# Biotechnology: Science for the New Millennium

# Biotechnology: Science for the New Millennium

Editor: Sophia Martin

STATES
ACADEMIC PRESS
www.statesacademicpress.com

**States Academic Press,**
109 South 5th Street,
Brooklyn, NY 11249, USA

Visit us on the World Wide Web at:
www.statesacademicpress.com



ISBN: 978-1-63989-081-1 (Hardback)



**Cataloging-in-Publication Data**

Biotechnology : science for the new millennium / edited by Sophia Martin.
p. cm.
Includes bibliographical references and index.
ISBN 978-1-63989-081-1
1. Biotechnology. 2. Genetic engineering. I. Martin, Sophia.
TP248.2 .B564 2021
660.6--dc23

# Table of Contents

# Preface

The world is advancing at a fast pace like never before. Therefore, the need is to keep up with the latest developments. This book was an idea that came to fruition when the specialists in the area realized the need to coordinate together and document essential themes in the subject. That's when I was requested to be the editor. Editing this book has been an honour as it brings together diverse authors researching on different streams of the field. The book collates essential materials contributed by veterans in the area which can be utilized by students and researchers alike.

Biotechnology is the technology based on biology which involves living systems and life forms that are interdependent on each other to create items using some mechanical application. It is based on the basic biological sciences such as molecular biology, cell biology, biochemistry and genetics. The most important applications of biotechnology are nutrient supplementation, abiotic stress resistance, strength fibers, healthcare, food processing, and fuel from waste. It covers the fields of molecular biology, bio-engineering, biomedical engineering, bio-manufacturing, and molecular engineering. This book brings forth some of the most innovative concepts and elucidates the unexplored aspects of biotechnology. From theories to research to practical applications, case studies related to all contemporary topics of relevance to this field have been included in it. The readers would gain knowledge that would broaden their perspective about biotechnology.

Each chapter is a sole-standing publication that reflects each author´s interpretation. Thus, the book displays a multi-facetted picture of our current understanding of application, resources and aspects of the field. I would like to thank the contributors of this book and my family for their endless support.

**Editor**

# 1

# Homologous recombination-mediated targeted integration in monkey embryos using TALE nucleases

Chu Chu[1†], Zhaohui Yang[1†], Jiayin Yang[2], Li Yan[1], Chenyang Si[1], Yu Kang[1], Zhenzhen Chen[1], Yongchang Chen[1], Weizhi Ji[1] and Yuyu Niu[1*]

**Abstract**

**Background:** Non-human primate (NHP) models can closely mimic human physiological functions and are therefore highly valuable in biomedical research. Genome editing is now developing rapidly due to the precision and efficiency offered by engineered site-specific endonuclease-based systems, such as transcription activator-like effector nucleases (TALENs) and the clustered regularly interspaced short palindromic repeats (CRISPR)/CRISPR-associated protein-9 nuclease (Cas9) system. It has been demonstrated that these programmable nucleases can introduce genetic changes in embryos from many species including NHPs. In 2014, we reported the first genetic editing of macaques using TALENs and CRISPR/Cas9. Subsequently, we characterized the phenotype of a methyl CpG binding protein 2 (MECP2)-mutant cynomolgus monkey model of Rett syndrome generated using the TALEN approach. These efforts not only accelerated the advance of modeling genetic diseases in NHPs, but also encouraged us to develop specific gene knock-in monkeys. In this study, we assess the possibility of homologous recombination (HR)-mediated gene replacement using TALENs in monkeys, and generate preimplantation embryos carrying an EmGFP fluorescent reporter constructed in the *OCT4* gene.

**Result:** We assembled a pair of TALENs specific to the first exon of the *OCT4* gene and constructed a donor vector consisting of the homology arms cloned from the monkey genome DNA, flanking an *EmGFP* cassette. Next, we co-injected the TALENs-coding plasmid and donor plasmid into the cytoplasm of 122 zygotes 6–8 h after fertilization. Sequencing and immunofluorescence revealed that the *OCT4-EmGFP* knock-in allele had been successfully generated by TALENs-mediated HR at an efficiency of 11.3% (7 out of 62) or 11.1% (1 out of 9), respectively, in monkey embryos.

**Conclusion:** We have successfully, for the first time, obtained *OCT4-EmGFP* knock-in monkey embryos via HR mediated by TALENs. Our results suggest that gene targeting through TALEN-assisted HR is a useful approach to introduce precise genetic modification in NHPs.

**Keywords:** Non-human primate, Homologous recombination, TALENs, Gene editing

## Background

Although a wide range of organisms, including worms, fruit flies, zebrafish, and rodents, have been used as research tools for most discoveries [1], non-human primates (NHPs) are excellent models to study human disease and develop therapeutic strategies because of their genetic, physiological and neurological similarity to humans [1–5]. Over the years, genetic approaches have been widely applied to model organisms. However, genetic engineering in monkeys was not available until 2001, when retrovirus-based methods were successfully used to generate transgenic monkey models [6]. This seminal work demonstrated that the monkey genome could be genetically modified efficiently. Subsequently, taking advantage of the high efficiency of lentiviral infection, the groups of Li and Sasaki successfully obtained several transgenic monkeys carrying the mutant *HTT*

* Correspondence: niuyy@lpbr.cn
†Chu Chu and Zhaohui Yang contributed equally to this work.
[1]Yunnan Key Laboratory of Primate Biomedicine Research; Institute of Primate Translational Medicine, Kunming University of Science and Technology, Kunming 650500, China

and *GFP* transgenes, respectively [7, 8]. However, while lentiviral transduction has been a powerful and efficient tool in genetic modification of animal genomes, it has often failed to achieve precise editing. The 2011 emergence of new-generation genome-editing methods based on reprogrammable, site-specific endonucleases, especially the clustered regularly interspaced short palindromic repeats (CRISPR)/CRISPR-associated protein-9 nuclease (Cas9) system and transcription activator-like effector nuclease (TALEN), made genome engineering more effective and reliable. Relative construction simplicity as well as precise and efficient genetic modification of target sites associated with these technologies has dramatically advanced genome editing in diverse species [9, 10]. In particular, these methods substantially facilitated the generation of non-human primate (NHP) models of human diseases. Until now, a number of such models have been established, including those of autism spectrum disorder (ASD) [11], Huntington's disease (HD) [7], Parkinson's disease (PD) [12], Duchenne muscular dystrophy (DMD) [13], and, most recently, Rett syndrome [3].

TALENs are fusions of the endonuclease domain consisting of the FokI restriction enzyme, which is only active as a dimer, and the engineered DNA-binding domain comprised by TAL effector (TALE) repeats. These highly conserved repeats are derived from naturally-occurring TALEs of the plant pathogen *Xanthomonas*, which contribute to disease or trigger defense by binding the host DNA and activating effector-specific host genes [14] .

TALENs can easily modify genomic DNA by introducing double-strand breaks (DSBs) at the desired site, stimulating the cellular DNA repair system. The main components of this system are non-homologous end-joining (NHEJ) and homologous recombination (HR) that result in knock-outs and knock-ins, respectively. NHEJ is a stochastic and error-prone repair process that does not require a repair template and

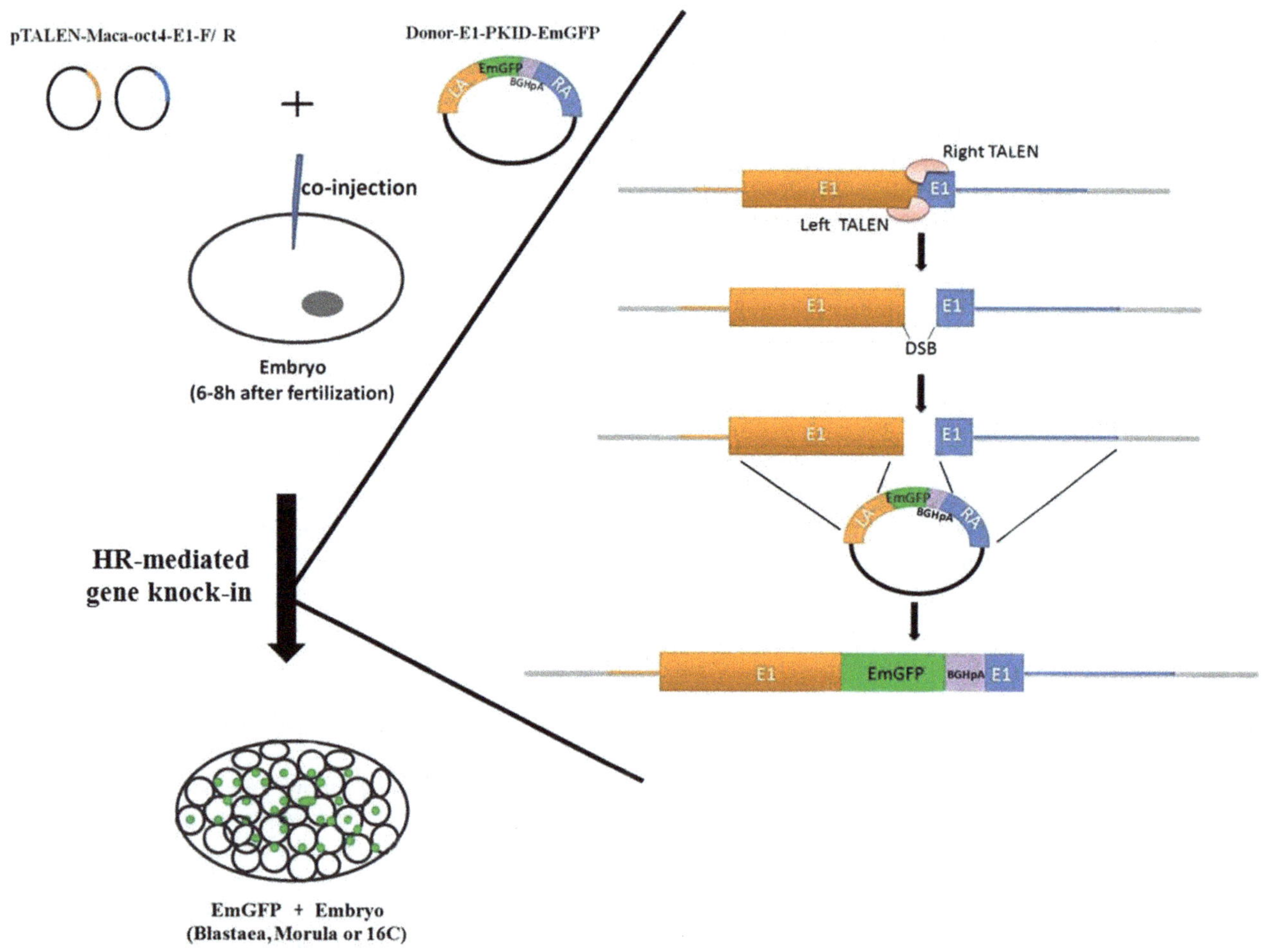


**Fig. 1** Workflow of TALEN-mediated generation of a monkey embryo carrying an EmGFP reporter in the OCT4 gene. TALENs-coding plasmids, pTALEN-Maca-oct4-E1-F/R, and the donor vector Donor-E1-PKID-EmGFP that targets exon 1 of the *OCT4* gene were designed and co-injected into the cytoplasm of a zygote 6–8 h after fertilization. Treated embryos at the blastocyst, morula and 16-cell stages were collected and analyzed

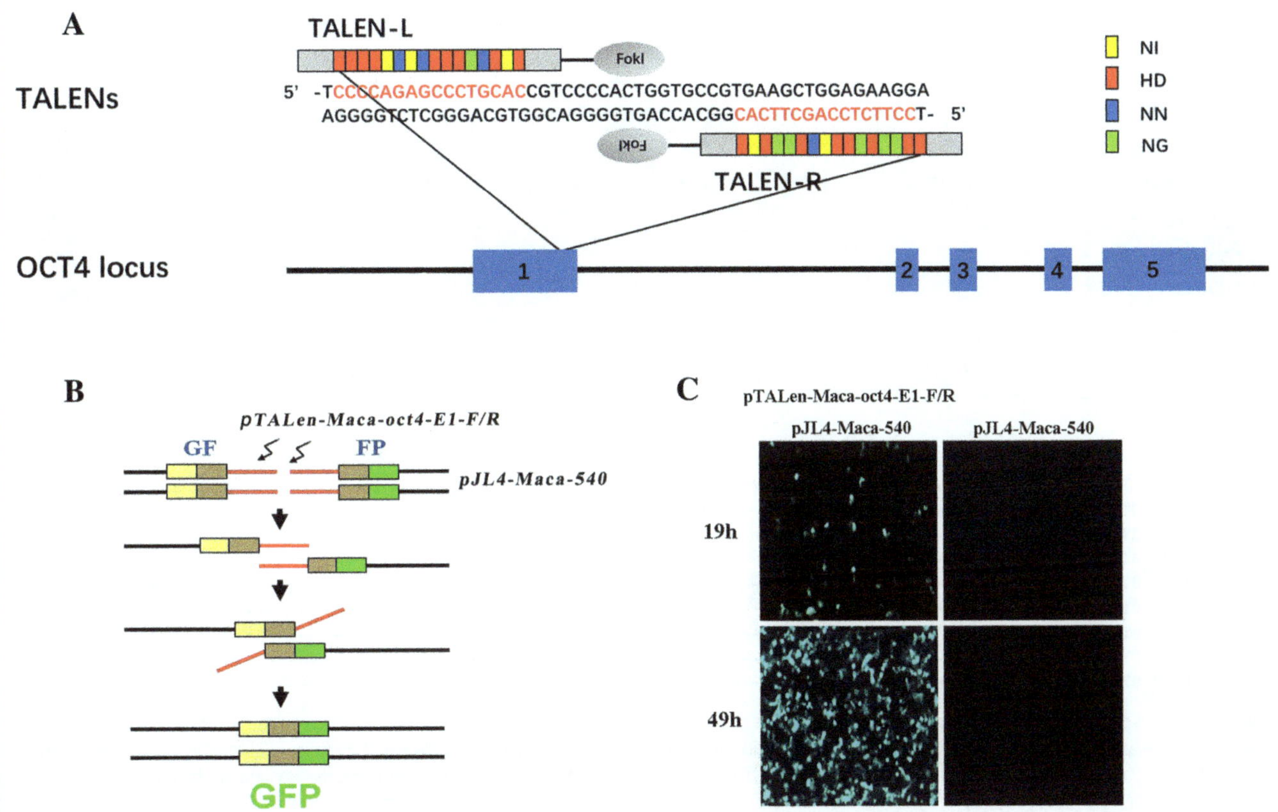


**Fig. 2** Schematic overview of the OCT4-EI-TALEN construction and the SSA assay for testing the OCT4-TALEN-coding plasmids. **a** The TALEN-targeted sequences within exon 1 of the *OCT4* gene are labeled in red. Assembled Repeat Variable Diresidue (RVD) repeats are represented schematically as boxes and labeled in yellow, red, blue and green. **b** The overview of the GFP-reporter system for SSA. Red lines represent 541 bp oligonucleotides near the TALEN-targeted sites amplified from genomic DNA. **c** Fluorescence intensity of GFP in 293 T cells 19 h or 49 h post transfection with OCT4-E1-TALEN-coding plasmids and the reporter vector pJL4-SSA (left) or controls transfected only with the reporter plasmid (right)

introduces random small insertions or deletions at DNA breakpoints [15]. Gene editing by this system has been used in mammalian cells to disrupt genes, delete chromosomal segments, or restore aberrant reading frames [14]. HR, on the other hand, is a precise repair pathway that requires a DNA template, which harbors a homologous sequence corresponding to the genomic DNA at the DSB site. HR has been used to correct certain mutations associated with diseases such as epidermolysis bullosa [16], sickle cell disease [17, 18], beta-thalassemia [19], and alpha-1 antitrypsin deficiency [20]. The efficiency of gene correction strategies based solely on HR is closely linked to the genomic target, cell type, cell-cycle state, organism species, and efficient delivery of an exogenous repair template. Unlike NHEJ, which occurs during the whole cell cycle, HR is only active during the late S/G2 phases, making HR-dependent gene knock-ins more difficult to achieve compared to the gene knock-outs mediated by NHEJ. Currently, HR-mediated knock-ins have been generated in many species including zebrafish [21], mice [22], rabbit [23], goat [24], sheep [25], cattle [26] and pig [27]. Recently, the groups of Yang and Niu successfully used CRISPR/Cas9-assisted HR to obtain knock-in cynomolgus monkeys with a reporter gene system [28, 29]. Here, we demonstrate TALEN-assisted, HR-mediated generation of monkey preimplantation embryos carrying an EmGFP fluorescent reporter constructed in the *OCT4* gene (Fig. 1). Our results demonstrate that TALEN-based HR is a feasible approach to generate knock-in NHP models.

**Table 1** In vitro developmental potential of the TALENs-injected embryos

| Group | Number of embryos | Development stage reached (%) | | | | |
|---|---|---|---|---|---|---|
| | | 2-cell | 8-cell | 16-cell | morula | blastocyst |
| Injected | 122 | 115 (94.3) | 96 (78.7) | 71 (58.2) | 57 (47.7) | 49 (40.2) |
| Control | 34 | 34 (100) | 30 (88.2) | 23 (67.64) | 17 ( 50) | 13 (38) |

A

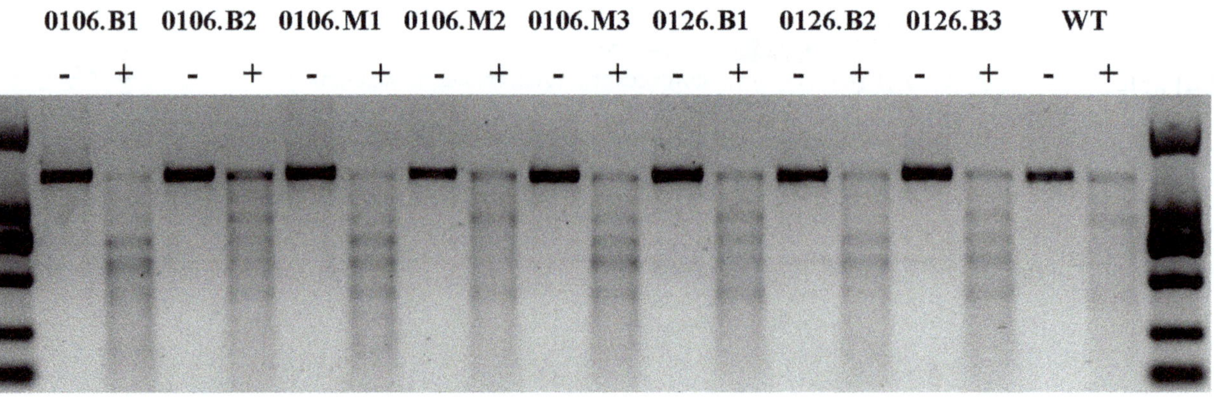


B

```
               OT-B-F/R GGAGAGCAACTCCGATGGGGCCTCCCCAGAGCCCTGCACCGTCCCCACTGGTGCCGTGAAGCTGGAGAAGGAGAAGCTGGAGCAAAACCC
OT-0106.B1_OT-B-F_TSS ------------------------------a---------------t.........----------------------------------  (-9)
OT-0106.B2_OT-B-F_TSS ------------------------------a-----------ctagt.........----------------------------------  (-14)
OT-0106.M1_OT-B-F_TSS ----------------------------------------.................---------------------------------  (-17)
OT-0106.M2_OT-B-F_TSS ------------------------------------------------------------------------------------------
OT-0106.M3_OT-B-F_TSS ------------------------------a---------------t.......------------------------------------  (-7)
OT-0126.B1_OT-B-F_TSS --------------------------------------................caccgtac----------------------------  (-23)
OT-0126.B2_OT-B-F_TSS ------------------------------a---------------t...----------------------------------------  (-3)
OT-0126.B3_OT-B-F_TSS ----------------------------------------------t.... --------------------------------------  (-4)
```

**Fig. 3** Detection of TALEN-induced mutations in the genomes of monkey embryos. **a** The T7EN1 cleavage assay for on-target efficiency of TALENs .The positive cleavage bands were visible in 7 samples in addition to 0106.M2. **b** Sequencing results showed different indels at the target site. The two short base sequences in red are the target sites for the TALEN pair, while the 17 bp between them comprise the region cleaved by FokI

## Results

### Construction of TALENs and evaluation of their activity in human cells

The monkey *OCT4* gene contains 5 exons, similar to that of humans and mice (monkey *OCT4* Gene ID: 714760, human *OCT4* Gene ID: 5460, mice *OCT4* gene ID: 18999, from https://www.ncbi.nlm.nih.gov/gene/?term=OCT4). We designed a pair of TALEN constructs to target exon 1 (Fig. 2a), which is the efficient site for inserting a reporter gene to indicate the expression of OCT4 [30, 31]. These sites are conserved in humans, so we initially used the single strand annealing (SSA) assay in 293 T cells to evaluate the cutting efficiency of Oct4-E1-TALENs based on the GFP expression level,

**Table 2** Primers used for genotyping

| Primer | Sequence (5′ to 3′) | Usage | Amplicon(bp) |
|---|---|---|---|
| L1-F | AGGATCGCTTGAGGACAGGAATT | PCR | 3259 bp (5′gDNA~ 3′gDNA) |
| L1-R | GGAGCCATCACAGGAGACAGAAAA | PCR/sequencing | |
| S1-F | GATGCATTGAGGGATAGCTCCACACACACATTC | PCR/sequencing | 1795 bp (5′ gDNA ~EmGFP) |
| S1-R | AGGACCATGTGATCGCGCTTCT | PCR | |
| S2-F | CAGAAGAACGGCATCAAG | PCR | 2109 bp (EmGFP~ 3′gDNA) |
| S2-R | AGTAGGGAGGAAGCAGTAGAA | PCR | |
| G-F | CACAAGTTCAGCGTGTCCG | PCR | 421 bp (EmGFP) |
| G-R | AGTTCACCTTGATGCCGTTC | PCR | |
| LI-W1R-G08 | ATCCCTCAGCAGCTCTACATTT | sequencing | |
| L1-W2R-H11 | CAGCTTCACGGCACCAGTAC | sequencing | |
| L1-W3R-C04 | GTAGCGGGCGAAGCACTG | sequencing | |
| S12-W1R-C01 | TCCGAGGATCAACCCAGC | sequencing | |
| OT-F | CACCACCATTAGGCAAAC | PCR | 1201 bp target site included |
| OT-R | GAGGCTGAAGTCAATCAAAA | PCR | |

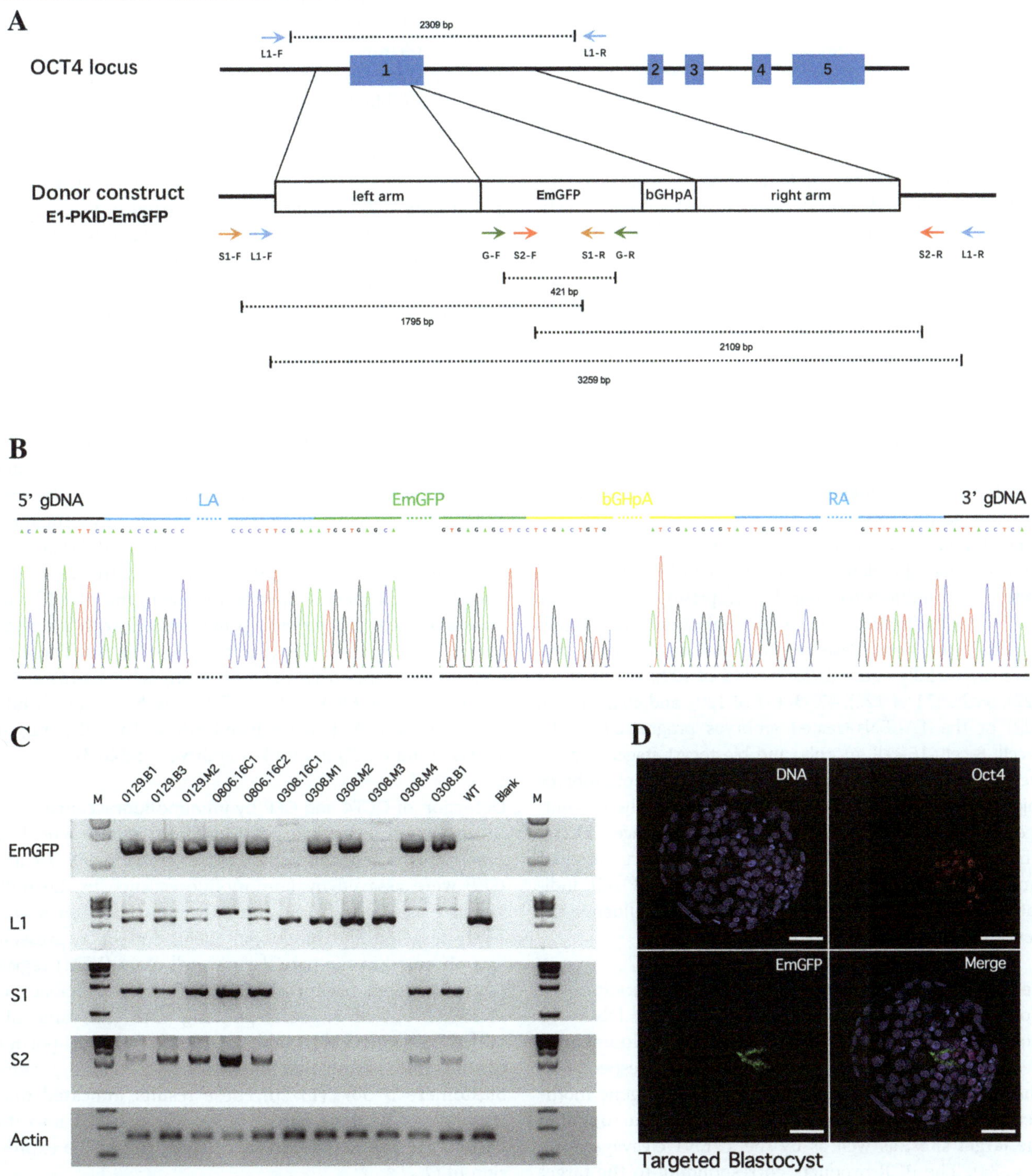


**Fig. 4** Genotyping and immunofluorescence of embryos generated by TALEN-mediated genome engineering. **a** Schematic overview of the strategy to generate an *OCT4-EmGFP* knock-in allele. The homologous arms of the donor vector are indicated as LA (839 bp) and RA (1001 bp). PCR primers used for PCR genotyping are shown as arrows in different colors. **b** PCR products obtained using primers L1-F and L1-R were sequenced. Sequence across the targeted region confirmed the correct fusion of EmGFP to the first exon of OCT4. **c** *EmGFP*: PCR genotyping using primers G-F and G-R produced bands of 421-bp fragment in the samples from nine embryos, suggesting the possible insertion or precise knock-in of the *EmGFP* gene. L1: PCR genotyping using primers L1-F and L1-R produced large products (3259 bp) in seven embryo samples, indicating the *EmGFP* sequence was integrated. The samples 0806.16C1, 0308 M4 and 0308.B1 only contain the larger product, suggesting either both alleles were targeted, or one allele failed to amplify. S1 and S2: PCR genotyping using primers S1-F and S1-R, S2-F and S2-R produced bands with correct size (1795 bp and 2109 bp) in the samples from the targeted embryos. **d** Immunostaining of targeted blastocysts using anti-GFP antibody showed a signal in the ICM. Scale bar, 50 μM

**Table 3** The positive rates of *EmGFP* knock-in demonstrated by PCR analysis or GFP detection

| Method | Number of zygote injected with plasmid | Number of positive embryos | positive rate |
|---|---|---|---|
| PCR analysis | 62 | 7 | 11.3% |
| GFP detection | 9 | 1 | 11.1% |

namely the percentage of GFP positive cells (Fig. 2b). Expression of GFP could be observed in 293 T cells transfected with the Oct4-E1-TALEN-coding plasmids, pTALEN-Maca-oct4-E1-F /R, and the reporter vector pJL4-Maca-540 both after 19 h and 49 h, but not in the control group transfected only with the reporter plasmid (Fig. 2c). These results showed that the TALENs were designed and constructed to effectively induce DSBs at the target sites in the human genome, suggesting that this pair of TALENs would work in a monkey as well because of high genomic homology.

#### Development of monkey embryos microinjected with the constructed plasmids

The plasmids coding for TALENs, pTALEN-Maca-oct4-E1-F /R, and the donor vector, Donor-E1-PKID-EmGFP, were mixed and injected into the cytoplasm of 122 zygotes 6–8 h after fertilization using a microinjection system under standard conditions. Based on the number of zygotes that were successfully injected, 94.3% (115 of 122), 78.7% (96 of 122), 58.2% (71 of 122), 47.7% (57 of 122), and 40.2% (49 of 122) of the TALENs-treated embryos progressed to the 2-cell, 8-cell, 16-cell, morula, and blastocyst stage, respectively (Table 1). The developmental competence of embryos injected with TALENs was normal, judging by the percentage of embryos that reached the blastocyst stage (40.2%). These results are similar to our previous report for those embryos derived from in vitro fertilization (IVF) [2], indicating that this system is safe and does not influence the development of NHP embryos.

#### Genotyping to confirm the targeted gene integration

To test the efficiency of TALENs, the genomic DNA was isolated from the embryos with normal development at the 16-cell, morula and blastocyst stages, respectively, and screened for the presence of site-specific gene modification by PCR amplification of the regions surrounding the target sites, as well as by the T7EN1 cleavage assay (Fig. 3a). The PCR products corresponding to the target sites were amplified using the OT-F/R primers (Table 2) from individual embryos ($n = 8$) in the morula or blastocyst stage and then sequenced. The genomes of up to 87% (7/8) of monkey embryos injected with TALEN-coding plasmids were modified at the target site, visualized by the cleavage bands in the target gene (Fig. 3a). Further characterization of the cleavage by sequencing indicated different indels at the target site with variable mutation size (Fig. 3b). This data demonstrated that the TALENs could also function effectively in monkey embryos.

To determine whether the *EmGFP* sequence was inserted into the target *OCT4* locus, four primers (Fig. 4a, Table 2) were used to amplify the different PCR products from the individual embryos ($n = 62$) in the 16-cell, morula or blastocyst stages, and these genes were then sequenced. PCR using a pair of primers for *EmGFP* (G-F/R) amplified a specific band in the genomic DNA of the embryos (Fig. 4c), which suggested the possible insertion or precise knock-in of the *EmGFP* gene. Using the forward primer outside of the left homology arm (LA) region (L1-F) and the reverse primer outside the right homology arm (RA) region (L1-R), we detected a large PCR product (3259 kb) for the EmGFP-knock-in in the *OCT4* gene locus of seven embryos (Fig. 4c). The PCR products of samples 0806.16C1 and 0806.16C2 were sequenced, and both 5′ and 3′ boundaries between *OCT4* and *EmGFP* were detected as designed (Fig. 4b, Additional file 1). To further confirm the precise modification, we were then able to amplify a fragment of the expected size from the seven *EmGFP*-positive embryos using a forward primer outside of the LA region (S1-F) with an *EmGFP*-specific reverse primer (S1-R), and an *EmGFP*-specific forward primer (S2-F) with a reverse primer outside of the RA region (S2-R) (Fig. 4c). These results of electrophoresis and sequencing demonstrated that the *OCT4-EmGFP* knock-in allele has been successfully generated by TALENs-mediated HR at the efficiency of 11.3% (7 out of 62) in monkey embryos (Table 3).

#### Detection of OCT4 and GFP by immunofluorescence

In order to investigate the function of the knock-in *EmGFP* as a reporter for *OCT4* expression, *OCT4* and *GFP* were assayed by immunofluorescence in the individual blastocysts ($n = 9$). As expected based on a correctly targeted and functional allele, EmGFP expression was observed in the partial inner cell mass (ICM) region of the targeted blastocysts (Fig. 4d) at an efficiency of 11.1% (1 out of 9), corresponding with the value obtained by sequencing (Table 3). Besides, the proportion of the EmGFP-positive blastomere in OCT4-positive blastomere is 50% (11/20).These results indicated that, although with mosaicism, *EmGFP* was inserted into the desired site and could also successfully report the expression of *OCT4.*

#### Assessment of the specificity of TALEN-mediated cleavage

To test the specificity of the cleavage by TALENs, seven highly-ranked potential off-target sites for *OCT4* TALENs were predicted and selected (Additional file 2: Table S1). The off-target effects were comprehensively assessed in the same way as the on-target effect, using genomic DNA from 10 embryos that were modified at the target site.

Thus, the fragments around the seven potential off-target gene loci were amplified by PCR and sequenced. Sequencing results (Additional file 3) showed that no off-target mutations were found at the predicted sites of genomic DNA from the TALEN-injected embryos, suggesting that the TALENs we designed and constructed had high specificity and thus were sufficiently safe to use in monkeys.

## Discussion

With the emergence of reliable and precise gene editing tools in the recent years, gene modification has been dramatically brought to a new level. Among the engineered site-specific nucleases, CRISPR/Cas9 system has developed most rapidly and is now the most widely-used tool in research labs because of its construction simplicity and high efficiency. However, this system is still controversial in terms of safety because of the associated frequency of off-target effects [32, 33]. In comparison, TALEN is safer to use due to its requirement for being a dimer to be active and because it consists only of proteins. Therefore, TALEN is still an important and attractive genetic tool for functional genomics. In this study, we used the classic method of Golden Gate cloning for TAL assembly to construct a TALEN pair that could effectively work in both humans and monkeys at exon 1 of the *OCT4* gene (Fig. 2, Fig. 3).

Mammalian animal models have contributed significantly to studies in both basic biology and human diseases. In this regard, genome engineering in mammals is highly relevant, especially targeted gene knock-in methods. For example, allowing reporter genes to be successfully inserted into specific genes important in embryonic stem cell mechanisms, such as *OCT4, NANOG* and *SOX2*, can facilitate developmental biology research [27] [34]. Furthermore, insertion of certain functional genes can be helpful in the generation of animal models or in animal husbandry [3, 24]. In addition, gene knock-in could be used to replace the endogenous base sequences for repairing mutations. Many knock-in animals of different species have been generated to date. In this work, we obtained monkey embryos with precise *EmGFP* knock-in at the *OCT4* gene locus by TALEN-mediated HR at an efficiency of 11% (Fig. 4, Table 3), which corresponds to the general achievable efficiency of gene knock-in in mammalian systems [22] [35]. According to the results of immunofluorescence experiments, detection of GFP not only confirmed the expression of OCT4, but also reliably indicated the ICM regions of blastocysts. Therefore, this system could be a useful tool for embryonic development research in monkeys.

NHPs are the most suitable models to study human disease because of their high homology to humans. Our results are significant for the genetic modification research in monkeys, and are therefore highly valuable for the construction of reliable human disease models. Similar to the work of Yang and Niu groups on gene knock-in monkeys using CRISPR/Cas9 system [28, 29], our results confirm the successful gene insertion but show a degree of mosaicism, indicated by the immunofluorescence result (Fig. 4d) and electrophoretic images from four samples, 0129.B1, 0129.B3, 0129.M2 and 0806.16C2. Nevertheless, there are also three genetically modified samples (0806.16C1, 0308 M3 and 0308B1) that seem to be homozygous (Fig. 4c), encouraging us to undertake further study to overcome mosaicism in monkey genome editing.

Recently, J.R. et al. [36] combined 2C-HR-CRISPR with a modified biotin–streptavidin system to localize the repair templates to target sites, resulting in a more-than-tenfold increase (up to 95%) in knock-in efficiency over standard methods. Although their result was based on the CRISPR/Cas9 system unlike our study, it prompts us to carefully consider the time window for the injection of the TALEN pair and the donor template. The structure of chromatin is one of the determining factors for achieving high-efficiency of gene knock-in. In this research, we injected TALENs-coding plasmids and the donor vector 6–8 h after fertilization, when chromatin is most opened. Therefore, in the future, we need to find the optimum conditions of introducing the editing system to obtain more efficient gene knock-in results in monkeys.

## Conclusions

In this work, for the first time, we have successfully obtained *OCT4-EmGFP* knock-in monkey embryos by TALEN-assisted HR. Even though mosaicism was detected, the GFP could faithfully report the expression of OCT4. Our results suggest that gene targeting through HR mediated by TALENs is a feasible approach to introduce precise genetic modification in NHPs.

## Methods

### Construction of TALEN-coding plasmids and the donor plasmid

TALEs were designed using online software (https://tale-nt.cac.cornell.edu/node/add/talen). Repeat Variable Diresidue (RVD)-containing repeats were assembled using the Golden Gate TAL Assembly Kit, and then were sub-cloned into FokI nuclease expression vectors containing modified ELD/KKR domains, ptCMV-153/47-VR-HD (Addgene #50703). Assembled RVD repeats were as follows: HD HD HD HD NI NN NI NN HD HD HD NG NN HD NI HD for the left TALEN targeting CCCCAGAGCCCTGCAC; HD HD NG NG HD NG HD HD NI NN HD NG NG HD NI HD for the right TALEN targeting CCTTCTCCAGCTTCAC.

HR donor vector E1-PKID-EmGFP was constructed by standard molecular-cloning methods. For the *OCT4* locus, the donor vector consists of the 839-bp

5′-homology arm and the 1001-bp 3′-homology arm cloned from monkey (050139) gDNA, flanking an *EmGFP* cassette followed by the BGH polyA signal (Fig. 4, Additional file 4: Figure S1).

#### SSA assay in human cells

For construction of the reporter plasmid pJL4-Maca-540, the 541-bp oligonucleotides near the TALENs target site was amplified with primer Maca-540-F1/R1 (Additional file 5) and cloned into the backbone vector pJL4-SSA(Cell Inspire Bio)linearized with BglII and EcoRI (NEB, R0144 and R0101). In this way the GFP-coding sequence could be separated into two parts, GF and FP. The two parts could be annealed to express intact GFP only following successful application of TALENs. The TALEN plasmids and the reporter plasmids were transfected into HEK293T cells using polyethylenimine (PEI,Polysciences 23,966–2) by applying standard protocol [37]. At 19 h and 47 h post transfection, respectively, the fluorescence intensity of GFP was measured using confocal microscopy.

#### Animals

Healthy female cynomolgus monkeys (*Macaca fascicularis*), ranging in age from 5 to 8, were selected for use in this study. All animals were housed at Yunnan Key Laboratory of Primate Biomedical Research (LPBR). These animals were only used for oocytes collection. After a seven-day observation period, no euthanasia needed, these animals returned to the outdoor animal colony of LPBR.

#### Embryo collection

Embryo collection and transfer were performed as previously described [38]. In brief, 11 healthy female cynomolgus monkeys aged 5–8 years with regular menstrual cycles were selected as oocyte donors for superovulation, which was performed by intramuscular injection with rhFSH (recombinant human follitropin alfa, GONAL-F, Merck Serono) for 8 days, then rhCG (recombinant human chorionic gonadotropin alfa, OVIDREL, Merck Serono) on day 9. 32–35 h after rhCG administration, Atropine(0.02 mg/kg,Tianjin Pharmaceutical Group, China), Benzyl penicillin sodium (20 mg/kg, Harbin Pharmaceutical Group, China) and morphine hydrochloride (0.2 mg/kg, Northeast Pharmaceutical Group, China) were intramuscular administered for preparing for minimally invasive surgery. Animals were anesthetized using Ketamine (10 mg/kg, Jiangsu Zhongmu Beikang Pharmaceutical CO., LTD, China).Oocytes were collected by laparoscopic follicular aspiration. Mature oocytes(MII oocytes,first polar body present) were used to perform intracytoplasmic sperm injection (ICSI) and the fertilization was confirmed by the presence of two pronuclei. The zygotes(including those subsequently microinjected with plasmids and) were cultured in the chemically defined, protein-free hamster embryo culture medium-10 (HECM-10) containing 10% fetal calf serum (Hyclone Laboratories, SH30088.02) at 37 °C in 5% CO2.

#### Microinjection of plasmids

TALENs-coding plasmids (50 ng/μl) and donor plasmid (10 ng/μl) were mixed and microinjected into the cytoplasm of zygotes 6–8 h after fertilization using a microinjection system under standard conditions. The zygotes were then cultured in the chemically defined, protein-free hamster embryo culture medium-10 (HECM-10) containing 10% fetal calf serum (Hyclone Laboratories, SH30088.02) at 37 °C in 5% CO2.

#### Genomic DNA extraction and genotyping by PCR

The individual embryos at the stage of 16-cell, morula or blastocyst were collected, and their genomic DNA was amplified using REPLI-g Single Cell Kit (QIAGEN, 150345). All PCR experiments were performed with Takara PrimeSTAR GXL DNA Polymerase (Takara, R050B), and the products were analyzed by sequencing. Primers OT-F/R were used to amplify the target site region at the following conditions: 95 °C for 5 min; 35 cycles of 95 °C for 30 s, 58 °C for 30 s, 72 °C for 1 min; 72 °C for 10 min. Primers G-F/R localizing to *EmGFP* were used to confirm possible insertion of *EmGFP*, under the following conditions: 95 °C for 5 min; 35 cycles of 95 °C for 30 s, 56 °C for 30 s, 72 °C for 30 s; 72 °C for 5 min. Primers S1-F/R amplified the fragment over outside the LA region and *EmGFP* under the conditions: 95 °C for 3 min; 16 cycles of 95 °C for 30 s, 66 °C for 30 s, 72 °C for 1.5 min, − 0.5 °C/cycle; 30 cycles of 95 °C for 30 s, 58 °C for 30s, 72 °C for 1.5 min; 72 °C for 5 min. Primers S2-F/R amplified the fragment over outside the RA region and *EmGFP* under the conditions: 95 °C for 5 min; 35 cycles of 95 °C for 30 s, 54 °C for 30 s, 72 °C for 2 min; 72 °C for 5 min. Primers L-F/R detected the large PCR product for the intact *EmGFP* knock-in at the *OCT4* locus, under the following conditions: 95 °C for 5 min; 35 cycles of 95 °C for 30 s, 60 °C for 30 s, 72 °C for 3 min; 72 °C for 5 min. The detailed information for all the primers is given in Table 2.

#### Immunofluorescence

Immunofluorescence experiments were performed as described previously (Zhou et al., 2006). After fixation in 4% paraformaldehyde in PBS for 30 min and permeabilization in 0.2% Triton X-100/PBS for 45 min at room temperature, all embryo samples were blocked in PBS buffer containing 3% BSA for 45 min before incubation with the primary antibodies: anti-GFP antibody (1:1200, Abcam,ab13970) and anti-OCT4 antibody (1:100, Abcam,ab27985) overnight at 4 °C. After rinsing with PBS, embryos were incubated with the secondary antibodies and DAPI in PBS for

2 h at room temperature. EmGFP was labeled with Goat anti-Chicken IgY (H + L) Secondary Antibody, Alexa Fluor 488 (1:500, Invitrogen, A-11039). Endogenous OCT4 was stained by Donkey anti-Goat IgG (H + L) Cross-Adsorbed Secondary Antibody, Alexa Fluor 594 (1:500, Invitrogen, A-11058). DNA was counterstained with a DAPI solution (1:500, Solarbio, C0060). Finally, embryos were mounted on glass slides and examined using confocal microscopy.

### Off-target analysis

All potential off-target sites with a degree of homology to the pair of TALENs were predicted by Paired Target Finder in TAL Effector Nucleotide Targeter 2.0 [39] using NCBI GenBank assembly Mmul_051212 (GCA_000002255.2). We chose the 7 highest-ranked off-target sites for analysis. The fragments around these gene sites were amplified using the corresponding primers (Additional file 6) and analyzed by sequencing (Additional file 3).

## Additional files

**Additional file 1: Supplementary information 1.** Sequencing and Blast results of large PCR products of samples 0806.16C1 and 0806.16C2. The sequencing result indicated that *EmGFP* was successfully inserted into exon 1 of *OCT4* as expected. (PDF 1720 kb)

**Additional file 2: Table S1.** List of putative off-target sites. The top 7 potential off-target sites (OTS1-OTS7) were selected for analysis. (XLSX 15 kb)

**Additional file 3: Supplementary information 2.** Results of off-target sequencing analysis. (PDF 123 kb)

**Additional file 4: Figure S1.** Plasmid profile of the donor vector Donor-PKID-E1-EmGFP. (PDF 200 kb)

**Additional file 5: Supplementary information 3.** Construction of the reporter plasmid for the SSA assay. (PDF 226 kb)

**Additional file 6: Table S2.** Primers used for off-target analysis. (XLSX 13 kb)

#### Abbreviations

ASD: Autism spectrum disorder; CRISPR/Cas9: Clustered regularly interspaced short palindromic repeats; DMD: Duchenne muscular dystrophy; DSBs: Double-strand breaks; HD: Huntington's disease; HDR: Homology-directed repair; HR: Homologous recombination; ICM: Inner cell mass; NHP: Non-human primate; PD: Parkinson's diseases; TALENs: Transcription activator-like effector nucleases

#### Acknowledgements

We appreciate the technical support for this study by the Shenzhen Cell Inspire Biotechnology. We also appreciate the professional support about animal management and treatment from senior veterinary surgeon Hong Wang.

#### Funding

This works is supported by The National Key R&D Program (2016YFA0101401 and 2016YFC0105703). The former were used for the purchase, raising, and welfare of animals and the funding from the later contributed to experiment in vitro. Besides, the funding bodies were not involved in the design of the study and collection, analysis, and interpretation of data or in writing the manuscript.

#### Author contributions

YYN and WZJ designed the project. CC, ZHY, JYY, LY, CYS, YK, ZZC and YCC performed or analyzed experiments. CC, YYN wrote the paper. All authors read and approved the final manuscript.

#### Ethics approval

The cynomolgus monkey (*Macaca fascicularis*) involved in this study is accredited by AAALAC International and all experimental protocols were approved in advance by the Institutional Animal Care and Use Committee of Yunnan Key Laboratory of Primate Biomedical Research.

#### Consent for publication

Not applicable.

#### Competing interests

The authors declare that they have no competing interests.

#### Author details

[1]Yunnan Key Laboratory of Primate Biomedicine Research; Institute of Primate Translational Medicine, Kunming University of Science and Technology, Kunming 650500, China. [2]The Cardiology Division, Department of Medicine, Li Ka Shing Faculty of Medicine, University of Hong Kong, Hong Kong, SAR, China.

#### References

1. Izpisua Belmonte JC, et al. Brains, genes, and primates. Neuron. 2015;86(3): 617–31.
2. Niu Y, et al. Generation of gene-modified cynomolgus monkey via Cas9/RNA-mediated gene targeting in one-cell embryos. Cell. 2014;156(4):836–43.
3. Chen, Y., et al., Modeling Rett syndrome using TALEN-edited MECP2 mutant Cynomolgus monkeys. Cell, 2017. 169(5): p. 945–955.e10.
4. Liu Z, et al. Generation of a monkey with MECP2 mutations by TALEN-based gene targeting. Neurosci Bull. 2014;30(3):381–6.
5. Kang Y, et al. CRISPR/Cas9-mediated Dax1 knockout in the monkey recapitulates human AHC-HH. Hum Mol Genet. 2015;24(25):7255–64.
6. Chan AWS, Chong KY, Martinovich C, Simerly C, Schatten G. *Transgenic Monkeys Produced by Retroviral Gene Transfer into Mature Oocytes*. Science. 2001;291(12):309–12.
7. Yang SH, et al. Towards a transgenic model of Huntington's disease in a non-human primate. Nature. 2008;453(7197):921–4.
8. Chan AW, Yang SH. Generation of transgenic monkeys with human inherited genetic disease. Methods. 2009;49(1):78–84.
9. Nemudryi AA, Valetdinova KR, Medvedev SP, Zakian SM. *TALEN and CRISPRCas Genome Editing Systems Tools of Discovery*. Acta Nat. 2014;6(3):22.
10. Pu J, et al. Utilization of TALEN and CRISPR/Cas9 technologies for gene targeting and modification. Exp Biol Med (Maywood). 2015;240(8):1065–70.
11. Zhao H, Jiang YH, Zhang YQ. Modeling autism in non-human primates: opportunities and challenges. Autism Res. 2018;11(5):686–94.
12. Ding F, et al. Development of a stable, early stage unilateral model of Parkinson's disease in middle-aged rhesus monkeys. Exp Neurol. 2008; 212(2):431–9.
13. Chen Y, et al. Functional disruption of the dystrophin gene in rhesus monkey using CRISPR/Cas9. Hum Mol Genet. 2015;24(13):3764–74.
14. Ousterout DG, Gersbach CA. The development of TALE nucleases for biotechnology. Methods Mol Biol. 2016;1338:27–42.
15. Sakuma T, et al. MMEJ-assisted gene knock-in using TALENs and CRISPR-Cas9 with the PITCh systems. Nat Protoc. 2016;11(1):118–33.
16. Osborn MJ, et al. TALEN-based gene correction for epidermolysis bullosa. Mol Ther. 2013;21(6):1151–9.
17. Sun N, et al. Optimized TAL effector nucleases (TALENs) for use in treatment of sickle cell disease. Mol BioSyst. 2012;8(4):1255–63.
18. Voit RA, et al. Nuclease-mediated gene editing by homologous recombination of the human globin locus. Nucleic Acids Res. 2014;42(2):1365–78.

19. Ma N, et al. Transcription activator-like effector nuclease (TALEN)-mediated gene correction in integration-free beta-thalassemia induced pluripotent stem cells. J Biol Chem. 2013;288(48):34671–9.
20. Choi SM, et al. Efficient drug screening and gene correction for treating liver disease using patient-specific stem cells. Hepatology. 2013;57(6):2458–68.
21. Shin J, Chen J, Solnica-Krezel L. Efficient homologous recombination-mediated genome engineering in zebrafish using TALE nucleases. Development. 2014;141(19):3807–18.
22. Chu VT, et al. Efficient generation of Rosa26 knock-in mice using CRISPR/Cas9 in C57BL/6 zygotes. BMC Biotechnol. 2016;16:4.
23. Yang D, et al. Identification and characterization of rabbit ROSA26 for gene knock-in and stable reporter gene expression. Sci Rep. 2016;6:25161.
24. Cui C, et al. Gene targeting by TALEN-induced homologous recombination in goats directs production of beta-lactoglobulin-free, high-human lactoferrin milk. Sci Rep. 2015;5:10482.
25. Wu M, et al. Rosa26-targeted sheep gene knock-in via CRISPR-Cas9 system. Sci Rep. 2016;6:24360.
26. Haibo Wu , Y Wang , Yan Zhang , Mingqi Yang , Jiaxing Lv , Jun Liu , and Yong Zhang, *TALE nickase-mediated SP110 knockin endows cattle with increased resistance to tuberculosis.* PNAS march 2, 2015 p. March 2, 2015.
27. Lai S, et al. Generation of Knock-in pigs carrying Oct4-tdTomato reporter through CRISPR/Cas9-mediated genome engineering. PLoS One. 2016;11(1): e0146562.
28. Yao X, et al. Generation of knock-in cynomolgus monkey via CRISPR/Cas9 editing. Cell Res. 2018;28(3):379–82.
29. Cui Y, et al. Generation of a precise Oct4-hrGFP knockin cynomolgus monkey model via CRISPR/Cas9-assisted homologous recombination. Cell Res. 2018;28(3):383–6.
30. Hockemeyer D, et al. Efficient targeting of expressed and silent genes in human ESCs and iPSCs using zinc-finger nucleases. Nat Biotechnol. 2009; 27(9):851–7.
31. Quan L, et al. Establishment of a rabbit Oct4 promoter-based EGFP reporter system. PLoS One. 2014;9(10):e109728.
32. Kosicki M, Tomberg K, Bradley A. Repair of double-strand breaks induced by CRISPR-Cas9 leads to large deletions and complex rearrangements. Nat Biotechnol. 2018; 36(8):765-71.
33. Haapaniemi E, Botla S, Persson J, Schmierer B, Taipale J. CRISPR-Cas9 genome editing induces a p53-mediated DNA damage response. Nat Med. 2018;24(7):927-30.
34. Yang H, et al. One-step generation of mice carrying reporter and conditional alleles by CRISPR/Cas-mediated genome engineering. Cell. 2013; 154(6):1370–9.
35. Certo MT, et al. Tracking genome engineering outcome at individual DNA breakpoints. Nat Methods. 2011;8(8):671–6.
36. Gu B, Posfai E, Rossant J. Efficient generation of targeted large insertions by microinjection into two-cell-stage mouse embryos. Nat Biotechnol. 2018; 36(7):632–7.
37. Longo PA, et al. Transient mammalian cell transfection with polyethylenimine (PEI). Methods Enzymol. 2013;529:227–40.
38. Niu Y, et al. Transgenic rhesus monkeys produced by gene transfer into early-cleavage-stage embryos using a simian immunodeficiency virus-based vector. Proc Natl Acad Sci U S A. 2010;107(41):17663–7.
39. Cermak T, et al. Efficient design and assembly of custom TALEN and other TAL effector-based constructs for DNA targeting. Nucleic Acids Res. 2011; 39(12):e82.

2

# Enhancement of using combined packing materials on the removal of mixed sulfur compounds in a biotrickling filter and analysis of microbial communities

Xiang Tu[1,2,3], Meiying Xu[1,2,3], Jianjun Li[1,2,3*], Enze Li[1,2,3], Rongfang Feng[1,2,3], Gang Zhao[1,2,3], Shaobin Huang[4] and Jun Guo[1,2,3]

**Abstract**

**Background:** Packing materials is a critical design consideration when employing biological reactor to treat malodorous gases. The acidification of packing bed usually results in a significant drop in the removal efficiency. In the present study, a biotrickling filter (BTF2) packed with plastic balls in the upper layer and with lava rocks in the bottom layer, was proposed to mitigate the acidification.

**Results:** Results showed that using combined packing materials efficiently enhanced the removal performance of BTF2 when compared with BTF1, which was packed with sole lava rocks. Removal efficiencies of more than 92.5% on four sulfur compounds were achieved in BTF2. Average pH value in its bottom packing bed was about 4.86, significantly higher than that in BTF1 (2.85). Sulfate and elemental sulfur were observed to accumulate more in BTF1 than in BTF2. Analysis of principal coordinate analysis proved that structure of microbial communities in BTF2 changed less after the shutdown but more when the initial pH value was set at 5.5. Network analysis of significant co-occurrence patterns based on the correlations between microbial taxa revealed that BTF2 harbored more diverse microorganisms involving in the bio-oxidation of sulfur compounds and had more complex interactions between microbial species.

**Conclusions:** Results confirmed that using combined packing materials effectively improved conditions for the growth of microorganisms. The robustness of reactor against acidification, adverse temperature and gas supply shutdown was greatly enhanced. These provided a theoretical basis for using mixed packing materials to improve removal performance.

**Keywords:** Biotrickling filter, Packing materials, Malodorous gases, Microbial community, Co-occurrence network

## Background

Many kinds of malodorous pollutants are generated from various industrial activities [1]. Among them, $H_2S$ and volatile organic sulfur compounds include methyl mercaptan, ethyl thioether and dimethyl disulfide appear most frequently in malodorous gases and are recognized as the key components due to their extremely low odor threshold [2]. Their release into ambient air cannot be ignored for their threats on health and living quality of nearby residents [3]. Besides, wastewater treatment plant can suffer severely economic loss every year from sulfide-induced corrosion [4].

Biofiltration is commonly considered as a cost-saving and environmentally friendly approach for the treatment of malodorous gases. High removal efficiencies of single malodorous substance were reportedly obtained by biological reactors. However, biological purification of the mixed gases containing multi-component malodorous pollutants remains challenging, predominantly because

* Correspondence: lijianjun1672@163.com
[1]Guangdong Institute of Microbiology, Guangdong Academy of Sciences, Guangzhou 510070, People's Republic of China
[2]State Key Laboratory of Applied Microbiology Southern China, Guangzhou 510070, People's Republic of China

of the accumulation of sulfate in packing materials arising from the bio-oxidization of sulfur compounds [5]. Severe acidification of packing materials in bioreactor could greatly inhibit the degrading activities of microorganisms, and thus lower the overall removal performance [6].

Strategies available to mitigate the impact of acidification on removal efficiencies of multi-components malodorous gases include washing packing bed [7], inoculating specific degrader [8], using two-stage bioreactor [9] or adding alkaline to neutralize acidity [10]. Here, another method of using combined packing materials was proposed. Previous study found that the bottom bed accounted for major removal of odorous pollutants [11]. However, removal of the upper packing materials from filter bed significantly lowered the overall removal performance of bioreactor. So, it was inferred that the upper packing materials played an important role in buffering the spatial gradient of environmental parameters including pH and sulfate, and then in shaping microbial communities. This has not been elucidated well from the ecological point of view. There are still large uncertainties about interactions among different microbial taxa during the biofiltration process of the mixed malodorous gases.

In the present study, two biotrickling filters (BTFs) using different packing materials were set up. The control was packed with sole lava rocks while the treatment had lava rocks in the bottom layer and plastic balls in the upper layer. The two BTFs were used (1) to assess the effects of combined packing materials on synchronous removal of $H_2S$, dimethyl sulfide, ethyl mercaptan and dimethyl disulfide; (2) to prove that combined packing materials help accelerate the leaching of metabolites from the filter bed; (3) to determine the robustness of the BTF employing combined packing materials against adverse environmental conditions; (4) to identify the functional microbial taxa that were responsible for the biotransformation of reduced sulfur compounds; and (5) to decipher the possible mechanisms that underlie the enhancement of combined packing materials on the synchronous removal of mixed sulfur compounds.

## Results

### Removal performances of biotrickling filters for mixed sulfur compounds

Two BTFs were operated in parallel in an effort to compare the effect of using combined packing materials on the removal of mixed sulfur compounds. After the startup, removal efficiencies (REs) of $H_2S$ rapidly increased near to 100% in both BTFs within four days, and steadily maintained at this level throughout the experiment regardless of varied inlet loading rates. The rapid acclimation BTFs to $H_2S$ within such short time can be attributed to the inoculums, which originated from a previous reactor treating $H_2S$-containing gases.

As to volatile organic sulfur compounds (VOSCs), both BTFs took much longer time to adapt. Figure 1 showed the overall removal performances of both BTFs on three VOSCs. Comparatively, BTF2 performed better. REs of ethyl mercaptan, dimethyl sulfide and dimethyl disulfide gradually increased in BTF2. On Day 80, these VOSCs were efficiently removed in BTF2 with 97.8, 92.3 and 96.7% of REs, respectively. BTF1 turned out to be less efficient, and REs of ethyl mercaptan and dimethyl disulfide were 84.5 and 75.1%, respectively. Dimethyl sulfide was among the most difficult to be removed in BTF1. Only 48.2% of dimethyl sulfide was removed on Day 80. Besides, the increase of inlet concentration with up to two times on Day 32 significantly lowered the removal efficiency of dimethyl sulfide in BTF1, which never exceeded 52% throughout the experiment.

There are two possible reasons that both BTFs required longer times to adapt the three organic sulfur compounds. Most likely, acidification of filter bed arising from production of sulfate might lower the activities of some methyl trophic microbes which play crucial roles in the biotransformation of organic sulfur compounds and grow well in neutral environment. Besides, as an important intermediate, the presence of $H_2S$ could inhibit the bio-oxidation of the three organic sulfur compounds.

### Evaluation of robustness of both BTFs against adverse conditions

Both BTFs were also compared in terms of their robustness against adverse environment conditions including low temperature, shutdown and low pH values of nutrient solution. Temperature is an important factor influencing the removal performance of BTF, particularly in winter. To ensure the successful acclimation, the recycling nutrient solution was heated to about 28 °C from Day 32 to 81. Results showed that higher temperature significantly promoted the removal of all VOSCs in both BTFs. Once the heating was halted from Day 82 to 96, drops of 19.4, 13.8 and 17.1% in REs of ethyl mercaptan, dimethyl sulfide and dimethyl disulfide were observed in BTF2, compared to 20.9, 15.5 and 28.4% seen in BTF1. Apparently, BTF2 was more robust to lower temperature than BTF1 was.

Both BTFs were temporally shut down from Day 109 to 137 due to national holidays. Results showed that about a month idle phase had different effects on the two BTF. No obvious decrease in REs of ethyl mercaptan and $H_2S$ were observed in both BTFs after shutdown. However, REs of DMS and DMDS decreased drastically in BTF1 from 51.9 and 72.0% to 23.1 and 21.8%

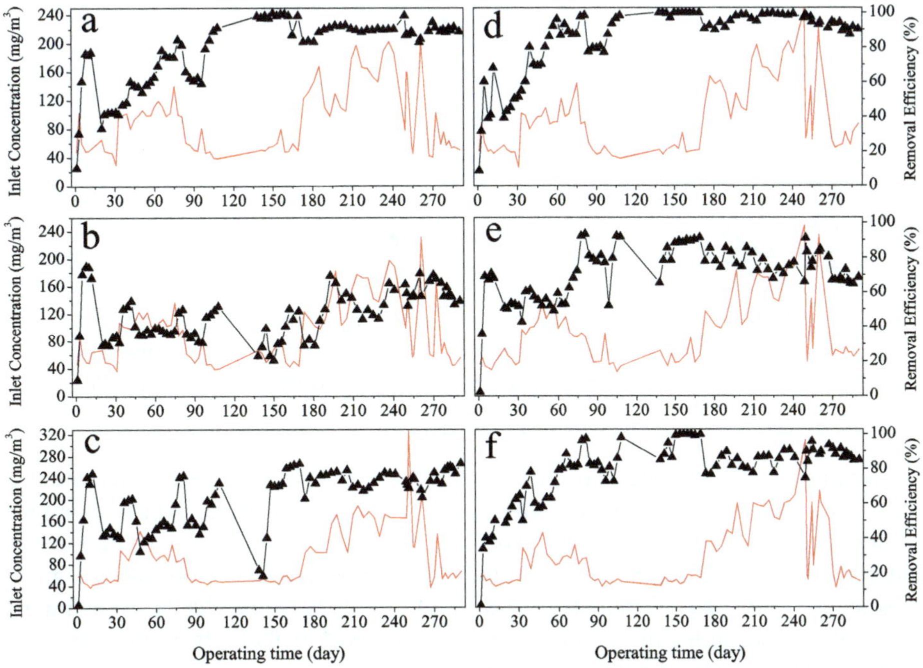


**Fig. 1** Performance of the biotrickling filters for the removal of volatile organic sulfur compounds: (**a**) EM removal in BTF1, (**b**) DMS removal in BTF1, (**c**) DMDS removal in BTF1, (**d**) EM removal in BTF2, (**e**) DMS removal in BTF2, (**f**) DMDS removal in BTF2

respectively. Their restoration to prior REs took about 15 and 30 days, respectively. As to BTF2, only 13% drop in the RE of DMDS was observed, and its full recovery took about one week. The RE of DMS was also affected, decreased from 91.8 to 65.3%. So again, BTF2 was superior to BTF1 against shutdown.

During the last phase of experiment (Day 270–290), pH values of nutrient solution were procedurally reduced from 7.0 to 5.5, resulting in significantly lowered removal performance of BTF2 on all organic sulfur compounds. On the contrary, BTF1 behaved more robust, possibly because it had been acidified already and the community inside got used to low pH.

### Overall removal capacities of both BTFs on four organic sulfur compounds

To further assess the overall performances, experimental data from Day 139 to 272 were used to perform the macro kinetics analysis. During this phase, the inlet concentration of individual component changed in a relatively wider range. Figure 2 showed that the Michaelis-Menten type model well fitted to the experimental EC ($P < 0.001$), indicating the two BTFs were not inhibited during this stage. The total $EC_{max}$ calculated from Michaelis-Menten model was 95.0 g/m$^3$ h for BTF1 and was 124.3 g/m$^3$ h for BTF2. Both estimated total ECs values are very higher than the actual, which was 54.9 for BTF1 and 68.8 g/m$^3$.h for BTF2 at inlet loading rates of 74.5 and 77.6 g/m$^3$.h, respectively. Macro kinetics showed that neither BTF reached its maximum EC, particularly for $H_2S$ since the REs of $H_2S$ in both BTFs were close to 100%. The $EC_{max}$ was the maximum elimination capacity that the BTFs can theoretically achieve in a given range of inlet concentration. However, to accurately predict the behavior of the BTFs in practical application, a pilot-scale experimental should be performed to obtain more empirical data for constructing a reliable kinetic model. Table 1 listed the removal performances of some bioreactors reported in literature. By comparison, the total maximum EC obtained in BTF2 was higher than the BTFs reported by Chen et al. [12–15]. However, the maximum EC varied in a much wide range, depends on the type of packing materials, the composition of waste gases or the configuration of bioreactor.

### Structure and composition of microbial communities in two BTFs

High-throughput sequencing of bacterial 16S rRNA genes generated an average of 16230 quality sequence reads per sample. All sequence reads were clustered into 1619 operational taxonomic units (OTUs) at 97% similarity, 96.4% of which could be assigned to bacterial or archaeal phylum. The most sequences belonged to bacterial domain, while Archaeal domain only accounted for less than 1.0% of all sequence. Bacterial sequences primarily comprised phylum Proteobacteria (68.7%), followed by phyla Bacteroidetes (15.3%), Actinobacteria

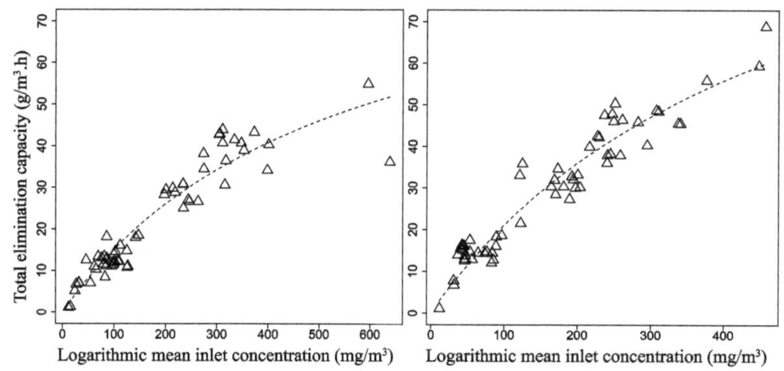


**Fig. 2** Fitting of Michaelis-Menten equation to the experimental elimination capacities recored in BTF1 (left) and BTF2 (right). Experimental data (triangle) and model data (dash-dot line)

(4.9%) and Planctomycetes (1.4%). Among the phylum Proteobacteria, class Beta proteobacteria had highest average relative abundance (30.9%), followed by class Gamma proteobacteria (18.4%) and Alpha proteobacterial (18. 1%). Besides, bacterial Firmicutes and archaeal phylum Euryarcheotic suddenly occurred in BTF1 on Day 150 with very higher relative abundances.

Principal co-ordinate analysis (PCoA) showed that data could be reduced into two principal components (PC1 and PC2) with combined eigenvalues explaining 45.0% of total variation. Figure 3 demonstrated that microbial communities inside both BTFs dynamically changed with the operating conditions. Samples taken on Day 150 were far distant from others in both BTFs, indicating the structure of microbial communities was seriously affected by the shutdown operation. However, it is evident that using combined packing material introduced a positive effect on the robustness of the microbial community, as seen in BTF2. More than two months later after the shutdown, samples on Day 220 and 250 taken from BTF2 closely grouped with those from Day 75, especially for those from the bottom layer. Comparatively, microbial communities inside BTF1 still sharply diverged even long after BTF1 was restored to the normal operation. Such community changes reconciled with the tendency of removal efficiencies observed in both BTFs. However, BTF2 was much more vulnerable to lower pH conditions than BTF1. Samples collected from BTF2 during acidification period were more distant from those collected on Day 75 and 220 compared to BTF1. It is most likely that microorganisms acclimated within BTF2 favored a neutral condition.

#### Correlation network analysis

Network analysis of taxon co-occurrence patterns can offer new insight into the interconnection of complex microbial communities. Co-occurrence networks were constructed for each BTF based on the Spearman's correlation ($r > 0.8$, adjusted $P$ values$< 0.001$). BTF1 network comprised 360 significant associations (edges) of 110 nodes with a clustering coefficient of 0.506 and an overall diameter of 7 (Fig. 4). BTF2 network exhibited 435 strongly positive associations of 108 nodes with an average clustering coefficient of 0.587 and an overall diameter of 9 (Fig. 4).

Both networks could be modularized into several major modules. The modularity of BTF1 and BTF2 network was 0.5590 and 0.6297, respectively. The modularity of more than 0.4 indicates that a network has modular structure [16]. BTF1 network contained 11 modules. Among them, module 2 was the biggest, which consisted of 34 highly interconnected nodes. BTF2 network mainly comprised of six modules. Module 1, 2 and

**Table 1** Removal performances of some bioreactors reported in literature

| Packing materials | Pollutants | Maximum EC (g/m$^3$.h) | Removal efficiency (%) | EBRT (s) | Reference |
|---|---|---|---|---|---|
| Polyurethane foam | DMDS | 86 | 100 | 40 | [2] |
| | $H_2S$ | 17 | 100 | | |
| Maifanite | DMDS | 19 | 100 | 123 | [12] |
| Polyurethane foam | $H_2S$ | 55 | 80 | 150 | [13] |
| Polyurethane foam | DMS | 23 | 65 | 40 | [14] |
| Polyethylene rings | DMS | 58 | 88 | 60 | [15] |
| | $H_2S$ | 83 | 100 | | |

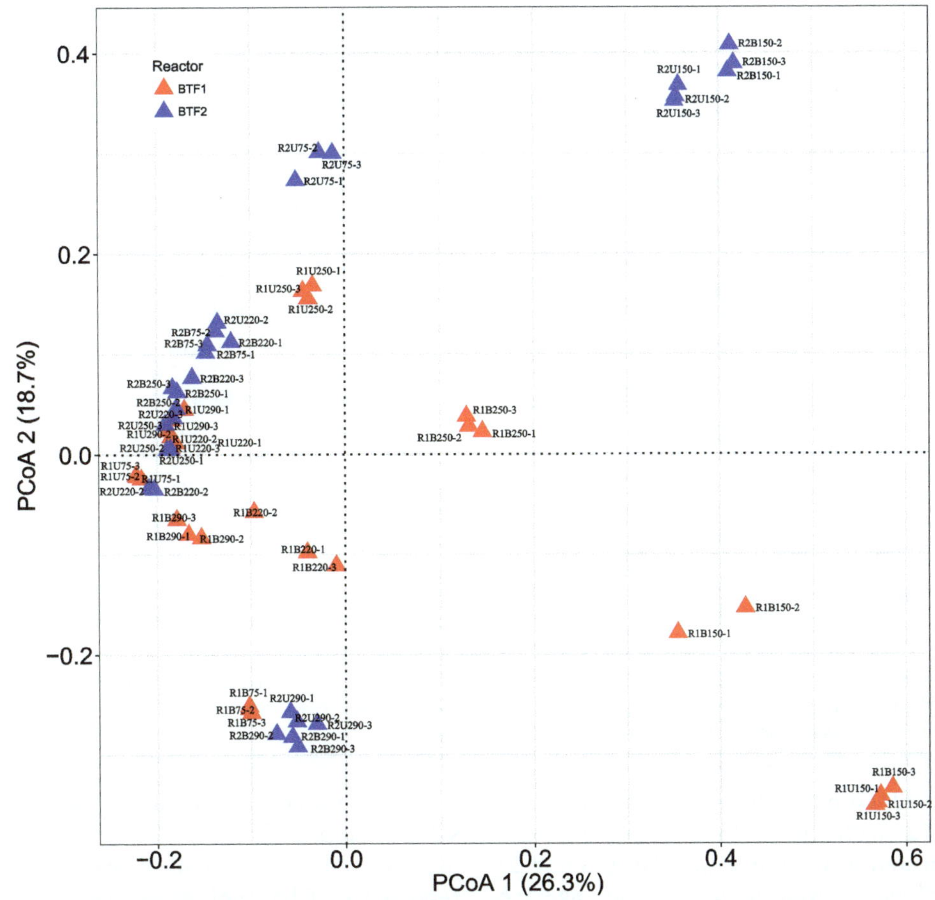


**Fig. 3** Principal coordinate analysis (PCoA) of microbial communities during different conditions

6 were three major highly dense modules. Although more nodes were included in BTF1 network, BTF2 network had denser connections. So, it was hypothesized that functional groups concerning sulfur oxidation may be enriched in BTF2, which resulted in a highly interconnected network.

## Discussion

Malodorous gases usually contain a variety of pollutants such as hydrogen sulfide, mercaptan and thioether. The complete bio-transformation of these compounds in biofiltration reactor generates sulfate as the main end product [17]. The accumulation of sulfate and the consequent drop in pH values inside packing materials can heavily affect the removal performance. When hydrogen sulfide co-exists with organic sulfur compounds in the odorous gases, some researchers suggested employing two-stage bioprocess with which hydrogen sulfide was removed in the first stage and organic sulfur compounds were in the second stage [10]. However, this would undoubtedly increase the footprint of bioreactor and the consequent construction cost.

Packing materials provide a surface for the biofilm formation. Different packing materials have different physico-chemical properties. Plastic balls and lava rocks are two kind of packing materials commonly used in bioreactor, and greatly differ in many aspects such as roughness, porosity and bulk density. The former features higher porosity and smooth surface by which metabolic product of sulfur compounds could be easily poured out from packing bed. Comparatively, lava rocks have rough surfaces, which retain microbial cells and metabolic products within the micropores.

It has been reported that the bio-degrading activities mainly occurred in the bottom layer of packing bed [11, 18]. As a result, acidification in the bottom layer is more severe, and the growth rates of sulfur-oxidizing microorganisms inhabiting there can be largely reduced. In the present study, lava rocks and plastic balls were used in combination in an effort to improve environmental conditions, and then to create different niches for more diverse microorganisms which could co-operated with each other to degrade the mixed pollutants. Results confirmed that BTF2 packed with combined materials outperformed BTF1

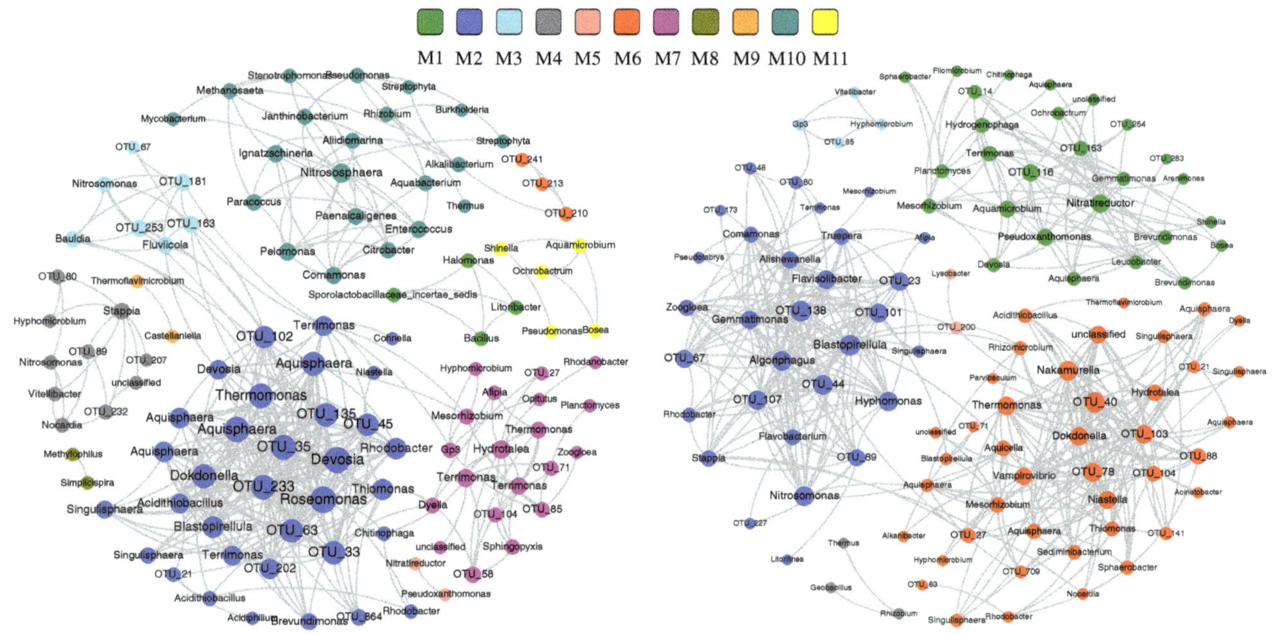


**Fig. 4** Co-occurrence network showing the correlation between bacterial OTUs in BTF1 (left) and BTF2 (right). A connection stands for a strong (Spearman's r > 0.8) and significant (Ajusted *p*-value< 0.001) correlation. The size of each node is proportional to the number of connections. Labels according to the taxonomic affiliations. The thickness of each connection between two nodes is proportional to the values of Spearman's correlation coefficients. Nodes were colored by modularity

the control, and all target pollutants were efficiently removed in BTF2 with more than 92.5% of REs. Paired t-test indicated a difference ($p = 0.004$, Table 2) in pH values in the bottom layer, with an average 2.85 in BTF1 and 4.86 in BTF2. Mainly owing to the low pH condition, the growth of heterotrophic microbes to some extent may be limited. The pressure drops across the filter bed did not exceed 50 Pa and clogging problem did not occur in both BTFs. Determination of metabolites from Day 251 to 288 also showed that the bottom packing bed of BTF1 accumulated more sulfate than BTF2 (32.3 mg/L versus 21.1 mg/L, $p < 0.05$) (Table 3), and sulfate has been reported to negatively impact the removal performance of biological reactor [18]. Besides, elevated elemental sulfur in BTF1 bottom layer (10.5 versus 2.1 mg/cm$^3$-packing) suggested an incomplete bio-oxidation of sulfur compounds in BTF1. It is most likely that low pH reduced the solubility of oxygen and pollutants as reported by Charnnok et al. [19].

**Table 2** pH values tested inside different filter layers of the two BTFs

| Sampling sites | BTF | Operating time (day) | | | | | | | *p*-value |
|---|---|---|---|---|---|---|---|---|---|
| | | 12 | 42 | 75 | 108 | 164 | 251 | 273 | |
| Upper | BTF1 | 6.74 | 6.68 | 7.05 | 7.34 | 7.51 | 6.53 | 6.99 | 0.5581 |
| | BTF2 | 6.71 | 7.66 | 7.03 | 7.43 | 6.90 | 6.99 | 6.93 | |
| Middle | BTF1 | 4.13 | 4.27 | 4.67 | 4.89 | 6.80 | 6.05 | 6.60 | 0.3443 |
| | BTF2 | 5.14 | 7.07 | 6.01 | 6.51 | 6.05 | 5.86 | 4.94 | |
| Bottom | BTF1 | 1.92 | 2.33 | 2.58 | 2.61 | 3.55 | 3.48 | 3.47 | 0.0004 |
| | BTF2 | 4.07 | 5.5 | 4.14 | 5.01 | 5.80 | 5.22 | 4.26 | |

Sulfate and elemental sulfur were the main end products from sulfur compounds biodegradation since no other intermediates were detected in both BTFs. Given that the average concentrations of sulfate in both BTFs were noticeably higher than that of elemental sulfur, it is likely that most sulfur compounds were bio-converted into sulfate. The occurrence of local anoxic zone within filter bed was the leading cause of the slight accumulation of elemental sulfur. Sulfur compounds could be incompletely bio-oxidized to elemental sulfur when oxygen supply is limited. Theoretical Gibbs free energy changes of DMS, EM and DMDS bio-oxidation were calculated according to stoichiometric Eqs. 1, 2, 3. Results indicate that the bio-oxidation of the three organic

**Table 3** Sulfate concentration tested inside different filter layers of the two BTFs (mg/L)

| location | day | 251 | 264 | 274 | 288 | *p*-value |
|---|---|---|---|---|---|---|
| Upper | BTF1 | 12.1 | 16.9 | 19.3 | 39.8 | 0.021 |
| | BTF2 | 9.5 | 11.5 | 14.9 | 31.6 | |
| Middle | BTF1 | 13.3 | 12.8 | 17.4 | 25.4 | 0.018 |
| | BTF2 | 8.4 | 10.8 | 10.6 | 20.5 | |
| Bottom | BTF1 | 18.7 | 21.0 | 27.0 | 62.5 | 0.031 |
| | BTF2 | 10.5 | 14.8 | 15.9 | 43.0 | |

sulfur compounds is thermodynamically favorable. BTF heavily relies on microorganisms to degrade various gaseous pollutants. In this study, all reduced sulfur compounds can serve as carbon or energy source to support the growth of microorganisms. Most energy generated in these processes is captured by microorganisms in the form of ATP for their anabolisms.

$$CH_3\text{-}S\text{-}CH_3 + 5O_2 \rightarrow 2CO_2 + 2H_2O + H_2SO_4 \ \Delta G^0 = -1948.9 \ KJ/mol \quad (1)$$

$$CH_3\text{-}CH_2\text{-}SH + 5O_2 \rightarrow 2CO_2 + 2H_2O + H_2SO_4 \ \Delta G^0 = -2021.0 \ KJ/mol \quad (2)$$

$$CH_3\text{-}S\text{-}S\text{-}CH_3 + 13/2 \ O_2 \rightarrow 2CO_2 + H_2O + 2H_2SO_4 \ \Delta G^0 = -2391.3 \ KJ/mol \quad (3)$$

Using oxygen as an electron acceptor, these compounds could be biodegraded to sulfate by phylogenetically diverse microorganisms following an oxidative pathway. For instance, DMS can be oxidized by methyltransferase or DMS monooxygenases to methanethiol. Methanethiol is subsequently degraded by methanethiol oxidase to produce sulfide. Finally, Sulfide is converted to sulfate via sulfite by the well-known SOX system [20].

The biodegradation pathway of EM and DMDS has not been well deciphered in literature. A few reports found that the EM degradation was initiated by formation of diethyl disulfide [21, 22]. Although the main end products were confirmed as elemental sulfur and sulfate based on the calculation of sulfur mass balance, the subsequent oxidation route of diethyl disulfide remains unclear, and enzymes and genes involved were not reported.

Together with these results confirmed that using combined packing materials successfully prevented severe acidification in the packing materials. Consequently, both BTFs differed in the structure of microbial diversity. Apart from samples taken on Day 150, the shannon index was significantly higher in BTF2 than in BTF1 (3.13 vs 2.69, $p < 0.01$). Microorganisms able to degrade sulfur compounds are extremely diverse, including autotrophs such as *Thiobacillus* and *Acidithiobacillus*, and heterotrophs microorganisms such as *Bacillus*, *Xanthobacter* and *Hyphomicrobium* [14, 15, 23, 24]. Most of them prefer neutral conditions. For instance, the drop pH value could lower the specific growth rate of genus *Hyphomicrobium* [25]. Only members of genus *Acidithiobacillus* could oxidize hydrogen sulfide in extremely acidic conditions [15].

OUT_6, OUT_7 and OUT_8 were the three key OUTs in BTF1 communities with relative abundance of 1.78, 2.86 and 2.48% on average. OUT_6 was assigned to genus *Thiomonas*, while OUT_7 and 8 were assigned to genus *Acidithiobacillus*. Both genera are (i) well-known sulfur-oxidizers, (ii) capable of chemo lithotrophic growth on various sulfur compounds [26, 27], and (iii) often reported to occur inside bioreactors treating sulfur-containing waste gases [17, 28]. Co-occurrence network analysis demonstrated that the three co-occurred in Module 2 of BTF1 network. So, it can be inferred that this module had potential links with the removal of sulfur compounds in BTF1. It is worth noting that the most abundant OUT (OUT_1) with relative abundance of 26.29% on average, were not constructed into BTF1 networks, possibly because it is autotrophs whose survival rarely relies on microbial incorporations. OUT_1 was assigned to genus *Thiobacillus*, and most of members of *Thiobacillus* are obligate autotrophs using elemental sulfur and reduced inorganic sulfur compounds as energy source [29]. The consistent predominance of *Thiobacillus* might account for the effective removal of $H_2S$ in both BTFs. $H_2S$ removal prior to other organic sulfur compounds is critical since that $H_2S$ is the main intermediate metabolite of organic sulfur compounds and has an inhibitive effect on the oxidation of the latter [30, 31]. OUT_2 and OUT_4 were other two predominant OTUs of not being constructed into BTF1 network. Their average relative abundances in BTF1 reached to 8.89 and 3.66%, respectively. OUT_2 belonged to the genus *Halothiobacillus*, while OUT_4 was classified as member of genus *Hyphomicrobium*. Members of both genera were reportedly the important sulfur-oxidizing microbes [32, 33].

Sulfur-degraders inside BTF2 network were mainly clustered in Module 6, which contained 42 highly interconnected nodes. This module resembles Module 2 of BTF1 network in ecological function but is much denser, suggesting that BTF2 harbored more diverse microorganisms involving in the removal of sulfur compounds. OTU_6 (*Thiomonas*) and OTU_8 (*Acidithiobacillus*) also predominated in BTF2 with average relative abundance of 2.52 and 0.69%, respectively. In addition, OTU_15 (*Hyphomicrobium*) co-occurred with OTU_6 and OTU_8 in Module 6, which was another bacterial taxon that dominated inside BTF2 (4.03%). Methylotrophic *Hyphomicrobium* is considered as an important degrader of organic sulfur compounds [34], so its co-occurrence with other sulfur-oxidizing bacteria might contribute a lot to the simultaneous removal of $H_2S$ and organic sulfur compounds. OTU_1 (*Thiobacillus*) was also the most predominant microbial taxon (28.9%) but failed to be constructed in BTF2 network. Unlike BTF1, other top abundant taxa in BTF2 were successfully included in the network, indicating that more type of microorganisms took actively part in the degradation of mixed sulfur compounds in BTF2, resulting in a much-enhanced co-operative network.

Together, this study proved that using combined packing materials efficiently enhanced the synchronous removal of mixed sulfur compounds by facilitating the elimination of acidic product out from filter bed. This strategy offers a promising alternative for the control of malodorous gases. Nonetheless, there is still a pressing need to further understand the ecological mechanisms that underlie the synergistic removal of complex sulfur compounds in one-stage bioreactor. To achieve the goal, functional genes that are responsible for degradation of each sulfur compound need to be quantitatively analyzed and should be linked to specific microbial species. Besides, the understanding of the removal profile of each malodorous pollutant and the spatial distribution of microbial community along the filter bed will aid in identifying the roles played by the upper and bottom packing materials in removing complex pollutants.

Malodorous gas that generated from a real emission source is usually complicated. The composition and the concentration of malodorous gases often change. To better use this strategy, pilot-scale experiments are required to be conducted in some industrial emission source. For industrial application, the volume ratio of the upper/bottom packing materials should be adjusted correspondently, depending on the composition and the concentration of malodorous gas. Other materials having similar physico-chemical properties also can be used as packing materials of biotrickling filters in the same way.

## Conclusions

The removal of malodorous pollutants mostly occurred in the bottom packing materials of biotrickling filter. However, the roles of upper packing materials in buffering the distribution of pH values along the packing bed are usually neglected. Results confirmed that using different packing materials in combination greatly enhanced the removal of mixed sulfur compounds. More favorable environmental condition for the growth of microorganisms was created in BTF2, resulting in a denser microbial correlation network in which diverse microorganisms took part in the degradation of mixed sulfur compounds. The robustness of microbial community against environmental stress such as shutdown and low temperature, was also enhanced in BTF2. These findings shed some light on the biofiltration of complex malodorous gases.

## Methods

### Biotrickling filters setup and operation

Two identical laboratory-scale biotrickling filters (BTFs) were employed, named BTF1 and BTF2, respectively. The schematic diagram and the actual photograph of the BTFs was showed in Figs. 5 and 6, respectively. Each was composed of three columns in the stacked configuration, with an inner diameter of 10 cm and a total height of 38 cm. The resulted packing volume of each BTF was about 3 L. BTF1 was packed with lava rocks, which have particle sizes of 6–10 mm and bulk density of 0.57 g/cm$^3$. BTF2 was packed with lava rocks in the bottom layer, while with plastic ball in the upper layer. Plastic balls are polyhedral and hollow spheres which are made of polypropylene and are 2.5 cm in diameter. The bulk density, porosity, and specific surface area of plastic ball are 80 kg/m$^3$, 92% and 460 m$^2$/m$^3$, respectively. For generating the malodorous gases, compressed air was split into a major and a minor airstream. Mixed liquid of three organic sulfur compounds including dimethyl sulfide (DMS), ethyl mercaptan (EM) and dimethyl disulfide (DMDS), was injected into the minor air stream via a syringe pump

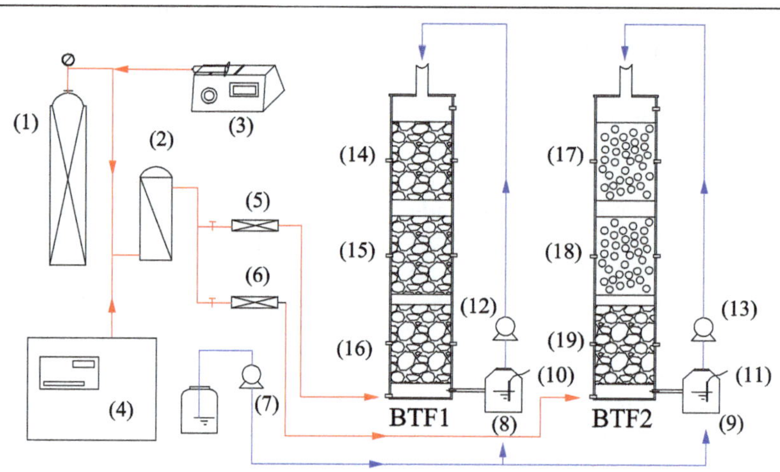


**Fig. 5** Schematic diagram of the biotrickling filters (1) $H_2S$ cylinder, (2) mixing chamber, (3) syringe pump, (4) air compressor, (5 and 6) gas flowmeter, (7) NaOH dosing pump, (8 and 9) nutrient tank, (10 and 11) pH probe, (12 and 13) peristaltic pump, (14–19) filter material sampling ports

**Fig. 6** Actual photograph of the biotrickling filters (1) nutrient solution, (2) bottom filter bed, (3 and 5) filter materials sampling ports, (4) upper filter bed, (6) peristaltic pump, (7) syringe pump, (8) gas flowmeter

(ShenChen, SPLab02, China). $H_2S$ vapors were also directly introduced from a gas cylinder into the minor air stream. Two air streams were finally mixed in a chamber and were then fed to both BTFs in the up-flow mode.

Both BTFs were operated at an empty bed residence time (EBRT) of 60 s across the experiment. The experimental schedule and operating conditions were listed in Table 4. Initial inoculum originated from a previous biotrickling filter treating $H_2S$-containing gases [35]. Nutrient solution was stored in a holding-tank, and evenly sprayed into packing column by using a peristaltic pump (ShenChen, BT300, China). Nutrient solution contained $NH_4Cl$ 2.5 g, $Na_2HPO_4$ 1.0 g, $KH_2PO_4$ 0.7 g, $MgSO_4$ 0.05 g, $CaCl_2$ 0.015 g per liter in water. The pH values of nutrient solution were adjusted at 7.0 by automatically adding alkaline by using a dosing pump (SKEO, AKS600, Italy). Nutrient solution was periodically renewed to ensure sufficient nutrient and moisture for the growth of microorganisms.

### Gas-phase determination

Gas-phase concentrations were measured with a gas chromatograph (Shimadzu GC-2010, Japan) equipped with an FPD detector. The column used was a GS-Gas Pro capillary column (30 m × 0.32 mm × 1.0 μm, Agilent Technologies, USA). Gas samples were taken using Tedlar bags of 2 L. Total volume of 100 μL was injected into the GC using a gastight syringe. The injector and detector temperature were set at 70 °C and 250 °C, respectively. The GC oven temperature was programmed as follows: initial temperature of 80 °C for 2 min, increase to 250 °C at 10 °C $min^{-1}$ and maintain for 5 min.

**Table 4** Experimental schedule and operating conditions

| Phase | Time | Average inlet concentration (mg/m$^3$) | | | | Temperature (°C) | pH |
|---|---|---|---|---|---|---|---|
| | day | $H_2S$ | EM | DMS | DMDS | | |
| 1 | 1–31 | 101.0 ± 10.8 | 58.2 ± 17.2 | 59.3 ± 17.2 | 57.4 ± 15.5 | 19.1 ± 4.1 | 7.0 |
| 2 | 32–81 | 188.4 ± 56.1 | 102.5 ± 14.5 | 106.1 ± 13.9 | 100.9 ± 16.7 | 28.0 ± 0 | 7.0 |
| 3 | 82–96 | 109.7 ± 21.4 | 56.3 ± 9.6 | 62.0 ± 15.5 | 51.0 ± 5.4 | 8.0 ± 4.9 | 7.0 |
| 4 | 97–108 | 98.7 ± 1.4 | 44.0 ± 5.0 | 43.5 ± 4.1 | 48.8 ± 3.1 | 28.0 ± 0 | 7.0 |
| 5 | 109–138* | – | – | – | – | – | – |
| 6 | 139–272 | 174.2 ± 93.1 | 96.3 ± 74.5 | 99.1 ± 78.5 | 105.5 ± 83.6 | 26.9 ± 1.3 | 7.0 |
| 7 | 273–290 | 118.1 ± 16.5 | 59.1 ± 13.2 | 60.7 ± 11.4 | 59.8 ± 10.9 | 26.9 ± 1.3 | 5.5 ± 0.8 |

*Both BTFs were shut down during this phase

Macro-kinetic model was used to evaluate the elimination capacities (ECs) of each BTF. The ECs commonly follow a behavior that can be adequately described by a Michaelis-Menten model type [36]. This model is constructed based on ECs and is relatively simple and useful in predicting the removal performance of practical engineering.

$$EC = EC_{max}\frac{Cln}{Ks + Cln}$$

Where $EC_{max}$ (g.m$^3$/h) is the maximum elimination capacity, Cln (g/m$^3$) is the logarithmic average of the inlet and outlet concentrations of pollutants in the gas phase and Ks (g/m$^3$) is the saturation constant.

### DNA extraction and high-throughput sequencing

2 g of packing materials were sterilely collected in triplicate from the upper and bottom filter layer of each BTF on Day 75, 150, 220, 250 and 290. Packing materials were mixed with 10 mL phosphate buffer (NaCl 8 g/L, KCl 0.2 g/L, $Na_2HPO_4$ 1.42 g/L, $KH_2PO_4$ 0.27 g/L). After gently shaking, biofilms were completely detached from the surface by ultrasonication for 10 min. The suspensions were collected and centrifuged for 10 min at 10, 000 g, and the resulted pellet was collected for the DNA extraction.

Genomic DNA was extracted using the MO-BIO PowerSoil® DNA Isolation Kit (MO BIO Laboratories, Inc., USA) according to the manufacturer's instructions. Universal primers 515F (5′-GTG CCA GCM GCCGCG GTA A-3′) and 909R (5′- CCC CGY CAA TTC MTT TRA GT – 3′) with 12 nt unique barcodes were used to amplify the V4 and V5 hypervariable regions of the 16S rRNA gene for pyrosequencing using a MiSeq sequencer.

### Data analysis

The raw data were firstly quality-filtered with QIIME Pipeline to remove reads that did not meet the desired quality [37]. All sequence reads were trimmed and assigned to the corresponding samples based on their barcodes. Multiple steps were taken to trim the sequences, such as the removal of sequences < 200 bp, and average base quality score Q < 25. Any chimeric sequences were identified and removed using Uchime algorithm. Sequences were clustered into operational taxonomic units (OTUs) using a 97% similarity. Taxonomic assignment was performed using the RDP classifier at a confidence level of 80%. All samples were randomly rarefied to an equal sequence number to neutralize the bias of varied sequencing depth.

Principal coordinate analysis (PCoA) was performed using vegan and ggplot2 packages in R software. PCoA provides information regarding the largest source of variation in the data and allows the observation of similarities and differences between samples.

Co-occurrence networks were inferred for each BTF based on the Spearman correlation matrix between bacterial OTUs. Valid connection indicates a strong ($r > 0.8$) and significant ($p < 0.001$) Spearman's correlation. To minimize pairwise comparisons and reduce the network complexity, only OTUs with more than 0.05% of relative abundance were considered. Both the correlation matrix and significance matrix were calculated using Hmisc package in R. To reduce the chances of obtaining false-positive results, *P*-values were adjusted by FDR using the BH method with the multtest package in R [38]. The network was visualized using Cytoscape 3.5.1 [39]. Modular structure of highly interconnected nodes was detected with multilevel algorithm using igraph package in R. A number of parameters (e.g. number of nodes and edges, betweenness centrality, clustering and modularity coefficient) were calculated using igraph too. OTUs with maximum betweenness centrality score were considered as keystone species [40].

#### Abbreviations

BTF: Biotrickling filter; DMDS: Dimethyl disulfide; DMS: Dimethyl sulfide; EBRT: Empty bed residence time; EC: Elimination capacity; EM: Ethyl mercaptan; OTUs: Operational taxonomic units; PCoA: Principal co-ordinate analysis; RE: Removal efficiency

**Acknowledgements**
Not applicable.

**Authors' contributions**
XT, RFF and JJL performed the experiments. JJL, XT and EZL designed the project and performed the bioinformatics studies. MYX, SBH, JJL and JG interpreted the experimental data and supervised the researching process. GZ performed the kinetics studies. JJL, XT and EZL wrote the manuscript. All authors read and approved the final manuscript.

**Funding**
This work was financially supported by the National Natural Science Foundation of China (U1701243, 31270169 and 41501244), the Science and Technology Projects of Guangdong (2016A030310314 and 2017B090901049), and the Science and Technology Projects of Guangzhou (201504010014, 201704020204 and 201707010377).

**Ethics approval and consent to participate**
Not applicable.

**Consent for publication**
Not applicable.

**Competing interests**
The authors declare that they have no competing interests.

**Author details**
[1]Guangdong Institute of Microbiology, Guangdong Academy of Sciences, Guangzhou 510070, People's Republic of China. [2]State Key Laboratory of Applied Microbiology Southern China, Guangzhou 510070, People's Republic of China. [3]Guangdong Provincial Key Laboratory of Microbial Culture Collection and Application, Guangzhou 510070, People's Republic of China. [4]School of Environment and Energy, South China University of Technology, Guangzhou 510006, People's Republic of China.

**References**

1. Barbusinski K, Kalemba K, Kasperczyk D, Urbaniec K, Kozik V. Biological methods for odor treatment-a review. J Clean Prod. 2017;152:223–41.
2. Arellano-García L, Le Borgne S, Revah S. Simultaneous treatment of dimethyl disulfide and hydrogen sulfide in an alkaline biotrickling filter. Chemosphere. 2018;191:809–16.
3. Wing S, Lowman A, Keil A, Marshall SW. Odors from sewage sludge and livestock: associations with self-reported health. Public Health Rep. 2014;129:505–15.
4. Jiang GM, Sun J, Sharma KR, Yuan ZG. Corrosion and odor management in sewer systems. Curr Opin Biotech. 2015;33:192–7.
5. Zhang YF, Liss SN, Allen DG. Effect of methanol on pH and stability of inorganic biofilters treating dimethyl sulfide. Environ Sci Technol. 2007;41:3752–7.
6. Barzgar S, Hettiaratchi JP, Pearse L, Kumar S. Inhibitory effects of acidic pH and confounding effects of moisture content on methane biofiltration. Bioresour Technol. 2017;245:633–40.
7. Rojo N, Munoz R, Gallastegui G, Barona A, Gurtubay L, Prenafeta-Boldu FX, Elias A. Carbon disulfide biofiltration: influence of the accumulation of biodegradation products on biomass development. J Chem Technol Biot. 2012;87:764–71.
8. Shu CH, Chen CK. Enhanced removal of dimethyl sulfide from a synthetic waste gas stream using a bioreactor inoculated with *Microbacterium* sp NTUT26 and *Pseudomonas putida*. J Ind Microbiol Biot. 2009;36:95–104.
9. Ramírez M, Fernández M, Granada C, Le Borgne S, Gómez JM, Cantero D. Biofiltration of reduced Sulphur compounds and community analysis of Sulphur-oxidizing bacteria. Bioresour Technol. 2011;102:4047–53.
10. Jover J, Ramírez M, Rodríguez I, Gómez JM, Cantero D. Strategies for pH control in a biofilter packed with sugarcane bagasse for hydrogen sulfide removal. J Environ Sci Heal A. 2012;47:990–6.
11. Tu X, Li JJ, Feng RF, Sun GP, Guo J. Comparison of removal behavior of two biotrickling filters under transient condition and effect of pH on the bacterial communities. PLoS One. 2016;11:5.
12. Chen XQ, Liang ZS, An TC, Li GY. Comparative elimination of dimethyl disulfide by maifanite and ceramic-packed biotrickling filters and their response to microbial community. Bioresour Technol. 2016;202:76–83.
13. Ramírez M, Gómez JM, Aroca G, Cantero D. Removal of hydrogen sulfide by immobilized *Thiobacillus thioparus* in a biotrickling filter packed with polyurethane foam. Bioresour Technol. 2009;100:4989–95.
14. Arellano-García L, Revah S, Ramírez M, Gómez JM, Cantero D. Dimethyl sulphide degradation using immobilized *Thiobacillus thioparus* in a biotrickling filter. Environ Technol. 2009;30:1273–9.
15. Sercu B, Núñez D, Van Langenhove H, Aroca G, Verstraete W. Operational and microbiological aspects of a bioaugmented two-stage biotrickling filter removing hydrogen sulfide and dimethyl sulfide. Biotechnol Bioeng. 2005;90:259–69.
16. Newman ME. Modularity and community structure in networks. PNAS. 2006;103:8577–82.
17. Tang K, Baskaran V, Nemati M. Bacteria of the Sulphur cycle: An overview of microbiology, biokinetics and their role in petroleum and mining industries. Biochem Eng J. 2009;44:73–94.
18. Lebrero R, Rodríguez E, García-Encina PA, Muñoz R. A comparative assessment of biofiltration and activated sludge diffusion for odour abatement. J Hazard Mater. 2011;190:622–30.
19. Charnnok B, Suksaroj T, Boonswang P, Chaiprapat S. Oxidation of hydrogen sulfide in biogas using dissolved oxygen in the extreme acidic biofiltration operation. Bioresour Technol. 2013;131:492–9.
20. Schäfer H, Myronova N, Boden R. Microbial degradation of dimethylsulphide and related C-1-Sulphur compounds: organisms and pathways controlling fluxes of Sulphur in the biosphere. J Exp Bot. 2010;61(2):315–34.
21. Wang XQ, Wu C, Liu N, Li SJ, Li W, Chen JM, Chen DZ. Degradation of ethyl mercaptan and its major intermediate diethyl disulfide by *Pseudomonas* sp strain WL2. App Microbiol Biotechnol. 2015;99(7):3211–20.
22. Sedighi M, Zamir SM, Vahabzadeh F. Cometabolic degradation of ethyl mercaptan by phenol-utilizing Ralstonia eutropha in suspended growth and gas-recycling trickle-bed reactor. J Environ Manag. 2016;165:53–61.
23. Giri BS, Mudliar SN, Deshmukh SC, Banerjee S, Pandey RA. Treatment of waste gas containing low concentration of dimethyl sulphide (DMS) in a bench-scale biofilter. Bioresour Technol. 2010;101:2185–90.
24. Giri BS, Pandey RA. Biological treatment of gaseous emissions containing dimethyl sulphide generated from pulp and paper industry. Bioresour Technol. 2013;142:420–7.
25. Hayes AC, Liss SN, Allen DG. Growth kinetics of *Hyphomicrobium* and *Thiobacillus* spp. in mixed cultures degrading dimethyl sulfide and methanol. Appl Environ Microbiol. 2010;76:5423–31.
26. Vésteinsdóttir H, Reynisdóttir DB, Örlygsson J. *Thiomonas islandica* sp. nov., a moderately thermophilic, hydrogen- and sulfur-oxidizing betaproteobacterium isolated from a hot spring. Int J Syst Evol Microbiol. 2011;61:132–7.
27. Nuñez H, Moya-Beltrán A, Covarrubias PC, Issotta F, Cárdenas JP, González M, Atavales J, Acuña LG, Johnson DB, Quatrini R. Molecular systematics of the genus *Acidithiobacillus*: insights into the phylogenetic structure and diversification of the taxon. Front Microbiol. 2017;8:30.
28. Pol A, van der Drift C, Op den Camp HJ. Isolation of a carbon disulfide utilizing *Thiomonas* sp. and its application in a biotrickling filter. Appl Microbiol Biotechnol. 2007;74:439–46.
29. Agha ABA, Kahrizi D, Ahmadvand A, Bashiri H, Fakhri R. Development of PCR primer systems for amplification of 16S-rDNA to detect of *Thiobacillus* spp. Cell Mol Bio. 2017;63:37–41.
30. Cáceres M, Silva J, Morales M, Martín RS, Aroca G. Kinetics of the bio-oxidation of volatile reduced Sulphur compounds in a biotrickling filter. Bioresour Technol. 2012;118:243–8.
31. Ben Jaber M, Anet B, Amrane A, Couriol C, Lendormi T, Le Cloirec P, Cogny G, Fillières R. Impact of nutrients supply and pH changes on the elimination of hydrogen sulfide, dimethyl disulfide and ethanethiol by biofiltration. Chem Eng J. 2014;258:420–6.
32. Boden R. Reclassification of *Halothiobacillus hydrothermalis* and *Halothiobacillus halophilus* to *Guyparkeria gen*. Nov. in the *Thioalkali*

*bacteraceae fam*. Nov., with emended descriptions of the genus *Halothiobacillus* and family Halothiobacillaceae. Int J Syst Evol Microbiol. 2017;67:3919–28.
33. Fernández HT, Rico IR, Prida JJ, Van Langenhove H. Dimethyl sulfide biofiltration using immobilized *Hyphomicrobium* VS and *Thiobacillus thioparus* TK-m in sugarcane bagasse. Environ Technol. 2013;34:257–62.
34. Boden R, Borodina E, Wood AP, Kelly DP, Murrell JC, Schafer H. Purification and characterization of dimethylsulfide monooxygenase from *Hyphomicrobium sulfonivorans*. J Bacteriol. 2011;193:1250–8.
35. Tu X, Guo J, Yang YG, Feng RF, Sun GP, Li JJ. Biofilms formed within the acidic and the neutral biotrickling filters for treating $H_2S$-containing waste gases. RSC Adv. 2017;7:25475–82.
36. Vergara-Fernández A, Revah S, Moreno-Casas P, Scott F. Biofiltration of volatile organic compounds using fungi and its conceptual and mathematical modeling. Biotechnol Adv. 2018;36(4):1079–93.
37. 454 Overview Tutorial: de novo OTU picking and diversity analyses using 454 data. http://qiime.org/tutorials/tutorial.html. Accessed 12 Feb 2018.
38. Benjamini Y, Hochberg Y. Controlling the false discovery rate - a practical and powerful approach to multiple testing. J Roy Stat Soc B Met. 1995;57:289–300.
39. Shannon P, Markiel A, Ozier O, Baliga NS, Wang JT, Ramage D, Amin N, Schwikowski B, Ideker T. Cytoscape: a software environment for integrated models of biomolecular interaction networks. Genome Res. 2003;13:2498–504.
40. Banerjee S, Kirkby CA, Schmutter D, Bissett A, Kirkegaard JA, Richardson AE. Network analysis reveals functional redundancy and keystone taxa amongst bacterial and fungal communities during organic matter decomposition in an arable soil. Soil Biol Biochem. 2016;97:188–98.

# 3

# Characterisation of three novel α-L-arabinofuranosidases from a compost metagenome

Brent Fortune[1], Sizwe Mhlongo[1], Leonardo Joaquim van Zyl[1], Robert Huddy[1,2], Mariette Smart[1,2] and Marla Trindade[1*]

**Abstract**

**Background:** The importance of the accessory enzymes such as α-L-arabinofuranosidases (AFases) in synergistic interactions within cellulolytic mixtures has introduced a paradigm shift in the search for hydrolytic enzymes. The aim of this study was to characterize novel AFase genes encoding enzymes with differing temperature optima and thermostabilities for use in hydrolytic cocktails.

**Results:** Three fosmids, pFos-H4, E3 and D3 were selected from the cloned metagenome of high temperature compost, expressed in *Escherichia coli* and subsequently purified to homogeneity from cell lysate. All the AFases were clustered within the GH51 AFase family and shared a homo-hexameric structure. Both AFase-E3 and H4 showed optimal activity at 60 °C while AFase-D3 had unique properties as it showed optimal activity at 25 °C as well as the ability to maintain substantial activity at temperatures as high as 90 °C. However, AFase-E3 was the most thermostable amongst the three AFases showing full activity even at 70 °C. The maximum activity was observed at a pH profile between pH 4.0–6.0 for all three AFases with optimal activity for AFase H4, D3 and E3 at pH 5.0, 4.5 and 4.0, respectively. All the AFases showed $K_M$ range between 0.31 mM and 0.43 mM, $K_{cat}$ range between 131 $s^{-1}$ and 219 $s^{-1}$ and the specific activity for AFase-H4, AFases-E3 and was 143, 228 and 175 U/mg, respectively. AFases-E3 and D3 displayed activities against pNP-β-L-arabinopyranoside and pNP-β-L-mannopyranoside respectively, and both hydrolysed pNP-β-D-glucopyranoside.

**Conclusion:** All three AFases displayed different biochemical characteristics despite all showing conserved overall structural similarity with typical domains of AFases belonging to GH51 family. The hydrolysis of cellobiose by a GH51 family AFase is demonstrated for the first time in this study.

**Keywords:** Thermostability, Arabinofuranosidase, Compost, Metagenomics

## Background

Hemicellulose is a highly branched hetero polymer which represents approximately one third of the total dry weight of plant biomass [1]. Heteroxylan forms the backbone of hemicellulose with xylopyranosyl residues linked by β-1, 4-glycosidic bonds which are mainly substituted with arabinose or glucuronic acid at their 2-O and/or 3-O positions [2, 3]. L-arabinosyl residues are mainly associated with arabinan, arabinoxylan, gum Arabic and arabinogalactan [4]. These residues participate in the crosslinking within the plant cell wall structure and thus strongly inhibit the degradation of xylan to simple xylose sugar units by xylan-degrading enzymes [5]. It is estimated that 40 million tons of xylan-rich agro-industrial biomass is generated globally on an annual basis. The potential of xylan-rich biomass feedstock has triggered an increasing interest in bioprospecting efforts for enzymes to degrade them to monomeric sugars which can be converted or used as part of raw materials in the production of value added products such as bioplastics, biodiesel and bioethanol [6].

The composition of enzyme cocktails largely depends on the type of biomass to be hydrolysed, the complexity of the bonds and carbohydrate structures [3]. Enzyme cocktails for biomass degradation typically include

* Correspondence: ituffin@uwc.ac.za
[1]Institute for Microbial Biotechnology and Metagenomics, University of the Western Cape, Bellville, South Africa

endoglucanases (EC 3.2.1.4), cellobiohydrolases (EC 3.2.1.91), β-glucosidases (EC 3.2.1.21) and a variety of hemicellulases. The core hemicellulases include endo-β-xylanases (EC 3.2.1.8), β-1,4-D-xylosidases (EC 3.2.1.37) and α-L-arabinofuranosidases (EC 3.2.1.55). These hemicellulases synergistically hydrolyse hemicellulose where α-L-arabinofuranosidase is involved in the hydrolysis of α-1,2 and α-1,3 glycosidic bonds that link α-arabinofuranoyl side moieties while endo-β-xylanases and β-1,4-D-xylosidases act on the β-1,4-bonds that link D-xylosyl residues [7].

Extreme environments such as hot springs, deep sea hydrothermal vents and organic composts are reservoirs of unique microbial diversity, providing the potential for isolating novel enzymes with desirable industrial properties [8]. The adaptation of microbial communities to these environmental conditions explains their high genomic and metabolic flexibility, and they often encode enzymes with properties suitable for many industrial applications [9]. Frequently, enzymes from these environments have been shown to be robust catalysts able to withstand high temperatures which are used to promote opening of the structure of lignocellulosic material. This not only leads to better enzyme penetration and therefore cell wall degradation but reduces or eliminates the need for cooling of the material prior to enzymatic pre-treatment [8]. This has the potential of increasing reaction rates by several orders of magnitude resulting in reduced enzyme loading and time required for efficient hydrolysis and saccharification, thus reducing the cost of the overall production process [8, 10, 11].

Using function-based metagenomic screening approaches, novel sequences from both cultivable and uncultivable microorganisms can be exploited without prior sequence information [6]. This screening approach thus provides access to previously unknown genes and their novel enzymes with unique structural and kinetic properties [12, 13]. Ohlhoff et al. [14] constructed a fosmid metagenomic library from high temperature compost which was subsequently screened for several classes of lignocellulases [15]. In this study, three AFases were chosen from these initial screens and were characterized with the aim of bioprospecting for unique biochemical properties that could be of use in industrial applications.

## Methods

### Bacterial strains and plasmids used in this study

Bacterial strains and plasmids used in this study are listed in Table 1 and primers are listed in Table 2. Restriction endonucleases and T4 DNA ligase were purchased from Fermentas (ThermoFisher Scientific) and used following the manufacturer's instructions. Unless otherwise stated, all *E. coli* strains were maintained in Luria broth (LB) or agar (LA) (Sigma Aldrich, South Africa) and cultivated at 37 °C with shaking at 160 rpm. *E. coli* strains transformed with pJET1.2 and pET21a plasmids were cultured in medium supplemented with 100 μg/mL ampicillin. *E. coli* strains transformed with

**Table 1** Bacterial strains, plasmid vectors and their recombinant versions used in this study

| Strain | Genotype[a] | Reference |
|---|---|---|
| *E. coli* Epi300 | F-mcrA D(mrr-hsdRMS-mcrBC) f80dlacZDM15 DlacX74recA1 endA1 araD139 (ara, leu) 7697 galK1-rspLnupGtrfr | Invitrogen, USA |
| Genehog | F-mcrA Δ(mrr-hsdRMS-mcrBC) $_{φ}$80dlacZΔM15 ΔlacX74recA1 araD139 Δ(ara, leu) 7697 galUgalKrpsL (StrR(endA1 nupGfhuAIS2r | Epicentre Biotechnology, USA |
| BL21 (DE3) | F-ompThsdSB(rB-mB-)gal dcm gal Λ(DE3) | Invitrogen, USA |
| pCC1Fos™ | pCC1Fos™L-Arabinose inducible promoter Copy Control; $Cam^R$, F factor ori, oriV high copy ori, λcos site for λ packaging, Bacteriophage T7 RNA polymerase promoter | Invitrogen, USA |
| pFos-H4 | pCC1FOS containing 17.5 kb of cloned metagenomic DNA as an insert with AFase activity, $Cam^R$ | IMBM |
| pFos-E3 | pCC1FOS containing 20.7 kb of cloned metagenomic DNA as an insert with AFase activity, $Cam^R$ | Dr C. Ohlhoff, IMBM, UWC, SA |
| pFos-D3 | pCC1FOS containing 10. 7 kb of cloned metagenomic DNA as an insert with AFase activity, $Cam^R$ | Dr C. Ohlhoff, IMBM, UWC, SA |
| pJET 1.2/blunt | Suicide cloning vector (eco47IR), blunt DNA ends for ligation with insert, T7 promoter, $Amp^R$ | Fermentas, USA |
| pJET-H4 | 1467 bp AFase-H4 gene amplicon blunt-end ligated into pJet1.2 | This study |
| pJET-E3 | 1547 bp AFase-E3 gene amplicon blunt-end ligated into pJet1.2 | This study |
| pJET-D3 | 1482 bp AFase-D3 gene amplicon blunt-end ligated into pJet1.2 | This study |
| pET21a(+) | Expression vector with a C- terminal His-tag, AmpR, T7 promoter and terminator, MCS. | Novagen, USA |
| pET21a-H4 | 1467 bp NdeI-XhoI fragment from pJET-H4 cloned in pET21a. | This study |
| pET21a-E3 | 1547 bp NdeI- HindIII fragment from pJET-H4 cloned in pET21a. | This study |
| pET21a-D3 | 1482 bp NdeI-XhoI fragment from pJET-H4 cloned in pET21a. | This study |

[a] Plasmid encoding Ampicillin and Chloramphenicol resistance is indicated as $Amp^R$ and $Cam^R$, respectively. Ori is used as an abbreviation for origin of replication and MCS is the acronym for multiple cloning site. The Hexa-histidine tag is abbreviated to His-tag that was used to purify the AFase proteins of this study through nickel-ion affinity liquid chromatography

**Table 2** Primers used in this study

| Primer | 5′ to 3′ sequence[a] | Reference |
|---|---|---|
| T7 Promoter forward | TAATACGACTCACTATAGGG | Epicentre® Biotechnologies |
| pCC1Fos reverse | CTCGTATGTTGTGTGGAATTGTGAGC | Epicentre® Biotechnology |
| MUKAN-1 FP-1 | CTGGTCCACCTACAACAAAGG | Epicentre® Biotechnologies |
| MUKAN-1 RP-1 | AGAGATTTTGAGACAGGATCCG | Epicentre® Biotechnologies |
| M13 Forward | CCCAGTCACGACGTTGTAAAACG | Inqaba Biotec |
| M13 Reverse | AGCGGATAACAATTTCACACAGG | Inqaba Biotec |
| H4-pET-Fwd | GTT**CATATG**AATCACATCAAGATTGATTTAGAT | This Study |
| H4-pET-Rev | CG**CTCGAG**TAAGTCAAAGCTGAGC | This Study |
| E3-pET-Fwd | AT**CATATG**GACGGAGGCGCATGCG | This Study |
| E3-pET-Rev | C**AAGCTT**GGACGGTCGGCGG | This Study |
| D3-pET-Fwd | GAT**CATATG**AACAATGTCGTCATCAATGTGG | This Study |
| D3-pET-Rev | CTT**CTCGAG**ACCTAATCTTAGAATGCCGAC | This Study |

[a]The restriction endonuclease sites incorporated into the PCR primers include *XhoI*, *NdeI* and *HindIII* indicated as bold nucleotides

pCC1Fos™ were cultured in medium supplemented with 15 μg/mL chloramphenicol while a double selection with 15 μg/mL chloramphenicol and 50 μg/mL kanamycin was applied for the transposon treated fosmid constructs.

### Fosmid selection

AFase positive clones identified during primary screening in [15] were first subjected to thermostability assays to identify those with desirable properties. Briefly, *E. coli* harbouring recombinant fosmids were inoculated into microtiter plates with LB containing 0.01% (*w*/*v*) L-arabinose and 15 μg/mL chloramphenicol and incubated for 16 h. A 200 μL aliquot of the overnight cultures was lysed with 10 μL Bugbuster™ protein extraction reagent (Novagen, USA) and the extracts incubated at 25, 40, 50, 60, 70, 80 or 90 °C for 60 min. Fosmid cell-free extracts incubated on ice served as controls. After incubation, the thermally treated extracts were placed on ice for 30 min before centrifugation at 13000 x g for 5 min at 4 °C to remove cell debris. Aliquots of 100 μL were then transferred into a flat bottom 96 well microtiter plate (Sterilin®) and *p*NP-arabinofuranose in 100 mM $NaPO_4$ (pH 7.0) was added to the extracts at a final concentration of 2 mM. The microtiter plate was incubated at 37 °C for 60 min and AFase activity was measured using a spectrophotometer at 410 nm (SPECTROstar Nano; BMG Labtech). All assays were performed in triplicate. From these assays, pFos-H4, pFos-E3, and pFos-D3, were identified as expressing the most thermostable AFase activities and selected for further study.

### Transposon mutagenesis for identification of AFase encoding open reading frames

Transposon mutagenesis was performed using the HyperMu™ < KAN-1 > Insertion Kit (Epicentre® Biotechnologies, USA) according to the manufacturer's instructions. Thereafter, the transposon-treated fosmid library together with respective control fosmids lacking the AFase insert were transformed into electrocompetent *E. coli* Epi300 cells and cultured on LA containing 15 μg/mL chloramphenicol and 50 μg/mL kanamycin. Single colonies growing on double selection plates were sub-cultured onto LA medium supplemented with chloramphenicol and kanamycin and cultured for 16 h at 37 °C. Subsequently, the mutated fosmid library and the respective controls were inoculated into individual wells of 96-well microtitre plates (Sterilin®) containing LB supplemented with kanamycin and chloramphenicol. The microtitre plates were duplicated using a 96-pin QPix2 automated colony picker (Genetix) and were incubated as described above. Thereafter, the cells were lysed to release soluble cell-free extract using BugBuster™ protein extraction reagent (Novagen). Enzyme activities of the mutants and their respective untreated AFase controls were detected by the addition of 1 mM *p*NP-arabinofuranose in 50 mM $NaPO_4$ buffer (pH 7.0) to each well of the microtiter plate containing the cell-free extracts and incubated at 37 °C until the development of a yellow colour indicating the release of *p*NP from the synthetic substrate. Transposon-treated fosmid clones that did not develop the distinct yellow colour or were observed to produce significantly reduced levels of yellow colour relative to that of the untreated fosmid controls were identified and chosen for further characterization.

Sequencing of mutated fosmids was conducted by the University of Stellenbosch's Central Analytical Facility (CAF). The sequences were manually edited using Chromas version 2.01 (Technelysium DNA sequencing software, Australia) and DNAMAN version 4.13 (Lynnon Corp., San Ramon, CA, USA). Sequence identity and similarity searches of DNA sequences were performed using the basic local alignment search tool (BLAST) programs as provided by the National Centre for Biotechnology Information (NCBI) [16, 17]. Putative ORFs were identified within the consensus sequence using GeneMark for prokaryotes (https://tinyurl.com/ya78gkes) and Interproscan [18] was used to identify conserved protein domains. The gene sequences for the three AFases were submitted to the GenBank database with accession numbers (QBG80847-QBG80849).

#### Expression and purification of AFases

The AFase genes, AFase-H4, AFase-E3 and AFase-D3 were amplified from the purified fosmid DNA with each primer bearing a specific restriction site for cloning into pET vectors (Table 1). The polymerase chain reactions (PCRs) were performed using Phusion DNA polymerase (ThermoFisher Scientific™) with AFase-H4 and AFase-D3 genes amplified from the fosmid DNA using the following conditions; initial denaturation at 98 °C for 30 s followed by 35 cycles of 10 s at 98 °C for denaturation, 30 s at 70 °C for annealing, 45 s at 72 °C for elongation and the final elongation at 72 °C for 10 min. Similar conditions were employed for amplifying the AFase-E3 gene except for annealing and elongation times which were reduced to 15 s and 30 s, respectively. The resulting PCR products were purified using the gel extraction kit from Machery Nagel. These were cloned into the pJET1.2/blunt cloning vector to create the recombinant constructs pJET1.2-AFase-H4, pJET1.2-AFase-D3 and pJET1.2-AFase-E3. The cloned inserts were sequenced to confirm that PCR errors were not introduced. The cloned genes were subsequently excised from the recombinant pJET-AFase constructs and cloned into the pET21a (+) vector digested with the restriction enzymes engineered into the primer sequences (Table 2). Following plasmid miniprep, the DNA sequence of constructs observed to contain the correct sized inserts was determined to ensure that the putative AFase genes had been cloned in-frame with the promoter and C-terminal histidine tag. The recombinant constructs, pET21a-H4, pET21a-E3 and pET21a-D3 were transformed into chemical competent *E. coli*-BL21 (DE3) (Table 1) cells for protein expression.

Two millilitres from overnight starter cultures of *E. coli* harbouring the respective pET21 constructs were inoculated in 500 mL Erlenmeyer flasks containing 50 mL LB with appropriate antibiotics and incubated to an $OD_{600nm}$ of approximately 0.6–0.8. Protein expression was induced by treating two experimental cultures with 0.5 and 1 mM isopropyl β-D-1-thiogalactopyranoside (IPTG, Fermentas) respectively. The third culture served as an uninduced control. Following incubation for three hours all the cultures were centrifuged at 4000 x g for 20 min at 4 °C. The pellets were resuspended in 10 mL of 1 x binding buffer (250 mM NaCl, 20 mM Tris-HCl, 5 mM imidazole, pH 7.9) and then disrupted by sonication (Bandelin Sono plus Ultrasonic Homogenizer, Germany) on ice with 6 pulses of 20 s. Following sonication, the lysates were centrifuged at 6000 x g for 20 min at 4 °C to remove cell debris. Proteins were purified by nickel affinity chromatography using the His-Bind® resin and buffer kit (Novagen, USA) following the manufacturer's instructions and dialysed against 200 volumes 20 mM sodium phosphate; 50 mM sodium chloride buffer (pH 7.0) using a 10 kDa MW cut-off membrane (Thermo Fischer Scientific, USA) for 16 h at 4 °C. The concentrations of purified AFases were determined by the method of Bradford [19] using the Bio-Rad protein assay kit with bovine serum albumin as a standard. Protein purity was assessed by sodium dodecyl sulphate-polyacrylamide gel electrophoresis (SDS-PAGE) under denaturing conditions as described by Laemmli [20].

#### Biochemical characterization

Thermal stability assays with purified AFases were performed by incubating them for 90 min or 24 h at various temperatures (25, 40, 50, 60, 70, 80 and 90 °C). For the AFases incubated for 24 h, samples were removed at 2 h intervals until 12 h and at 24 h. AFase activities were measured as described for cell-free thermal stability assays in preliminary screening described above with the respective control reactions incubated on ice.

The optimum temperature of the purified AFases (0.5 μg per assay) was assessed by measuring the activity of each AFase against 2 mM *p*NP-arabinofuranosyl at 6, 20.5, 25, 40, 50, 60, 70, 80 and 90 °C after 2 min of incubation with the substrate at the indicated temperatures. The temperature at which the highest activity was recorded was selected as the optimum temperature for each AFase tested. The optimal pH of the AFases was evaluated at 40 °C in 50 mM citrate and/or phosphate buffers with pH varying from 2 to 8 and the buffers without AFase added were used in control reactions. The pH of the buffer containing the highest AFase activity was selected as the optimal pH for each AFase tested.

#### Substrate specificity and enzyme kinetics

For substrate specificity the following substrates were assayed: *p*NP-β-D-cellobioside, *p*NP-α-D-mannopyranoside, *p*NP-β-D-fucopyranoside, *p*NP-β-D-glucuronide, *p*NP-α-D-glucopyranoside, *p*-NP- β-D-glucopyranoside *p*NP-α-L-arabinopyranoside and *p*NP-β-L-arabinopyranoside from Sigma-Aldrich (SA) and *p*NP-α-L-arabinofuranoside, *p*NP-β-D-xylopyranoside, *p*-methylumbelliferyl-β-D-xylopyranoside and *p*NP-β-D-mannopyranoside from Carbosynth Ltd. All the substrates were prepared following their respective manufacturer's instructions. The enzymatic assays were performed in a total reaction volume of 250 μL containing 2 mM substrate in 50 mM citrate buffer (pH 7.0) with 0.5 μg enzyme for 2 min. The reactions were stopped by raising the pH with the addition of 750 μL of 0.4 M $Na_2CO_3$. A 250 μL reaction volume was aliquoted into a single well of a 96-well microtiter plate (Sterilin®). The enzyme activities were measured spectrophotometrically at 410 nm and the standard curve was generated using 1–10 mM pNP under assay conditions. One unit (U) of the enzyme is defined as the amount of enzyme that can liberate 1 μmol of *p*NP per minute. All assays were performed in triplicate with the addition of the appropriate controls. Complex substrates were hydrolyzed for four hours at each enzyme's apparent temperature optimum and pH. Reducing sugars resulting from the hydrolysis of complex α-L-arabinose polysaccharide substrates: arabinoxylan, arabinan and linear arabinan (Megazyme, 0.5% *w*/*v*) were measured using the dinitrosalicylic acid (DNS) assay as described by Miller [21]. Briefly, 150 μL of DNS solution was added to 50 μL reactions and boiled for 5 min before being rapidly cooled on ice. The volume was made up to 1000 μL with $dH_2O$ and the absorbance determined at 510 nm (OD510; SPECTROstar Nano; BMG Labtech). The reducing sugars were determined using a L-arabinose standard curve generated under the same reaction conditions. Kinetic parameters ($K_M$, $V_{max}$ and $k_{cat}$) were evaluated by measuring the enzyme activity in the presence of varying *p*NP-arabinofuranosyl concentrations after 1 min. The Michealis-Menten plot was generated using GraphPad Prism 4 (GraphPad Software).

### Structural analysis of AFases

Prediction of the quaternary structures was performed by fast protein liquid chromatography (AKTA FPLC, Amersham Biosciences) using a Superdex G200 column run at 0.5 ml/min with running buffer (50 mM NaCl, 20 mM $Na_2PO_4$ (pH 7.0)). The proteins were resuspended in running buffer and loaded onto the column according to the manufactures instructions. Proteins used as molecular weight markers were β-amylase (200 kDa), alcohol dehydrogenase (150 kDa), albumin (66 kDa) and carbonic anhydrase (29 kDa) which were also resuspended in the same running buffer. These were used to plot the log of their MW against retention time to generate a standard curve. Protein molecular weights were approximated based on the retention time using the standard curve generated.

## Results

### Functional screening and selection of AFase encoding fosmids

In a previous study by Ohlhoff et al. [14], a metagenomic library consisting of more than 150,000 fosmid clones with an average insert size of 31 kb was constructed. A total of 46,000 clones were screened for α-L-arabinofuranosidase (AFase) activity and 13 putative AFase positive clones were identified [15]. Here we performed preliminary thermostability screening by incubating cell-free extracts at a range of temperatures for 60 min. Cell-free extract from pFos-D3 AFase retained close to 100% activity when incubated at 25 and 40 °C. The enzyme activity decreased by approximately 40 and 90% when incubated at 50 and 60 °C, respectively, while at 70 °C and higher, negligible AFase activity was detected. AFase pFos-H4 retained 100% activity following incubation at 25, 40, 50 and 60 °C. The relative residual activity decreased sharply following incubation at temperatures greater than 60 °C. In contrast to pFos-D3, the AFase activity encoded by cell-free extract from pFos-H4 and pFos-E3 displayed an increase in relative activity following incubation at temperatures greater than 30 °C, with the highest activity observed between 60 and 70 °C respectively. At these temperatures, the recombinant AFase-E3 maintained full activity, relative to the untreated control, and was the most thermostable of the three enzymes. Based on these preliminary thermostability profiles, the crude protein extracts of pFos-D3, pFos-H4 and pFos-E3 clones were identified as thermolabile, moderately thermostable and thermostable respectively, and were selected for further characterization.

### Sequence and phylogenetic analysis of AFase-encoding ORFs

Transposon mutagenesis was employed to identify the ORFs encoding the α-L-arabinofuranosidase activities. Fosmids with loss of, or decreased activity, compared to the non-mutated constructs, indicative of gene disruption, were selected for sequence analysis. Overlapping sequences were generated from 4 to 6 mutants for each fosmid (pFos-H4, pFos-E3 and pFos-D3) to assemble putative AFase encoding regions of 2693, 2406 and 2505 bp for each fosmid, respectively.

The putative ORF (489 amino acids; 55 kDa) from pFos-H4 shared 100% amino acid identity with a sequence derived from a Brazilian compost metagenome (QGUR01000024.1) which, although reported on earlier [22], was only recently added to the NCBI database (June 2018). One of the pFos-H4 end sequences also matched 100% at nucleotide level to the metagenomic contig this gene was located on, suggesting that AFase-H4 originates from a closely related bacterium. The next closest hit was to an AFase from *Truepera radiovictrix* DSM 17093 at 61% identity. Conserved catalytic and C-terminal domains were identified within the predicted AFase amino acid sequence, and no evidence of a signal peptide was detected. This could indicate that a "non-classical protein secretion" mechanism may be involved in exporting the AFase-H4 protein [23]. The catalytic domain overlaps the AFase C-terminal domain and is positioned from P41-A346 and E261-V482, respectively. A putative cellulase-like domain is located within the catalytic domain from V161-W240. The catalytic domain has sequence identity to that of Glycoside Hydrolase (GH) family 51 (GH51).

For pFos-E3, the predicted ORF consists of 510 amino acids (58 kDa) and the sequence shared 99% identity over the full length of the protein to a putative AFase from *Paenibacillaceae* bacterium JTherm, also isolated from a compost heap in Vacaville, CA, USA [24]. Nucleotide sequence from both ends of pFos-E3 also matched 100% at the nucleotide level to this genome, indicating that AFase-E3 likely originates from an identical or very closely related bacterium. This organism has been proposed to be the first member of a new genus *Candidatus Reconcilibacillus cellulovorans* responsible for the initial breakdown of cellulosic material in a compost heap allowing other community members to proliferate and hydrolyse the material further [24].

The 494 amino acid (56 kDa) sequence of pFos-D3 shared 77% identity (87% similarity) with a *Paenibacillus taihuensis* ORF. Although not yet published, this organism's genome was sequenced as part of the third phase of the Joint Genome Institutes' Genomic Encyclopedia of Type Strains program which covers genomes from soil and those that are plant associated. It could be argued that these are mesophilic environments and may be the reason AFase-D3 displays mesophilic characteristics.

All the key amino acids in the catalytic site of AFases have previously been identified [25] and these were perfectly conserved in both AFase-H4 and AFase-E3 (F23, E25, R65, N70, W95, N170, E171, H239, Y241, E289, W293, and Q340; AFase-H4 numbering). In AFase-D3 however, Asn70 is replaced by a cysteine residue. This residue is thought to be important to substrate binding as a hydrogen bonding partner to Glu25. Glu25 may not be directly involved in catalysis but is thought to be crucial in coordinating the hydrogen bonding network at the non-reducing end of the substrate and in this way enable substrate binding [26]. Despite this substitution the enzyme is still functional and this may be part of the reason why a broad apparent temperature optimum is observed as explained in the discussion.

Phylogenetic analysis of the three AFase sequences showed that they all clustered together with representative AFase sequences of the GH51 family which contains several thermostable proteins (Fig. 1). The sequences of the more thermostable AFase-H4 and AFase-E3 are more related as compared to the less thermostable AFase-D3 (Fig. 1). Both AFase-E3 and AFase-D3 show similarity to AFases from *Bacillus subtilis*, a group that contains both mesophilic and thermophilic variants of this protein [27, 28].

### Biochemical characterization of the recombinant proteins

The three selected AFase ORFs were cloned and expressed in *E. coli* and purified to near homogeneity as judged by SDS-PAGE analysis using nickel-ion affinity chromatography (results not shown). All three proteins showed molecular masses close to their in silico predicted sizes of 55 kDa, 56 kDa and 58 kDa for AFase-H4, –D3 and -E3 respectively. The quaternary structures of the native AFase-H4, AFase-E3 and AFase-D3 proteins were determined by gel filtration. The respective retention times of the three AFases, relative to that of molecular weight standards suggested that all three proteins are homo-hexamers or a trimer of dimers.

AFase-H4 and AFase-E3 had an apparent temperature optimum of 60 °C while for AFase-D3 the apparent optimal temperature was 25 °C (Fig. 2). Thermal stability analysis showed that AFase-H4 maintained 100% residual activity even after a prolonged incubation time of 12 h at 50 °C. The residual activity started to decrease when the protein was incubated at 60 °C. AFase-D3 maintained 100% residual activity after incubation for 1 h at 40 °C, whereas AFases H4 and E3 displayed classic temperature dependence profiles. Interestingly, AFase-D3 maintained approximately 50% activity even at 80–90 °C. AFase-E3 maintained 100% residual activity even after incubation for 24 h at 70 °C, the most thermostable of the three enzymes identified (Additional file 1: Figure S1). All three AFases showed best activity in the pH range 4.0–6.0 (Fig. 3).

Due to the various domains identified for these enzymes based on sequence analysis, the substrate specificities of these AFases were assessed on various pNP-linked substrates. All three AFases showed strong activity towards pNP-α-L-arabinofuranoside, with AFase-E3 displaying the highest activity at about 130 U/mg, followed by AFase-D3 (~ 90 U/mg) and AFase-H4 (~ 50 U/mg) (Additional file 2: Figure S2). AFase-H4 also showed substantial activity on pNP-β-D-cellobiose (~ 10 U/mg) while relatively low but detectable activities were displayed by all AFases when assayed on other pNP-linked substrates. However, none of the AFases were able to hydrolyse pNP-β-D-xylopyranoside substrate. The hydrolytic preferences of these enzymes were also examined on more complex substrates including arabinoxylan, arabinan, and linear arabinan (Fig. 4). These polysaccharide substrates were chosen to represent differing internal bonds between arabinose subunits or sidechains. All three AFases displayed different effectiveness for hydrolysing arabinofuranosyl bonds within these three substrates. AFase-E3 showed highest activity on these substrates (arabinan and linear arabinan) when compared to AFase-D3 while AFase-H4 was incapable of hydrolyzing any of these substrates and none of the AFases released reducing sugars from arabinoxylan.

All three enzymes displayed Michaelis-Menten kinetics when assayed on pNP-α-L-arabinofuranoside (Additional file 3: Figure S3). AFase-H4 and AFase-E3 displayed slightly lower $K_M$ values when compared to AFase-D3 indicating that AFase-D3 has a slightly lower

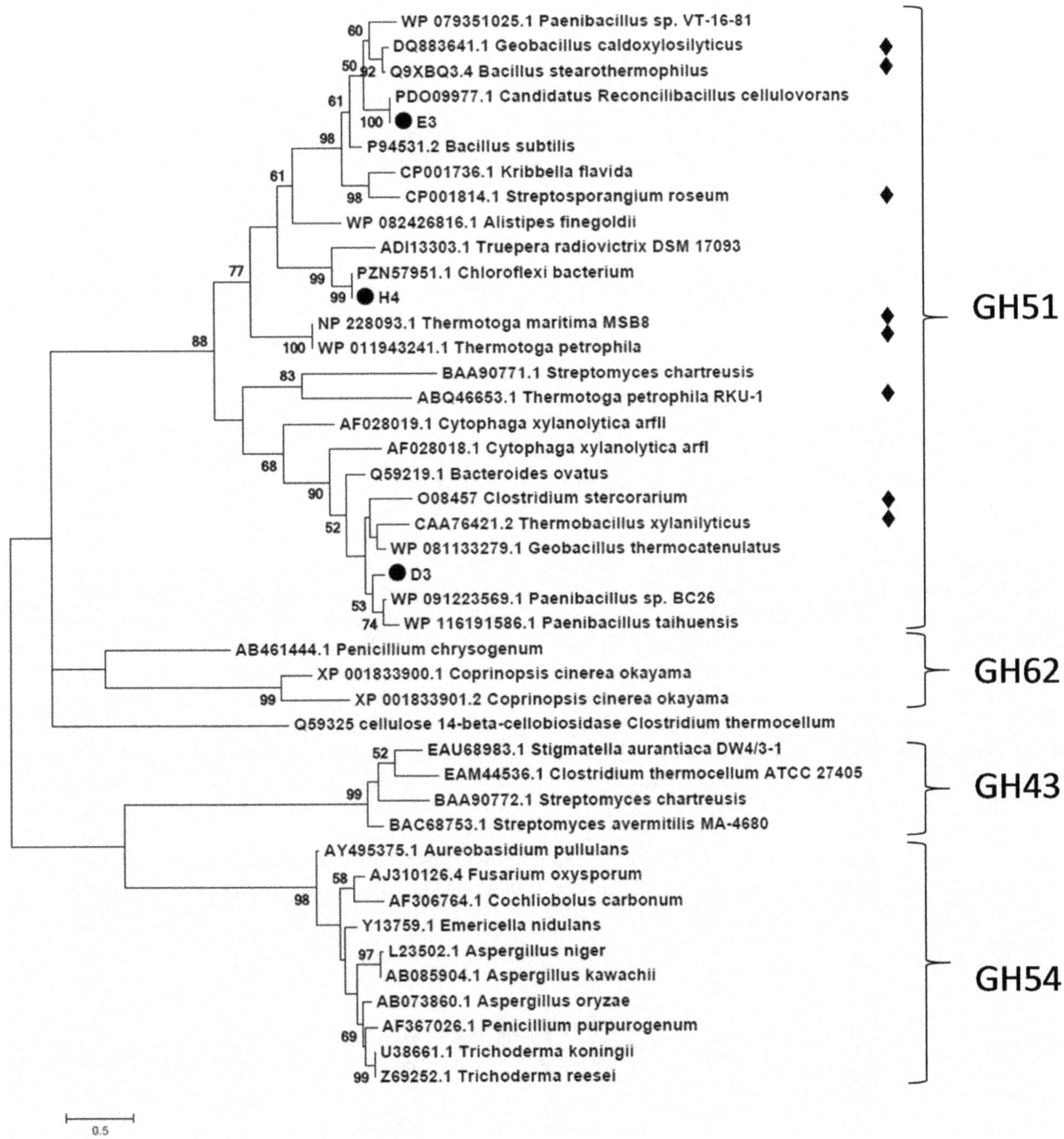


**Fig. 1** Maximum Likelihood (JTT model) tree of the phylogenetic relationship of AFase-H4, –E3 and -D3 to characterised AFases from four glycoside hydrolase families. The functionally characterized thermophilic AFases are indicated with the black diamond symbol (♦). The scale bar indicates the number of amino acid substitutions

affinity for this substrate compared to the other two enzymes. AFase-E3 showed the highest specific activity and the highest turnover number of the three enzymes (Table 3).

## Discussion

Here we present three novel AFases, identified from a compost metagenomic library. Although all three sequences have close relatives on the GenBank database, owing to the recent addition of metagenomic and bacterial genome sequence, none of these closely related enzymes have been characterized previously. The presence of nearly identical AFase-H4 and AFase-E3 sequences in compost heaps that are biogeographically separated speaks to them being selected for and particularly

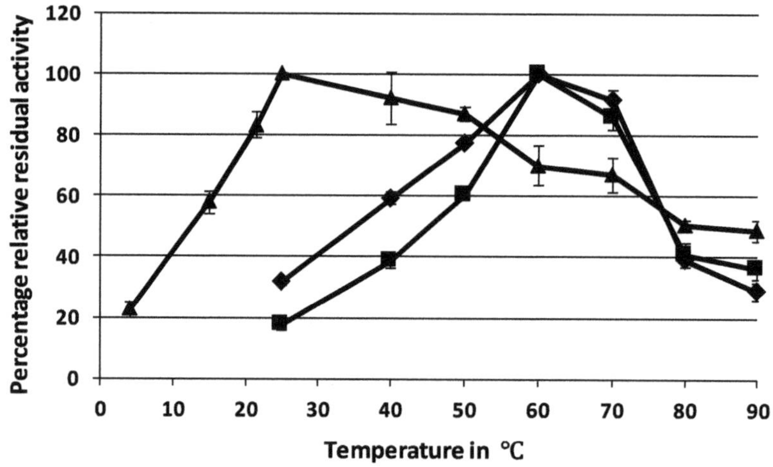


**Fig. 2** The temperature optima of AFase H4 (♦), E3 (■) and D3 (▲). Data represents the mean ± standard error ($n = 3$)

successful in compost environments. All bioinformatic and biochemical characterization data presented supports the classification of these three enzymes as belonging to the GH51 family [EC 3.2.1.55]. The AFases shared a homo-hexameric structure or possibly consist of a dimer of trimers. Although hexameric structures for AFase proteins have been reported, specifically for the GH51 family of AFases, tetramers and octamers have also been reported from this family. Homo-tetrameric structures have been reported for enzymes from a *Streptomyces* species, *Geobacillus caldoxylolyticus* TK4 and *Anoxybacillus kestanbolensis* AC26Sari [29–31].

AFase-H4 and AFase-E3 displayed optimal activity at 60 °C while AFase-D3 displayed a mesophilic thermal profile with an optimum activity at 25 °C yet retaining up to 50% activity at temperatures as high as 80–90 °C. The thermostability profiles of AFase-H4 and AFase-E3 were both similar to that of functionally characterized AFases from *Streptomyces* sp. S9 [29], *Paenibacillus* sp. TH51 [7] and *G. caldoxylolyticus* TK4 [30]. AFase-E3 shares sequence similarity with *B. subtilis* derived AFase and this species is known to encode both mesophilic and thermophilic AFases. The presence of AFases with both mesophilic and thermophilic properties in compost could be explained by the stages involved during the composting process. In the early stages of composting mesophilic microorganisms dominate and their metabolic activities cause an increase in temperature. This stimulates the growth and activity of thermophilic microorganisms which is subsequently followed by a cooling stage also referred to as the maturation stage [32, 33]. Since the AFases described here are derived from compost that reaches only

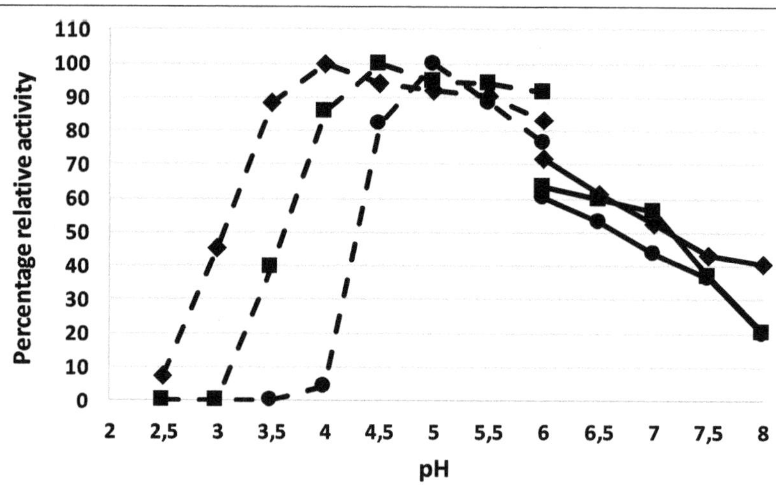


**Fig. 3** pH optima of AFase H4 (●), E3 (♦) and D3 (■) determined in phosphate (dashed line) and citrate buffers (dotted line) between pH 2.5 and 6.0 and pH 6.0 to 8.0, respectively. Data represents the mean ± standard error ($n = 3$)

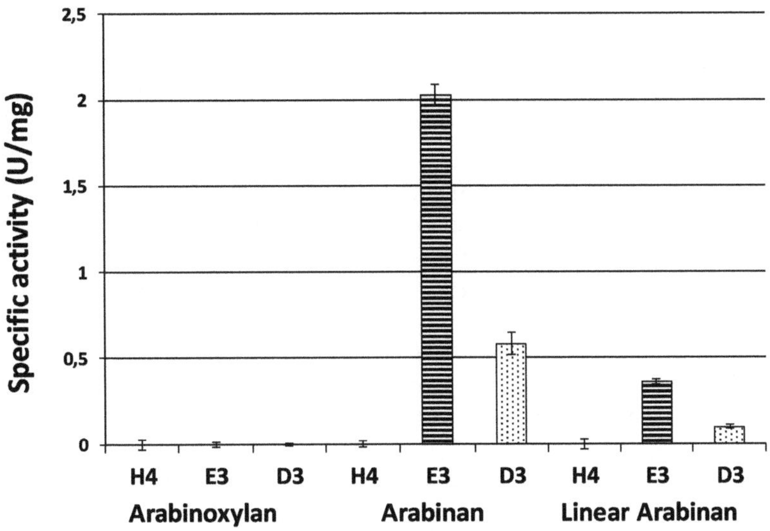


**Fig. 4** AFase activity on complex substrates, namely arabinoxylan, arabinan and linear arabinan. The specific activity was determined in U/mg. ($n = 3$) ± standard error

70 °C, this may put an upper limit on the thermal stability and apparent temperature optimum of enzymes we could have identified in this study.

The mesophilic/thermophilic characteristic displayed by AFase-D3 has only been reported for AFases that belong to the GH43 family previously isolated from *Paenibacillus* species [34] with no characterized GH51 family enzyme displaying this thermal profile. Broad temperature optima have been observed for several different enzymes isolated from a range of organisms including plants [35–38]. Of particular interest are enzymes (phophoribosyl anthranilate isomerase, indoleglycerol phosphate synthase and l-isoaspartyl (D-aspatyl) O-methyltransferase) from *T. maritima* that show highest catalytic efficiency at 25 °C as opposed to the organism's optimum growth temperature [39–41]. *T. maritima* has a wide growth temperature range up to its optimum of 90 °C. Although a thermally robust GH51 family AFase which could maintain full activity even at temperatures as high as 90 °C has been isolated from *T. maritima* MSB8, there was no evidence of a thermophilic/mesophilic property for this enzyme [42]. While the thermal behavior of AFase-H4 and AFase-E3 correspond to enzymes which have a large $\Delta H_{eq}$ for the $E_{act}/E_{inact}$ transition according to the equilibrium model for dependence of enzyme activity on temperature, AFase-D3 likely has a small $\Delta H_{eq}$ resulting in the broad temperature optimum [43]. A small $\Delta H_{eq}$ may be the result of the substrate used to assay for the apparent temperature optimum as the $E_{act}/E_{inact}$ transition appears to be linked to the active site conformation. It could be that the *p*NP-arabinofuranosyl substrate locks the active site, and by extension other parts of the protein, in a stable conformation. Taken together, this leaves open the possibility that AFase-D3 is a thermophilic enzyme. The *Thermotogae* phylum contains microorganisms with temperature ranges that span both thermophilic to mesophilic temperatures and they are interesting models for studying evolutionary changes [39, 44]. This flexibility has

**Table 3** Enzyme kinetic comparison of AFases characterised in this study and thermostable AFases obtained from literature

| AFase | Micro-organism | $K_M$ in mM (Std. error) | Sp. act. in U/mg (Std. error) | $k_{cat}$ ($s^{-1}$) | $k_{cat}/K_M$ ($M^{-1}\ s^{-1}$) | Reference |
|---|---|---|---|---|---|---|
| *AFase-H4* | Unknown | 0.31 (±0.03) | 143.1 (±3.7) | 131 | $4.2 \times 10^5$ | This study |
| *AFase-E3* | Unknown | 0.33 (±0.04) | 228.6 (±9.7) | 219 | $6.6 \times 10^5$ | This study |
| *AFase-D3* | Unknown | 0.38 (±0.05) | 166.6 (±6.9) | 155 | $4.1 \times 10^5$ | This study |
| abf51S9 | *Streptomyces* sp. S9 | 1.45 | 221 | 203 | $1.4 \times 10^5$ | [29] |
| AbfATK4 | *Geobacillus caldoxylolyticus* TK4 | 0.17 | 588 | 568 | $3.3 \times 10^6$ | [30] |
| AbfAC26Sari | *Anoxybacillus kestanbolensis* | 0.14 | 1019 | 968 | $7 \times 10^6$ | [31] |
| Tm-Afase | *Thermotoga maritima* | 0.42 | 23.5 | 22 | $5.2 \times 10^4$ | [42] |
| AFase | *Caldocellulosiruptor sacchrolyticus* | 1.29 | 295 | 285 | $2.2 \times 10^5$ | [60] |
| abfB | *Auriobasidium pullulans* | 6.27 | 78.1 | 96.3 | $1.5 \times 10^4$ | [62] |

been ascribed to lateral gene transfer where most of the acquired genes in this phylum are involved in carbohydrate metabolism [44]. These gene acquisitions afford the bacteria in this phylum a selective advantage to transit between the mesophilic and thermophilic environments and of metabolizing across a broad spectrum of temperatures [45, 46]. It would be of interest to isolate the host of the AFase-D3 to assess whether the thermostability profile observed for this protein extends to the entire organism as an adaptation mechanism as described for *T. maritima* and other *Thermotogae* lineages.

Studies conducted to discriminate between mesophilic and thermophilic proteins have shown that there are no significant differences in these protein orthologues as they often share a similar structure, sequence identity and the same catalytic mechanisms [45]. There appears to be no uniform property that confers increased protein thermostability across the proteome, but rather could be a result of minor differences in sequence and structure due to point mutations, increased numbers of salt bridges and the presence of specific amino acid residues on the protein surface such as fewer thermolabile amino acid residues observed in thermophilic proteins [46]. This seems to be evident in this study as cysteine residues which are often associated with increased protein thermostability were identified in the protein sequence of AFase-E3, which showed the highest thermostability among the AFases described in this study. Cysteine residues are known to form disulphide bridges with alternate cysteine residues within the catalytic domain and to increase thermostability by 10 to 20 °C [29, 47]. The contribution of cysteine residues to the thermostability has been previously shown through substitution of cysteine residues with alanine, resulting in a decreased thermostability of AFases from *G. caldoxylolyticus*, *Geobacillus stearothermophilus*, *Thermobacillus xylanilyticus* and *B. subtilis* [30, 48]. Further evidence of the contribution of subtle changes in the amino acid sequence to thermostability of the proteins has been shown through comparison of amino acid composition of thermophilic and mesophilic protein homologues. The diguanylate cyclase and glutamate dehydrogenase enzymes from the hyperthermal *T. maritima* had a smaller hydrophobic accessible surface and a greater charged surface area when compared to their mesophilic homologues from *Pseudomonas aeruginosa* and *Clostridium symbiosum*, respectively [44]. This conformation results in stronger hydrophobic interactions in the interior of the protein and increased ion pairing on the surface conferring greater thermostability to these thermophilic proteins compared to their respective mesophilic counterpart [45].

The three AFases described here had optimum activity within a slightly acidic pH range of 4.0–5.0. A pH optimum in this range is typical for AFases belonging to the GH51 family as previously shown for AFases from *Paenibacillus* sp., *Aureobasidium pullulans*, *Aspergillus oryzae* and *Streptomyces coelicolor* [7, 49]. Industrial processes often operate at high temperatures and slightly acidic pH which make the AFases identified in this study suitable for application processes conducted under these conditions [50].

All three AFases displayed the greatest hydrolytic activity towards pNP-α-L-arabinofuranoside compared to the activity on the other pNP-glycosides. Of the 145 GH families reported to date only families GH2, 3, 43, 51, 54 and 62 are known to encode α-L-arabinofuranosidase activity with family GH51 having the largest number of α-L-arabinofuranosidases [6, 51]. Lower activity was observed on the other pNP-glycosides except for AFase-H4 which showed substantial activity on pNP-β-D-cellobioside. The activity on pNP-β-D-cellobioside shown by AFase-H4 is the first report for a GH51 family AFase. None of the AFases could hydrolyse the pNP-β-D-xylopyranoside substrate. The enzymatic hydrolysis of pNP-β-D-xylopyranoside is a characteristic of the GH43 family of AFases, which are known for their dual activity, possessing both β-xylosidase and the conventional α-L-arabinofuranosidase activity [52]. Only two GH51 AFases have been shown to hydrolyse this bond and these were isolated from *G. caldoxylolyticus* TK4 [30] and *Paenibacillus sp.* TH51 [7]. AFase-D3 had very low activity on pNP-β-D-glucopyranoside and pNP-α-L-mannopyranose. No other GH51 AFase has been reported to have the capability of hydrolyzing these synthetic substrates and together with its exceptionally wide thermostability profile, it makes it a unique enzyme.

All three AFases showed different levels of activity for hydrolysis of the arabinofuranosyl bonds within the natural substrates tested. AFase-E3 indicated a higher affinity for these substrates when compared to AFase-H4 and AFase-D3. The hydrolysis of arabinan has been a common functional characteristic for GH51 AFases [53] but, AFase-H4 showed no hydrolysis of arabinan. Similarly, the AFases from *Streptomyces* sp. and *Penicillium purpurogenum* were unable to hydrolyse this substrate [54, 55]. AFase-E3 and AFase-D3 had a lower activity on linearized arabinan and it is known that GH51 AFases weakly hydrolyze the α-1,5-L arabinofuranosyl bonds, whereas this is a capability of the GH43 family AFases [52, 53]. A similar linear arabinan hydrolysis profile has been reported previously [54] suggesting that the AFases described here are typically exo-acting enzymes contrary to GH43 AFases which readily hydrolyze the α-1,5-L arabinofuranosyl bonds [34, 52]. None of the AFases were able to release reducing sugars from arabinoxylan. This activity is characteristic of some GH51 AFases, but not all [29–31, 47, 56]. Substrate specificity towards arabinoxylan has been attributed to specific residues in the catalytic domain in the *Thermotoga* species and *Thermobacillus* species of AFases [47, 57]. A tryptophan located at approximately the 96-100th position in the catalytic domain [47, 57] is a key xylan-binding residue. However only AFase-E3 has two of these residues required for an electrostatic active site to liberate L-arabinose from the xylan backbone. Figure 3.20 shows no liberation of L-arabinose from the xylan backbone and this could be due to the absence of the residues required for an adequate electrostatic surface within the active site [26]. The inability to liberate the L-arabinose from the xylan backbone has also been correlated to the ratio of the α-1,3-L and α-1,2-L- arabinofuranosyl bonds that differs within different arabinoxylan-containing materials. There are particular arabinoxylans that are highly saturated with either α-1,3-L or the α-1,2-L bonds [58]. Therefore, it is plausible that the arabinoxylan used in this experiment is saturated with more of α-1,2-L bonds. This conclusion is consistent owing to the AFases showing an affinity for the *p*-nitrophenyl-α-L-arabinofuranoside bond which resembles the α-1,3-L arabinofuranosyl bonds [58, 59].

Multifunctional enzymes with two or more activities are highly desirable for the hydrolysis of complex polymers and offer the possibility of reduced complexity of enzyme cocktails and better synergies [7]. However, as evidenced in this study, the shortfall is that these additional activities are always minor and are not displayed when GH51 AFases are employed to hydrolyse heteroxylan or xylo-oligosaccharides in more complex and natural substrates [7]. The relatively good cellobioside activity of AFase-H4, together with its moderate thermostability, could make this enzyme a good starting point for engineering of a dual functional enzyme for use in thermophilic ethanologenesis processes.

Although we have highlighted the unique features and differences that exist between these and previously described enzymes, overall, the kinetic data for these three AFases demonstrate that they are not exceptional in this regard and have characteristics that are in the same range to that of previously characterized AFases [30, 31, 60]. Thus, even though these AFases are phylogenetically distinct, their biochemical characteristics are similar to those that have been described previously. This recurring theme in enzymology, once again reiterates the power of selective pressure to ensure that the enzymes' activity falls within a limited range [61]. However, the hydrolysis of bonds other than the α-1,5-L-arabinofuranosyl linkage makes the AFases identified in this study novel and they could be used as a starting point for engineering to optimise their suitability for various industrial applications.

## Conclusions

Here we describe three GH51-related α-arabinofuranosidases that are novel at the amino acid level, substrate specificity and thermostability profile (AFase-D3) with the potential to be engineered to be used in industrial processes. Their discovery once again demonstrates the power of functional metagenomics and evolutionary pressure to explore novel sequence space.

## Additional files

**Additional file 1: Figure S1.** Thermostability profiles for the purified AFase H4 (A), E3 (B) and D3 (C) at 25 (✚), 40 (▲), 50 (♦), 60 (■), 70 (╳) and 80 °C (●). Data represents the average of three replicates ± standard error ($n = 3$). (DOCX 30 kb)

**Additional file 2: Figure S2.** Substrate range of AFases measured on pNP linked glycosides. A) H4, B) E3 and C) D3. Data represents the average of three replicates ± standard error ($n = 3$). (DOCX 46 kb)

**Additional file 3: Figure S3.** Michaelis-Menten plots of AFase-D3, −E3 and -H4. The kinetics of each enzyme was determined at their respective pH and temperature optima with increasing amounts of pNP-α-L-arabinofuranoside (mM). ($n = 3$) ± standard error. (DOCX 72 kb)

### Abbreviations

AFases: Arabinofuranosidase; GH: Glycosyl hydrolases; IPTG: Isopropyl-β-D-1-thiogalactopyranoside; NGS: Next generation sequencing; ORF: Open reading frame; pNP: *para*-nitrophenyl

### Acknowledgements

We'd like to thank the NRF for their financial support.

### Funding

The study was funded by the National Research Foundation (NRF) of South Africa under the Thuthuka program (UID: 69116). The funders sole role was providing funding for the study and all other aspects of the project (design of the study, collection, analysis, and interpretation of data and writing of the manuscript) was executed by the authors.

### Authors' contributions

BF, RH and MS conducted the experiments, analyzed the data and edited the manuscript, SM prepared the manuscript, LJVZ generated the figures and contributed to writing of and editing the manuscript, the study was conceived and supervised by MT who also edited the manuscript. All authors read and approved the manuscript.

### Ethics approval and consent to participate

Not applicable.

### Consent for publication

Not applicable.

### Competing interests

The authors declare that they have no competing interests.

### Author details

[1]Institute for Microbial Biotechnology and Metagenomics, University of the Western Cape, Bellville, South Africa. [2]Centre for Bioprocess Engineering Research, University of Cape Town, Cape Town, Western Cape, South Africa.

**References**


1. Pathak VM, Navneet. Review on the current status of polymer degradation: a microbial approach. Bioresour Bioprocess. 2017;4:15.
2. Mohanram S, Amat D, Choudhary J, Arora A, Nain L. Novel perspectives for evolving enzyme cocktails for lignocellulose hydrolysis in biorefineries. Sustain Chem Process. 2013;1:1–12.
3. Jonathan MC, DeMartini J, Van Stigt TS, Hommes R, Kabel MA. Characterisation of non-degraded oligosaccharides in enzymatically hydrolysed and fermented, dilute ammonia-pretreated corn Stover for ethanol production. Biotech Biofuels. 2017;10:112.
4. Ahmed S, Luis AS, Bras JLA, et al. A novel α-L-arabinofuranosidase of family 43 glycoside hydrolase (Ct43Araf) from *Clostridium thermocellum*. PLoS One. 2013;8:9.
5. Hatfield RD, Rancour DM, Marita JM. Grass cell walls: a story of cross linking. Front Plant Sci. 2016;7:2056.
6. Maruthamuthu M, van Elsas JD. Molecular cloning, expression, and characterization of four novel thermo-alkaliphilic enzymes retrieved from a metagenomic library. Biotechnol Biofuels. 2017;10:142.
7. Bouraoui H, Desrousseaux ML, Ioannou E, et al. The GH51 α-l-arabinofuranosidase from *Paenibacillus sp.* THS1 is multifunctional, hydrolysing main-chain and side-chain glycosidic bonds in heteroxylans. Biotechnol Biofuels. 2016;9:140.
8. Turner P, Mamo G, Karlsson EN. Potential and utilization of thermophiles and thermostable enzymes in biorefining. Microb Cell Factories. 2007;6:9.
9. DeCastro ME, Rodríguez-Belmonte E, González-Siso MI. Metagenomics of thermophiles with a focus on discovery of novel thermozymes. Front Microbiol. 2016;7:1521.
10. Viikari L. Hydrolysis of amorphous and crystalline cellulose by heterologously produced cellulases of *Melanocarpus albomyces*. J Biotechnol. 2007;136:140–7.
11. Karnaouri A, Matsakas L, Topakas E, Rova U, Christakopoulos P. Development of thermophilic tailor-made enzyme mixtures for the bioconversion of agricultural and forest residues. Front Microbiol. 2016;7:177.
12. Dougherty MJ, Dhaeseleer P, Hazen TC, Simmons BA, Adams PD, Hadi MZ. Glycoside hydrolases from a targeted compost metagenome, activity-screening and functional characterization. BMC Biotechnol. 2012;12:38.
13. Li LL, McCorkle SR, Monchy S, Taghavi S, van der Lelie D. Bioprospecting metagenomes: glycosyl hydrolases for converting biomass. Biotechnol Biofuels. 2009;2:10.
14. Ohlhoff CW, Kirby BM, van Zyl L, Mutepfa DLR, Casanueva A, Huddy RJ, et al. An unusual feruloyl esterase belonging to family VIII esterases and displaying a broad substrate range. J Mol Catal B Enzym. 2015;118:79–88.
15. Smart M, Huddy RJ, Cowan DA, Tuffin M. Liquid phase multiplex high-throughput screening of metagenomics libraries using p-Nitrophenyl-linked substrates for accessory lignocellulosic enzymes. Methods Mol Biol. 2017; 1539:219–28.
16. Altschul SF, Gish W, Miller W, Myers EW, Lipman DJ. Basic local alignment search tool. J Mol Biol. 1990;215:403–10.
17. Besemer J, Borodovsky M. GeneMark: web software for gene finding in prokaryotes, eukaryotes and viruses. Nucleic Acids Res. 2005;33:W451–4.
18. Jones P, Binns D, Chang H-U, Fraser M, Li W, McAnulla C, et al. InterProScan 5: genome-scale protein function classification. Bioinformatics. 2014;30: 1236–40.
19. Bradford MM. A rapid and sensitive method for the quantitation of microgram quantities of protein utilizing the principle of protein-dye binding. Anal Biochem. 1976;72:248–54.
20. Laemmli UK. Cleavage of structural proteins during the assembly of the head of bacteriophage T4. Nature. 1970;227:680–5.
21. Miller GL. Use of dinitrosalicylic acid reagent for determination of reducing sugar. Anal Chem. 1959;31:426–8.
22. Martins LF, Antunes LP, Pascon RC, de Oliveira JCF, Digiampietri LA, et al. Metagenomic analysis of a tropical composting operation at the São Paulo zoo park reveals diversity of biomass degradation functions and organisms. PLoS One. 2013;8:10.
23. Bendtsen JD, Jensen LJ, Blom N, Von Heijne G, Brunak S. Feature-based prediction of non-classical and leaderless protein secretion. Protein Eng Des Sel. 2004;17:349–56.
24. Kolinko S, Wu Y-W, Tachea F, Denzel E, Hiras J, Gabriel R, Bäcker N, Chan LJG, Eichorst SA, Frey D, Chen Q, Azadi P, Adams PD, Pray TR, Tanjore D, Petzold CJ, Gladden JM, Simmons BA, Singer SW. A bacterial pioneer produces cellulase complexes that persist through community succession. Nat Microbiol. 2018;3:99–107.
25. Hövel K, Shallom D, Niefind K, Belakhov V, Shoham G, Baasov T, Shoham Y, Schomburg D. Crystal structure and snapshots along the reaction pathway of a family 51 alpha-L-arabinofuranosidase. EMBO J. 2003;22:4922–32.
26. Souza TACB, Santos CR, Souza AR, Oldiges DP, Ruller R, Prade RA, Squina FM, Murakami MT. Structure of a novel thermostable GH51 α-Larabinofuranosidase from *Thermotoga petrophila* RKU-1. Protein Sci. 2011;20:1632–7.
27. Inacio JM, Correia IL, de Sá-Nogueira I. Two distinct arabinofuranosidases contribute to arabino-oligosaccharide degradation in *Bacillus subtilis*. Microbiology. 2008;154:2719–29.
28. Hoffman ZB, Oliveira LC, Cota J, Alvarez TM, Diogo JA, Neto MDO, Citadini APS, Leite VBP, Squina FM, Murakami MT, Ruller R. Characterization of a hexameric exo-acting GH51 α-L-arabinofuranosidase from the mesophilic *Bacillus subtilis*. Mol Biotechnol. 2013;55:260–7.
29. Shi P, Li N, Yang P, Wang Y, Luo H, Bai Y, Yao B. Gene cloning, expression, and characterization of a family 51 alpha-L-arabinofuranosidase from *Streptomyces* sp. S9. Appl Biochem Biotechnol. 2010;162:707–18.
30. Canakci S, Belduz A, Saha BC, Yasar A, Ayaz FA, Yayli N. Purification and characterization of a highly thermostable alpha-L-arabinofuranosidase from *Geobacillus caldoxylolyticus* TK4. Appl Microbiol Biotechnol. 2007;75:813–20.
31. Canakci S, Kacagan M, Inan K, Belduz AO, Saha BC. Cloning, purification, and characterization of a thermostable α-L arabinofuranosidase from *Anoxybacillus kestanbolensis* AC26Sari. Appl Microbiol Biotechnol. 2008;81:61–8.
32. Goyal S, Dhull SK, Kapoor KK. Chemical and biological changes during composting of different organic wastes and assessment of compost maturity. Bioresour Technol. 2005;96:1584–91.
33. Sundberg C, Smars S, Jonsson H. Low pH as an inhibiting factor in the transition from mesophilic to thermophilic phase in composting. Bioresour Technol. 2004;95:145–50.
34. Wagschal K, Heng C, Lee CC, Wong DWS. Biochemical characterization of a novel dual-function arabinofuranosidase/xylosidase isolated from a compost starter mixture. App Microbiol Biotechnol. 2009;(5):855–63.
35. Deb P, Talukdar SA, Mohsina K, Sarker PK, Sayem SMA. Production and partial characterization of extracellular amylase enzyme from *Bacillus amyloliquefaciens* P-001. Springer Plus. 2013;2:154.
36. Zhang T, Liang J, Wang P, Xu Y, Wang Y, Wei X, Fan M. Purification and characterization of a novel phloretin-2′-O-glycosyltransferase favouring phloridzin biosynthesis. Sci Rep. 2016;6:35274.
37. Yildirim V, Baltaci MO, Ozgencli I, Sisecioglu M, Adiguzel A, Adiguzel G. Purification and biochemical characterization of a novel thermostable serine alkaline protease from *Aeribacillus pallidus* C10: a potential additive for detergents. J Enz Inhib Med Chem. 2017;32:468–77.
38. Scapin SMN, Souza FHM, Zanphorlin LM, de Almeida TS, Sade YB, Cardoso AM, et al. Structure and function of a novel GH8 endoglucanase from the bacterial cellulose synthase complex of *Raoultella ornithinolytica*. PLoS One. 2017;12:e0176550.
39. Sterner R, Kleemann GR, Szadkowski H, Lustig A, Hennig M, Kirschner K. Phosphoribosyl anthranilate isomerase from *Thermotoga maritima* is an extremely stable and active homodimer. Protein Sci. 1996;5:2000–8.
40. Ichikawa JK, Clarke S. A highly active protein repair enzyme from an extreme thermophile: the Lisoaspartyl methyltransferase from *Thermotoga maritima*. Arch Biochem Biophys. 1998;358:222–31.
41. Merz A, Knöchel T, Jansonius JN, Kirschner K. The hyperthermostable indoleglycerol phosphate synthase from *Thermotoga maritima* is destabilized by mutational disruption of two solvent-exposed salt bridges. J Mol Biol. 1999;288:753–63.
42. Miyazaki K. Hyperthermophilic a-L-arabinofuranosidase from *Thermotoga maritima* MSB8: molecular cloning, gene expression, and characterization of the recombinant protein. Extremophiles. 2005;9:399–406.
43. Peterson ME, Daniel RM, Danson MJ, Eisenthal R. The dependence of enzyme activity on temperature: determination and validation of parameters. Biochem J. 2007;402:331–7.
44. Pollo SMJ, Zhaxybayeva O, Nesbø CL. Insights into thermoadaptation and the evolution of mesophily from the bacterial phylum *Thermotogae*. Can J Microbiol. 2015;61:655–70.
45. Taylor TJ, Vaisman II. Discrimination of thermophilic and mesophilic proteins. BMC Struct Biol. 2010;10:1–10.

46. Zeigler DR. The *Geobacillus* paradox: why is a thermophilic bacterial genus so prevalent on a mesophilic planet? Microbiol. 2014;160:1–11.
47. Dumbrepatil A, Park J, Jung TY, Song H, Jang M, Han N, Kim T, Woo E. Structural analysis of α-Larabinofuranosidase from *Thermotoga maritima* reveals characteristics for thermostability and substrate specificity. J Microbiol Biotechnol. 2012;22:1724–30.
48. Nurcholis M, Nurhayati N, Helianti I, Ulfah M, Wahyuntari B, Wardani AK. Cloning of α-Larabinofuranosidase genes and its expression in *Escherichia coli*: a comparative study of recombinant arabinofuranosidase originating in *Bacillus subtilis* DB104 and newly isolated *Bacillus licheniformis* CW1. Microbiol Indones. 2012;6:1–8.
49. Shi P, Chen X, Meng K, Huang H, Bai Y, Luo H, Yang P, Yao B. Distinct actions by *Paenibacillus* sp. strain E18 α-L-arabinofuranosidases and xylanase in xylan degradation. Appl Environ Microbiol. 2013;79:1990–5.
50. Patel H, Chapla D, Divecha J, Shah A. Improved yield of α-Larabinofuranosidase by newly isolated *Aspergillus niger* ADH-11 and synergistic effect of crude enzyme on saccharification of maize Stover. Bioresour Bioprocess. 2015;2:11.
51. Wilkens C, Andersen S, Dumon C, Berrin JG, Svensson B. GH62 arabinofuranosidases: structure, function and applications. Biotechnol Adv. 2017;35:792–804.
52. Zhou J, Bao L, Chang L, Zhou Y, Lu H. Biochemical and kinetic characterization of GH43 β-D: - xylosidase/α-L: -arabinofuranosidase and GH30 α-L-arabinofuranosidase/ β-D-xylosidase from rumen metagenome. J Industrial Microbiol Biotechnol. 2012;39:143–52.
53. McKee LS, Pena MJ, Rogowski A, Jackson A, Lewis RJ, York WS, Krogh KBRM, Vikso-Nielsen A, Skjot M, Gilbert HJ, Marles-Wright J. Introducing endoxylanase activity into an exo-acting arabinofuranosidase that targets side chains. Proc Natl Acad Sci U S A. 2012;109:6537–42.
54. Fritz M, Ravanal MC, Braet C, Eyzaguirre J. A family 51 alpha-larabinofuranosidase from *Penicillium purpurogenum*: purification, properties and amino acid sequence. Mycol Res. 2008;112:933–42.
55. Yan Q, Tang L, Yang S, Zhou P, Zhang S, Jiang Z. Purification and characterization of a novel thermostable α-L-arabinofuranosidase (α-LAFase) from *Chaetomium* sp. Process Biochem. 2012;47:472–8.
56. Cartmell A, Mckee LS, Pen MJ, Larsbrink J, Brumer H, Kaneko S, Ichinose H, Lewis RJ, Vikso-Nielsen A, Gilbert HJ, Marles-Wright J. The structure and function of an arabinan-specific alpha-1,2-arabinofuranosidase identified from screening the activities of bacterial GH43 glycoside hydrolases. J BiolChem. 2011;286:15483–51495.
57. Paës G, Skov LK, O'Donohue MJ, Rémond C, Kastrup JS, Gajhede M, Mirza O. The structure of the complex between a branched pentasaccharide and *Thermobacillus xylanilyticus* GH-51 arabinofuranosidase reveals xylan-binding determinants and induced fit. Biochemistry. 2008;47:7441–51.
58. Caffall KH, Mohnen D. The structure, function, and biosynthesis of plant cell wall pectic polysaccharides. Carbohydr Res. 2009;344:1879–900.
59. Saha BC. Alpha-l-arabinofuranosidases: biochemistry, molecular biology and application in biotechnology. Biotechnol Adv. 2000;18:403–23.
60. Lim YR, Yoon RY, Seo ES, Kim YS, Park SC, Oh DK. Hydrolytic properties of a thermostable a-Larabinofuranosidase from *Caldicellulosiruptor saccharolyticus*. J Appl Microbiol. 2010;109:1188–97.
61. Tuffin M, Anderson D, Heath C, Cowan DA. Metagenomic gene discovery: how have we moved into novel sequence space? Biotechnol J. 2009;4: 1671–83.
62. Ohta K, Fujii S, Higashida C. Characterization of a glycoside hydrolase family-51 α-larabinofuranosidase gene from *Aureobasidium pullulans* ATCC 20524 and its encoded product. J Biosci Bioeng. 2013;116:287–92.

# 4

# Production of alkaline pectinase: a case study investigating the use of tobacco stalk with the newly isolated strain *Bacillus tequilensis* CAS-MEI-2-33

Ge Zhang[1,2,3], Shugui Li[1], Yingbo Xu[5], Juan Wang[3], Fan Wang[2], Yuhua Xin[3], Zhong Shen[3], Haibo Zhang[2], Ming Ma[4*] and Haobao Liu[1,3*]

**Abstract**

**Background:** Tobacco stalk (TS), a major agricultural waste abundant in pectin, has resulted in concerns about the need for its reuse. The nicotine in TS is considered a chemical that is to\xic and hazardous to the environment.

**Results:** In this study, *Bacillus tequilensis* CAS-MEI-2-33 was isolated from cigar wrappers to produce alkaline pectinase using TS. Subsequently, the medium and fermentation conditions for the production of pectinase by *B. tequilensis* CAS-MEI-2-33 were optimized. The optimal fermentation period, pH of the initial fermentation medium, concentration of TS, and inoculum amount for *B. tequilensis* CAS-MEI-2-33 were 40 h, 40 g/L, 7.0, and 3%, respectively. Under optimal conditions, the pectinase activity was 1370 U/mL. Then, the enzymatic properties, such as the optimum pH, reaction temperature, temperature stability, and effects of metal ions, were studied. The optimal pH was determined to be 10.0, indicating that the enzyme was an alkaline pectinase. The optimal temperature was 40 °C, and pectinase activity was stable at 40 °C. The $Ag^{+}$ metal ions were shown to remarkably promote enzyme activity. The pectinase was partly purified by ammonium sulfate precipitation, ion exchange chromatography, and Sephacryl S-100 chromatography. Sodium dodecyl sulfate-polyacrylamide gel electrophoresis (SDS-PAGE) and LC-MS/MS analyses were utilized to analyze the pectinase.

**Conclusions:** This study provided a new alkaline pectinase candidate and a new strategy for the use of TS.

**Keywords:** Alkaline pectinase, Tobacco stalk, Screening *Bacillus tequilensis*, Fermentation, Enzymatic properties, Purification

## Background

Extremely large quantities of agro-industrial waste residues are generated from the processing of raw materials, such as tobacco stalk (TS), which is an abundant crop residue in China. It has been estimated that more than one hundred thousand metric tons of TS are discarded annually in China ([1] #1136). Stricter environmental legislation for tobacco industries in China has forced these industries to seek an environmentally friendly process for the disposal of tobacco waste. The use of TS would not only decrease the soil pollution caused by its improper disposal but also provide additional income for the tobacco producers ([2] #1155). Some progress has been made in the use of TS, such as the extraction of chemicals ([3] #1227), reconstituted tobacco or filler ([4] #1051), and nanofibrillated cellulose ([5] #1210). However, new approaches are necessary for the safe and sustainable use of TS.

Pectinase is widely used in food ([6] #1172), paper, pulp ([7] #1095), and textile industries ([8] #918). This enzyme can be used for the degradation of pectin

* Correspondence: 228532536@qq.com; liuhaobao@caas.cn; l88702236@163.com
[4]China Tobacco Standardization Research Center, Zhengzhou Tobacco Research Institute of CNTC, Zhengzhou 450001, People's Republic of China
[1]Key Laboratory of Tobacco Biology and Processing, Tobacco Research Institute of Chinese Academy of Agricultural Sciences, Qingdao 266101, People's Republic of China

substances in plant tissues. Several microorganisms, including bacteria and fungi, have been shown to produce pectinase ([9] #1173). To reduce production costs, agro-waste sources, such as fruit waste ([10] #1101; [11] #1092), carrot waste ([12] #1098), and onion wastes ([13] #1211), have been used to produce pectinase using microbial solid-state fermentation. Furthermore, bacterial pectinase is preferred over fungal pectinase because of the ease of fermentation and modern techniques for improving production yield ([11] #1092). In addition, pectinases can be classified into acidic and alkaline enzymes according to the optimal pH for enzymatic activity. Alkaline enzymes, which are almost exclusively produced by bacteria, have numerous applications, such as textile processing; pharmaceutical uses; and leather, detergent and paper production ([6] #1172; [14] #1304).

Based on these criteria, we screened bacteria that could produce pectinase using TS, which is abundant in pectin (approximately 10%) ([15] #1141). Furthermore, the morphological and molecular characteristics of the isolated strain were analyzed. The enzyme properties, production processes with TS, and the purification of the pectinase were studied. This study provides a new strategy for the use of TS and helps lay the foundation for the production of pectinase.

## Results

### Screening of isolates for pectinase activity

Pectinase has many applications in various industrial processes ([16] #1170; [17] #1311). To screen excellent strains for pectinase production, some isolates from the primary screening were spot inoculated on pectin agar plates (PAPs) in triplicate. An additional image was shown this in more detail [see Additional file 1: Figure S1]/(see Additional file 1). The Hc value of 33 for this isolate was significantly higher than the values for other isolates ($8.87 \pm 0.15$, Fig. 1). This bacterium was labeled as isolate *B. tequilensis* CAS-MEI-2-33, and further characterization and identification were carried out.

### Morphological characteristics of *B. tequilensis* CAS-MEI-2-33

*B. tequilensis* CAS-MEI-2-33 was a gram-positive bacterium that contained spores, and the shape of the bacterial cells was clubbed. When grown at 37 °C for 8–12 h in LB agar, colonies of *B. tequilensis* CAS-MEI-2-33 were round, smooth, and cream colored, and the margin was entire.

### Biochemical characterization of *B. tequilensis* CAS-MEI-2-33

The biochemical characterization of *B. tequilensis* CAS-MEI-2-33 was performed using a variety of tests, including the Voges–Proskauer, nitrate reduction, glucose utilization, catalase, motility, lysozyme tolerance, phenylalanine, gelatin, starch, lactose, casein, and mannitol tests. The results of the phenylalanine, starch, and mannitol tests were negative, while the results of the remaining tests were positive (Table 1).

### Identification based on phylogenetic analysis

The phylogenetic tree generated using the 16S rRNA gene sequence of the bacterial isolate showed the highest homology (100%) with *B. tequilensis*. Furthermore, the biochemical characterization of the ability of CAS-MEI-2-33 to metabolize starch and mannitol was different from that of *Bacillus subtilis* but similar to that of *B. tequilensis*. According to Bergey's Manual of Determinative Bacteriology, this strain was named *B. tequilensis* CAS-MEI-2-33 ([18] #1309). The constructed phylogenetic tree indicated that this strain had the closest genetic relationship with the *B. tequilensis* strain P12Pb (Fig. 2). The tree was inferred by the neighbor-joining method using MEGA 7.0 software. The numbers at the nodes of the tree are indications of the levels of bootstrap support based on neighbor-joining analysis of 1000 inferred replications.

### Toxicity of nicotine to *B. tequilensis* CAS-MEI-2-33

TS, contains nicotine, which is harmful to many bacteria, and serves as a distinct source of nutrition. The average nicotine content in TS is as high as 1900–3800 mg/kg ([1] #1136), which is equal to 76–152 mg/L in the fermentation medium. In this study, 500–2000 mg/L nicotine in the medium was tested. The highest concentration of nicotine in the test medium was approximately 13 times that in the TS fermentation medium. However, the growth of *B. tequilensis* CAS-MEI-2-33 was less affected under the experimental conditions (Fig. 3).

### Production of pectinase

The production of high titers of pectinase by optimizing the growth parameters is of prime importance in industry ([19] #1145). In the present study, the one-factor-a-t-a-time method was implemented to optimize the components of the medium and conditions. The enzymatic activity was detected using 3,5-dinitrosalicylic acid (DNS) reagent (Beijing Leagene Biotech. Co., Ltd.) based on the production of D-galacturonic acid ([20] #1161). The curve for the production of pectinase during the fermentation of *B. tequilensis* CAS-MEI-2-33 was pivotal for indicating the optimal fermentation. The activity of pectinase increased gradually until 40 h, except for 32 h, and then began to decline (Fig. 4a). The highest pectinase activity for the process of fermentation using TS as the feedstock was 293 U/mL at 40 h. Parameters such as the pH of the fermentation medium have a substantial influence on the growth of strains and pectinase production. Pectinase activity ($618 \pm 9$

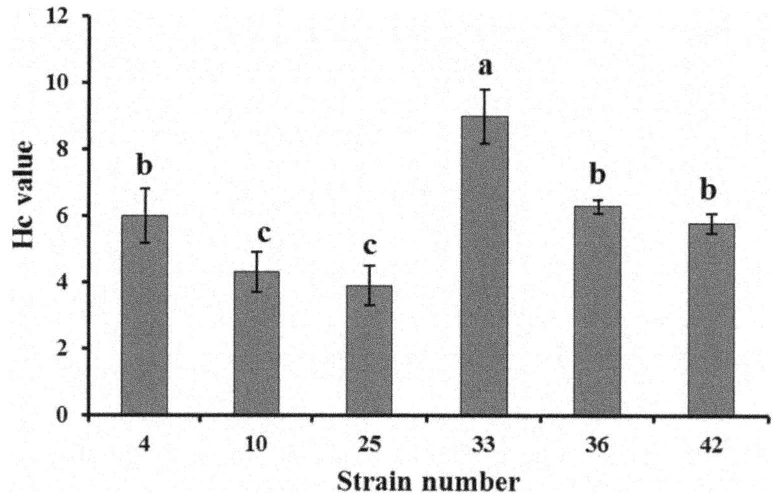


**Fig. 1** The Hc values of screened strains. Values are given as the means ± standard deviation ($n = 3$). Different letters indicate significant differences at 5%

U/mL, Fig. 4b) was significantly increased at pH 7.0 compared to that under other conditions.

After the determination of pectinase activity with different concentrations of TS in the fermentation medium, it was found that at the concentration of 40 g/L TS and the optimized pH, pectinase activity was the highest (730 ± 38 U/mL, Fig. 4c), as supported by ANOVA ($p = 0.05$).

The inoculum amount played a key role in the initial fermentation. Under the optimal pH and TS concentration, inoculum concentrations of 1, 3, 5, 7, and 9% were tested. The highest pectinase activity (1370 ± 126 U/mL, Fig. 4d) with 3% inoculum was not significantly different from that with 5% inoculum. At higher levels, such as 7 and 9%, enzyme production declined, which could be due to competition for nutrients among the population of bacteria, as has been observed in *Thermomucor indicae-seudaticae* ([21] #1230).

**Table 1** Biochemical characterization of the isolate *B. tequilensis* CAS-MEI-2-33. (+, positive; –, negative)

| Biochemical test | CAS-MEI-2-33 |
|---|---|
| Voges–Proskauer | + |
| Nitrate reduction | + |
| Glucose utilization | + |
| Catalase | + |
| Motility | + |
| Lysozyme tolerance | + |
| Phenylalanine | – |
| Gelatin | + |
| Starch | – |
| Lactose | + |
| Casein | + |
| Mannitol | – |

### Pectinase properties

The results of the pectinase property analysis are shown in Fig. 5. The pH of the reaction system can affect pectinase activity. Pectinase activity increased when the pH of the reaction systems was increased from 6.0 to 10.0, but this activity was nearly undetectable at pH 11.0. Our results showed that 791 ± 42 U/mL pectinase activity at pH 10.0 was the highest recorded value (Fig. 5a). Subsequently, the influence of temperature on pectinase activity was investigated. Pectinase activity increased when the reaction temperature was increased from 30 °C to 40 °C, reached maximum activity at 40 °C, and then decreased rapidly as the temperature increased beyond 40 °C. The pectinase activity at 40 °C was higher than that at other temperatures (Fig. 5b). Thus, pectinase was more active at an alkaline pH and intermediate temperature. The effects of different metal ions on pectinase activity are shown in Fig. 5c. $Ag^{+}$, $Li^{+}$, $Cu^{2+}$, $Ca^{2+}$, $Ba^{2+}$, and $Mn^{2+}$ ions increased enzyme activity; in particular, $Ag^{+}$ ions increased pectinase activity by 193.95%, which was approximately 1.94 times higher than that of the control. $K^{+}$, $Co^{2+}$, $Ni^{2+}$, $Mg^{2+}$, $Zn^{2+}$, $Cd^{2+}$, and $Fe^{3+}$

**Table 2** The partial purification results of alkaline pectinase

| | Total activity (U) | Total protein (ug) | Specific activity | Purification fold |
|---|---|---|---|---|
| Fermentation liquid | 4,074,513 | 7,859,964 | 0.5 | 1 |
| Ammonium sulfate | 284,144 | 166,842 | 1.7 | 3.3 |
| High-Q-8.0 | 77,611 | 15,066 | 5.2 | 10.2 |
| Sephacryl S-100 | 41,357 | 1580 | 26.2 | 51.4 |

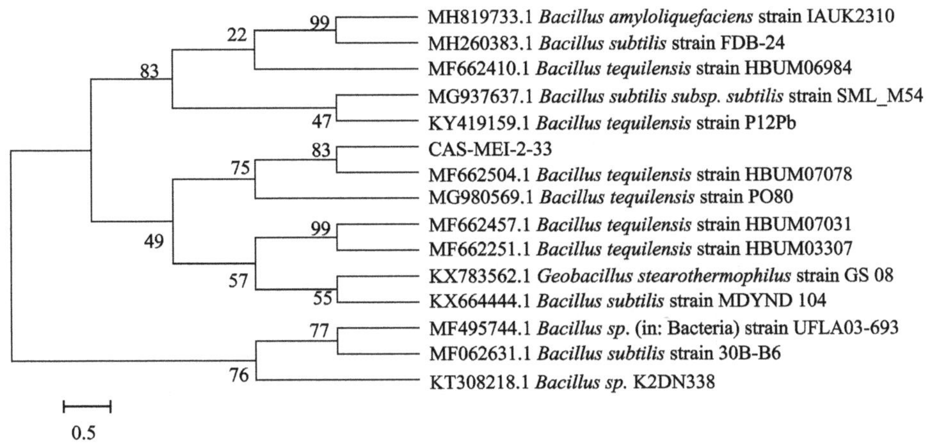


**Fig. 2** Phylogenetic tree based on 16S ribosomal RNA sequence analysis showing the position of the strain CAS-MEI-2-33 using MEGA 7.0 software. Numbers at branching points refer to bootstrap values (1000 resamplings) with 0.50 as the sequence divergence

ions, especially $Zn^{2+}$, inhibited pectinase activity. Figure 5d shows the thermal stability of pectinase produced by *B. tequilensis* CAS-MEI-2-33. However, pectinase activity decreased when the cell-free supernatant was placed at 60 °C. Pectinase activity was stable when the cell-free supernatant was incubated at 40 °C; however, pectinase could not tolerate the high temperature for a long time.

### Partial purification of pectinase from CAS-MEI-2-33

Under the optimal fermentation conditions, 2.30 L supernatant was obtained by centrifuging the bacteria in a 3.0 L fermentation flask. The pectinase activity reached 1771 U/mL, and the total pectinase activity was 4,074,513 U (Table 2).

According to the ammonium sulfate fractionation curve, suitable saturation was selected for salting out. The fractionation curve is shown in Fig. 6a. When ammonium sulfate was saturated to 70–80%, the pectinase activity of the precipitate increased, while that of the supernatant decreased significantly. When the saturation rate reached 80–90%, the pectinase activities (pH 10.0, Gly-NaOH) of the precipitate and supernatant did not change significantly. The target protein was

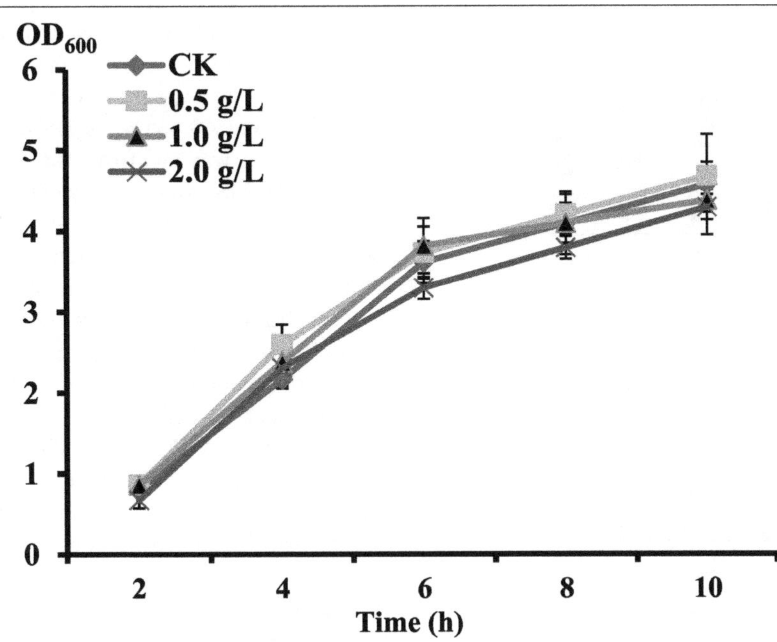


**Fig. 3** The growth of *B. tequilensis* CAS-MEI-2-33 under different concentrations of exogenous nicotine. Values are given as the means ± standard deviation (n = 3)

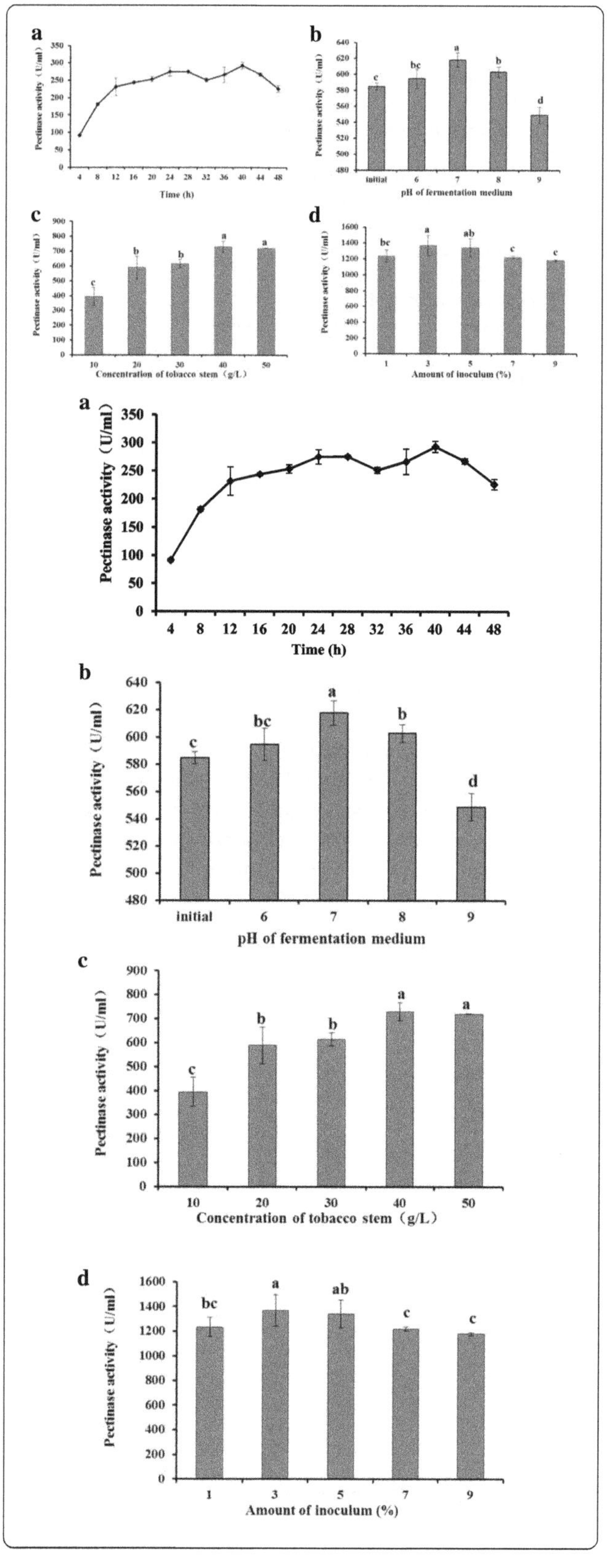


**Fig. 4** Pectinase activity of *B. tequilensis* CAS-MEI-2-33 during fermentation. **a** Pectinase activity during *B. tequilensis* CAS-MEI-2-33 growth. **b** Effect of the initial pH of fermentation medium on enzyme activity. **c** Effect of tobacco stalk concentration in the fermentation medium on enzyme activity. **d** Effect of the amount of inoculum on enzyme activity. Values are given as the means ± standard deviation (n = 3). Different letters indicate significant differences at 5%

dissolved with 30 mL buffer (pH 8.0, Tris-HCl) after salting out. Then, the target protein was further renatured with 55.0 mL buffer (pH 8.0, Tris-HCl) through a dialysis bag. The pectinase activity was 5166 U/mL, and the total activity was 284,144 U.

After dialysis, the supernatant was purified with a Mini Macro-Prep High-Q ion exchange column. The results are shown in Fig. 6b. The pectinase was eluted with pH 8.0 Tris-HCl buffer, and when the column was gradiently eluted with pH 8.0 Tris-HCl buffer containing 0–1 mol/L NaCl, peaks eluted. By detecting pectinase activity, it was found that the first elution peak was active. The first elution peak was collected with pectinase activity of 862,352 U/mL. The total activity was 77,611 U. Sephacryl S-100 was equilibrated with ultrapure water, sampled, and then eluted with pH 7.2 PBS buffer, with a flow rate of 0.8 mL/min, and the effluent was detected online by a UV detector at a wavelength of 280 nm to record the ultraviolet absorption peak curve (Fig. 6c). The components were collected and used for the determination of enzyme activity and protein content. The pectinase activity was 13,786 U/mL, and the total activity was 41,357 U.

The molecular weight of the purified alkaline pectinase was detected with SDS-polyacrylamide gel electrophoresis (PAGE) as described by Mehrnoush et al. ([22] #1306). The molecular weight of the pectinase, which was approximately 45.4 kDa, is shown in Fig. 6d. Then, the protein band was cut from the SDS-PAGE gel, subjected to LC-MS/MS analysis by Shanghai Applied Protein Technology ([23] #1320; [24] #1321), and identified by searching the UniProt database. The additional figure about protein base peak was shown this in more detail [see Additional file 2: Figure S2]/(see Additional file 2 for protein peak). Finally, the protein was identified as pectate lyase, the sequence was K.ASSSNVYTVSNR.N (Fig. 7). The molecular weight was 45.4 kDa, which was consistent with the SDS-PAGE results. The results indicated that the enzyme was a good candidate for pectate lyase. Furthermore, this study was a new attempt to recycle and reuse TS in agricultural production.

## Discussion

This study made a successful primary attempt to produce pectinase using TS with a new strain. In previous studies, a variety of microorganisms were identified as

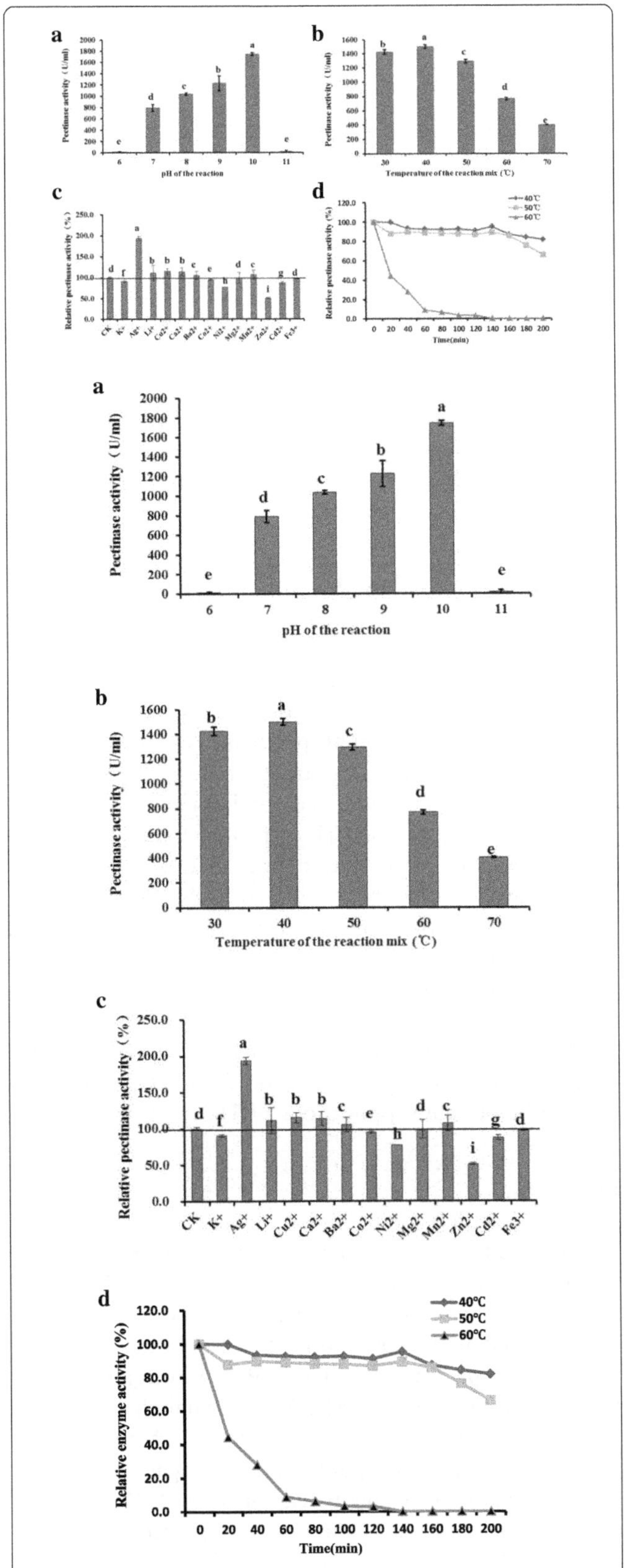


**Fig. 5** Enzymatic properties of pectinase. **a** Effect of substrate pH on pectinase activity. **b** Effect of reaction temperature on pectinase activity. **c** Effect of metal ions on pectinase activity. **d** Temperature stability of pectinase. Values are given as the means ± standard deviation (n = 3). Different letters indicate significant differences at 5%

pectinase producing, including *B. subtilis* ([8] #918; [25] #921), *B. megaterium* ([26] #1152), *Rhizopus oryzae* ([27] #1147), *Aspergillus oryzae* ([10] #1101), and *B. tequilensis* ([28] #1307).

In addition, it was critical to determine the fermentation period of *B. tequilensis* CAS-MEI-2-33 during optimization experiments. The one-factor-at-a-time method was implemented to optimize the components of the medium and conditions, especial the pH of medium. The optimum pH for the growth of and pectinase production by other bacterial strains, such as *B. subtilis* ([8] #918), has been recorded to vary from 7.0 to 10.0. Meneghel et al. ([29] #1193) and Fawole and Odunfa ([30] #1231) have reported that pectinase production by *Aspergillus* fungi occurs at an acidic pH. Different strains have species-specific acid-base conditions for their growth and fermentation processes. And then, as the primary carbon source in the medium, the concentration of TS exerted a profound effect on pectinase production by *B. tequilensis* CAS-MEI-2-33. However, pectinase activity with 50 g/L TS in the fermentation medium was not significantly different from that with 40 g/L TS, and this similarity might be caused by the other components, such as tobacco-specific nitrosamines (TSNAs) ([31] #1310).In the optimized fermentation (40 h, pH 7.0, 40 g/L TS, 3% inoculum), the highest pectinase activity was 1370 U/ml. By optimizing the inoculum amount, pectinase activity was improved 1.88 times. The results indicated that the *B. tequilensis* CAS-MEI-2-33 strain was a good candidate for biorefining TS. In subsequent experiments, optimization to improve the efficiency of pectinase production using TS and exploration of applications in industry, as conducted by Swarupa Rani Chiliveri ([28] #1307), would attract more interest.

The properties of pectinase are important in the application of this enzyme ([16] #1170); in particular, enzyme deactivation and stability are considered major constraints. The pectinase enzymes from different strains exhibited different properties ([16] #1170). In this study, the enzyme was an alkaline pectinase, which has various environmental and economic applications, such as textile processing; pharmaceutical uses; and leather, detergent and paper industry ([6] #1172). Pectinase is produced by both prokaryotic microorganisms, which almost exclusively synthesize alkaline pectinases, and eukaryotic microorganisms, which synthesize acid pectinases ([6] #1172). Ping & Chaochao reported that the

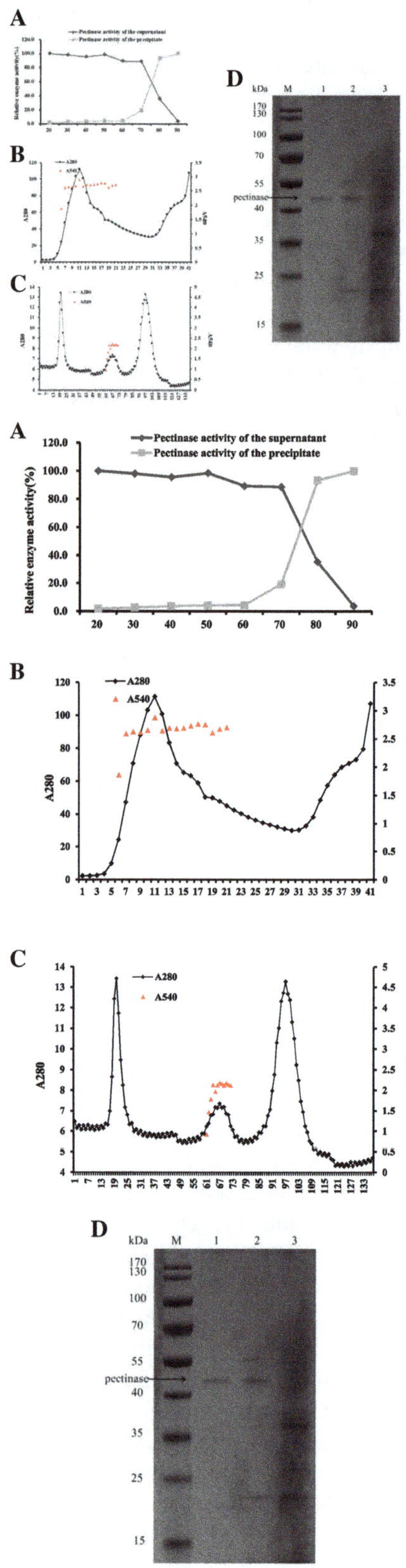


**Fig. 6** The purification of alkaline pectinase from *B. tequilensis* CAS-MEI-2-33. **a** Ammonium sulfate fractionation curve; **b** Elution curve of Mini Macro-Prep High-Q ion exchange chromatography; **c.** Elution curve of Sephacryl S-100 column chromatography; **d.** The partial purification of alkaline pectinase from *B. tequilensis* CAS-MEI-2-33 using TS. M: molecular weight makers; 1. Sephacryl S-100 column chromatography with pH 7.2; 2: Mini Macro-Prep High-Q ion exchange chromatography with pH 8.0; 3. Concentrated ammonium sulfate salting out solution

optimal pH for pectinase production by *Bacillus* sp. was 5.0 ([32] #1221). The difference in optimal pH might be caused by the strain or fermentation medium. This is the first report of the ability of *B. tequilensis* strain to produce pectinase using TS and provides a direction for further investigation of the possible applications of TS.

## Conclusions

This study successfully determined that the strain isolated from the cigar wrapper was a potential candidate organism that was useful in the production of pectinase using TS. The *B. tequilensis* CAS-MEI-2-33 strain will not only aid in addressing TS waste but also increase the possible sources of alkaline pectinase. The medium and culture conditions for this strain were optimized in this study. Based on the optimized conditions (pH 7.0, 40 g/L TS, 3% inoculation, and 40 h cultivation), the highest enzymatic activity of the strain was 1370 U/mL. The enzymatic properties were studied, including the pH of the reaction buffer, reaction temperature, effects of metal ions, and thermal stability. The alkaline pectinase was relatively stable at 40 °C, and the molecular weight was approximately 45.4 kDa. This is the first report of the ability of the *B. tequilensis* strain to produce pectinase using TS and provides a direction for further investigation of the possible applications of TS.

## Methods

### Primary screening

Samples collected from the cigar wrapper leaf from the Haikou Cigar Research Institute, Hainan Provincial Branch of the China National Tobacco Corporation, were diluted with 0.9% sterile NaCl solution and then plated onto pectin agar plates (PAPs) containing 0.5% pectin (galacturonic acid≥ ≥ 74.0%, Shanghai Macklin Biochemical Co., Ltd.), 0.1% $K_2HPO_4$, 0.3% $(NH_4)_2SO_4$, 0.001% $Fe_2SO_4$, 0.1% $MgSO_4$, and 1.8% agar powder. Inoculated plates were incubated for 48 h at 37 °C to obtain bacterial colonies. Colonies were picked, inoculated on fresh PAPs, and incubated at 37 °C for 48 h to obtain pure cultures. The pure cultures were maintained on Luria–Bertani (LB) agar plates at 4 °C and stored at − 80 °C with 15% glycerol for further use.

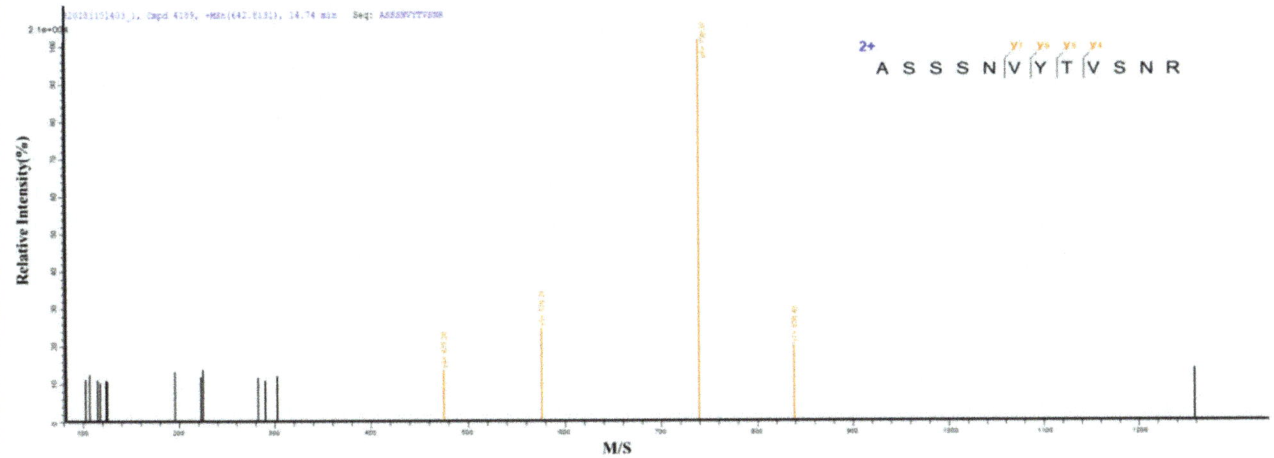


**Fig. 7** LC-MS/MS analysis of protein bands by SDS-PAGE

### Secondary screening

The bacterial isolates from the primary screening were tested for their ability to produce pectinases. The isolates were spot inoculated on PAPs in triplicate and incubated at 37 °C for 36–48 h. Depending on the zone of clearance surrounding the colonies (Hc value, the ratio of the diameter between clearance and the strain) using Congo red (Shanghai Aladdin Bio-Chem Technology Co., Ltd.) ([10] #1101), the strain used for further experiments was selected.

### Identification of the isolated bacterial strain

The isolated bacterial strain was identified according to morphological, physiological, and biochemical characteristics, as well as phylogenetic analysis of the 16S ribosomal RNA (rRNA) gene. Morphological characteristics of the isolated bacterial strain, e.g., colony morphology (color, shape, margin, and surface) and cell morphology (shape and Gram staining), were studied ([25] #921). The biochemical characteristics of the strain were studied with microbial biochemical identification tubes (Guangdong Huankai Microbial Sci. & Tech. Co., Ltd.). These biochemical and characterization tests include the Voges–Proskauer, nitrate reduction, glucose utilization, catalase, motility, lysozyme tolerance, phenylalanine, gelatin, starch, lactose, casein, and mannitol tests. The 16S rRNA gene fragment of the strain was obtained using polymerase chain reaction (PCR) amplification with universal primers (BGI Life Tech Co., Ltd., Qingdao, China): 27F (5′-3′): AGAGTTTGATCCTGGCTCAG and 1492R (5′-3′): ACGGCTACCTTGTTACGACTT. The reaction mixture was prepared in a final reaction volume of 50 μl consisting of 25.0 μl Premix Taq™ (LA Taq™ Version 2.0, Takara Bio, Japan), 1.0 μl 10 μM primers, 1.0 μl genomic DNA template (57 ng/μl), and 22.0 μl dd$H_2O$. The genomic DNA of *B. tequilensis* CAS-MEI-2-33 was extracted with a TIANamp Bacteria DNA Kit (TIANGEN Biotech, Beijing) according to the manufacturer's instructions. The PCR products were detected using 1.0% agarose gel electrophoresis. The product of the appropriate size (1.5 kbp) was sequenced (BGI Life Tech, China), and the sequence was compared with the NCBI nucleotide sequence database (https://blast.ncbi.nlm.nih.gov/Blast.cgi?PROGRAM=blastn&PAGE_TYPE=BlastSearch&LINK_LOC=blasthome) and deposited under accession number MH806356.1. The phylogenetic tree was constructed using MEGA (version 7) ([33] #1324). The phylogenetic tree was inferred using the neighbor-joining method. Bootstrap analysis was based on 1000 resamplings.

### Pectinase activity assay

Pectinase activity production was assessed using the 3,5-dinitrosalicylic acid (DNS) method ([29] #1193; [9] #1173), and the absorbance was measured at 540 nm with a UV-VIS spectrophotometer (Cary 50 UV-Vis, Varian, Inc., North America). Cell-free supernatant (20 μl) was mixed with 1 mL pectin (0.2%, galacturonic acid≥74.0%, Shanghai Macklin Biochemical Co., Ltd.) as the substrate using NaOH-glycine buffer (50 mmol/L, pH 9.0) and incubated at 40 °C for 30 min. DNS solution (1 mL; Beijing Leagene Biotechnology Co., Ltd., China) was added, and the reaction mixture was boiled for 5 min. The absorbance of the cooled reaction mixture was read at 540 nm, while the supernatant was boiled for 5 min to inactivate the pectinase.

A standard curve was established using D-galacturonic acid as the reducing sugar. The final concentrations of the solutions in each tube with 5 mL DNS solution were 0.008, 0.016, 0.024, 0.032, 0.040, and 0.048 mg/mL D-galacturonic acid (Beijing Leagene Biotechnology Co., Ltd.). The tubes were boiled for 5 min and cooled. A blank sample was prepared with distilled water instead of D-galacturonic acid. One unit (U) of polygalacturonase activity was defined as the amount of enzyme that generated 1 μmol galacturonic acid per min under the assay conditions. Pectinase activity was calculated using the following formula:

$$\text{Enzyme Activity}\left(\frac{\text{U}}{\text{ml}}\right)=\frac{1000\times\Delta\text{A}\times\text{N}}{\text{K}\times\text{M}\times\text{T}}$$

where ΔA is the absorbance of samples subtracted from control absorbance, N is the dilution factor, K is the slope of the standard curve, T is 30 min, and M is the molecular weight of D-galacturonic acid (M = 194.14 Da).

**Optimization of the fermentation conditions**

It is critical to determine the fermentation period of *B. tequilensis* CAS-MEI-2-33 during optimization experiments. We took samples and detected the pectinase activity of the fermentation medium every 4 h to confirm the maximum activity of pectinase during fermentation. The initial fermentation medium contained 20 g/L TS, 0.5 g/L $MgSO_4$, 3 g/L $(NH_4)_2SO_4$, and 0.01 g/L $Fe_2SO_4$ at pH 7.0. Fermentation was performed in shake-flask experiments using 500 mL shake flasks containing 100 mL fermentation medium incubated with the strain *B. tequilensis* CAS-MEI-2-33. The culture was grown at 37 °C and 180 rpm. All experiments were performed in triplicate. The reagents were purchased from Sinopharm Chemical Reagent Co., Ltd.

Inocula were prepared using the culture of *B. tequilensis* CAS-MEI-2-33 incubated for 8 h at 37 °C in LB. The $OD_{600}$ of *B. tequilensis* CAS-MEI-2-33 was approximately 4.52, and the inoculum volume was 1% of the fermentation volume. All experiments were performed in triplicate. One-way ANOVA with Tukey's test was used to determine any significant differences ($p < 0.05$) between treatments using SAS (Statistical Analysis System, SAS Institute Inc., USA).

To test the toxicity of nicotine to *B. tequilensis* CAS-MEI-2-33, growth of this strain in LB containing nicotine was evaluated in shake-flask experiments using 150 mL shake flasks containing 50 mL medium that was incubated with 1% (v/v) inocula. Nicotine was added to the LB medium at final concentrations of 0, 0.5, 1.0, and 2.0 g/L. The optical density of the incubation was monitored at 600 nm every two hours.

Optimization of the pH was performed in a similar manner for the selected isolate. The initial fermentation medium was prepared with pH ranging from 6.0–8.0 using NaOH and the initial medium prior to sterilization at 121 °C for 20 min. The medium was incubated at 37 °C and 180 rpm for 40 h.

TS provided the carbon and nitrogen sources in the medium and was important during fermentation. The fermentation medium was prepared with TS concentrations ranging from 10 to 50 g/L at the optimal pH. The medium was incubated at 37 °C and 180 rpm for 40 h.

Effect of the inoculum size on enzyme production was studied with different concentrations of inocula. Different amounts, i.e., 1, 3, 5, 7, and 9% (v/v), were inoculated into the fermentation medium with 40 g/L TS at pH 7.0. The medium was incubated at 37 °C and 180 rpm for 40 h.

**Enzymatic properties**

The enzyme activity was highly influenced by the pH, temperature, and metal ions. The pH and temperature are important characteristics of a biocatalyst for use in industrial applications. The optimal pH of pectinase in the cell-free supernatant was assayed using citrate–phosphate (pH 6.0), phosphate (pH 7.0–8.0), and glycine–NaOH (pH 9.0–11.0) as buffers (50 mmol/L) and 0.2% pectin (Shanghai Macklin Biochemical Co., Ltd., China) as a substrate at 40 °C. The optimal temperature for pectinase activity was determined by incubating the reaction mixture from 30 °C to 70 °C at the optimal pH. The effects of metal ions ($Ag^+$, $K^+$, $Li^+$, $Cu^{2+}$, $Ca^{2+}$, $Ba^{2+}$, $Co^{2+}$, $Ni^{2+}$, $Mg^{2+}$, $Mn^{2+}$, $Zn^{2+}$, $Cd^{2+}$, and $Fe^{3+}$) on pectinase activity were assayed at a concentration of 1.0 mM in the reaction mixture. Thermal stability was investigated by measuring pectinase activity every 20 min following incubation of the enzyme solution at 40 °C, 50 °C, and 60 °C.

**Purification of pectinase**

According to the optimal fermentation parameter, we obtained approximately 2.30 L fermentation liquid. After fermentation, the fermentation liquid was centrifuged at 4 °C and 5000×g for 30 min to collect the supernatant, which contained the alkaline pectinase. Crude protein was obtained after separation with ammonium sulfate fractionation, ion exchange chromatography and column chromatography.

Fifty milliliters of supernatant was placed in an ice bath, and 20, 30, 40, 50, 60, 70, 80, and 90% saturation with ammonium sulfate was conducted overnight for protein precipitation. Then, the solution was centrifuged in a high-speed refrigerated centrifuge at 5000×g for 15 min to obtain the precipitate and supernatant. By detecting the pectinase activity of the precipitate and supernatant with different levels of saturation with ammonium sulfate, the ammonium sulfate fractionation curve and suitable saturation were determined. Based on the results from the ammonium sulfate fractionation curve, 70% saturation with ammonium sulfate was conducted, and the precipitate was removed by centrifugation. Then, the supernatant was brought to 80% saturation with ammonium sulfate to precipitate the target enzyme. The precipitate was crude pectinase. By dialyzing the precipitate a few times in equilibrated buffer, the crude enzyme was obtained.

The crude enzyme was subjected to High-Q ion exchange chromatography (Bio-Rad Laboratories (Shanghai Co., Ltd.)). Then, the column was equilibrated with Tris-HCl buffer (pH 8.0) and eluted with a linear salt gradient (0–1 M NaCl in 200 mL Tris-HCl buffer, 0.4 mM, pH 8.0). The eluent was collected automatically at 1 min/tube.

Sephacryl S-100 was equilibrated with ultrapure water, sampled, and eluted with pH 7.2 PBS buffer, with a flow rate of 0.8 mL/min, and the effluent was detected by an online UV detector at a wavelength of 280 nm to record the ultraviolet absorption peak curve. The components were collected, and the enzyme activity and protein content were tested.

### Determination of protein content and SDS-polyacrylamide gel electrophoresis (SDS-PAGE)

The protein content was determined using a BCA (bovine serum albumin) standard curve, and SDS-PAGE ([34] #1297) was conducted to identify the enzyme purity.

### LC-MS/MS identification of protein

Protein bands were cut from the SDS-PAGE gel and subjected to LC-MS/MS (liquid chromatographic-tandem mass spectrometric) analysis by Shanghai Applied Protein Technology ([23] #1320; [24] #1321). Protein samples were digested using trypsin for 20 h at 37 °C. Then, peptides were trapped and desalted on Zorbax 300SB-C18 peptide traps (Agilent Technologies, Wilmington, DE, USA) followed by separation on a C18-reversed-phase column (0.15 × 150 mm, Column Technology Inc., Fremont, CA, USA). An Easy nLC system (Thermo Fisher Scientific) was used to deliver mobile phases A (0.1% formic acid) and B (0.1% formic acid in 84% acetonitrile) ([35] #1322) according to routine methods. The mass spectrometer was operated in MS/MS mode scanning from 380 to 1800 amu. The top 20 multiply charged ions were selected from each scan for MS/MS analysis ([36] #1323). Then, the MS/MS spectra were searched using MASCOT 2.2 (Matrix Science, London, UK), and the protein was identified by searching the *B. tequilensis* database in UniProt (https://www.uniprot.org/).

## Additional files

**Additional file 1: Figure S1.** The transparent zone around the strain of *B. tequilensis* CAS-MEI-2-33 on PAPs. (DOCX 65 kb)

**Additional file 2: Figure S2.** Protein peak detected by IC-MS/MS. The protein was isolated from a gel. (DOCX 78 kb)

### Abbreviations

16S rRNA: 16S ribosomal RNA; ANOVA: Analysis of variance; *B. tequilensis*: *Bacillus tequilensis*; DNS method: 3,5-dinitrosalicylic acid method; LB: Luria–Bertani; LC-MS/MS: Liquid chromatographic-tandem mass spectrometric; PAPs: Pectin agar plates; PCR: Polymerase chain reaction; TS: Tobacco stalk

### Acknowledgments

We thank International Science Editing (http://www.internationalscienceediting.cn/) and American Journal Experts (https://www.aje.cn/) for help with English usage.

### Funding

We greatly appreciate the financial support from the Key Program for Science and Technology of Hainan Provincial Branch of the CNTC (2017000024052), the Key Program for Science and Technology of Hainan Provincial Branch of the CNTC (2017000024056), and the Youth Innovation Promotion Association (CAS NO. 2017252). The funders role was providing funding for the study and all other aspects of the project (design, collection, analysis, and interpretation of data and writing of the manuscript) was executed by the authors.

### Authors' contributions

GZ carried out the experiments in this study and drafted the manuscript. HZ, MM and HL conducted the design of the experiment and helped draft and finalize the manuscript. YX and ZS revised the manuscript. SL, JW and YX assisted with the screening and fermentation experiments. FW participated in the pectinase activity detection and data analysis work. All authors read and approved the final manuscript.

### Ethics approval and consent to participate

Not applicable.

### Consent for publication

Not applicable.

### Competing interests

The authors declare that they have no competing interests.

### Author details

[1]Key Laboratory of Tobacco Biology and Processing, Tobacco Research Institute of Chinese Academy of Agricultural Sciences, Qingdao 266101, People's Republic of China. [2]CAS Key Laboratory of Biobased Materials, Qingdao Institute of Bioenergy and Bioprocess Technology, Chinese Academy of Sciences, Qingdao 266101, People's Republic of China. [3]Haikou Cigar Research Institute, Hainan Provincial Branch of CNTC, Haikou 571100, People's Republic of China. [4]China Tobacco Standardization Research Center, Zhengzhou Tobacco Research Institute of CNTC, Zhengzhou 450001, People's Republic of China. [5]Research and Development Center, China Tobacco Anhui Industrial Co., Ltd., Hefei 230088, People's Republic of China.

### References

1. Su Y, Xian H, Shi S, Zhang C, Nuruzzaman MSM, Mao J, Zhang G, Liao W, Wang Q, Liu H. Biodegradation of lignin and nicotine with white rot fungi for the delignification and detoxification of tobacco stalk. BMC Biotechnol. 2016;16(1):81–9.
2. Zhong W, Zhu C, Shu M, Sun K, Zhao L, Wang C, Ye Z, Chen J. Degradation of nicotine in tobacco waste extract by newly isolated *Pseudomonas sp.* ZUTSKD. Bioresour Technol. 2010;101(18):6935–41.

3. Hu R, Wang J, Li H, Ni H, Chen Y, Zhang Y, Xiang S, Li H. Simultaneous extraction of nicotine and solanesol from waste tobacco materials by the column chromatographic extraction method and their separation and purification. Sep Purif Technol. 2015;146:1–7.
4. Yuan Y, Lu Z, Huang L, Li Y, Lu F, Bie X, Teng Y, Lin Q. Biodegradation of nicotine from tobacco waste extract by *Ochrobactrum intermedium* DN2. J Ind Microbiol Biotechnol. 2007;34(8):567–70.
5. Tuzzin G, Godinho M, Dettmer A, Zattera AJ. Nanofibrillated cellulose from tobacco industry wastes. Carbohyd Polym. 2016;148:69–77.
6. Kohli P, Gupta R. Alkaline pectinases: a review. Biocatal Agr Biotechnol. 2015;4(3):279–85.
7. Ahlawat S, Mandhan RP, Dhiman SS, Kumar R, Sharma J. Potential application of alkaline pectinase from *Bacillus subtilis* SS in pulp and paper industry. Appl Biochem Biotechnol. 2008;149(3):287–93.
8. Ahlawat S, Dhiman SS, Battan B, Mandhan RP, Sharma J. Pectinase production by *Bacillus subtilis* and its potential application in biopreparation of cotton and micropoly fabric. Process Biochem. 2009;44(5):521–6.
9. Jayani RS, Saxena S, Gupta R. Microbial pectinolytic enzymes: a review. Process Biochem. 2005;40(9):2931–44.
10. Akbar S, Prasuna RG. Exploitation of fruit wastes for pectinase production using *Aspergillus oryzae*. Int J Pharm Bio Sci. 2012;3(3):756–65.
11. Kumar A, Sharma R. Production of alkaline pectinase by bacteria (*Cocci sps.*) isolated from decomposing fruit materials. J Phytopathol. 2012;4(1):01–5.
12. Patil RC, Murugkar TP, Shaikh SA. Extraction of pectinase from pectinolytic bacteria isolated from carrot waste. Int J Pharma Bio Sci. 2012;3(1):B261–6.
13. Pereira GS, Cipriani M, Wisbeck E, Souza O, Strapazzon JO, Gern RMM. Onion juice waste for production of *Pleurotus sajor-caju* and pectinases. Food Bioprod Process. 2017;106:11–8.
14. Hoondal GS, Tiwari RP, Tewari R, Dahiya N, Beg QK. Microbial alkaline pectinases and their industrial applications: a review. Appl Microbiol Biotechnol. 2002;59(4–5):409–18.
15. Xiao R, Chen Y. Research progress on comprehensive utilization of tobacco stalk. Sci Technol Inform. 2012;15:55.
16. Gummadi SN, Panda T. Purification and biochemical properties of microbial pectinases a review. Process Biochem. 2003;38(2003):987–96.
17. El-Enshasy HA, Elsayed AE, Suhaimi N, Malek RA, Esawy M. Bioprocess optimization for pectinase production using *Aspergillus niger* in a submerged cultivation system. BMC Biotechnol. 2018;18(2018):71.
18. R. EB, N. EG. Bergey's manual of determinative bacteriology eight edition(in Chinese). Beijing: Science Press; 1984.
19. Reddy MP, Saritha KV. Effects of the culture media optimization on pectinase production by *Enterobacter sp.* PSTB-1. Biotech. 2016;6(2):207–17.
20. Uzuner S, Cekmecelioglu D. Enhanced pectinase production by optimizing fermentation conditions of *Bacillus subtilis* growing on hazelnut shell hydrolyzate. J Mol Catal B-Enzym. 2015;113:62–7.
21. Kaur P, Satyanarayana T. Production and starch saccharification by a thermostable and neutral glucoamylase of a thermophilic mould *Thermomucor indicae-seudaticae*. World J Microb Biot. 2004;20(4):419–25.
22. Amid M, Manap Y, Zohdi K. Purification and characterisation of thermo-alkaline pectinase enzyme from *Hylocereus polyrhizus*. Eur Food Res Technol. 2014;239(1):21–9.
23. Gao L, Li Z, Xia C, Qu Y, Liu M, Yang P, Yu L, Song X. Combining manipulation of transcription factors and overexpression of the target genes to enhance lignocellulolytic enzyme production in *Penicillium oxalicum*. Biotechnol Biofuels. 2017;10(1):100–17.
24. Qian W, Gang X, Zhang T, Wei L, Yang X, Li Z, Yang Y, Song L, Wang P, Peng J, et al. Protein kinase a-mediated phosphorylation of the broad-complex transcription factor in silkworm suppresses its transcriptional activity. J Biol Chem. 2017;292(30):12460–70.
25. El-Sayed MH. Thermoalkali-stable pectinase from *Bacillus subtilis* strain NVFO 19 isolated from agricultural waste dump soil. Curr Res Microbiol Biotechn. 2015;3(6):805–15.
26. Tosin AB, Maria BO, Omowunmi OO, Oladipo AO. Pectinase production by *Bacillus megaterium*, *Bacillus bataviensis* and *Paenibacillus sp.* isolated from decomposing wood residues in the Lagos lagoon. J Trop Life Sci. 2017;7(3):204–7.
27. Chowdhury TI, Jubayer F, Uddin B, Aziz G. Production and characterization of pectinase enzyme from *rhizopus oryzae*. Slovak J Food Sci. 2017;11(1):641–51.
28. Chiliveri SR, Koti S, Linga VR. Retting and degumming of natural fibers by pectinolytic enzymes produced from *Bacillus tequilensis* SV11-UV37 using solid state fermentation. Springerplus. 2016;5(1):559–75.
29. Meneghel L, Reis G, Reginatto C, Malvessi E. Silveira MMd. Assessment of pectinase production by *Aspergillus oryzae* in growth-limiting liquid medium under limited and non-limited oxygen supply. Process Biochem. 2014; 49(11):1800–7.
30. Fawole OB, Odunfa SA. Some factors affecting production of pectic enzymes by *Aspergillus niger*. Int Biodeterior Biodegradation. 2003;52(4):223–7.
31. Guo P, Xia Y, Li R, Lü Y, Qiu M, Zhao W, Yu Y. Assessment of tobacco specific nitrosamines in tobacco genotypes. Adv Mater Res. 2013;641-642(1):871–4.
32. Yu P, Xu C. Production optimization, purification and characterization of a heat-tolerant acidic pectinase from *Bacillus sp.* ZJ1407. Int J Biol Macromol. 2017;108:972–80.
33. Kumar S, Stecher G, Tamura K. MEGA7: molecular evolutionary genetics analysis version 7.0 for bigger datasets. Mol Biol Evol. 2016;33(7):1870–4.
34. Celestino SMC, Maria dFS, Javier MF, MVd S, Filho EXF. Purification and characterization of a novel pectinase from *Acrophialophora nainiana* with emphasis on its physicochemical properties. J Biotechnol. 2006;123(1):33–42.
35. Martin P, Julie K, Ri-T A, Vasiliki B, George M, Antonia V, William M, Harald M, Vera J. Comparison of higher energy collisional dissociation and collision-induced dissociationMS/MS sequencing methods for identification of naturally occurring peptides in human urine. Proteomics Clin Appl. 2015;9:531–42.
36. Huang J, Liu T, Li K, Song X, Yan R, Xu L, Li X. Proteomic analysis of protein interactions between Eimeria maxima sporozoites and chicken jejunal epithelial cells by shotgun LC-MS/MS. Parasite Vector. 2018;11(1):226–35.

# 5

# Primary allogeneic mitochondrial mix (PAMM) transfer/transplant by MitoCeption to address damage in PBMCs caused by ultraviolet radiation

Francisco Cabrera[1,5,9,10], Mayra Ortega[2,9,10†], Francesca Velarde[9,10†], Eliseo Parra[6,7,8], Stephany Gallardo[2], Diego Barba[3], Lina Soto[3], Gabriela Peña[3], Luis Alberto Pedroza[3,4], Christian Jorgensen[5,10], Maroun Khoury[6,7,8,10] and Andrés Caicedo[3,9,10,11*]

**Abstract**

**Background:** Artificial Mitochondrial Transfer or Transplant (AMT/T) can be used to reduce the stress and loss of viability of damaged cells. In MitoCeption, a type of AMT/T, the isolated mitochondria and recipient cells are centrifuged together at 4 °C and then co-incubated at 37 °C in normal culture conditions, inducing the transfer. Ultraviolet radiation (UVR) can affect mitochondria and other cell structures, resulting in tissue stress, aging, and immunosuppression. AMT/T could be used to repair UVR cellular and mitochondrial damage. We studied if a mitochondrial mix from different donors (Primary Allogeneic Mitochondrial Mix, PAMM) can repair UVR damage and promote cell survival.

**Results:** Using a simplified adaption of the MitoCeption protocol, we used peripheral blood mononuclear cells (PBMCs) as the recipient cell model of the PAMM in order to determine if this protocol could repair UVR damage. Our results showed that when PBMCs are exposed to UVR, there is a decrease in metabolic activity, mitochondrial mass, and mtDNA sequence stability as well as an increase in p53 expression and the percentage of dead cells. When PAMM MitoCeption was used on UVR-damaged cells, it successfully transferred mitochondria from different donors to distinct PBMCs populations and repaired the observed UVR damage.

**Conclusion:** Our results represent an advancement in the applications of MitoCeption and other AMT/T. We showed that PBMCs could be used as a PAMM source of mitochondria. We also showed that these mitochondria can be transferred in a mix from different donors (PAMM) to UVR-damaged, non-adherent primary cells. Additionally, we decreased the duration of the MitoCeption protocol.

**Keywords:** Mitochondria, MitoCeption, Artificial mitochondria transfer / transplant (AMTT), Primary allogeneic mitochondrial mix (PAMM), Ultraviolet radiation (UVR), Cellular damage, p53, Primary immune cells, Cell repair

* Correspondence: acaicedo@usfq.edu.ec
[†]Mayra Ortega and Francesca Velarde contributed equally to this work.
[3]Colegio de Ciencias de la Salud, Escuela de Medicina, Universidad San Francisco de Quito, 17-12-841 Quito, Ecuador
[9]Instituto de Investigaciones en Biomedicina, Universidad San Francisco de Quito, 17-12-841 Quito, Ecuador
Full list of author information is available at the end of the article

## Background

A substantial number of in vitro and in vivo assays have demonstrated the natural ability of cells to transfer mitochondria amongst each other [1]. This phenomenon is most commonly observed in mitochondrial transfer from healthy mesenchymal stem/stromal cells (MSCs) to damaged cells [2–7]. The transfer replaces or repairs damaged mitochondria and thereby reduces the percentage of dead cells and restores normal functions [3, 4, 8]. In 1982, Clark and Shay introduced a type of AMT/T model using a co-incubation step between the recipient cell and exogenous mitochondria [9]. Their pioneering study demonstrated for the first time that the mitochondrial DNA (mtDNA) of donor cells could be integrated into recipient cells and subsequently transmit hereditary traits and induce functional changes. AMT/T mimics the natural process of mitochondrial transfer, reprograms cellular metabolism, and induces proliferation [10–13]. The introduction of this model elucidated the possible use of mitochondria as an active therapeutic agent.

Since 1982, numerous adaptations of AMT/T have been developed for in vitro and in vivo applications [10–12]. Among all available methods, the use of a centrifugation during co-incubation seems to reduce the quantity of mitochondria needed to facilitate successful mitochondrial internalization by the recipient cells [11, 14, 15]. In-vitro cultured cells, especially MSCs, have been used as one of the most common sources of mitochondria for AMT/T [11, 12, 14]. However, using stem cells or other cultured cells, which require an extensive time to proliferate, increases the cost and reduces time-effectiveness of the process. Furthermore, a large number of cells are needed to successfully obtain high yields of mitochondria for transfer. As an advancement in AMT/T, McCully et al. successfully transplanted autologous mitochondria from skeletal muscle and injected them into damaged myocardium after ischemic injury, which lead to an improvement in ventricular function in humans [16].

Our study tests a modification of the original MitoCeption protocol which reduces the time and complexity of the protocol. We sought to determine if primary allogenic mitochondrial mix (PAMM) MitoCeption could be used to repair peripheral blood mononuclear cells (PBMCs) damaged by ultraviolet radiation (UVR) (UVC-UVR wavelength of 254 nm). PAMM is composed of the PBMCs of at least three donors. A secondary objective was to provide further evidence as to how UVR affects mitochondria and cell viability. To first determine the effects of UVR on cells and mitochondria, we created a cellular model in which human PBMCs were irradiated with UVR. Mitochondrial damage was assessed according to changes in mitochondrial mass, metabolic activity estimated by the 3-(4,5-Dimethylthiazol-2-yl)-2,5-diphenyltetrazolium bromide (MTT) assay, and percentage of dead cells; these indicators were examined 30 min to 120 min after (early time point) and 18 h after (late time point) exposure to radiation. Then, we selected a standard exposure time of 3 min for the protocol, because this level of UVR exposure resulted in harm but not complete cell death. Irradiated cells were rescued with varying doses of mitochondria isolated from different PBMC donors (PAMM) using the updated MitoCeption protocol. Using this approach, we showed that PAMM transfer from PBMC donors can repair UVR damage in recipient PBMCs. PBMCs can internalize PAMM and decrease the percentage of dead cells together with the repairing effect of immune cells' respiratory burst (RB) after 18 h. Here, we describe a new method to repair UVR-damaged mitochondria using PAMM MitoCeption; 1 h to 2 h after PAMM MitoCeption, we determined that exogenous mitochondria had been internalized by recipient cells and normal cell viability and mitochondrial mass, metabolic function, DNA patrimony, p53 expression, and RB had been recuperated.

## Results

### UVR exposure decreases PBMC mitochondrial mass and metabolic function and increases the percentage of dead cells

To understand the effects of UVR on cell physiology, we developed an in vitro model that used fresh PBMCs and exposed them to a gradient of increasing doses of UVR during different lengths of time. For these assays, we isolated PBMCs from healthy donors of different ages and genders. PBMCs were exposed to increasing doses of UVC from 2 min (48 mJ/cm$^2$) to 6 min (144 mJ/cm$^2$). The effects of UVR on mitochondrial mass, function, and percentage of dead cells were measured at two separate times: immediately after exposure (early time point) and 18 h after culture (late time point). To test the effects of UVR on mitochondrial mass, we labeled the mitochondria with MitoTracker Green® and measured changes using flow cytometry with mean fluorescence intensity values (MFI). The shape, size, and granularity of treated and untreated PBMCs were analyzed (Additional file 1: Figure S1). The metabolic activity of PBMCs was measured with spectrophotometry using the MTT assay. Mitochondrial mass data was analyzed using flow cytometry and MTT assay for metabolic activity and subsequently normalized using sample/control-average transformation [17]. Finally, we stained PBMCs with Trypan Blue to determine the percentage of dead cells.

We observed a statistically significant reduction (ANOVA test, $^*p < 0.05$) in the mitochondrial mass of the lymphocytes (PBMCs) immediately after UVR exposure especially from 3 min (72 mJ/cm$^2$) to 6 min (144 mJ/cm$^2$) (Fig. 1a). Other specific PBMC populations differed in their mitochondrial masses and varied responses to UVR damage were observed (data not shown). After 18 h of culture, lymphocytes in the PBMCs that survived the UVR exposure did not recover their mitochondrial mass in most of the conditions tested (ANOVA test, $^{***}p < 0.001$) (Fig. 1b). The MTT

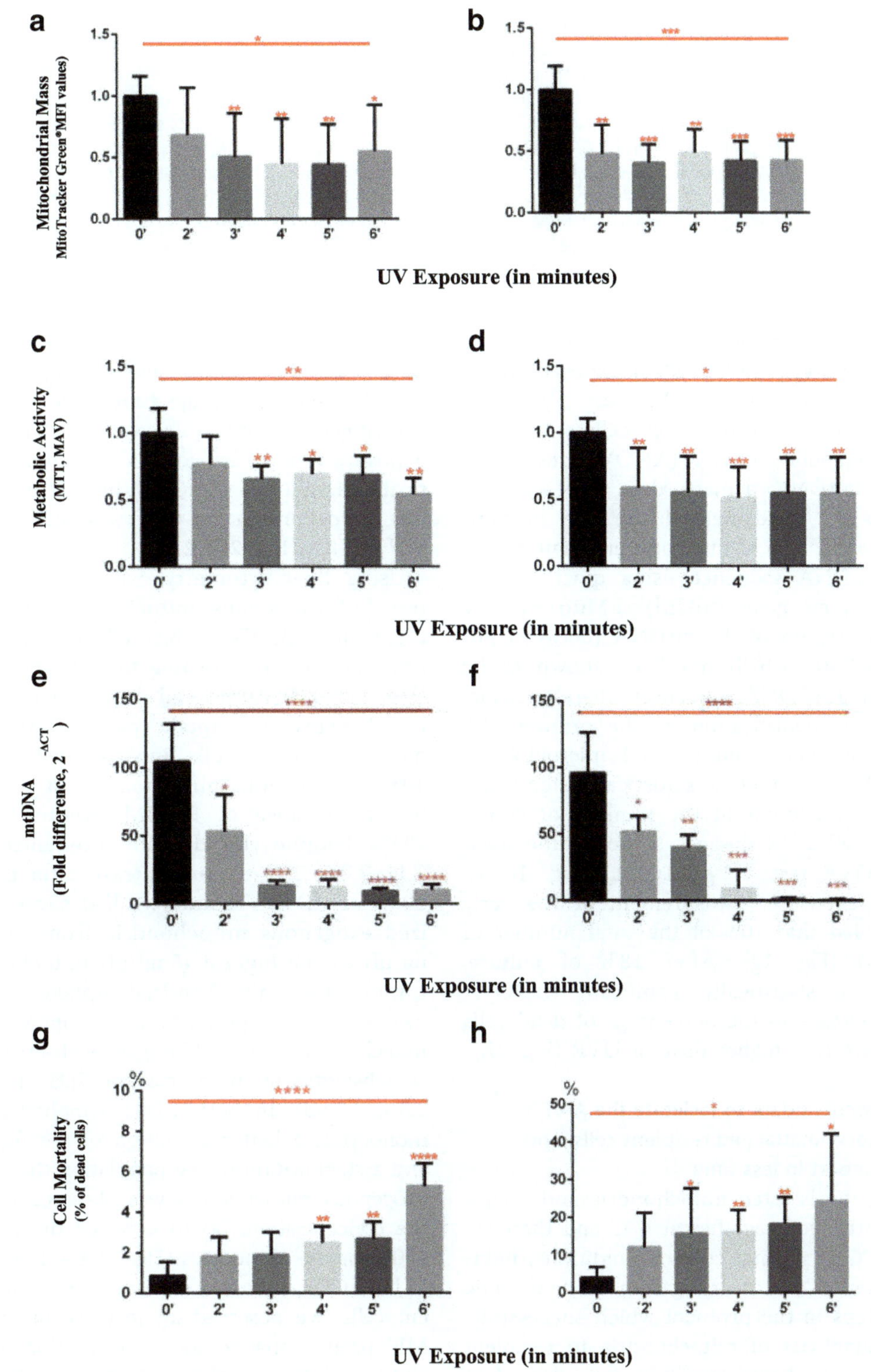


**Fig. 1** (See legend on next page.)

(See figure on previous page.)

**Fig. 1** PBMCs' mitochondrial mass, metabolic activity, mortality, and mtDNA quantification after UVR exposure. **a**, **b** Mitochondria mass ($n = 8$, 8 PBMC donors) determined by the mean fluorescence intensity values (MFI) of mitochondria labelled with MitoTracker® Green. **a** 1 h culture after exposure and 30'of MitoTracker® Green incubation. **b** 18 h culture and 30'of MitoTracker® Green incubation. Analysis, the sample/control-average transformation was used, Anderson-Darling normality test was applied. Un-paired, ANOVA test and Student's t-test are shown in the figure with an alpha-value of 0.05 (*$p < 0.05$, **$p < 0.01$, ***$p < 0.001$, ****$p < 0.0001$) to observe statistically significant differences. **c**, **d** Metabolic activity measured by MTT ($n = 5$, 5 PBMC donors). Mean absorbance values (MAV) were measured by spectrophotometry of PBMCs treated with MTT after UVR exposure. **c** After 2 h incubation with MTT. **d** After 18 h in culture and 2 h incubation with MTT. Analysis: the sample/control-average transformation was used, Anderson-Darling normality test was applied. Un-paired, ANOVA test and Student's t-test are shown in the figure with an alpha-value of 0.05 (*p < 0.05, **p < 0.01, ***p < 0.001, ****p < 0.0001) to observe statistically significant differences. **e**, **f** Determination of the effects on the mtDNA sequence of PBMCs after the exposure to UVR by qPCR (n = 8, 8 PBMC donors). The qPCR $2^{-\Delta CT}$ fold method was used. Primers: HB2M as the housekeeping gene sequence (Human B2M Beta-2-microglobulin, NCBI AH002619.1.), and HMito (designed for the mitochondrial genome, between positions 241 and 390, NCBI NC_012920.1) from Ajaz et al.(2015) [18] **e** 1 h culture after exposure. **f** After 18 h culture. Analysis: Un-paired, ANOVA test and Student's t-test are shown in the figure with an alpha-value of 0.05 (*$p < 0.05$, **$p < 0.01$, ***$p < 0.001$, ****$p < 0.0001$) to observe statistically significant differences. **g**, **h** Mortality of PBMCs induced by UVR ($n = 7$, 7 PBMC donors). The percentage of dead cells was estimated by counting those positive for Trypan Blue staining and then dividing them by the total; non-labelled viable cells were considered as well. **g** 1 h culture after exposure. **h** After 18 h culture. Statistical analysis: Un-paired, ANOVA test and Student's t-test are shown in the figure with an alpha-value of 0.05 (*$p < 0.05$, **$p < 0.01$, ***$p < 0.001$, ****$p < 0.0001$) to observe statistically significant differences

assay showed a reduction of metabolic activity of the PBMCs in all conditions tested when exposed to increasing UVR exposure of (ANOVA test, **$p < 0.01$) (Fig. 1c). After 18 h of culture, viable cells still exhibited reduced metabolic activity (ANOVA test, *$p <$ 0.05) (Fig. 1d). mtDNA damage data was analyzed using the qPCR $2^{-\Delta CT}$ fold method; the level of damage was estimated by hMito primers' failure to recognize the mtDNA sequence using qPCR with a nuclear housekeeping gene (hB2M). hMito primers bind to a unique region of the mtDNA (between positions 241 and 390), which has been shown to be less similar to nuclear DNA sequences, thereby ensuring more accurate identification by the primers [18, 19]. hB2M is the nuclear human β-2 microglobulin gene [18, 19]. We observed a statistically significant and proportional reduction in the number of copies of mtDNA after all UVR dosages at both time measurements (ANOVA test, ****$p < 0.0001$) (Fig. 1e, f). Following UVR exposure, measurement at the early time point revealed that 10% of the total number of cells were dead (Fig. 1g). After 18 h of culture, PBMCs showed a statistically significant (ANOVA test, *p < 0.05) increase in the percentage of dead cells following exposure to a higher dose of UVR (Fig. 1h).

### MitoCeption protocol update to facilitate the AMT/T using PAMM donors, unattached recipient cells, and alternative cell support in less time

In MitoCeption, the isolated mitochondria and recipient cells are centrifuged together at 4 °C and then co-incubated at 37 °C at normal culture conditions, inducing the transfer. The centrifugation and thermic shock are key steps in the protocol which successfully transfer small quantities of mitochondria to recipient cells [11]. The protocol was modified and adapted to transfer mitochondria to unattached PBMCs without using a plate or coating of a well surface as cell support. The cell support was changed from plates to 1.5 mL micro-centrifuge tubes (Fig. 2.1). We reduced centrifugation force (500 x g for 5 min at 4 °C) and exposure time to exogenous mitochondria (1 h); statistically significant (ANOVA test, ***$p < 0.001$) exogenous mitochondria up-take was still observed at 1 h and at 18 h (Fig. 2.2; 2.3).

Using flow cytometry, we observed that the patterns of exogenous mitochondria internalization by recipient PBMCs subjected to the MitoCeption protocol varied according to cell type and population (Fig. 2.2; 2.3; Additional file 1: Figure S1). In CD3+ lymphocytes, two uptake patterns were observed. In one population, cells incorporated between 5 and 40% of exogenous mitochondria in proportion to the increasing dose of PAMM. Another population of CD3+ lymphocytes did not internalize mitochondria (Fig. 2.2.a, 2.2c-d). In contrast, after one hour, 60 to 100% of CD14+ monocytes had successfully internalized exogenous mitochondria from the lowest (1.25 ng/μl) to the highest (5 ng/μl) mitochondrial concentration; the mitochondrial uptake rate in CD14+ monocytes was proportional to increasing exogenous mitochondrial doses, although the observed effect was not as substantial as in the case of CD3+ lymphocytes (Fig. 2.3.a, 2.3c-d). In both CD3+ lymphocytes and CD14+ monocytes, a better definition of the fluorescence peak and a stabilization of the population that internalized the exogenous mitochondria was observed independently of the concentration, 30 to 40% for the lymphocytes and 100% for the monocytes (Fig. 2.2.b – 2.2.e-f; 2.3.b-2.3.e-f). In both CD3 + lymphocytes and CD14+ monocytes recipient cells, we observed an increase in fluorescence and MFI proportional to the concentration of mitochondria and successful mitochondria internalization after 1 h and 18 h (Fig. 2.2a-b; 2.2.d; 2.2.f; 2.3a-b; 2.3.d; 2.3.f).

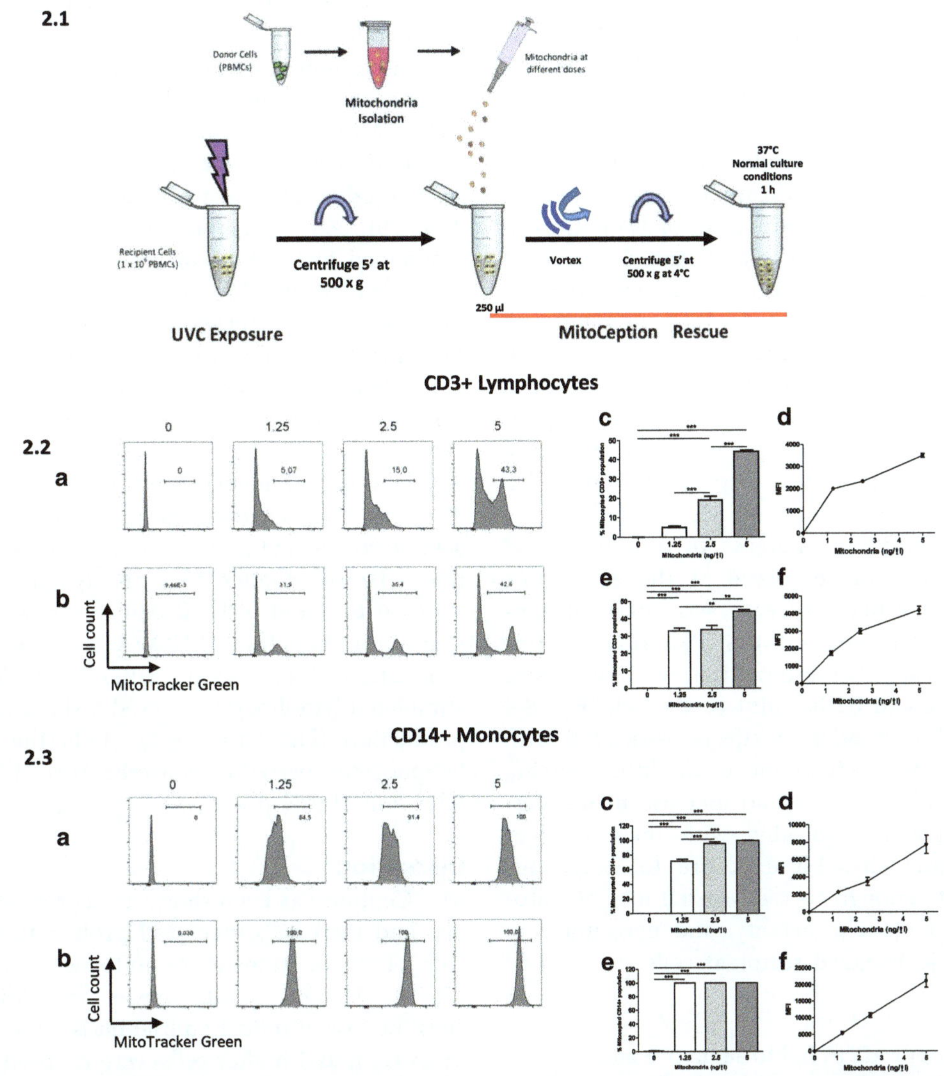


**Fig. 2** MitoCeption protocol update. (1) Schema of the updated protocol. Unattached recipient cells can be MitoCepted in a microcentrifuge tube of 1.5 ml and then centrifuged at 500 x g at during 5 min and immediately put in incubation for at least 1 h with mitochondria. Then, cells can be washed of excess mitochondria and used for downstream applications. Graphs show the proportional increase of the transfer of fluorescent mitochondria in percentages and MFIs due to the uptake in relation to the control of non-MitoCepted mitochondria. PAMM is composed of the PBMCs of at least three donors. (2, 3) MitoCepted CD3+ Lymphocytes and CD14+ Monocytes ($n = 3$, 3 PBMC donors, 3 PBMC donors for PAMM). Fresh PBMCs were MitoCepted with isolated mitochondria labeled with MitoTracker® Green from PBMCs; lymphocytes and monocytes were selected by their size, granularity, singles, alive cells (– for Annexin and 7AAD), and CD3+ and CD14+ identification. (2.2a, 2.2c, 2.2d) Flow cytometry of CD3+ cells after 1 h of MitoCeption. (2.2b, 2.2e, 2.2f) CD3+ after wash of excess mitochondria and 18 h of MitoCeption. (2.2c) CD3+ percentage of MitoCeption cells after 1 h. (2.2d) CD3+ MFI of the MitoCepted cells after 1 h. (2.2e) CD3+ Percentage of MitoCeption cells after 18 h. (2.2f) CD3+ MFI of the MitoCepted cells after 18 h. (2.3a, 2.3c, 2.3d) Flow cytometry of CD14+ cells after 1 h of MitoCeption. (2.3b, 2.3e, 2.3f) Flow cytometry of CD14+ cells after 18 h of MitoCeption. (2.3c) CD14+ percentage of MitoCeption cells after 1 h. (2.3d) CD14+ MFI of the MitoCeption cells after 1 h. (2.3e) CD14+ Percentage of MitoCeption cells after 18 h. (2.3f) CD14+ MFI of the MitoCepted cells after 18 h. Statistical analysis for all conditions: Mean ± SEM one-way ANOVA and Tuckey's pot-test. (* $p < 0.05$, ** $p < 0.01$, *** $p < 0.001$)

### PAMM transfer of mitochondria by MitoCeption rescues mitochondrial mass, function, and viability in PBMCs damaged by UVR

We demonstrated that UVR exposure in PBMCs cause a reduction of mitochondrial mass and function, loss of or changes in mtDNA, and an increase in the percentage of dead cells (Fig. 1a-h). Among all UVR doses tested, we determined that 3 min (72 mJ/cm$^2$) was the minimal dose required to induce statistically significant cell and mitochondrial damage (ANOVA test, ** $p < 0.01$); we used this UVR dose in all of the tested conditions. PBMCs damaged by 3 min of UVR exposure (72 mJ/cm$^2$) were MitoCepted during 1 h with allogeneic mitochondria from different PBMCs donors (PAMM) and subsequently washed to remove excess exogenous organelles that were not successfully internalized (Fig. 3). Isolated mitochondria were MitoCepted in doses of 1.25, 2.5, and 5 ng/µl to $1 \times 10^6$ UVR-damaged PBMCs.

Additionally, we used 5 ng/µl of isolated mitochondria to MitoCept PBMCs that had not been exposed to UVR in order to observe cytotoxicity. As expected, controls irradiated with a 3 min dose of UVR (72 mJ/cm$^2$) exhibited a statistically significant decrease in their mitochondrial mass and function and increase in percentage of dead cells (Fig. 3a-f). Treatment with PAMM induced a complete recovery of mitochondrial mass and function and reduced percentage of dead cells at the early and late time points in the lymphocytes population (PBMCs) (Fig. 3a-f). In both the non-exposed PBMCs and those exposed to 3 min (72 mJ/cm$^2$) of UVR which were then MitoCepted with 5 ng/µl of mitochondria, an increase of dead cells and mitochondrial mass was observed at the early time point; mitochondrial mass increase was more substantial in the UVR-damaged PBMCs. Additionally, this finding may indicate that internalization of excessive of mitochondria could be harmful to the cell (Fig. 3e). We did not observe adverse effects associated with the use of allogeneic mitochondria in doses ranging from 1.25 to 2.5 ng; there was no increase in the percentage of dead cells associated with these dosages (Fig. 3e-f). Additionally, 1 h exposure to exogenous mitochondria was enough to recover the loss of mitochondrial mass, metabolic activity, and percentage of dead cells in UVR-damaged recipient cells.

#### Rescue of the loss of mtDNA hMito sequence and normalization p53 mRNA expression in UVR-damaged PBMCs using PAMM MitoCeption

After determining the appropriate UVR dose (3 min) in our PBMC model and the quantity of exogenous mitochondria (2.5 ng / µl by $1 \times 10^6$ cells) needed to achieve statistically significant improvements in damaged cells, we assessed if the UVR damage of mtDNA could induce changes in p53 gene expression and if this could be repaired by PAMM MitoCeption. UVR caused a statistically significant loss of the mtDNA recognition sequence for hMito primers by qPCR, which was reversed by PAMM MitoCeption at both the early and late time points (Fig. 4a-b). UVR exposure caused a statistically significant increase (ANOVA test, $^{***}p < 0.001$) in p53 gene expression, as measured by qRT-PCR at 2 h after exposure and subsequent MitoCeption; this increase was then reversed by MitoCeption (Fig. 4c). We chose to measure the mRNA of p53 at 2 h because in preliminary assays the p53 expression at 1 h or after 18 h was minimally detected.

#### Rescue of RB (ROS production) of damaged PBMCs by MitoCeption

We used flow cytometry analysis of lymphocyte and monocyte size and granularity (Additional file 1:Figure S1) in order to determine the level of ROS production by Phorbol 12-myristate 13-acetate (PMA); we detected ROS production using Dihydro-rhodamine-123 (DHR-123) in all conditions tested. PBMCs monocytes responded to PMA by increasing ROS production (also known as RB) in all tested conditions (Fig. 5a-b). Monocytes that were MitoCepted increased their ROS production quantity at the early and late time points. After the UVR damage, monocytes decreased their ROS production and recovered it after MitoCeption; the statistical significance of these results was confirmed using ANOVA and Student's t-test (ANOVA test, $^{*}p < 0.05$). The lymphocytic population did not exhibit statistically significant change in ROS production at the early time point. However, after 18 h, the lymphocyte population exposed to 3 min of UVR damage showed a statistically significant reduction (ANOVA test, $^{**}p < 0.01$) in ROS production (Fig. 5c-d). After MitoCeption, PMA-stimulated lymphocytes showed a slight increase in ROS production (Fig. 5a-b). After 18 h, the UVR-damaged lymphocytes partially recovered their ROS production following MitoCeption.

### Discussion

MitoCeption has been shown to reprogram cell metabolism and increase cancer cell proliferation and invasion 24 h after the transfer of mitochondria isolated from MSCs [11]. Our results showed that MitoCeption rescues the loss of mitochondrial mass, function, and viability as soon as 1 h after cells were exposed to UVR. Here, we adapted the MitoCeption protocol to be used with fresh non-adherent immune cells such as PBMCs and changed the exogenous mitochondria source, centrifugation time and force, co-incubation time, and cell support. Some protocols use MSCs as a mitochondria source; however, the reduced proliferation capacity of MSCs makes obtaining sufficient cells impractical and expensive [14, 20, 21]. Therefore, we tested the suitability of PBMCs as the mitochondria source, which constitutes a novel approach in the field and is a more accessible mitochondria source because a greater number of mitochondria donor cells can be obtained from multiple donors. We decreased the force and time of centrifugation (500 x g for 5 min at 4 °C) and changed the original protocol's volume of medium from 1 mL to 250 µL in order to maximize recipient cell and mitochondria interaction. Finally, we used a different supporting material by changing the culture plate to 1.5 mL microcentrifuge tubes, thereby adapting this protocol for use with unattached cells like PBMCs.

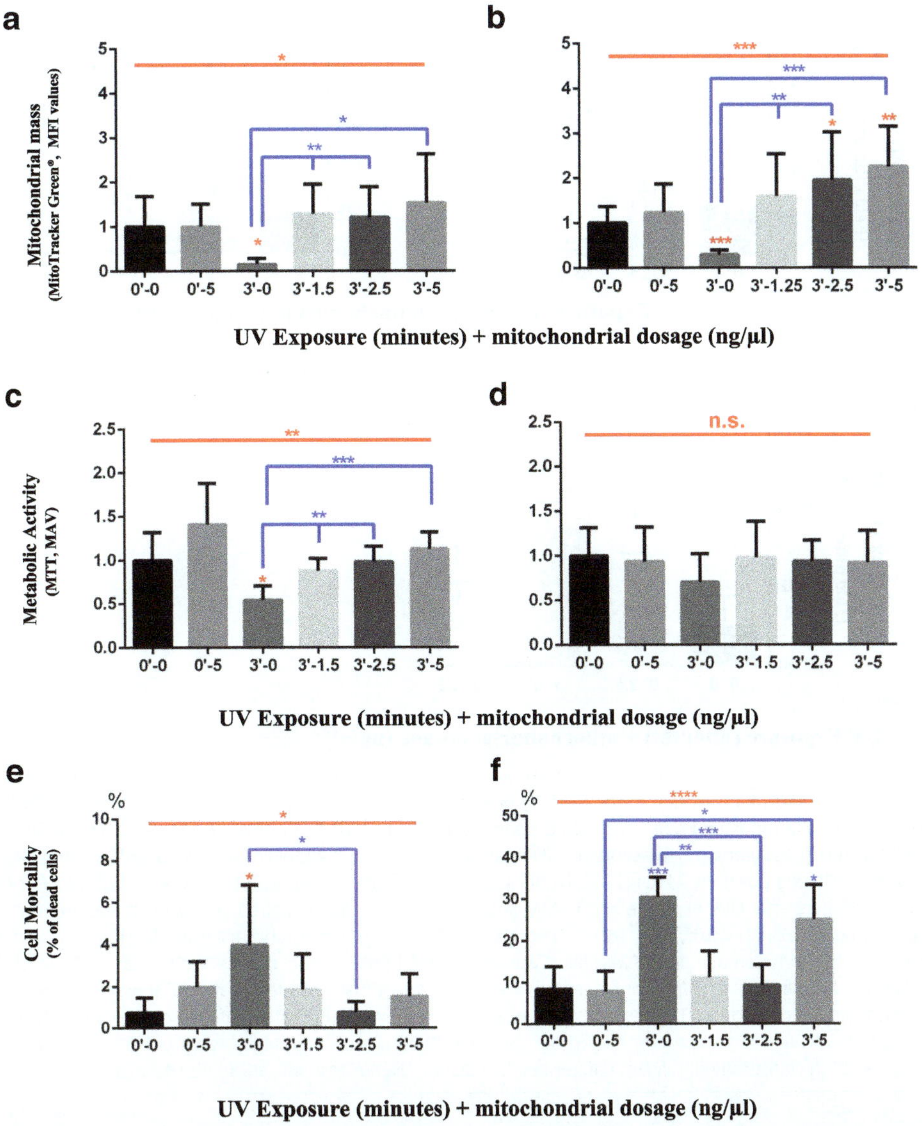


**Fig. 3** Estimation of the rescue of mitochondrial mass, metabolic activity, and mortality of damaged PBMCs using PAMM MitoCeption. PBMCs were isolated, exposed to UVR (3 min = 72 mJ/cm$^2$), and treated with 3 allogeneic mitochondria doses (1,25-2,5–5,0 µg of mitochondria). **a**, **b** Mitochondria mass (n = 7, 7 PBMC donors exposed to UVR and PAMM of 3 donors). MFI of PBMCs labelled with MitoTracker Green® was analyzed by flow cytometry. **a** 1 h after exposure to UVR, 1 h after MitoCeption, wash, and 30` incubation with MitoTracker® Green. **b** After 1 h exposure to UVR, 1 h MitoCeption, wash and 18 h in culture. Analysis, the sample/control-average transformation was used, Anderson-Darling normality test was applied. Un-paired, ANOVA test and Student's t-test are shown in the figure with an alpha-value of 0.05 (*$p < 0.05$, **$p < 0.01$, ***$p < 0.001$, ****$p < 0.0001$) to observe statistically significant differences. **c**, **d** Metabolic activity measured by MTT ($n = 5$, 5 PBMC donors exposed to UVR and PAMM of 3 donors). MAV were measured by spectrophotometry of PBMCs treated with MTT after UVR exposure and MitoCeption at the early and late time points. **c** 1 h after exposure to UVR, 1 h of MitoCeption, wash and 2 h of incubation with MTT. **d** 1 h after exposure to UVR, 1 h of MitoCeption, wash, 18 h in culture and 2 h of incubation with MTT. Analysis: the sample/control-average transformation was used, Anderson-Darling normality test was applied. Un-paired, ANOVA test and Student's t-test are shown in the figure with an alpha-value of 0.05 (*$p < 0.05$, **$p < 0.01$, ***$p < 0.001$, ****$p < 0.0001$) to observe statistically significant differences. **e**, **f** Mortality of PBMCs caused by UVR (n = 5, 5 PBMC donors exposed to UVR and PAMM of 3 donors), the percentage was estimated by counting dead cells positive for Trypan Blue staining and dividing that number by viable cells. **e** Early time point measured 1 h after exposure to UVR, 1 h MitoCeption and wash **f** Late time point 1 h after exposure to UVR, 1 h MitoCeption, wash and 18 h in culture. Analysis: Anderson-Darling normality test was applied to the resulting data. Un-paired, ANOVA test and Student's t-test are shown in the figure with an alpha-value of 0.05 (*p < 0.05, **p < 0.01, ***p < 0.001, ****p < 0.0001) to observe statistically significant differences

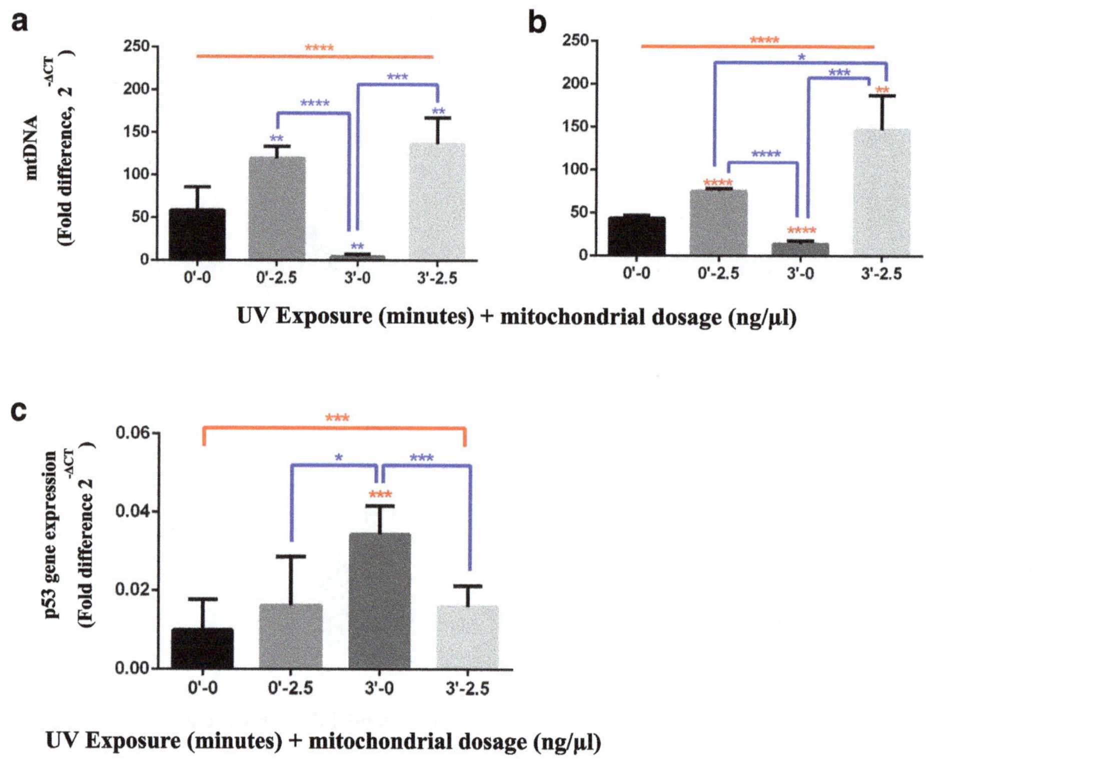


**Fig. 4** Determination of the effects on the mtDNA sequence, changes in the expression of the p53 gene induced by UVR, and rescue by adding exogenous mitochondria using PAMM MitoCeption in the PBMC model. **a**, **b** mtDNA damage quantification by the $2^{-\Delta CT}$ fold by qPCR after 3 min UVR exposure and rescue by PAMM Mitoception (n = 5, 5 PBMC donors exposed to UVR and PAMM of 3 donors). The qPCR $2^{-\Delta CT}$ fold method was used. Primers: HB2M as the housekeeping gene sequence (Human B2M Beta-2-microglobulin, NCBI AH002619.1.), and HMito (designed for the mitochondrial genome, between positions 241 and 390, NCBI NC_012920.1) from Ajaz et al.(2015) [18] **a** 1 h culture after UVR exposure and PAMM MitoCeption. **b** 18 h culture after UVR exposure and PAMM MitoCeption. Analysis: Un-paired, ANOVA test and Student's t-test are shown in the figure with an alpha-value of 0.05 (*p < 0.05, **p < 0.01, ***p < 0.001, ****p < 0.0001) to establish statistically significant differences. **c** Analysis of the p53 gene expression after UVR damage and PAMM MitoCeption (*n* = 6, 6 PBMC donors exposed to UVR and PAMM of 3 donors). Primers for detecting p53 mRNA and for the nuclear PUM (housekeeping) were used to estimate p53 levels after UVR exposure and PAM MitoCeption in receptor cells with or without UVR damage. p53 gene expression was measured by qRT-PCR at 2 h after exposure and subsequent MitoCeption. Statistical analysis: Un-paired, ANOVA test and Student's t-test are shown in the figure with an alpha-value of 0.05 (*p < 0.05, **p < 0.01, ***p < 0.001, ****p < 0.0001) to establish statistically significant differences. Un-paired, ANOVA test and Student's t-test (p < 0.05)

The MitoCeption protocol uses centrifugation to put recipient cells in close contact with the exogenous mitochondria and a thermic shock to attempt to facilitate mitochondrial uptake. It has been reported that MitoCeption requires 1.25 to 20 µg/ml of mitochondria (concentration measured using Bradford protein quantification assay) per 100,000 cancer cells to achieve phenotypical changes in the recipient cells [11]. MitoCeption allows minimal quantities of mitochondria to be used. The use of additional steps in co-incubation may depend on the user needs, cell type, and application. Studies performed by Lorberboum-Galski and her team showed that simple incubation between receiver cells and exogenous mitochondria was sufficient to facilitate transfer in a process called "Mitochondrial Transformation". In most assays, her team used 300 µg / mL of mitochondria with 50,000 HepG2 cells, MCF-7, 30,000 HEK-293, or 75,000 fibroblasts, did not use centrifugation, and completed data collection 1 h and 24 h after combining the recipient cells and exogenous motochondria [15]. In our study, we found that we were able to obtain yields of 50 µg / mL of mitochondria using only $16 \times 10^6$ PBMCs (4 donors), which was sufficient for all conditions tested. Interestingly, other groups have also found that performing a centrifugation step prior to the co-incubation of the isolated mitochondria with the recipient cells greatly improves the mitochondria uptake; for example, Mi Jin Kim et al. found that 0.05, 0.5 and 5 µg of exogenous mitochondria were successfully transferred to 10,000 recipient cells after 24 h [14].

To our knowledge, this study is the first to demonstrate in-vitro that MitoCeption can be used to re-establish mitochondrial function loss caused by UVR exposure. Additionally, we successfully transferred a mix of different PBMC donors to one PAMM that was used to repair damaged cells. Other research groups have

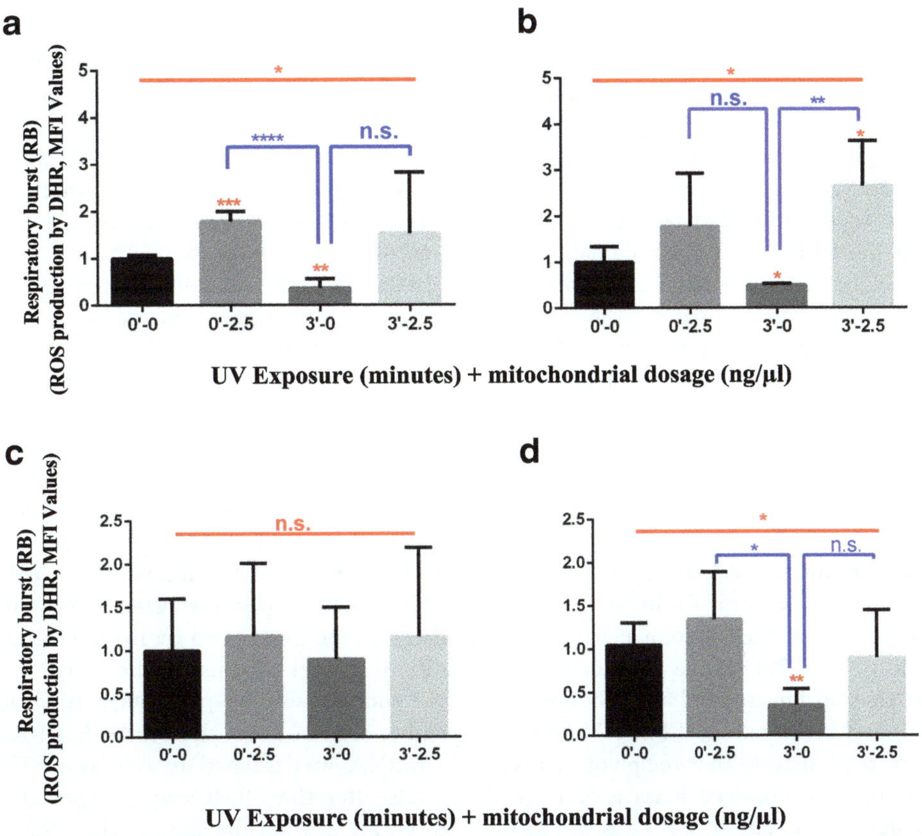


**Fig. 5** Estimation of the RB (ROS production) of UVR-exposed PBMCs (monocytes and lymphocytes) and assessment of PBMC rescue by PAMM MitoCeption. **a**, **b**, **c**, **d** ROS production measured with DHR-PMA treatment and flow cytometry (n = 6, 6 PBMC donors exposed to UVR and PAMM of 3 donors) **a**. Monocytes behavior 1 h after 3 min exposure to UVR and rescue by PAMM MitoCeption. **b**. Monocytes behavior after 18 h culture and rescue by PAMM MitoCeption. **c**. Lymphocytes behavior 1 h after 3 min exposure to UVR and rescue by PAMM MitoCeption. **d**. Lymphocytes behavior after 18 h culture and rescue by PAMM MitoCeption. Analysis: the sample/control-average transformation was used, Anderson-Darling normality test was applied. Un-paired, ANOVA test and Student's t-test are shown in the figure with an alpha-value of 0.05 (*$p < 0.05$, **$p < 0.01$, ***$p < 0.001$, ****$p < 0.0001$) to observe statistically significant differences

successfully transferred mitochondria from one cell donor type to others [10, 11, 20]; however, none of them have mixed mitochondria isolated from different donors for the transfer/transplant. This study elucidates the potential to use mitochondria from different donors (PAMM) to treat UVR stress and possibly other types of damage or metabolic malfunctions in cells, resulting in not only in-vitro but also ex-vivo applications. The repairing capacity of the PAMM and MitoCeption protocols require further investigation because the effects beyond 18 h of culture, the long-term effects of multiple mitochondria patrimonies inside the recipient cells, and the capacity to re-use them in-vivo for distinct applications is unknown.

In our study, assessment of mitochondrial state in the PAMM was limited, although the PAMM demonstrated reparative properties. Our results demonstrate that PAMM MitoCeption can repair some cell damage inflicted by UVR exposure. In order to promote mitochondrial stability during co-culture, in our protocol, we reduced the centrifugation and co-incubation time and the possible cellular stress induced by the long centrifugation and thermic shock included in the original MitoCeption methodology. Nonetheless, it is crucial that further assays be performed to determine if the mitochondria in PAMM are active, energetically coupled or uncoupled, and structurally complete in order to fully understand their repairing properties.

We observed that UVR-exposed PBMCs exhibited a decrease in mitochondrial mass and metabolic activity as measured by MTT. Likewise, UVR exposure induced changes or loss of mtDNA and increased the percentage of dead cells. Furthermore, increased p53 gene expression was associated with changes or loss in the mtDNA target sequence of hMito primers. UVR exposure also caused a decrease in ROS production (especially in monocytes), thus affecting the RB. The RB is part of the innate response to infection and host defense based on ROS production by the NADPH oxidase enzyme, which is situated in the membranes of monocytes and which may be sensitive to UVR [22]. We provided further insight into how mitochondria are affected by UVR in PBMCs in order to validate the use of MitoCeption to rescue cells exposed to UVR-induced stress (Additional file 2: Figure S2).

UVC was used to generate a standard damaging procedure for PBMCs to validate the use of the MitoCeption of PAMM to repair this damage. UVC greatly disturbs DNA and particularly mtDNA; damage especially accumulates in the latter molecule due to the lack of mtDNA repairing systems as compared to the nucleus [23]. Moreover, when mtDNA reaches a threshold limit for damage, it causes mitochondrial malfunction, increased ROS production, and induces apoptosis [23–26]. By using UVC, our study showed that increasing doses of radiation causes proportional loss of mtDNA, mitochondrial mass integrity, and metabolic function as soon as 1 h after exposure.

Measurement of the mitochondrial mass by MitoTracker Green® has been proven to be an accurate method to distinguish differences among cells and experimental conditions [27, 28]. Once we successfully used MitoCeption to transfer isolated mitochondria to undamaged PBMCs and UVR-damaged PBMCs, flow cytometry was used to demonstrate that recipient cells exhibited an increase in mitochondrial mass proportional to the dose of transferred mitochondrial. Thus, a statistically significant increase in mitochondrial mass was observed primarily in recipient cells that had been cocultured with exogenous mitochondria in concentrations greater than or equal to 2.5 ng / µl. After UVR damage, the decrease in percentage of dead cells was completely recovered by the adapted PAMM MitoCeption protocol.

In the MTT technique, the metabolic activity of living cells transforms the MTT chemical compound into formazan crystals, which appear blueish to violet in color. Once living cells have transformed the MTT in formazan crystals, dimethyl sulfoxide (DMSO) lysates the cells and dilutes the crystals. The viability and metabolic activity of the cells can be estimated by analyzing this bluish to violet solution with spectrophotometry [29]. The Trypan Blue dye exclusion assay is a widely-used technique to determine living cells (non-blue) from dead cells (blue); however, it has been reported that it overestimates cell viability and underestimates non-viable cell density [30]. In Fig. 3, at the early time point, the MTT results (c panel) and the percentage of dead cells (e panel) showed a congruent response: a decrease in MTT and an increase in the number of dead cells at the 3 min measurement. Exposure to 5 ng of mitochondria increased the MTT labeling, although it induced a slight increase in dead cells. This effect could be related to cytotoxicity. A recovery of the MTT staining, which reflects metabolic activity, was observed in the 3 min exposure conditions and repaired by PAMM MitoCeption. Cells exposed to 3 min of UVR and which were MitoCepted showed a greater decrease in the percentage of dead cells as compared with cells exposed to 3 min of UVR but which were not MitoCepted. At the early time point, even Trypan Blue differentiates living from dead cells, it does not provide information about cells that could have been weakened by UVR and MitoCeption and which could die after 18 h.

After 18 h (Fig. 3), cells continued to die and accumulated from the early time point in the conditions of 3 min UVR exposure and 3 min UVR plus MitoCeption, resulting in the increased percentage of positive Trypan Blue cells at the late time point (non-viable) (panel f). This result demonstrated that the number of dead cells increased after 18 h under the most stressful conditions. These conditions were cells exposed to 3 min of UVR. It seems that the initial Trypan Blue staining could have underestimated the number of weakened cells. The MTT conversion to formazan crystals in the cells at 18 h can only be performed by living cells under all the conditions (panel d); this reaction occurs independently of the increase in the percentage of dead cells (panel f), which did not affect the overall MTT assay result.

mRNA was isolated to measure p53 gene expression levels after the PBMCs were exposed to UVR and cultured for 2 h. Results showed that PBMCs exposed to radiation increased their p53 mRNA levels in a response that could be associated with the induction of cell arrest or repair of mtDNA [31]. The decreased recognition of mtDNA sequence by hMito primers in cells exposed to UVR and the increased p53 mRNA expression could imply that these effects are linked. However, when PBMCs were rescued by adding 2.5 ng of exogenous mitochondria, mtDNA was recovered and p53 expression levels were decreased. The mechanism by which exogenous mitochondria transfer interacts with p53 gene expression is unknown and could influence future research on how MitoCeption could inhibit damage or senescence response [31].

Our article showed that the in-vitro use of PAMM MitoCeption could repair UVR damage in our cellular model. Even if the MitoCeption technique is limited to an in-vitro AMT/T, our protocol allows an effective and proportional exogenous mitochondrial uptake in cell lines and primary cells in proportion to their dose. This technique and properties could be used for ex-vivo experimentation.

It is still a challenge to standardize and predict proportional uptake in-vivo AMT/T by damaged tissue; this is true whether exogenous mitochondria are delivered to the tissue by direct or systemic injection. Nevertheless, the therapeutic value of exogenous mitochondria for patients is supported by medical evidence [10, 16].

MitoCeption in combination with PAMM could have clinical applications in the ex-vivo transfer of mitochondria to target cells in order to repair or modify them prior to reintroducing them to a patient's body. In the future, the MitoCeption technique could be used as part of a protocol to isolate primary cells with direct or indirect mitochondrial malfunctions, transfer/transplant exogenous mitochondria, and reintroduce them into the source organism. Today, the ex-vivo modification of cells for therapeutic purposes (such as in CAR-T cell generation and others) has been applied in anti-cancer medicine [32–36]. MitoCeption or other mitochondria transfer/transplant techniques could be used as in CAR-T cell therapy design and clinically applied to patients in future studies.

## Conclusion

In conclusion, we showed for the first time that PAMM MitoCeption can repair the decrease in metabolic activity, mitochondrial mass, and mtDNA sequence stability, decrease p53 gene expression, and decrease percentage of dead PBMCs, all of which are caused by UVR damage. We used an adaptation of the MitoCeption technique in PBMCs, which not only allows an effective and proportional uptake in primary cells, but also reduces the time of the protocol for better efficiency in downstream applications. This study opens the possibility to use mitochondria from different donors (PAMM) to treat UVR stress and possibly other metabolic diseases or malfunctions in cells in ex-vivo applications.

## Methods

### PBMCs isolation

Blood was collected from healthy female and male donors between 20 and 30 years old, each of whom provided written informed consent in accordance with relevant guidelines and regulations for mitochondria studies approved by the Bioethics Committee (the Institutional Review Board) of the Universidad San Francisco de Quito (study code: 2017-026IN). After blood collection, PBMCs were isolated using Ficoll-Paque Premium (GE Healthcare Life Sciences; MA, U.S) and centrifuged at 400 x g for 30 min without break or acceleration in order to create a density gradient. PBMCs were washed by suspension with 1X phosphate buffered saline (PBS) (Gibco by Life Technologies, ThermoFisher Scientific; MA, U.S) in a standardized volume of 10 mL and centrifuged at 1500 x g for 20 min at 4 °C. Supernatant was discarded and the pellet was re-suspended in 10 mL of PBS. Cells were counted in a Neubauer cell counting chamber and diluted to $1 \times 10^6$ cell/mL in 1% FCS RPMI (Gibco by Life Technologies, ThermoFisher Scientific, MA, U.S.), resulting in a final volume of 2 mL in an Eppendorf tube; one such tube was prepared for each condition. Cells were stored for a maximum of 2 h at 4 °C until further assays were performed.

### UVR exposure

Isolated cells were incubated at room temperature for 20 min prior to UVR exposure. Cells stored at low temperature seemed to resist irradiation better, thus creating variability in the results (data not shown). PBMCs were exposed to a UVR wavelength of 254 nm (UVC) emitted by a BS-02 UVR system (Purifier Logic Class II, KS, U.S). The UVC exposure dosages were: 2 min (48 mJ/cm$^2$), 3 min (72 mJ/cm$^2$), 4 min (96 mJ/cm$^2$), 5 min (120 mJ/cm$^2$), and 6 min (144 mJ/cm$^2$).

### Mitochondrial isolation

Mitochondria was isolated from 10 to $20 \times 10^6$ of fresh PBMCs using the Mitochondria Isolation Kit for Tissue (ThermoFisher Scientific, MA, U.S) following manufacturer's instructions with a final recovery step using a centrifugation at 3000 x g for 15 min to obtain highly purified mitochondria isolates. A second wash was performed using the same centrifugation conditions in order to discard residue of Reactive C from the kit. 1 mL of DMEM High Glucose (Gibco by Life Technologies, ThermoFisher Scientific, MA, U.S) without serum (DMEM pure) was added to re-suspend the mitochondrial pellet. Mitochondrial concentration was determined using a Pierce™ Coomassie Plus (Bradford) Assay Kit (ThermoFisher Scientific, MA, U.S). A typical mitochondria yield was 30 to 50 µg/mL. The mitochondrial solution was stored at 4 °C until MitoCeption of PBMCs was performed.

### Adapted MitoCeption protocol

PBMCs from all conditions were centrifuged at 500 x g at 4 °C for 5 min in 2 mL of RPMI 1% FCS in a microcentrifuge tube. The supernatant was discarded and 250 µL of the isolated mitochondrial solution in pure DMEM was added at different concentrations. 1.5 mL Eppendorf tubes were vortexed at maximum speed for 5 s with the 250 µL of mitochondria and PBMCs. A final centrifugation was performed at 500 x g for 5 min at 4 °C. Finally, cells were co-incubated at 37 °C for 1 h, washed with 1 mL of PBS, and left to rest until downstream assays were performed. PAMM is composed of the PBMCs of at least three donors (Fig. 2.1).

### Percentage of dead cells assessment

After UVR exposure, MitoCeption, and after the early and late time points, cells were washed and resuspended in 250 µL of 1% FCS RPMI. A 10 µl aliquot of each exposed, non-exposed, treated, and un-treated cells with mitochondria was mixed through resuspension, with 10 µL of 0.02% Trypan Blue staining (Gibco by Life

Technologies, ThermoFisher Scientific, MA, U.S). Two different operators counted the cells in a Neubauer cell counting chamber using an inverted Olympus microscope. Percentage of dead cells was calculated by dividing the number of dead cells (positive for Trypan Blue) by the total number of cells in the chamber.

#### Mitochondria labeling with MitoTracker green®

PBMCs were incubated with 250 nm/mL of MitoTracker Green® (Molecular Probes by Life Technologies, ThermoFisher Scientific, MA, U.S) solution in 1% FCS RPMI at 37 °C during 30 min. Two washes were performed to eliminate excess dye and cells were re-suspended in 1 mL of PBS.

#### Determination of the mitochondria transfer

PBMCs tested in all conditions were washed with 1 mL of PBS and centrifuged at 500 x g for 5 min prior to the assay to eliminate un-transferred mitochondria. MitoCepted PBMCs passed directly to flow cytometry analysis to determine the transfer success. Cell viability was verified (selection of cells - for Annexin and 7AAD). During the flow cytometry analysis, we focused on the lymphocyte or monocyte population using size, granularity, and CD3+ or CD14+ labeling. PBMCs were pelleted by centrifugation at 500 x g for 5 min, supernatant was discarded, and fresh RPMI + 1% FCS was added. Cells were analyzed 1 h or 18 h after MitoCeption by flow cytometry, focusing on the CD3+ lymphocyte and CD14+ monocyte populations. Cells were stained with APC mouse anti-human CD3 (555,335; BD biosciences) and PE mouse anti-human CD14 (555,398; BD Biosciences) antibodies. To determine the transfer success, a Flow Cell Cytometer Canto™ II (BD Biosciences, NJ, U.S) was used and, for the mitochondrial mass quantitation, the assays were performed using a Flow Cell Cytometer BD Accuri™ C6 Plus (BD Biosciences, NJ, U.S.). Data was analyzed by the FlowJo™, BD Accuri™ CSampler and flow cytometry Diva BD Bioscience software.

#### Metabolic activity

A solution of MTT (3-(4,5-Dimethylthiazol-2-yl)-2,5-Diphenyltetrazolium Bromide) (M6494; ThermoFisher Scientific, MA, U.S) was prepared at a concentration of 20 µg/mL and filtered through a 0.45 µm filter. 100 µL of the MTT solution was added to each tested condition in 1 mL of PBMCs in 1% FCS RPMI (final MTT concentration was 200 nmol/mL), followed by a 2 h incubation. Centrifugation at 500 x g for 5 min was performed to concentrate the viable cells containing dark violet crystals. After centrifugation, we discarded RPMI plus MTT and allowed 50 µL of the medium mix to settle so as not disturb the pellet. Then, 200 µL of DMSO (sc-358,801; Santa Cruz Biotechnology, TX, U.S) was added to each condition and mixed carefully to dissolve the dark violet crystals. To perform the measurement in the microplate spectrophotometer (Epoch; BioTek Instruments), we used a 96-well ELISA plate with a flat bottom with 100 µL of each condition. The plate was read using a wavelength of 570 nm.

#### Mitochondrial DNA damage quantitation by qPCR

PBMCs tested in all conditions were washed with 1 mL PBS and centrifuged at 500 x g for 5 min prior to the assay to eliminate any kind of debris or un-transferred mitochondria. DNA was extracted from PBMCs using MagMax™-96 Mutli-Sample Kit (Applied Biosystems by ThermoFisher Scientific, MA, U.S). Real time quantitative PCR was performed in a StepOne Real-Time PCR System (Applied Biosystems by ThermoFisher Scientific, MA, U.S). Fast SYBR® Green Master Mix (4,385,610; Applied Biosystems by ThermoFisher Scientific, MA, U.S) was used according to manufacturer's instructions, resulting in a final DNA concentration of 10 ng/µL in a 10 µL reaction. mtDNA was identified and quantified using primers (hMitoF3 and hMitoR3), manufactured by Invitrogen™: hMitoF3 5′- CACTTTCCACACAGA-CATCA – 3′; hMitoR3 5′- TGGTTAGGCTGGTGT-TAGGG – 3′ corresponding to a unique region of the mitochondria DNA between positions 241 and 390, which has been shown to be less similar to nuclear DNA sequences, thereby ensuring more accurate identification by the primers [18, 19]. The nuclear human β-2 microglobulin gene was used as a housekeeping gene with the primers: hB2MF1 5′- TGT TCC TGC TGG GTA GCT CT – 3′; and hB2MR1 5′- CCT CCA TGA TGC TGC TTA CA – 3′, as suggested by Ajaz et al.(2015) [18]. The qPCR reaction was based on the Ajaz et al. (2015) protocol with a minor modification: we included pre-incubation at 95 °C for 5 min (1 cycle); denaturation at 95 °C for 10 min; annealing and extension at 63 °C for 30 min (denaturation and extension steps were repeated for 40 cycles); and melting at 95 °C for 5 min, 63 °C for 60 min, 95 °C for 5 min, and 40 °C for 30 min. Data were collected by estimating the $2^{-\Delta CT}$; for statistical analysis, we used the Anderson-Darling test for normality to determine the use of parametric or non-parametric analysis. After statistical normality assumptions, we used un-paired, parametric ANOVA and Student's t-test to analyze the differences among conditions, using an alpha-value of 0.05 ($^{*}p < 0.05$, $^{**}p < 0.01$, $^{***}p < 0.001$, $^{****}p < 0.0001$) to establish significant differences.

#### p53 gene expression

RNA was isolated from PBMCs using the PureLink™ RNA Mini Kit (12,183,025; Ambio by Life Technologies, ThermoFisher Scientific, MA, U.S). Real time quantitative PCR was performed in a StepOne Real-Time PCR

System. SuperScript® III Platinum® SYBER® Green One-Step qRT-PCR Kit with Rox (11,745,500; Invitrogen by Life Technologies, ThermoFisher Scientific, MA, U.S) was used according to the manufacturer's instructions with a final DNA concentration of 10 ng/µL in a 10 µL reaction. mtDNA was quantified using primers manufactured by Invitrogen: Tp53F 5′ - CCT CAG CAT CTT ATC CGA GTG G – 3′; Tp53R 5′ – TGG ATG GTG GTA CAG TCA GAG C – 3′; and the nuclear PUM gene was used as the housekeeping gene with the primers: PUM1F 5′- AGT GGG GGA CTA GGC GTT AG – 3; and PUM1R 5′- GTT TTC ATC ACT GTC TGC ATC C – 3′. The qPCR reaction was performed under the following conditions: pre-incubation at 95 °C for 5 min (1 cycle); denaturation at 95 °C for 15 min; annealing and extension at 60 °C for 30 min (repeat denaturation and extension steps for 45 cycles); melting at 95 °C for 15 min, 60 °C for 60 min, and 95 °C for holding. Data were collected by estimating the $2^{-\Delta CT}$.

### ROS quantification

Aliquots of 1 million PBMCs in 2 mL of RPMI medium with 1% FCS were exposed to 6 min UVR (dosage: 72 mJ/cm$^2$) in 2 mL Eppendorf tubes. After 1 h culture at 37 °C in a 5% CO2/humidified air atmosphere, exposed and non-exposed samples (controls) were MitoCepted with 2.5 ng of mitochondria, according to the method explained above. 10 µL of Dihydro-rhodamine-123 (DHR-123) (D23806; ThermoFisher Scientific, MA, U.S) was added to 400 µL of each sample (final concentration: 1 mmol/L) followed by 1 µL Phorbol 12-myristate 13-acetate (PMA) (final concentration of 1 mmol/L) (79, 346; SIGMA, Merck, U.S) to induce cellular stress. Cells were cultured in the dark for 2 h. Flow cytometry was performed with a BD AccuriTM C6 Plus device (BD Biosciences, NJ, U.S) with 488 nm laser excitation and analyzed with BD AccuriTM CSampler software. Five thousand cells were collected from each sample. Results were calculated as the mean fluorescence intensity (MFI) (arbitrary units) and normalized by folding. Our data are shown with mean fluorescence intensity values.

### Statistical analyses

Data were analyzed using Minitab 17 and GraphPad Prism 6 software. The results for the mitochondrial mass and ROS production (estimated by flow cytometry), and metabolic activity (estimated with MTT assay) were normalized by using sample/control-average transformation. The percentage of dead cells was estimated using the Trypan Blue stain. The qPCR $2^{-\Delta CT}$ fold method was applied to mtDNA quantification and expression of the p53 gene. Then, we performed an Anderson-Darling normality tests on all data sets to determine the use of parametric or non-parametric analysis. After statistical normality assumptions, we used un-paired, parametric ANOVA and Student's t-test to analyze the differences among conditions, using an alpha-value of 0.05 ($^*p < 0.05$, $^{**}p < 0.01$, $^{***}p < 0.001$, $^{****}p < 0.0001$) to observe significant differences. Red headlines with asterisks represent the overall ANOVA significance; blue headlines with asterisks represent Student's t-test significance between conditions and red asterisks over columns represent Student's t-test significance between the control and the individual condition as stated in each figure.

## Additional files

**Additional file 1: Figure S1.** PBMCs lymphocytes and monocytes populations. Representative images. (PPTX 81 kb)

**Additional file 2: Figure S2.** Schematic representation of the UVR damage and PAMM MitoCeption rescue of PBMCs. (PPTX 209 kb)

#### Abbreviations

AMT/T: Artificial Mitochondrial Transfer or Transplant; CD14: Cluster of differentiation- co-receptor detection Lipopolysaccharide (LPS); CD3: Cluster of differentiation- T-cell co-receptor; DHR-123: Dihydro-rhodamine-123; DMEM: Dulbecco's Modified Eagle Medium; DMSO: Dimethyl sulfoxide; ELISA: Enzyme-linked immunosorbent assay; FCS: Fetal calf serum; MAV: Mean absorbance values; MFI: Mean fluorescence intensity; MSCs: Mesenchymal stem/stromal cells; mtDNA: Mitochondrial DNA; MTT: 3-(4,5-Dimethylthiazol-2-yl)-2,5-diphenyltetrazolium bromide; NADPH: Nicotinamide adenine dinucleotide phosphate; p53: Tumor suppressor gene 53; PAMM: Primary allogeneic mitochondrial mix; PBMCs: Peripheral blood mononuclear cells; PBS: Phosphate buffered saline; PMA: Phorbol 12-myristate 13-acetate; qPCR: Real-time quantitative polymerase chain reaction; RB: Respiratory burst; ROS: Reactive oxygen species; RPMI: Roswell Park Memorial Institute; UVC: UVR wavelength of 254 nm; UVR: Ultraviolet radiation

#### Acknowledgements

We would like to thank Marisol Llerena, Veronica Castañeda and Emilia Morales, biotechnology students from the Universidad San Francisco de Quito (USFQ) for their assistance during the assays of the article. We are thankful to Dr. Marcela Bovera, Anita Franco, Nancy Tigasi, and Yolanda Calderón, members of the "Laboratorio de Análisis Clínicos" of the Hospital de los Valles in Quito for their help during blood collection from donors. We would like to thank Danny Navarrete, a USFQ colleague, for his help in the statistical analysis. The authors would also like to thank Esther L. Wisdom and Caley Mikesell for assistance in revising the manuscript.

#### Authors' contributions

FC, MO, FV, EP and AC planned and performed the experiments, wrote the manuscript, analyzed data and created figures. SG set-up the MTT assays and contributed to the manuscript. DB performed the qPCR experiments regarding p53 expression levels. LN, GP, LAP, and EP performed flow cytometry assays and contributed to the manuscript. ChJ and MK advised on data analysis and contributed to the manuscript. F.V. drew the images of Fig. 2.1. AC conceptualized the article and developed PAMM. All authors have read and approved the manuscript.

#### Funding

This research was funded by USFQ Collaboration Grant (institutional codes: HUBi 5501, USFQ Veterinary Medicine School Grants for Research; # 3281 and #10098) and SIME R&D. The funding bodies evaluated this project before its submission to BMC Biotechnology through grant proposals and bioethics review processes according to institutional criteria. Funding bodies did not play any role in the design, data collection, analysis, and interpretations of data or in writing the manuscript.

**Ethics approval and consent to participate**
Blood was collected from healthy female and male donors between 20 and 30 years old, each of whom provided written informed consent in accordance with relevant guidelines and regulations for mitochondria studies approved by the Bioethics Committee (the Institutional Review Board) of the Universidad San Francisco de Quito (study code: 2017-026IN).

**Consent for publication**
Not applicable.

**Competing interests**
The authors declare that they have no competing interests.

**Author details**
[1]Colegio de Ciencias de la Salud, Escuela de Medicina Veterinaria, Universidad San Francisco de Quito, 17-12-841 Quito, Ecuador. [2]Colegio de Ciencias Biológicas y Ambientales, Escuela de Biotecnología, Universidad San Francisco de Quito, 17-12-841 Quito, Ecuador. [3]Colegio de Ciencias de la Salud, Escuela de Medicina, Universidad San Francisco de Quito, 17-12-841 Quito, Ecuador. [4]Baylor College of Medicine: Department of Pediatrics, Section of Immunology, Allergy, and Rheumatology, Baylor College of Medicine, Houston, TX, USA. [5]IRMB CHU Saint Eloi, University of Montpellier, France, 80 rue Augustin Fliche, 34295 Montpellier, cedex 5, France. [6]Laboratory of Nano-Regenerative Medicine, Faculty of Medicine, Universidad de Los Andes, Santiago, Chile. [7]Consorcio Regenero, Chilean Consortium for Regenerative Medicine, Santiago, Chile. [8]Cells for Cells, Santiago, Chile. [9]Instituto de Investigaciones en Biomedicina, Universidad San Francisco de Quito, 17-12-841 Quito, Ecuador. [10]Mito-Act Research Consortium, Quito, Ecuador. [11]Sistemas Médicos, SIME, Universidad San Francisco de Quito, 17-12-841 Quito, Ecuador.

**References**

1. Miliotis S, Nicolalde B, Ortega M, Yepez J, Caicedo A. Forms of extracellular mitochondria and their impact in health. Mitochondrion. 2019. https://doi.org/10.1016/j.mito.2019.02.002.
2. Liu C-S, et al. Delivering healthy mitochondria for the therapy of mitochondrial diseases and beyond. Int J Biochem Cell Biol. 2014;53:141–6.
3. Jackson MV, et al. Mitochondrial transfer via tunneling nanotubes is an important mechanism by which mesenchymal stem cells enhance macrophage phagocytosis in the in vitro and in vivo models of ARDS: mitochondrial transfer from MSC to macrophages. Stem Cells. 2016;34: 2210–23.
4. Wang X, Gerdes H-H. Transfer of mitochondria via tunneling nanotubes rescues apoptotic PC12 cells. Cell Death Differ. 2015;22:1181.
5. Spees JL, Olson SD, Whitney MJ, Prockop DJ. Mitochondrial transfer between cells can rescue aerobic respiration. Proc Natl Acad Sci U S A. 2006;103:1283–8.
6. Islam MN, et al. Mitochondrial transfer from bone-marrow-derived stromal cells to pulmonary alveoli protects against acute lung injury. Nat Med. 2012; 18:759–65.
7. Acquistapace A, et al. Human mesenchymal stem cells reprogram adult cardiomyocytes toward a progenitor-like state through partial cell fusion and mitochondria transfer. Stem Cells. 2011;29:812–24.
8. Berridge MV, Schneider RT, McConnell MJ. Mitochondrial transfer from astrocytes to neurons following ischemic insult: guilt by association? Cell Metab. 2016;24:376–8.
9. Clark MA, Shay JW. Mitochondrial transformation of mammalian cells. Nature. 1982;295:605–7.
10. McCully JD, Levitsky S, del Nido PJ, Cowan DB. Mitochondrial transplantation for therapeutic use. Clin Transl Med. 2016;5(1):16.
11. Caicedo A, et al. MitoCeption as a new tool to assess the effects of mesenchymal stem/stromal cell mitochondria on cancer cell metabolism and function. Sci Rep. 2015;5:9073.
12. Caicedo A, et al. Artificial mitochondria transfer: current challenges, advances, and future applications. Stem Cells Int. 2017. https://doi.org/10.1155/2017/7610414.
13. Hsu Y-C, Wu Y-T, Yu T-H, Wei Y-H. Mitochondria in mesenchymal stem cell biology and cell therapy: from cellular differentiation to mitochondrial transfer. Semin Cell Dev Biol. 2016;52:119–31.
14. Kim MJ, Hwang JW, Yun C-K, Lee Y, Choi Y-S. Delivery of exogenous mitochondria via centrifugation enhances cellular metabolic function. Sci Rep. 2018;8:3330.
15. Kesner EE, Saada-Reich A, Lorberboum-Galski H. Characteristics of mitochondrial transformation into human cells. Sci Rep. 2016;6:26057.
16. Emani SM, Piekarski BL, Harrild D, del Nido PJ, McCully JD. Autologous mitochondrial transplantation for dysfunction after ischemia-reperfusion injury. J Thorac Cardiovasc Surg. 2017;154:286–9.
17. Dahlmanns M, et al. Chemotherapeutic xCT inhibitors sorafenib and erastin unraveled with the synaptic optogenetic function analysis tool. Cell Death Discov. 2017;3:17030.
18. Ajaz S, Czajka A, Malik A. Accurate measurement of circulating mitochondrial DNA content from human blood samples using real-time quantitative PCR. Methods Mol Biol Clifton NJ. 2015;1264:117–31.
19. Malik AN, Shahni R, Rodriguez-de-Ledesma A, Laftah A, Cunningham P. Mitochondrial DNA as a non-invasive biomarker: accurate quantification using real time quantitative PCR without co-amplification of pseudogenes and dilution bias. Biochem Biophys Res Commun. 2011;412:1–7.
20. Kitani T, Kami D, Matoba S, Gojo S. Internalization of isolated functional mitochondria: involvement of macropinocytosis. J Cell Mol Med. 2014;18: 1694–703.
21. Wu M, et al. Comparison of the biological characteristics of mesenchymal stem cells derived from the human placenta and umbilical cord. Sci Rep. 2018;8:5014.
22. Thomas DC, et al. Eros is a novel transmembrane protein that controls the phagocyte respiratory burst and is essential for innate immunity. J Exp Med. 2017;214:1111–28.
23. Birch-Machin M a, Russell E v, Latimer J a. Mitochondrial DNA damage as a biomarker for ultraviolet radiation exposure and oxidative stress. Br J Dermatol. 2013;169:9–14.
24. Gniadecki R, Thorn T, Vicanova J, Petersen A, Wulf HC. Role of mitochondria in ultraviolet-induced oxidative stress. J Cell Biochem. 2001;80:216–22.
25. Torregrosa-Muñumer R, Goffart S, Haikonen JA, Pohjoismäki JLO. Low doses of ultraviolet radiation and oxidative damage induce dramatic accumulation of mitochondrial DNA replication intermediates, fork regression, and replication initiation shift. Mol Biol Cell. 2015;26:4197–208.
26. Takai D, et al. UV-irradiation induces oxidative damage to mitochondrial DNA primarily through hydrogen peroxide: analysis of 8-oxodGuo by HPLC. Free Radic Res. 2006;40:1138–48.
27. Agnello M, Morici G, Rinaldi AM. A method for measuring mitochondrial mass and activity. Cytotechnology. 2008;56:145–9.
28. Marlein CR, et al. PGC-1α driven mitochondrial biogenesis in stromal cells underpins mitochondrial trafficking to leukemic blasts. Leukemia. 2018;1. https://doi.org/10.1038/s41375-018-0221-y.
29. van Meerloo J, Kaspers GJL, Cloos J. Cell Sensitivity Assays: The MTT Assay. In: Cree IA, editor. Cancer Cell Culture: Methods and Protocols: Humana Press; 2011. p. 237–45. https://doi.org/10.1007/978-1-61779-080-5_20.
30. Altman SA, Randers L, Rao G. Comparison of trypan blue dye exclusion and fluorometric assays for mammalian cell viability determinations. Biotechnol Prog. 1993;9:671–4.
31. Rufini A, Tucci P, Celardo I, Melino G. Senescence and aging: the critical roles of p53. Oncogene. 2013;32:5129–43.
32. Olweus J. Manufacture of CAR-T cells in the body. Nat Biotechnol. 2017;35:520.
33. Srivastava S, Riddell SR. Engineering CAR-T cells: design concepts. Trends Immunol. 2015;36:494–502.
34. Ren J, et al. A versatile system for rapid multiplex genome-edited CAR T cell generation. Oncotarget. 2017;8:17002–11.
35. Stewart MP, et al. *In vitro* and *ex vivo* strategies for intracellular delivery. Nature. 2016;538:183–92.
36. Schumann K, et al. Generation of knock-in primary human T cells using Cas9 ribonucleoproteins. Proc Natl Acad Sci. 2015;112:10437–42.

6

# Evaluation of strategies for improving the transgene expression in an oleaginous microalga *Scenedesmus acutus*

Anongpat Suttangkakul[1,2], Anchalee Sirikhachornkit[1,2], Piyada Juntawong[1,2], Wilasinee Puangtame[1], Thitikorn Chomtong[1], Suchada Srifa[1], Sukhita Sathitnaitham[1], Wasawat Dumrongthawatchai[1], Kanidtha Jariyachawalid[3] and Supachai Vuttipongchaikij[1,2*]

**Abstract**

**Background:** Genetic transformation of microalgae has been hampered by inefficient transgene expression, limiting the progress of microalgal biotechnology. Many vector tools and strategies have been developed in recent years to improve transgene expression in the model microalga *Chlamydomonas*, but these were hardly applied to other microalgae. In this work, naturally-isolated oleaginous microalgae were accessed for genetic transformation, and various expression systems were evaluated in a selected microalga to circumvent inefficient transgene expression.

**Results:** Initially, a strain of *Scenedesmus acutus* was selected from the oleaginous microalgal collection based on its highest transformation rate and transgene stability. This strain, which had very low or no GFP reporter expression, was first tested to improve transgene expression by using intron-containing constructs and the transcript fusion using *ble::E2A*. The intron-containing constructs yielded 2.5–7.5% of transformants with 2–4-fold fluorescence signals, while the majority of the transformants of the transcript fusion had the fluorescence signals up to 10-fold. Subsequently, three UV-induced *S. acutus* mutants were isolated with moderate increases in the level and frequency of transgene expression (2–3-fold and 10–12%, respectively). Finally, a transcript fusion system was developed using *psy* white mutants with an expression vector containing *PSY::E2A* for complementation and light selection. Transformants with green colonies were selected under light exposure, and the transgene expression was detected at protein levels. Although the improvement using *PSY::E2A* was only minor (1–2-fold increase and ~ 7% of transformants), this system provides an alternative selectable marker that is compatible with large-scale culture.

**Conclusions:** Here, the overall improvement of transgene expression using the *Chlamydomonas* tools was moderate. The most effective tool so far is the transcript fusion using *ble::E2A* system. This work demonstrates that, so far, genetic engineering of non-model microalgae is still a challenging task. Further development of tools and strategies for transgene expression in microalgae are critically needed.

**Keywords:** *Agrobacterium*-mediated transformation, Biofuel, Fluorescence reporter, Introns, Lipid, Microalgae, Mutagenesis, Transcript-fusion

* Correspondence: fsciscv@ku.ac.th
[1]Special Research Unit in Microalgal Molecular Genetics and Functional Genomics (MMGFG), Department of Genetics, Faculty of Science, Kasetsart University, 50 Ngam Wong Wan road, Chatuchak, Bangkok 10900, Thailand
[2]Center of Advanced studies for Tropical Natural Resources, Kasetsart University, 50 Ngam Wong Wan road, Chatuchak, Bangkok 10900, Thailand

## Background

Oleaginous microalgae emerged as a promising renewable feedstock for biofuel production [1–3]. These microalgae can be cultivated using wastewater, seawater or simple medium and, under certain conditions such as nutrient deprivation or stresses, they can accumulate high lipid contents [4]. For decades, microalgae have been isolated and screened from natural sources for properties beneficial to biofuel production including rapid cell growth, high lipid yields, tolerance to stresses, ease of harvest and ability to secrete lipid [5]. To date, however, these microalgae are unable to produce biofuels at competitive costs compared to those of fossil fuels. Genetic improvement is a major step for making algal biofuels economically feasible [6, 7].

Genetic transformation has not been widely used in microalgae, partly because of difficulties in nuclear gene transformation, as reported in the model microalga *Chlamydomonas reinhardtii* and other species for low transformation efficiencies [8], inefficient transgene expression [9], transgene cleavages [9, 10], transgene instability [11] and highly control transcriptional silencing [12, 13]. Genome editing was also proved to be very difficult as the homology-directed repair (HDR) is extremely inefficient [14, 15] and CRISPR/Cas9 system was shown to be toxic for the algal cell [16]. The only exception so far is a marine microalga *Nannochloropsis*, which was reported for an efficient nuclear transformation and its feasibility for genome editing by either HDR [17] and CRISPR/Cas9 [18]. Initially, many attempts were made for improving transgene expression in *Chlamydomonas* by codon optimization [19] and screening large numbers of transformants to minimize transgene positional effects [20], but little improvement was achieved. Recent developments have demonstrated marked improvements by a number of approaches including mutant isolations for reduced transcriptional silencing [21] and the incorporation of introns in the expression constructs [22] and transcript fusion system [23]. However, most of these tools were developed for *Chlamydomonas*, and only a few have been applied in other microalgae [24].

Various oleaginous microalgae have been tested for genetic transformation, and, indeed, they suffer from inefficient transgene expression as *Chlamydomonas* had in the past [25, 26]. Notably, most of the work reported so far relied on the gene or enzymatic activity to represent the transgene expression, without confirmation at protein levels [24, 27–29]. This shortcoming is problematic as reliable detection at protein levels is the key indicator for establishing stable transgene expression, before evaluating the effects of transgenes such as phenotypes or compound production. Along the line of improvements in *Chlamydomonas*, many developed tools and strategies are now available, and it would be proven fruitful if these could be used for improving other oleaginous microalgae.

Here, ten oleaginous microalgae isolated from natural sources in Thailand were tested for genetic transformation. Initially, some of these microalgae were tested by electro-transformation, but very few or no transformant was obtained. In contrast, *Agrobacterium*-mediated transformation yielded much more transformants and was used to access the transformation efficiency of the ten microalgae. A strain of *S. acutus*, which has rapid growth and high lipid accumulation, was selected based on its high transformation rate and transgene stability. As expected, this strain has a low capability for transgene expression as indicated by very low levels of GFP reporter. This strain was then used for evaluating various expression tools, which were previously developed in *Chlamydomonas*, whether they could mitigate the low transgene expression problem. These included the use of constructs containing introns, constructs with transcript fusion and mutant isolation for reduced transcriptional gene silencing. Furthermore, a transgene expression system using novel selectable marker was developed by generating *psy* white mutants and complementation using *PSY* gene and light as the mean of selection. Transgene expression was analyzed for both expression levels and the frequency of expressors among the transformants.

## Results

### *Agrobacterium*-mediated transformation of wild-isolated oleaginous microalgal strains

Ten wild-isolated freshwater unicellular microalgae with high lipid accumulations were tested for *Agrobacterium*--mediated transformation using a method based on Kumar et al. [30]. Before transformation, axenic cultures of the ten strains were verified for hygromycin B sensitivity to estimate the selective concentrations (Additional file 1). After transformation using *Agrobacterium* harboring pCXSN-*GFP*, hygromycin B resistant colonies were obtained from six out of ten strains with transformation rates approximately 10–200 CFU per $10^6$ cells (Table 1, see also Additional file 2 for an example of selection plates). Some background colonies conferring spontaneous resistant to hygromycin B were observed from TISTR8511, 8555, 8519, 8540 and 8447, accounting for 5–9% of total colonies counted in transformation plates. Two *S. acutus* strains with the highest transformation rates, TISTR8540 and 8447, were tested for assessing the transgene stability. After 10–12 rounds of subcultures on the non-selective medium before transferring onto the hygromycin B selective medium, the growth of the TISTR8447 transformants was more consistent on the selective medium than those of TISTR8540 transformants

**Table 1** Transformation rates of ten microalgal strains by *Agrobacterium*-mediated transformation

| Species | Algal media | Concentration of hygromycin B ($\mu g\ ml^{-1}$) | Selection period (days) | Transformation rate (CFU per $10^6$ cells) | Background control (CFU per $10^6$ cells) |
|---|---|---|---|---|---|
| *Coelastrum* sp. (TISTR 8511) | TAP | 10 | 15 | 0 ($n$ = 44) | 0 |
| *S. acutus* (TISTR 8555) | TAP | 30 | 15 | 20 ± 5.5 ($n$ = 12) | 1.5 ± 1.3 |
| *S. acutus* (TISTR 8540) | TAP | 50 | 10 | 72 ± 41 ($n$ = 8) | 3.7 ± 3.5 |
| *S. acuminatus* (TISTR 8519) | TAP | 50 | 10 | 31 ± 23 ($n$ = 11) | 2.7 ± 2.5 |
| *S. acutus* (TISTR 8447) | TAP | 30 | 15 | 217.5 ± 75 ($n$ = 12) | 19.5 ± 6.8 |
| *C. humicola* (TISTR 8434) | TAP | 20 | 15 | 10.8 ± 32 ($n$ = 4) | 0 |
| *M. braunii* (TISTR 8429) | BG-11 | 10 | 15 | 0 ($n$ = 8) | 0 |
| *A. falcatus* (TISTR 8557) | BG-11 | 10 | 15 | 0 ($n$ = 12) | 0 |
| *T. cumbricus* (TISTR 8480) | TAP | 50 | 18 | 9 ± 8.8 ($n$ = 5) | 0 |
| *A. densus* (TISTR 8505) | BG-11 | 10 | 15 | 0 ($n$ = 8) | 0 |

(Additional file 3). This result suggests the higher stability of transgenes in TISTR8447. This *S. acutus* TISTR8447 was used for subsequent studies.

### Evaluation of TISTR8447 as a platform for gene expression

TISTR8447 is a fast-growing strain, reaching its stationary phase within 5 days in TAP medium. It accumulates high lipid contents up to 3 ± 0.17 and 10 ± 1.04% dry weight in N-supplemented and N-deprived media, respectively, based on total lipid quantification using vanillin staining. The lipid content in this strain is somewhat in moderate levels for lipid production as compared to other potential feedstock microalgae [31]. Nile red staining also showed the abundance of lipid droplets in the cells (Additional file 4). TISTR8447 was further tested for transformation efficiency using four *Agrobacterium* strains: A41, EHA105, GV3101 and LBA4404. Despite no statistical difference among the *Agrobacterium* strains ($p > 0.05$), EHA105 and GV3101 appeared to provide the highest transformation rates (Additional file 5). The effect of Acetosyringone was tested, but no significant difference among the tested concentrations (0, 50, 100 and 200 μM) was observed ($p > 0.05$, data not shown), indicating that Acetosyringone is not required.

Transformed TISTR8447 were analyzed for the presence and expression of the transgene. Among 15 randomly selected potential-transformants subjected to either PCR or RT-PCR analysis, the *aphIV* gene (hygromycin B resistant gene) and its transcript was detected in 12 and 13 transformants, respectively (Additional file 6). This analysis indicates that most of the hygromycin B resistant colonies contained the transgene. However, after analyzing more than 60 transformants by confocal microscopy, little or no GFP signal was detected above the autofluorescence background (Additional file 6). RT-PCR of the *GFP* transcript and immunoblot analysis using anti-GFP antibody also failed to detect the presence of the transcript and protein (data not shown). These data demonstrate that TISTR8447 is a potential microalgal strain for biofuel production through its properties for growth, lipid accumulation and its ability to be efficiently transformed by *Agrobacterium*, but has a very low capacity for transgene expression.

### Transgene expression in TISTR8447 using *Chlamydomonas* introns

To test the effect of introns in transgene expression, four expression constructs with *Chlamydomonas* intron-carrying *mRuby2*, *mCerulean3*, *mVenus* and *Clover* reporters, which were previously shown to improve transgene expression in *Chlamydomonas* [22], were tested in the TISTR8447 through electro-transformation. Our electro-transformation rate for TISTR8447 was only 4.95 ± 5.18 CFU per $5 \times 10^6$ cells ($n$ = 20), and no background colony was observed when using paromomycin selection. Noting that electro-transformation of *S. obliquus* was previously reported as 494 ± 48 CFU per $10^6$ cells [25]. Fluorescence microplate analysis showed that most of the transformants had very low fluorescence signal ratios for the four reporters, similar to the levels observed in non-transformed controls (Fig. 1). Only a few showed 2–4-fold signal ratios indicating the expression of the reporter proteins above the background signals. The frequency of the expressers could be estimated 1–3 among 40 transformants (2.5–7.5%), which, nonetheless, is an improvement from the transformation using the pCXSN-*GFP*. This result indicates that the incorporation of *Chlamydomonas* introns can partly improve transgene expression in TISTR8447.

### Improving transgene expression in TISTR8447 using *ble::E2A* transcript fusion

Previously, transcript fusion constructs using a self-cleavage 2A peptide from foot-and-mouth disease virus (FMDV) [32] with *ble* selectable marker [33] were

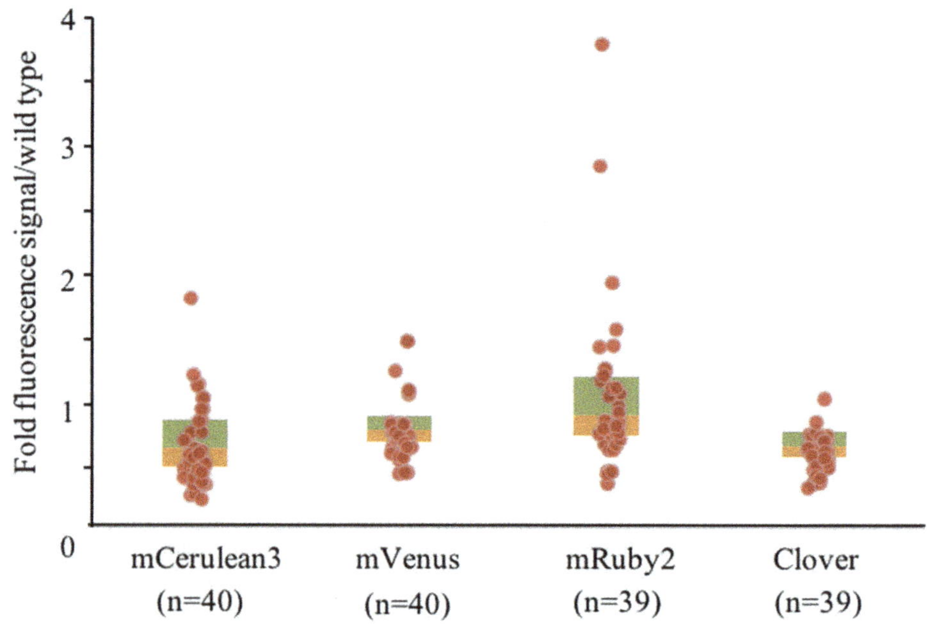


**Fig. 1** Analysis of TISTR8447 transformants expressing four fluorescence reporter genes harboring *Chlamydomonas* introns using fluorescence microplate reader. Data are presented as box plots overlaid with scatter plots for fold fluorescence reads over the wild type values. n indicates the number of transformants reads for each reporter

shown to be a strategy of choices for highly efficient transgene expression in *Chlamydomonas* [23, 34–37]. To test whether this could improve transgene expression in TISTR8447, *ble::E2A* fusion vectors were constructed using with- or without-introns *ble* marker and *mCherry* reporter (pCreZ-*intble::E2A-mCherry* and pCreZ-*ble::-E2A-mCherry*, Fig. 2 and Additional file 7). Transformants of these vectors were analyzed for the fluorescence reporter signal, and mCherry was readily observed under the confocal microscope (Fig. 2b). Fluorescence microplate analysis showed that most of the transformants from both constructs (19/24 or 79% for *ble* and 12/14 or 85% for *int-ble*) had more than a 2-fold increase of mCherry signal ratio to the wild type controls (Fig. 2c). The use of *RBSC2* promoter and introns in the *ble* gene (*int-ble*) showed higher signals compared to that using *PSAD* promoter and *ble* gene without introns (*ble*), though without significant difference by Kruskal-Wallis Test ($p > 0.05$). This result indicates that the *ble::E2A* transcript fusion system could provide an efficient transgene expression system for the TISTR8447 under selective pressure of zeocin.

Subsequently, the expression level of the *ble::E2A::mCherry* transcript fusion was tested whether it could be induced in a dose-dependent manner with elevated zeocin concentrations. Selected transformants with low, medium and high expression were cultured in at 10 μg ml$^{-1}$ zeocin, before transferring into the media supplemented with 0–50 μg ml$^{-1}$ zeocin for 2 days. However, no significant increase of the fluorescence signals following zeocin concentrations for both constructs was observed (Fig. 2d). At concentrations above 20 μg ml$^{-1}$ zeocin, the transformants grew poorly with low cell mass and became chlorosis. This observation demonstrates that increases in zeocin selective pressure could not provide a further improvement on transgene expression in the TISTR8447.

### Improving transgene expression in TISTR8447 by UV mutagenesis

To further improve the transgene expression, TISTR8447 mutants with enhanced transgene expression were generated using UV mutagenesis and a selection strategy using *CRY1-1* (conferring emetine resistant) as reported by Neupert et al. [21]. Transformation of the TISTR8447 using pCRE-*CRY1-1* was selected on the medium supplemented with 1 μg ml$^{-1}$ emetine. Fifty selected transformants were tested on media supplemented with 1–20 μg ml$^{-1}$ emetine to confirm that they could grow on the medium supplemented with at most 1 μg ml$^{-1}$ emetine. Three transformants (S14) were chosen for UV mutagenesis at 0.1% survival rate and subsequently selected using 2, 5 and 10 μg ml$^{-1}$ emetine. Finally, three mutants (SUV1, 2 and 3), which were able to grow on 10 μg ml$^{-1}$ emetine, were obtained (Fig. 3a). These mutants were then tested for any improvement on transgene expression by transformation using pOPT-*mCerulean3* followed by fluorescence microplate analysis (Fig. 3b). Although the highest levels of the fluorescence signal ratios were in the range of 2- to 3-fold, which were similar to transformants of control

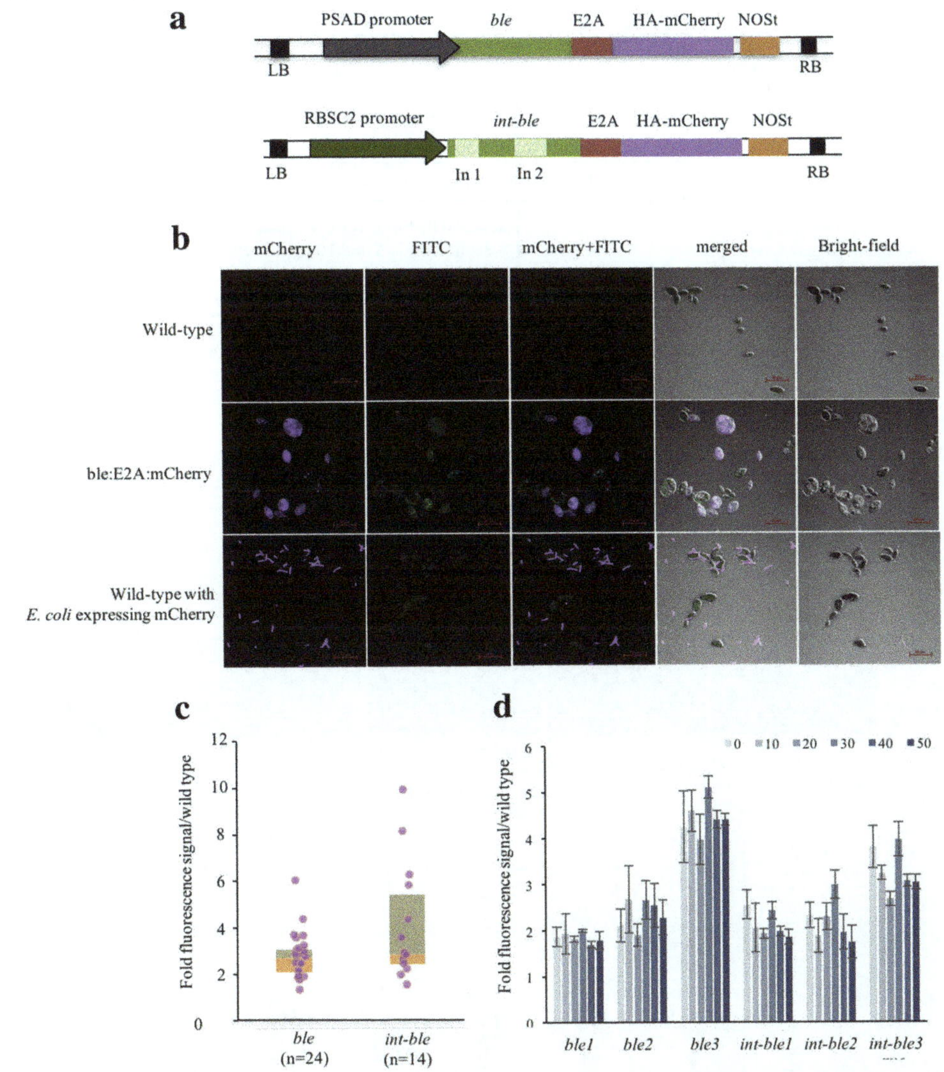


**Fig. 2** Analysis of TISTR8447 transformants expressing transcript fusions. **a** pCreZ-*ble::E2A::mCherry* (*ble*) and pCreZ-*intble::E2A::mCherry* (*int-ble*) vectors are illustrated. **b** TISTR8447 transformants analyzed by confocal microscopy showing mCherry signal and FITC channel-autofluorescence (scale bar = 20 μm). *E. coli* expressing *mCherry* was added to TISTR8447 wild type as a positive control. **c** mCherry signals of *ble* and *int-ble* vector transformants are presented as box plots overlaid with scatter plots. (n) indicates the number of transformants. **d** Fluorescence signals of low, medium and high mCherry expressors from each construct after subjected to media supplemented with 0–50 μg $ml^{-1}$ zeocin for 2 days. Error bars represent standard deviations ($n = 3$)

strains (S14), the SUV1-3 provided overall higher frequencies of transformants with high fluorescence signal ratios (8/81 or 10%, 5/43 or 12% and 2/17 or 12% for SUV1-3, respectively). This result demonstrates that although the level of transgene expression is only moderately improved, these TISTR8447 mutants provide increases in the frequency of transgene expressed transformants.

### Improving transgene expression by generating TISTR8447 white mutants and *PSY::E2A* transcript fusion

The transcript fusion system using *ble::E2A* have been shown so far to be the best improvement of transgene expression in TISTR8447, but the use of zeocin as a selective pressure is impractical for large-scale culture and affects the algal growth. To further improve this, an alternative expression strategy was developed by using *Phytoene synthase* (*PSY*) mutants, *PSY* gene complementation and light as a selective pressure. *Chlamydomonas* with *PSY* mutations were previously shown to exhibit pale-green or whitish colony, perish under the light as low as 8 μmol photons $m^{-2} s^{-1}$ and to be complemented by transformation using *PSY* gene [38]. By adopting this idea, white mutants of the TISTR8447 were generated by UV mutagenesis and selection of pale-green colonies grown in the dark. From 50 selection plates, each with

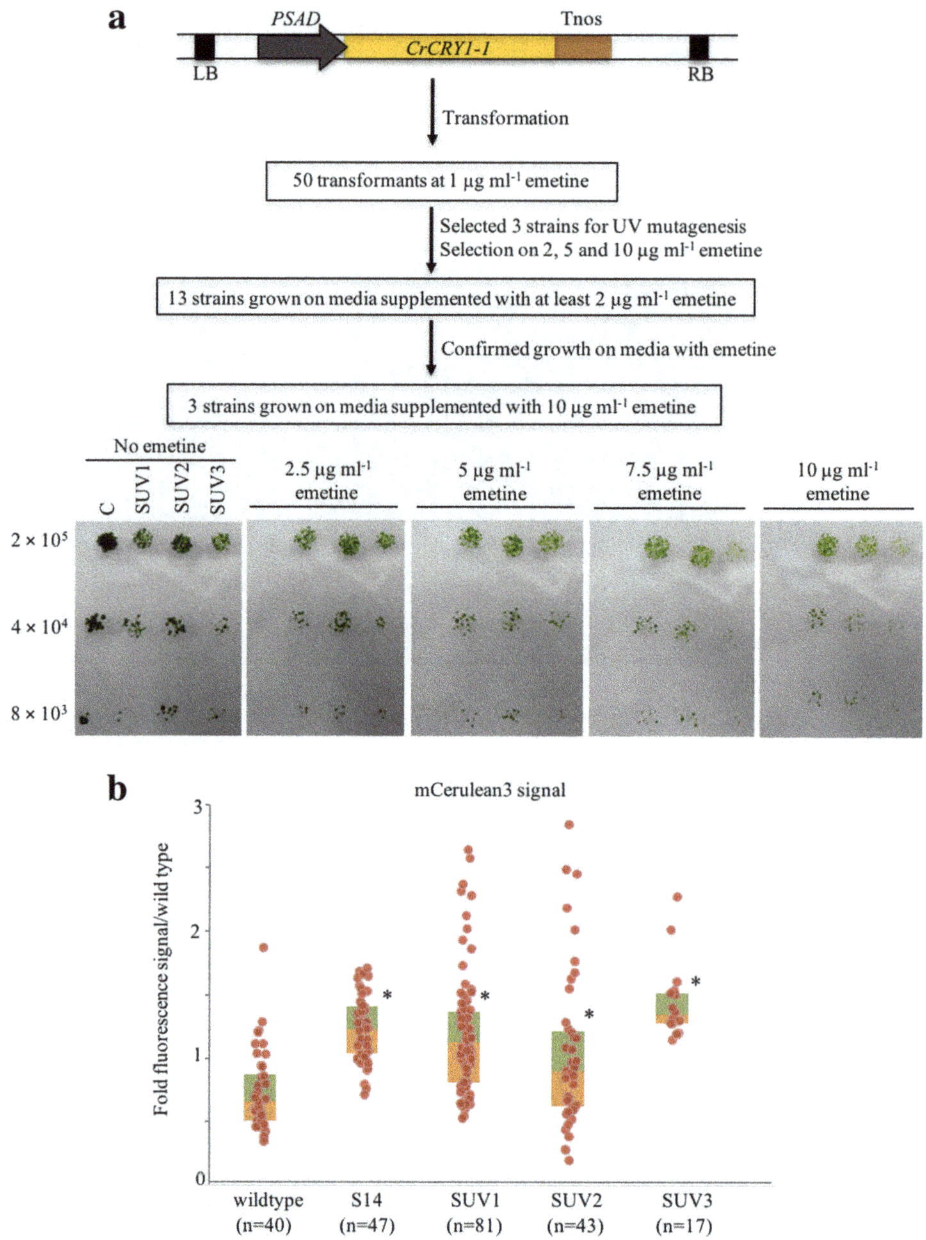


**Fig. 3** UV mutagenesis and selection of SUV strains. **a** *CrCRY1-1* expression construct and the experimental procedure for selecting transformants. The lower panel shows three selected strains, SUV1-3, that grew on the medium supplemented with emetine up to 10 μg $ml^{-1}$ compared to the TISTR8447 wild type. **b** Box plots overlaid with scatter plots of mCerulean3 fluorescence signals of transformants generated from wild type, S14 (a *CrCRY1-1* expressing strain) and SUV1-3. (n) indicates the number of transformants for each strain. Asterisks indicate significant differences between signals from transformants generated from wild type and others by Kruskal-Wallis Test ($p < 0.05$)

$10^6$ cells, six white mutants, named G1-G6, were obtained after three rounds of subcultures in the dark (Fig. 4a). Four white mutants (G1, G2, G3 and G5) ceased growth under light at 50 μmol photons $m^{-2} s^{-1}$, similar to the cc4113 and cc4109 *Chlamydomonas psy* white mutants [38], while G4 and G6 mutants were able to grow under the light. After sequencing the *PSY* gene in these mutants, point mutations were observed in the *PSY* coding sequence in G1, G2 and G5 mutants (Fig. 4b), but not in G3. The G1, G2 and G5 *psy* mutants were then tested for complementation using a construct containing *Chlamydomonas PSY* coding sequence fused with *E2A* and *HA-mCherry*. Transformation of these three white mutants using the *Agrobacterium* method was unsuccessful, and this may be because the mutant grew very slow and could not out-compete *Agrobacterium* growth. Electro-transformation yielded complementation from G1 and G2 mutants with green colonies after

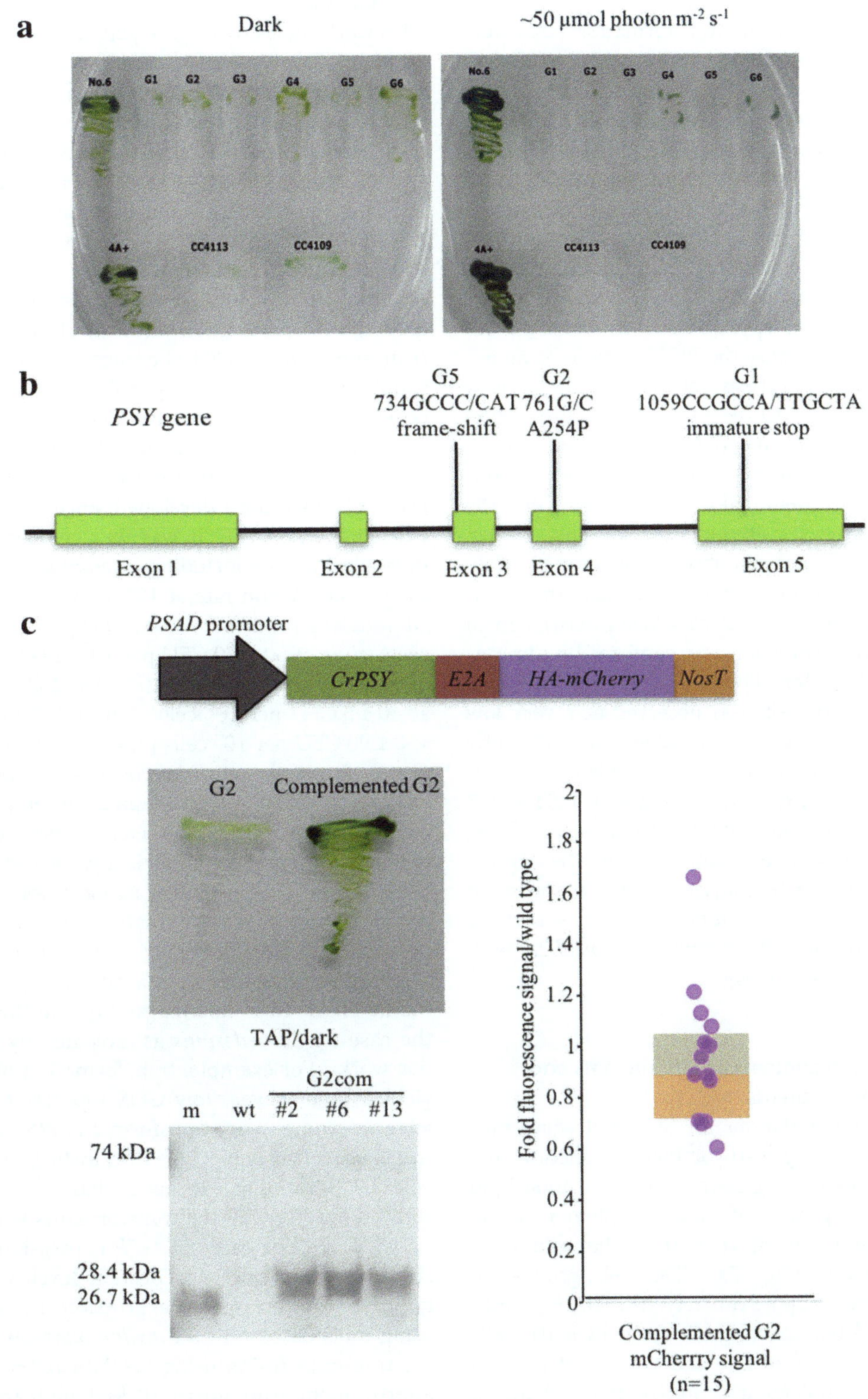


**Fig. 4** Generation of *psy* white mutants and complementation using *CrPSY*. **a** Six isolated white mutants (G1-6) on TAP medium under light and dark conditions (No.6: TISTR8447, 4A+: *Chlamydomonas*, cc4113 and cc4109: *Chlamydomonas psy* white mutants). **b** Genotyping of *psy* mutants by direct PCR and sequencing. **c** Complementation of *psy* white mutants using *CrPSY* coding sequence fused in-frame with *E2A* and *mCherry*. mCherry was detected by both fluorescence microplate reader and immunoblot analysis (m: mCherry from *E. coli*, wt: TISTR8447). mCherry carrying HA tag in the algae is 28.4 kDa, while that in *E. coli* is 26.7 kDa. A fainted band representing 74 kDa of CrPSY::E2A::mCherry fusion protein was observed (see Additional file 8 for longer exposures and Additional file 10 for the full length blot)

selection under light at 50 μmol photons $m^{-2} s^{-1}$ for 7–10 days (Fig. 4c). However, the transformation rates were extremely low, approximately 1–4 CFU per $5 \times 10^6$ cells per plate.

The G2 complemented strains were analyzed by fluorescence microplate reader and showed that most of the transformants had low levels of the fluorescence signals for mCherry (Fig. 4c). The frequency of the transgene expressers (up to a 2-fold signal ratio) was one among 15 transformants (~ 7%), somewhat similar to that of intron-containing vectors. Unlike that of *ble::E2A* vectors with the zeocin selection, the expression of *PSY::E2A* was still low. This may be because the PSY functions as an enzyme in carotenoid biosynthesis pathways, while ble protein functions by binding to zeocin in a direct proportion [23]. Therefore, light selection could not drive the *PSY* expression to the levels achieved by zeocin selection. Spontaneous revertants were also monitored in the transformation controls (without plasmid) and, so far, there were only two revertants observed among 40 selection plates ($5 \times 10^6$ cells per plate). To preclude the possibility of these complemented strains being derived from revertants, mCherry expression was confirmed by immunoblot analysis (Fig. 4c). The non-cleaved PSY::E2A::HA-mCherry fusion protein was observed at a very low level indicating that the E2A peptide was efficiently self-cleaved in the TISTR8447 (Additional file 8). This result demonstrates that *psy* white mutants of TISTR8447 can be used as host cells for transformation using *PSY* selectable marker and light selection. However, the expression levels of the *PSY::E2A* transgene were in the same range as that of constructs containing introns and far lesser than those using SUV mutants and *ble::E2A* with the zeocin selection, respectively.

### Growth and total lipid contents of SUV mutants and complemented white mutants

To examine whether mutations for SUV and white mutants had any effect on growth or lipid production, the mutants were analyzed for growth curves and total lipid contents. While the growth of three complemented G2 strains and S14 was similar to that of the wild type, SUV1-3 grew slower (Fig. 5a). The cell growth of SUV1-3 was limited to approximately $10 \times 10^6$ cells $ml^{-1}$, whereas the wild type could grow up to $15 \times 10^6$ cells $ml^{-1}$. The lipid contents of S14 and G2 complemented strains in both N-supplemented and N-deprived media were similar to those of wild type, but those of SUV1-3 were significantly lower for both media ($p < 0.05$) (Fig. 5b). This result indicates that mutations, which provide improved transgene expression in SUV mutants, also affect physiological processes of the TISTR8447 resulting in slow growth and less accumulated lipids.

## Discussion

This work shows that *Agrobacterium*-mediated transformation is a convenient method for simultaneously testing the transformation of many microalgae. In this work, ten strains were tested, and six of them were successfully transformed. The best transformation rate was obtained from *S. acutus* TISTR8447 at 217.5 ± 75 CFU per $10^6$ cells, and this was selected as our best candidate for evaluating the transgene expression. *Coelastrum, Chlorococcum, Monoraphidium* and *Tetradesmus* are reported for their first trial here for the *Agrobacterium* method. Noting that although *M. braunii, A. falcatus, A. densus* and *Coelastrum* spp. were unable to be transformed in this work, this should not preclude their possibility to be transformed by other means. *Agrobacterium*-mediated transformation of microalgae was first introduced by Kumar et al. [30] for a successful transformation in *Chlamydomonas*, and various protocols have been developed since then to improve its efficiency [9, 39, 40]. It was then applied to many other microalgal species including *Haematococcus pluvialis* with the transformation rate at 153 ± 4 CFU per $10^6$ cells [41], *Dunaliella bardawil* at 39–42 CFU per $10^6$ cells [42], *Schizochytrium* at 50–170 CFU per $10^6$ cells [43], *Scenedesmus almeriensis* at 90 CFU per $10^6$ cells [26], *Dunaliella salina* at 40 ± 5 CFU per $10^6$ cells [29] and *Tetraselmis chuii* at 150 ± 90 CFU per $10^6$ cells [27]. And, with a different unit of transformation rate, *Nannochloropsis* sp. [44], *Chlorella vulgaris* [45] and *Ankistrodesmus* sp. [46] were shown to be transformed based on *GUS* expression at approximately 21, 30 and 3.5% efficiency, respectively. Taken together, this work provides support for the use of the *Agrobacterium*-mediated method for transforming other potential industrial microalgae yet to be tested.

Extremely inefficiently transgene expression is commonly observed in microalgae [9], and this appears to be the case for *Scenedesmus* as indicated by this and previous works. For example, transformation of *S. almeriensis* yielded little or very low GUS reporter expression even after screening many transformants [26], that of *S. bajacalifornicus* had only 2.96% of transformants with detectable GUS staining [46] and that of *S. obliquus* also showed only 1.5% of the transformants with *GFP* expression [25]. In our case, no *GFP* expression was observed at neither transcript nor protein level. Our attempt to improve the transgene expression in TISTR8447 by using tools based on *Chlamydomonas*, so far, shows that the transcript fusion using *ble::E2A* provided the highest improvement with up to 10-fold increases of mCherry expression and up to 85% of transformants with observable mCherry. However, other strategies including the uses of intron-containing constructs, SUV1-3 mutants and white mutants with *PSY::E2A* complementation provided only minor improvements with at most 2–4-fold increases of the reporter expression. The majority of the

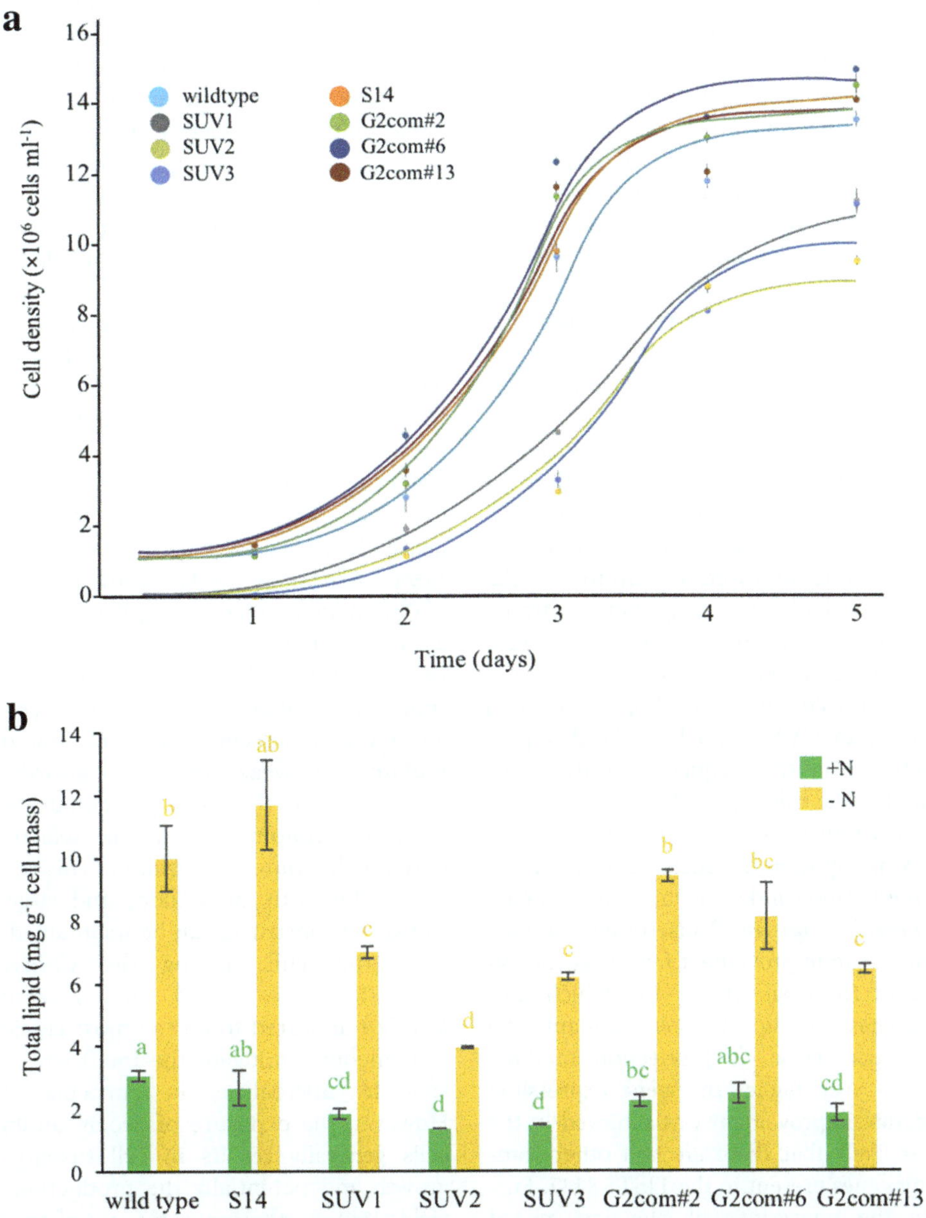


**Fig. 5 a** Growth curves of S14, SUV1-3 and complemented G2 strains compared to TISTR8447 wild type. **b** Total lipid content of the microalgae in N-supplemented and N-deprived media grown for 3 days. Error bars represent standard deviation ($n = 6$), and bars with the same letter are not significantly different (Tukey's test, $p < 0.05$)

transformants from the intron-containing constructs and the white mutants had very low transgene expression, while SUV mutants had increases in the frequency of transgene expressors.

Transcriptional gene silencing was proposed to be the key factor that causes the low transgene expression in *Chlamydomonas* [13, 19, 21, 36, 47] and most likely in many other microalgae [48–50]. Noting that a recent work by Mini et al. [9] reported that rearrangements of transgenes could also be another factor contributing to the inefficient expression in *Chlamydomonas.* Because gene expression tools that could circumvent the gene silencing effects were successfully developed in *Chlamydomonas*, these tools could be applied to other microalgae, particularly in the TISTR8447. However, the improvements were far less than those obtained in *Chlamydomonas.* The development of the *ble::2A* systems provided more than a 100-fold improvement of transgene expression in *Chlamydomonas* [23, 34, 35, 37], while it was only a 10-fold increase in the TISTR8447. Insertions of an endogenous intron(s) into transgenes or the surrounding 5′ and 3′ untranslated regions led to improved transgene expression in

*Chlamydomonas* [51–53]. In contrast, these *Chlamydomonas* introns only provided small increases in the transgene expression in the TISTR8447. A little improvement in the TISTR8447 observed here by the use of the introns could be because of the difference in intron splicing between the species, or these introns could only occasionally help the transgene escaping the silencing controls in the TISTR8447. Perhaps, *Scenedesmus* might have a more robust gene silencing mechanism than that in *Chlamydomonas*.

Neupert et al. [21] first reported the strategic isolation of *Chlamydomonas* mutants with improved transgene expression using *CRY1-1* and elevated emetine concentrations for selection. The selected mutants, UVM4 and UVM11, had demonstrated up to 100-fold increases in emetine resistant. Less than expected, our selections of mutagenized TISTR8447 yielded up to a 10-fold increase in emetine resistance, and the expression levels based on the fluorescence ratios were improved only up to a 2-fold. Nonetheless, the SUV1-3 had higher frequencies of transgene expressors, in which transformants with at least a 2-fold signal could be accounted for ~ 10–12% compared to ~ 2.5% observed in wild type. This improvement is somewhat comparable to UVM4 and UVM11 that provided 4.5-fold increases in the frequency of transgene expressors compared to the wild type [47].

MET1 (DNA methyltransferase1) is one of the key enzymes for the DNA methylation transgene silencing along with other DNA-methylation independent gene silencing enzymes [13]. *Chlamydomonas met1* insertional mutants were shown to provide an improvement for transgene expression. Work from Kurniasih et al. [13], which performed UV mutagenesis using the *met1* mutant and selection using Neupert et al. [21] selection method, yielded mutants with even higher transgene expression. Because only moderate improvement was achieved in the SUV mutants, it is likely that there are still other transcriptional silencing genes present in the TISTR8447. Further mutations in the transcriptional silencing related genes may be required for further improvement.

Mutations that provide improved transgene expression in *Chlamydomonas* are known to be associated with epigenetic controls, and these may affect cell growth, as demonstrated in *Chlamydomonas* UVM mutants of the *met-1* insertional mutant [13]. Indeed, mutations of SUV1-3 are most likely related to epigenetic control machinery, and they affect both cell growth and lipid production, as none of these effects was observed in the S14 control strain. This result shows that while the SUV1-3 mutants have gained some improvement on the transgene expression, their growth and ability to accumulate lipids were weakened, compromising cell mass and lipid production. This result confirms the previous observation of the negative aspect of using mutants related to epigenetic controls. Thus, cautions should be made before performing genetic screening for enhancing transgene expression, especially when the cell performances are the key to the application. Nonetheless, the white mutants, when complemented, could provide sufficiently intact growth and have only slight decreases in the lipid content. The result also suggests that while *CrPSY* can sufficiently complement the *SaPSY* knockout, yielding complemented strains with robust cell growth, it could not provide the lipid accumulation as efficient as the *SaPSY*.

Even though the *PSY::E2A* construct is not as effective as *ble::E2A* in forcing expression, this system provides an alternative selectable marker for microalgae. Current selectable markers used for microalgae transformation are based on antibiotic selection, for example, bleomycin/zeocin resistance gene [33] and paromomycin resistance gene [54]. Recently, the herbicide norflurazon was recently developed for transformation selection using *Phytoene desaturase* (*PDS*) with an amino substitution L505F in some microalgae [55–57]. The use of *PSY* gene as a selectable marker for microalgae, so far, has not been reported. This may be because of two main reasons: white mutants needed to be generated by mutagenesis and screened for each microalga, and white mutants are weak with slow growth and difficult to transform. For the work presented here, this is one of the most economical transgene selection approaches as only a light source is required. This selectable marker is compatible with an outdoor and large-scale culture as transgene constructs can be maintained by sunlight.

To this end, among the *Chlamydomonas* gene expression tools tested, the transcript fusion using *ble::E2A* appeared to be the most efficient and could instantaneously mitigate the inefficient transgene expression in microalgae, in particular, the TISTR8447. However, the exposure of zeocin antibiotic to the algal cells generally results in cell toxicity and affects cell growth and potentially the production of products and yields [58]. Furthermore, the use of zeocin is uneconomical for large-scale production, limiting the use of *ble::E2A* system. A consensus for inefficient transgene expression has been evidenced in many oleaginous microalgae and *Scenedesmus* species as reported here. The challenges in gaining high transgene expression in microalgae are the critical bottleneck for current development in Eukaryotic microalgal biotechnology. Further improvements and newly developed tools are critically needed to allow efficient use of microalgae for biotechnology application.

## Conclusions

This work demonstrates that oleaginous microalgae isolated from natural resources can be screened for selecting strains with high transformation efficiency, but they generally have a low capacity for transgene expression. A selected strain, *S. acutus* TISTR8447, was tested to improve the transgene expression using strategies developed in *Chlamydomonas*, but overall improvement was of moderate for both levels and frequency of expression. So far, the most effective tool giving the highest improvement is the transcript fusion using *ble::E2A* system. Furthermore, a new strategy for transgene expression has been developed using *psy* white mutants with an expression vector containing *PSY::E2A* for complementation and light selection. Finally, this work demonstrates that genetic engineering of non-model microalgae is still a challenging task. New tools and strategies are critically needed for transgene expression in promising industrial microalgae.

## Methods

### Microalgae and cultivation conditions

Microalgae were obtained from Thailand Institute of Science and Technological Research (TISTR) including *Coelastrum* sp. (TISTR8511), *Scenedesmus acutus* (TISTR8555), *S. acutus* (TISTR8540), *S. acuminatus* (TISTR8519), *S. acutus* (TISTR8447), *Chlorococcum humicola* (TISTR8434), *Monoraphidium braunii* (TISTR8429), *Ankistrodesmus falcatus* (TISTR8557), *Tetradesmus cumbricus* (TISTR8480), *Ankistrodesmus densus* (TISTR8505). These accessions were chosen based on their potential high accumulation of lipid droplets under the N deprivation condition and the variation of algal species. *Chlamydomonas* white mutants CC4113 (Its1-207 mt+) and CC4109 (Its1-203 mt+) were obtained from Chlamydomonas Resource Center. Microalgae were cultured in either Tris-acetate phosphate (TAP) or BG-11 media at 25 °C under continuous light (50 μmol photons $m^{-2} s^{-1}$), excepting that those white mutants were incubated in the dark. The N-deprived medium (TAP-N) was prepared by substituting $KNO_3$ with KCl. For growth analysis, three cultures of 50 ml were started at the of $10^6$ cells $ml^{-1}$ cell density, and 1 ml sample was collected daily to monitor for cell density.

### Plasmid constructions

pOPT vector series with reporter genes including *mRuby2, mCerulean3, mVenus* and *Clover* [22] were obtained from Chlamydomonas Resource Center. pCXSN-GFP was constructed by TA cloning using the *Agrobacterium* transformation vector pCXSN [59] and *GFP* coding sequence, which was PCR amplified using GFP-F and GFP-R primers (Additional file 9). Transcript fusion vectors were constructed using pPLV02 [60] as the vector backbone. The pPLV02 backbone was amplified using pplv-F and pplv-R primers. PSAD promoter, RBSC2 promoter-*ble* with introns and *ble* without intron fragments were amplified from pSL18 [61, 62] using psad-F and psad-R primers, pSP124S [51] using rbcs-F and Zeo-E2A-R primers and pICZ (Thermo Fisher) using Zeo-F and Zeo-E2A-R primers, respectively. Each primer has an adaptor sequence for subsequent sequence and ligation independent cloning (SLIC) [63]. All PCR cloning was performed using hi-fidelity Q5 DNA polymerase (New England Biolabs, USA). pCreZ-*E2A* and pCreZint-*E2A* were constructed by fusing the pPLV02 PCR fragment with PSAD promoter and *ble* without intron, and RBSC2 promoter-*ble* with introns, respectively, by SLIC technique. *HA-mCherry* was subsequently inserted into both vectors at NcoI site by PCR amplification using HA-mcherry-F and mcherry-R primers and SLIC.

For *CrCRY1-1* expression construct, the coding region of *CrCRY1* was PCR amplified from *Chlamydomonas* cDNA using gene specific primers, before being re-amplified using a reverse primer with a modified base at the last codon from CTG to CCG (cry1-1-F and cry1-1-TtoC-R primers) to generate *CrCRY1-1* coding sequence. The *CrCRY1-1* fragment with adaptor sequences was then inserted into pCreZ-*E2A* at EcoRI and BamHI sites by SLIC, removing *ble-E2A*. For *CrPSY::E2A::mCherry* constructs, the *CrPSY* coding sequence was PCR amplified from *Chlamydomonas* cDNA and re-amplified using crpsy-F and crpsy-R primers, which contain adaptor sequences for SLIC. The *CrPSY* fragment was then ligated to a vector fragment, which was amplified from pCreZ-*E2A* by E2A-F and psad-crpsy-R primers, replacing *ble* coding sequence and in frame with *E2A*. Plasmids for *Agrobacterium*-mediated transformation were transferred into *Agrobacterium* cells by electroporation.

### Transformation

The microalgal strains were tested for their sensitivity to hygromycin B before performing an *Agrobacterium*-mediated transformation. Briefly, $10^6$ cells were spread onto TAP or BG-11 agar media supplemented with 10–100 μg $ml^{-1}$ of hygromycin B for 14 days. This sensitivity test was performed in triplicate. For *Agro*-transformation, microalgae were cultured in liquid media for 3–4 days, and $10^6$ cells were grown on solid media for 2–3 days until forming a lawn. *A. tumefaciens* EHA105 harboring transformation plasmids were grown in LB media supplemented with 25 μg $ml^{-1}$ of rifampicin and 50 μg $ml^{-1}$ of kanamycin under continuous shaking at 28 °C for 2 days. For evaluation of *Agrobacterium* strains, A41, GV3101 and LBA4404 strains were used in replacement of EHA105. The algal lawn was mixed with 100 μl of *Agrobacterium* ($OD_{600}$ = 1) using a spreader and co-cultivated in the dark for 2 days before transferring onto solid TAP or BG-11 selective media supplemented with 250 μg $ml^{-1}$ cefotaxime and hygromycin B at specific concentrations. Colonies obtained from selection plates were subcultured twice on the selection media before being designated as transformants. Transformation rates were calculated based on the numbers of resistant colonies per $10^6$ starting cells. *Agrobacterium* cells without expression plasmid were used for transformation as negative controls.

For electro-transformation, log phase cells at no more than $2 \times 10^6$ cells $ml^{-1}$ were harvested by brief centrifugation at 400×g and resuspended in TAP supplemented with 40 mM sucrose (TAP sucrose) at $2 \times 10^8$ cells $ml^{-1}$. Thirty-eight microlitres of the algal suspension were mixed with 2 µl of a plasmid vector (400 ng) before transferring to a 2 mm gap electroporation cuvette and pulsing using Bio-Rad Gene Pulser II with the following condition: 0.4 kV, 25 µF and 500 Ω. The algal cells were immediately transferred to 10 ml TAP sucrose and cultured at the normal condition with constant shaking at 100 rpm under low light for an overnight before spreading onto TAP agar supplemented with 30 µg $ml^{-1}$ of paromomycin or 10 µg $ml^{-1}$ of zeocin for 7–10 days.

For the transformation of white mutants, a white colony was cultured in TAP sucrose with constant shaking in the dark up to 10 days or until reaching $1–2 \times 10^6$ cells $ml^{-1}$. Cells were harvested and resuspended in 40 mM sucrose TAP at $2 \times 10^8$ cells $ml^{-1}$. Subsequently, 38 µl of the suspension was mixed with 2 µl of a plasmid before being electroporated using the same condition as above and cultured in 10 ml TAP sucrose for overnight with constant shaking in the dark. The cells were plated onto TAP agar and incubated in the dark for 5 days and then under light (50 µmol photons $m^{-2} s^{-1}$) for 7–10 days.

#### UV mutagenesis and mutant screening

UV mutagenesis was performed by irradiating the TISTR8447 at a density of $5 \times 10^6$ cells $ml^{-1}$, which were previously kept in the dark for 2 h, in a petri dish using a UV transilluminator for 2 h to obtain a 0.1% survival rate. Cells were cultured overnight with shaking in the dark before plating onto TAP agar medium at $10^6$ cells per plate. For SUV mutants, *CRY1-1* transformants were UV irradiated and plated onto TAP agar medium supplemented with emetine at 2, 5 and 10 µg $ml^{-1}$. For white mutants, irradiated cells were plated onto TAP agar medium and incubated in the dark for 14 days. Colonies with pale-green were subcultured in the dark.

#### Molecular characterization

For DNA and RNA isolation, the algal cells were subjected to a mechanical disruption in liquid nitrogen using a mortar and pestle. DNA was isolated using the CTAB method, whereas total RNA was isolated using GENEzol TriRNA Pure Kit (Geneaid, Taiwan). RNA was purified by using RNase-free DNase and RNA Cleanup Kit. cDNA was synthesized using one microgram of total RNA by MMuLV reverse transcriptase. PCR and RT-PCR reactions were performed using *Taq* DNA polymerase and primers listed in Additional file 9. For sequencing of the *PSY* coding sequence of TISTR8447 (Genbank accession MF401544), *SaPSY* of the white mutants was amplified from total DNA using hi-fidelity Q5 DNA polymerase and sapsy-seq-F and sapsy-seq-R primers and sequenced.

For western analysis, a total protein was isolated by grinding the cells with liquid nitrogen and resuspended in a protein extraction buffer [20 mM Tris-HCl pH 8.0, 150 mM NaCl, 2 mM EDTA, 10% (*v*/v) glycerol, 0.5% IGEPAL CA-360 and 1 mM PMSF]. The total protein was quantified by Bradford assay before subjecting to SDS-PAGE and blotting onto a PVDF membrane. mCherry was detected using rabbit anti-mCherry as a primary antibody at 1:1000 dilution and anti-rabbit-HRP as a second antibody at 1:10,000 dilution. The western signal was detected using WesternSure Premium Chemiluminescence Substrate (Li-COR, USA) and C-DiGit Blot Scanner (Li-COR, USA).

#### Fluorescence detection

Fluorescent reporter signals were analyzed using a microplate reader. The excitation and emission wavelengths are as follows; mRuby2 (ex = 558 nm, em = 605 nm), mCerulean3 (ex = 445 nm, em = 503 nm), mVenus (ex = 515 nm, em = 550 nm), Clover (ex = 477 nm, em = 515 nm), mCherry (ex = 562 nm, em = 607 nm). Chlorophyll autofluorescence was measured at 440 nm for excitation and 685 nm for emission. Standard curves for the conversion of fluorescence signal to protein contents were generated from recombinant GST-mCerulean and GST-mCherry purified from *E. coli* and quantified using Bradford assay. The protein concentrations used was 1.25–400 ng $µl^{-1}$ in the TISTR8447 algal culture at $10^6$ cells $ml^{-1}$. The chlorophyll fluorescence was used for calculating the cell number based on a standard curve between the chlorophyll signals and cell numbers. Selected transformants with a high level of expression of transgenes were then assessed for fluorescence signal using Nikon C2 Si confocal microscopy.

#### Stability test

Transformants were maintained on media without hygromycin B for 2 weeks before subculturing onto media with and without hygromycin B to determine the retention of the selectable marker gene without the selection pressure. For the G2 complemented lines, the dark and light conditions were used instead of the antibiotic. These subcultures were repeated every 2 weeks.

#### Lipid quantification using vanillin method

This lipid quantification using vanillin was modified from Mishra et al. [64]. Cells were grown in a starting culture using liquid TAP medium for 2–3 days ($\sim 2 \times 10^6$ cells $ml^{-1}$

concentration) before dividing in half, and the medium of each half was replaced by liquid TAP or TAP-N at $2 \times 10^6$ cells $ml^{-1}$ concentration before culturing for 3 days. Approximately $1.5 \times 10^6$ cells with known mass was mixed with 200 µl sulfuric acid, incubated at 95 °C for 10 min before placing on ice for 5 min, added with 500 µl phospho-vanillin [0.12% (*w*/*v*) vanillin in 80% (*v*/v) phosphoric acid] and incubated at 37 °C for 15 min. The solution was briefly centrifuged at 3000 rpm for 5 min, and the absorbance at 530 nm of the sample was assessed. Total lipid was quantified against the standard curve generated from canola oil.

### Nile red staining

Microalgae were stained using Nile red (10 µg $ml^{-1}$ in DMSO) and Calcofluor (1%) for 30 min in the dark at 25 °C. The samples were then fixed in 4% paraformaldehyde before washing twice with PBS solution and analyzed using Nikon C2 Si confocal microscopy (543 nm excitation, 525–605 nm emission).

### Statistical analysis

The statistical tests used for analyzing the data varied based on the distribution of each data set. Kruskal-Wallis test was conducted to assess the difference in the transformation efficiency of different *Agrobacterium* strains and the levels of fluorescence signals among populations of microalgal strains. For datasets with a normal distribution, one-way analysis of variance (ANOVA) using an alpha of 0.05 and a multiple range test using Tukey's method was conducted.

## Additional files

**Additional file 1:** Survival of ten microalgal strains on different concentrations of hygromycin B. (PDF 37 kb)

**Additional file 2:** Representative cultured and selective plates for *Agrobacterium*-mediated transformation of TISTR8540. (PDF 552 kb)

**Additional file 3:** A transgene stability test of TISTR8540 and TISTR8447. (PDF 1536 kb)

**Additional file 4:** Nile red staining of the TISTR8447 after 3 days of nitrogen starvation. (PDF 101 kb)

**Additional file 5:** Transformation rates of the TISTR8447 using four *Agrobacterium* strains. (PDF 39 kb)

**Additional file 6:** *Agrobacterium*-mediated transformation of *S. acutus* TISTR8447. (PDF 446 kb)

**Additional file 7:** The schematic of pCreZ vectors. (PDF 64 kb)

**Additional file 8:** Western analysis of CrPSY::E2A::mCherry fusion protein with extended exposure times. (PDF 127 kb)

**Additional file 9:** Oligonucleotide primer sequences. (PDF 111 kb)

**Additional file 10:** The full length immuno blot of G2 complemented strains. (PDF 170 kb)

#### Abbreviations

ANOVA: Analysis of variance; aphIV: Hygromycin phosphotransferase; BG11: Blue-green Medium; ble: Bleomycin resistant protein; CaMV: Cauliflower mosaic virus; CFU: Colony forming unit; CrPSY: *Chlamydomonas reinhardtii* phytoene synthase; CRY1-1: Cryptochrome circadian clock 1-1; CTAB: Cetyltrimethylammonium bromide; DMSO: Dimethyl sulfoxide; EDTA: Ethylenediaminetetraacetic acid; GFP: Green fluorescence protein; GST: Glutathione S-transferase; GUS: Beta-glucuronidase; HA: Hemagglutinin; HRP: Horseradish peroxidase; MET1: DNA methyltransferase1; MMuLV: Reverse transcriptase: moloney murine leukemia virus reverse transcriptase; PDS: Phytoene desaturase; PMSF: Phenylmethylsulfonyl fluoride; PSY: Phytoene synthase; PVDF: Polyvinylidene fluoride; SaPSY: *Scenedesmus acutus* phytoene synthase; SDS-PAGE: Sodium dodecyl sulfate polyacrylamide gel electrophoresis; TAP: Tris-acetate phosphate buffer; TAP-N: Tris-acetate phosphate buffer without nitrogen; Taq polymerase: *Thermus aquaticus* polymerase; UV: Ultraviolet

#### Acknowledgments

We are grateful to the Thailand Institute of Scientific and Technological Research (TISTR) for providing the microalgae. We thank Pornthip Boonmahamongkol and Science Equipment center, Faculty of Science, Kasetsart University for equipment and technical support, particularly fluorescence microplate reader and confocal microscopy. We thank Dr. Thidaporn Supapakorn for support on statistical analysis. We thank Dr. Kunn Kangwansaichol, Chanakarn Chaikaeodang, Jureeporn Konpiewlert and Chanarong Sartsanga for technical support.

#### Funding

This work was supported by Kasetsart University Research and Development Institute (KURDI), PPT Public Company Limited and Thailand Research Fund (TRF-RSA6080031). These funding bodies had no role in the design of this study or any role during the collection, analysis, and interpretation of the data, and in writing the manuscript.

#### Authors' contributions

ASu and SV conceived and designed the experiments. ASu, WP, TK, SSr, SSa and WD performed experiments. ASu, ASi, PJ, KJ and SV analyzed the data. ASu and SV prepared the Figures and Tables and wrote the manuscript. ASi, PJ and SV corrected and proofread the manuscript. All authors read and approved the final manuscript.

#### Ethics approval and consent to participate

Not applicable.

#### Consent for publication

Not applicable.

#### Competing interests

This research was partly supported by PPT Public Company Limited (Thailand). Kanidtha Jariyachawalid is employed by the PTT public company limited. Patent applications (Thailand patent No.1501003942 and No.1601007906) have been filed for some of the technology disclosed in this publication. All other authors declare no competing interests.

#### Author details

[1]Special Research Unit in Microalgal Molecular Genetics and Functional Genomics (MMGFG), Department of Genetics, Faculty of Science, Kasetsart University, 50 Ngam Wong Wan road, Chatuchak, Bangkok 10900, Thailand. [2]Center of Advanced studies for Tropical Natural Resources, Kasetsart University, 50 Ngam Wong Wan road, Chatuchak, Bangkok 10900, Thailand. [3]PTT Research and Technology Institute, PTT Public Company Limited, Ayuthaya 13170, Thailand.

**References**

1. Chisti Y. Biodiesel from microalgae beats bioethanol. Trends Biotechnol. 2008;26(3):126–31.
2. Beer LL, Boyd ES, Peters JW, Posewitz MC. Engineering algae for biohydrogen and biofuel production. Curr Opin Biotechnol. 2009;20(3): 264–71.
3. Jones CS, Mayfield SP. Algae biofuels: versatility for the future of bioenergy. Curr Opin Biotechnol. 2012;23(3):346–51.
4. Singh A, Olsen SI. A critical review of biochemical conversion, sustainability and life cycle assessment of algal biofuels. Appl Energy. 2011;88(10):3548–55.
5. Wijffels RH, Barbosa MJ. An outlook on microalgal biofuels. Science. 2010; 329(5993):796–9.
6. Radakovits R, Jinkerson RE, Darzins A, Posewitz MC. Genetic engineering of algae for enhanced biofuel production. Eukaryot Cell. 2010;9(4):486–501.
7. Georgianna DR, Mayfield SP. Exploiting diversity and synthetic biology for the production of algal biofuels. Nature. 2012;488(7411):329.
8. Potvin G, Zhang Z. Strategies for high-level recombinant protein expression in transgenic microalgae: a review. Biotechnol Adv. 2010;28(6):910–8.
9. Mini P, Demurtas OC, Valentini S, Pallara P, Aprea G, Ferrante P, Giuliano G. *Agrobacterium*-mediated and electroporation-mediated transformation of *Chlamydomonas reinhardtii*: a comparative study. BMC Biotechnol. 2018; 18(1):11.
10. Zhang R, Patena W, Armbruster U, Gang SS, Blum SR, Jonikas MC. High-throughput genotyping of green algal mutants reveals random distribution of mutagenic insertion sites and endonucleolytic cleavage of transforming DNA. Plant Cell. 2014;26(4):1398–409.
11. Radakovits R, Jinkerson RE, Fuerstenberg SI, Tae H, Settlage RE, Boore JL, Posewitz MC. Draft genome sequence and genetic transformation of the oleaginous alga Nannochloropis gaditana. Nat Commun. 2012;3:686.
12. Cerutti H, Johnson AM, Gillham NW, Boynton JE. Epigenetic silencing of a foreign gene in nuclear transformants of *Chlamydomonas*. Plant Cell. 1997; 9(6):925–45.
13. Kurniasih SD, Yamasaki T, Kong F, Okada S, Widyaningrum D, Ohama T. UV-mediated *Chlamydomonas* mutants with enhanced nuclear transgene expression by disruption of DNA methylation-dependent and independent silencing systems. Plant Mol Biol. 2016;92(6):629–41.
14. Sodeinde OA, Kindle KL. Homologous recombination in the nuclear genome of *Chlamydomonas reinhardtii*. Proc Natl Acad Sci U S A. 1993; 90(19):9199–203.
15. Gumpel NJ, Rochaix JD, Purton S. Studies on homologous recombination in the green alga *Chlamydomonas reinhardtii*. Curr Genet. 1994;26(5):438–42.
16. Jiang W, Brueggeman AJ, Horken KM, Plucinak TM, Weeks DP. Successful transient expression of Cas9 and single guide RNA genes in *Chlamydomonas reinhardtii*. Eukaryot Cell. 2014;13(11):1465–9.
17. Kilian O, Benemann CS, Niyogi KK, Vick B. High-efficiency homologous recombination in the oil-producing alga *Nannochloropsis* sp. Proc Natl Acad Sci U S A. 2011;108(52):21265–9.
18. Wang O, Lu Y, Xin Y, Wei L, Huang S, Xu J. Genome editing of model oleaginous microalgae *Nannochloropsis* spp. by CRISPR/Cas9. Plant J. 2016; 88(6):1071–81.
19. Fuhrmann M, Oertel W, Hegemann P. A synthetic gene coding for the green fluorescent protein (GFP) is a versatile reporter in *Chlamydomonas reinhardtii*. Plant J. 1999;19(3):353–61.
20. Schroda M, Blöcker D, Beck CF. The HSP70A promoter as a tool for the improved expression of transgenes in *Chlamydomonas*. Plant J. 2000; 21(2):121–31.
21. Neupert J, Karcher D, Bock R. Generation of *Chlamydomonas* strains that efficiently express nuclear transgenes. Plant J. 2009;57(6):1140–50.
22. Lauersen KJ, Kruse O, Mussgnug JH. Targeted expression of nuclear transgenes in *Chlamydomonas reinhardtii* with a versatile, modular vector toolkit. Appl Microbiol Biotechnol. 2015;99(8):3491–503.
23. Rasala BA, Lee PA, Shen Z, Briggs SP, Mendez M, Mayfield SP. Robust expression and secretion of Xylanase1 in *Chlamydomonas reinhardtii* by fusion to a selection gene and processing with the FMDV 2A peptide. PLoS One. 2012;7(8):e43349.
24. Zorin B, Grundman O, Khozin-Goldberg I, Leu S, Shapira M, Kaye Y, Tourasse N, Vallon O, Boussiba S. Development of a nuclear transformation system for oleaginous green alga *Lobosphaera* (Parietochloris) *incisa* and genetic complementation of a mutant strain, deficient in arachidonic acid biosynthesis. PLoS One. 2014;9(8):e105223.
25. Guo SL, Zhao XQ, Tang Y, Wan C, Alam MA, Ho SH, Bai FW, Chang JS. Establishment of an efficient genetic transformation system in *Scenedesmus obliquus*. J Biotechnol. 2013;163(1):61–8.
26. Dautor Y, Úbeda-Mínguez P, Chileh T, García-Maroto F, Alonso DL. Development of genetic transformation methodologies for an industrially-promising microalga: *Scenedesmus almeriensis*. Trends Biotechnol. 2014; 36(12):2551–8.
27. Úbeda-Mínguez P, Chileh T, Dautor Y, García-Maroto F, Alonso DL. Tools for microalgal biotechnology: development of an optimized transformation method for an industrially promising microalga—*Tetraselmis chuii*. J Appl Phycol. 2015;27(1):223–32.
28. Chen CY, Kao AL, Tsai ZC, Chow TJ, Chang HY, Zhao XQ, Chen PT, Su HY, Chang JS. Expression of type 2 diacylglycerol acyltransferase gene *DGTT1* from *Chlamydomonas reinhardtii* enhances lipid production in *Scenedesmus obliquus*. Biotechnol J. 2016;11(3):336–44.
29. Simon DP, Anila N, Gayathri K, Sarada R. Heterologous expression of β-carotene hydroxylase in *Dunaliella salina* by agrobacterium-mediated genetic transformation. Algal Res. 2016;18:257–65.
30. Kumar SV, Misquitta RW, Reddy VS, Rao BJ, Rajam MV. Genetic transformation of the green alga—*Chlamydomonas reinhardtii* by *Agrobacterium tumefaciens*. Plant Sci. 2004;166(3):731–8.
31. Gim GH, Kim JK, Kim HS, Kathiravan MN, Yang H, Jeong SH, Kim SW. Comparison of biomass production and total lipid content of freshwater green microalgae cultivated under various culture conditions. Bioprocess Biosyst Eng. 2014;37(2):99–106.
32. Ryan MD, King AM, Thomas GP. Cleavage of foot-and-mouth disease virus polyprotein is mediated by residues located within a 19 amino acid sequence. J Gen Virol. 1991;72(11):2727–32.
33. Stevens DR, Purton S, Rochaix JD. The bacterial phleomycin resistance gene *ble* as a dominant selectable marker in *Chlamydomonas*. Mol Gen Genet. 1996;251(1):23–30.
34. Rasala BA, Barrera DJ, Ng J, Plucinak TM, Rosenberg JN, Weeks DP, Oyler GA, Peterson TC, Haerizadeh F, Mayfield SP. Expanding the spectral palette of fluorescent proteins for the green microalga *Chlamydomonas reinhardtii*. Plant J. 2013;74(4):545–56.
35. Rasala BA, Chao SS, Pier M, Barrera DJ, Mayfield SP. Enhanced genetic tools for engineering multigene traits into green algae. PLoS One. 2014;9(4):e94028.
36. Kong F, Yamasaki T, Kurniasih SD, Hou L, Li X, Ivanova N, Okada S, Ohama T. Robust expression of heterologous genes by selection marker fusion system in improved *Chlamydomonas* strains. J Biosci Bioeng. 2015;120(3):239–45.
37. Plucinak TM, Horken KM, Jiang W, Fostvedt J, Nguyen ST, Weeks DP. Improved and versatile viral 2A platforms for dependable and inducible high-level expression of dicistronic nuclear genes in *Chlamydomonas reinhardtii*. Plant J. 2015;82(4):717–29.
38. McCarthy SS, Kobayashi MC, Niyogi KK. White mutants of *Chlamydomonas reinhardtii* are defective in phytoene synthase. Genetics. 2004;168(3):1249–57.
39. Rajam MV, Kumar SV. Green alga (*Chlamydomonas reinhardtii*). In: Wang K, editor. Methods in molecular biology: *Agrobacterium* protocols, vol. 2. New Jersey: Human Press; 2006. p. 421–33. Part VIII.
40. Pratheesh PT, Vineetha M, Kurup GM. An efficient protocol for the *Agrobacterium*-mediated genetic transformation of microalga *Chlamydomonas* reinhardtii. Mol Biotechnol. 2014;56(6):507–15.
41. Kathiresan S, Chandrashekar A, Ravishankar GA, Sarada R. *Agrobacterium*-mediated transformation in the green alga *Haematococcus pluvialis* (Chlorophyceae, Volvocales). J Phycol. 2009;45(3):642–9.
42. Anila N, Chandrashekar A, Ravishankar GA, Sarada R. Establishment of *Agrobacterium tumefaciens*-mediated genetic transformation in *Dunaliella bardawil*. Eur J Phycol. 2011;46(1):36–44.
43. Cheng R, Ma R, Li K, Rong H, Lin X, Wang Z, Yang S, Ma Y. *Agrobacterium tumefaciens* mediated transformation of marine microalgae *Schizochytrium*. Microbiol Res. 2012;167(3):179–86.
44. Cha TS, Chen CF, Yee W, Aziz A, Loh SH. Cinnamic acid, coumarin and vanillin: alternative phenolic compounds for efficient *Agrobacterium*-mediated transformation of the unicellular green alga, *Nannochloropsis* sp. J Microbiol Methods. 2011;84(3):430–4.
45. San Cha T, Yee W, Aziz A. Assessment of factors affecting *Agrobacterium*-mediated genetic transformation of the unicellular green alga, *Chlorella vulgaris*. World J Microbiol Biotechnol. 2012;28(4):1771–9.
46. Sanitha M, Radha S, Fatima AA, Devi SG, Ramya M. *Agrobacterium*-mediated transformation of three freshwater microalgal strains. Pol J Microbiol. 2014; 63:387–92.

47. Kong F, Yamasaki T, Ohama T. Expression levels of domestic cDNA cassettes integrated in the nuclear genomes of various *Chlamydomonas reinhardtii* strains. J Biosci Bioeng. 2014;117(5):613–6.
48. Babinger P, Kobl I, Mages W, Schmitt R. A link between DNA methylation and epigenetic silencing in transgenic *Volvox carteri*. Nucleic Acids Res. 2001;29(6):1261–71.
49. Bird A. DNA methylation patterns and epigenetic memory. Genes Dev. 2002;16(1):6–21.
50. Fuks F, Hurd PJ, Wolf D, Nan X, Bird AP, Kouzarides T. The methyl-CpG-binding protein MeCP2 links DNA methylation to histone methylation. J Biol Chem. 2003;278(6):4035–40.
51. Lumbreras V, Stevens DR, Purton S. Efficient foreign gene expression in *Chlamydomonas reinhardtii* mediated by an endogenous intron. Plant J. 1998;14:441–7.
52. Eichler-Stahlberg A, Weisheit W, Ruecker O, Heitzer M. Strategies to facilitate transgene expression in *Chlamydomonas reinhardtii*. Planta. 2009;229:873–83.
53. Lauersen KJ, Berger H, Mussgnug JH, Kruse O. Efficient recombinant protein production and secretion from nuclear transgenes in *Chlamydomonas reinhardtii*. J Biotechnol. 2013;167(2):101–10.
54. Sizova I, Fuhrmann M, Hegemann PA. *Streptomyces rimosusa phVIII* gene coding for a new type phosphotransferase provides stable antibiotic resistance to *Chlamydomonas reinhardtii*. Gene. 2001;277(1):221–9.
55. Steinbrenner J, Sandmann G. Transformation of the green alga *Haematococcus pluvialis* with a phytoene desaturase for accelerated astaxanthin biosynthesis. Appl Environ Microbiol. 2006;72(12):7477–84.
56. Liu J, Gerken H, Huang J, Chen F. Engineering of an endogenous phytoene desaturase gene as a dominant selectable marker for *Chlamydomonas reinhardtii* transformation and enhanced biosynthesis of carotenoids. Process Biochem. 2013;48(5):788–95.
57. Liu J, Sun Z, Gerken H, Huang J, Jiang Y, Chen F. Genetic engineering of the green alga *Chlorella zofingiensis*: a modified norflurazon-resistant phytoene desaturase gene as a dominant selectable marker. Appl Microbiol Biotechnol. 2014;98(11):5069–79.
58. Lin G, Wang Y, Guo L, Ding H, Hu Y, Liang S, Zhang Z, Yang G. Verification of mutagen function of Zeocin in *Nannochloropsis oceanica* through transcriptome analysis. J Ocean Univ China. 2017;16(3):501–8.
59. Chen S, Songkumarn P, Liu J, Wang GL. A versatile zero background T-vector system for gene cloning and functional genomics. Plant Physiol. 2009;150(3):1111–21.
60. De Rybel B, van den Berg W, Lokerse AS, Liao CY, van Mourik H, Möller B, Llavata-Peris CI, Weijers D. A versatile set of ligation-independent cloning vectors for functional studies in plants. Plant Physiol. 2011;156(3):1292–9.
61. Fischer N, Rochaix JD. The flanking regions of PsaD drive efficient gene expression in the nucleus of the green alga *Chlamydomonas reinhardtii*. Mol Gen Genet. 2001;265(5):888–94.
62. Pollock SV, Colombo SL, Prout DL, Godfrey AC, Moroney JV. Rubisco activase is required for optimal photosynthesis in the green alga *Chlamydomonas reinhardtii* in a low-CO2 atmosphere. Plant Physiol. 2003; 133(4):1854–61.
63. Jeong JY, Yim HS, Ryu JY, Lee HS, Lee JH, Seen DS, Kang SG. One-step sequence-and ligation-independent cloning as a rapid and versatile cloning method for functional genomics studies. Appl Environ Microbiol. 2012; 78(15):5440–3.
64. Mishra SK, Suh WI, Farooq W, Moon M, Shrivastav A, Park MS, Yang JW. Rapid quantification of microalgal lipids in aqueous medium by a simple colorimetric method. Bioresour Technol. 2014;155:330–3.

7

# Crosslinked flagella as a stabilized vaccine adjuvant scaffold

Casey M. Gries[1], Rohith R. Mohan[2], Dimitrios Morikis[2] and David D. Lo[1*]

## Abstract

**Background:** Engineered vaccine proteins incorporating both antigen and adjuvant components are constructed with the aim of combining functions to induce effective protective immunity. Bacterial flagellin is a strong candidate for an engineered vaccine scaffold as it is known to provide adjuvant activity through its TLR5 and inflammasome activation. Moreover, polymerized flagellin filaments can elicit a more robust immunoglobulin response than monomeric flagellin, and the multimeric antigen form can also promote T cell-independent antibody responses. Here, we aim to produce and test a covalently stabilized polymerized flagellar filament, providing additional immune efficacy through stabilization of its polymeric filament structure, as well as stabilization for long-term storage.

**Results:** Computational modeling of monomer packing in flagellin filaments helped identify amino acids with proximity to neighboring flagella protofilaments. Paired cysteine substitutions were made at amino acids predicted to form inter-monomer disulfide cross-links, and these substitutions were capable of forming flagella when transfected into a flagellin-negative strain of *Salmonella enterica* subspecies Typhimurium. Interestingly, each paired substitution stabilized different helical conformational polymorphisms; the stabilized filaments lost the ability to transition between conformations, reducing bacterial motility. More importantly, the paired substitutions enabled extensive disulfide cross links and intra-filament multimer formation, and in one of the three variants, permitted filament stability in high acidic and temperature conditions where wild-type filaments would normally rapidly depolymerize. In addition, with regard to potential adjuvant activity, all crosslinked flagella filaments were able to induce wild-type levels of epithelial NF-κB in a cell reporter system. Finally, bacterial virulence was unimpaired in epithelial adherence and invasion, and the cysteine substitutions also appeared to increase bacterial resistance to oxidizing and reducing conditions.

**Conclusions:** We identified amino acid pairs, with cysteine substitutions, were able to form intermolecular disulfide bonds that stabilized the resulting flagellar filaments in detergent, hydrochloric acid, and high temperatures while retaining its immunostimulatory function. Flagellar filaments with disulfide-stabilized protofilaments introduce new possibilities for the application of flagella as a vaccine adjuvant. Specifically, increased stability and heat tolerance permits long-term storage in a range of temperature environments, as well as delivery under a range of clinical conditions.

**Keywords:** Vaccine, Polymer, Innate immunity

## Background

Current vaccines and their adjuvants are restricted by a limited local and systemic immune induction and insufficient molecular stability due to a dependence on the storage and shipment "cold chain". In 2012 the World Health Organization proposed new programmatic suitability requirements and preferred characteristics for vaccines; ideally one that is high-efficacy, multi-antigen, and heat-stable to be stored for extended periods of time above 8 °C with the current target threshold temperature of 40 °C [1]. Indeed, the use of biosynthetic vaccines and adjuvants has risen from inactivated whole components due to increasing safety demands and better-defined vaccine elements. However, this has necessitated the inclusion of appropriate adjuvants to enhance the immune responses induced by engineered antigens and is of particular importance when used for individuals suffering from immunodeficiency disorders.

* Correspondence: david.lo@medsch.ucr.edu
[1]Division of Biomedical Sciences, School of Medicine, University of California, Riverside, USA

Flagellin is a highly-conserved bacterial protein required for motility and adhesion to host tissues that elicits strong inflammatory responses in both plants and animals. The flagellin filament (flagella) is composed of thousands of individual flagellin proteins polymerized into 11 protofilaments, linked by non-covalent bonds, to form whip-like helical appendages hundreds of nanometers long and anchored at one end to a membrane-bound motor assembly. The non-covalent nature of the flagella allows for flexibility and conformation polymorphism. The flagellin monomer is divided into four domains: two terminal; D0 and D1, and two central; D2 and D3. The terminal, alpha-helical, regions are highly conserved between bacterial species and indispensable for filament polymerization as they are involved in the protein-protein interactions. The D1 domain contains the small but highly conserved region of flagellin primarily involved in immune recognition. The central regions, dispensable for flagella polymerization and function, are exposed on the filament and are highly variable. These regions define the various H-antigen subtypes and allow for the evasion of adaptive immune responses [2].

Innate immune pattern recognition receptors (PRRs), including toll-like receptors (TLRs) and NOD-like receptors (NLRs) recognize pathogen-associated molecular patterns (PAMPs) and binding of these ligands to their cognate receptors initiates a signaling cascade that results in the activation and maturation of antigen-presenting cells (APCs) which, in turn, activate adaptive immune responses. Flagellin is the sole agonist of TLR5 [2] which mobilizes NF-κB via a MyD88-dependent pathway. Intracellular flagellin is recognized by two NLR family receptors, NLRC4 and NAIP5, which activate the inflammasome. As opposed to many other PAMPs, flagellin produces a mixed Th1 and Th2 immune response [3, 4], considered important to mounting an effective immune memory response. However, the immune recognition site of flagellin is hidden, buried in the flagella core [2], suggesting that flagellin can be recognized by PRRs only in its monomeric form.

Interestingly, several studies have demonstrated that polymeric flagellin has equal if not greater immunostimulatory effects as a vaccine adjuvant compared to monomeric flagellin. We previously employed a hybrid engineered flagellin substituting the central flexible domain with the Env protein from Dengue Virus [5]. The resulting hybrid flagellin was found to be a more potent vaccine in polymeric versus monomeric form. This effect was due in part to the ability of the filament to efficiently crosslink B cell receptors; moreover, these filaments were also able to induce a T helper cell independent immune response, adding T helper-independent B cells to the available antibody repertoire. In this study, our approach is to further stabilize the flagellar filament scaffold by introducing specific cysteine amino acid substitutions in adjacent protofilament monomers to introduce intermolecular disulfide bonds.

## Results

### Introducing intermolecular disulfide bonds in FliC filaments

The flagella of *Salmonella* (In this study, we used *Salmonella enterica* serovar Typhimurium, referred to briefly as *S.* Typhimurium in this manuscript) consist of a filament that is a supercoiled assembly of monomers of proteins such as flagellin (FliC). The monomers have a shape resembling the capital Greek letter gamma (Γ), with the tail of the monomer consisting of alpha-helical domains joined to the arm by a flexible central domain. The alpha-helical domains align to form protofilaments. Eleven protofilaments form the flagellar filament, through translation-rotation screw symmetry, with the flexible arm of each monomer presented to the external solvent-facing surface. No covalent links provide stability to the filament, so that filaments can bend or de-polymerize under certain conditions. Computational models of the packing of flagellin monomers within polymerized filaments allowed us to identify candidate amino acid pairs where neighboring monomers are capable of forming intermolecular disulfide bonds when replaced with cysteine.

The illustration in Fig. 1 depicts the D0-D1 domain interface of adjacent FliC protofilament monomers with the identified amino acids and their intermolecular distance indicated. Asparagine, glutamine, and glutamic acid residues were chosen because combinations of their rotameric states can assume the distance needed for the formation of the disulfide bridge upon replacement by cysteine. The interacting pairs of amino acids selected connect (see Fig. 1): Disulfide 1) the C'D1 domain of protofilament monomer 1 with the N'D0 domain of protofilament monomer 2; Disulfide 2) the C'D1 domain of protofilament monomer 1 with the N'D0-N'D1 linker of protofilament monomer 2; and Disulfide 3) the N'D1 domain of protofilament monomer 1 with the N'D1 domain of protofilament monomer 2. Incorporation of disulfide bridges through mutations to Cys at the aforementioned sites is expected to improve the stability of the multimeric complexes without significantly altering the local structures and leaving the D2 and antigenic D3 domain unrestrained. The domain D1 is more amphipathic (and more stable) than domain D0, thus the cross-link at the helical N-terminus and at the linker at the other end will preserve its flexibility. As has been suggested in previous studies, the D0 domain and lower parts of the D1 domain do not have a significant effect on the structure of the monomer [6]. Thus, we propose that the identified amino acid substitutions would permit stabilizing intermolecular disulfide bond formation, retain filament formation, and allow for TLR5 immune recognition. Moreover, these studies leave the antigenic D3 domain intact for future introduction of immunogenic vaccine antigens, including substitution of the D3

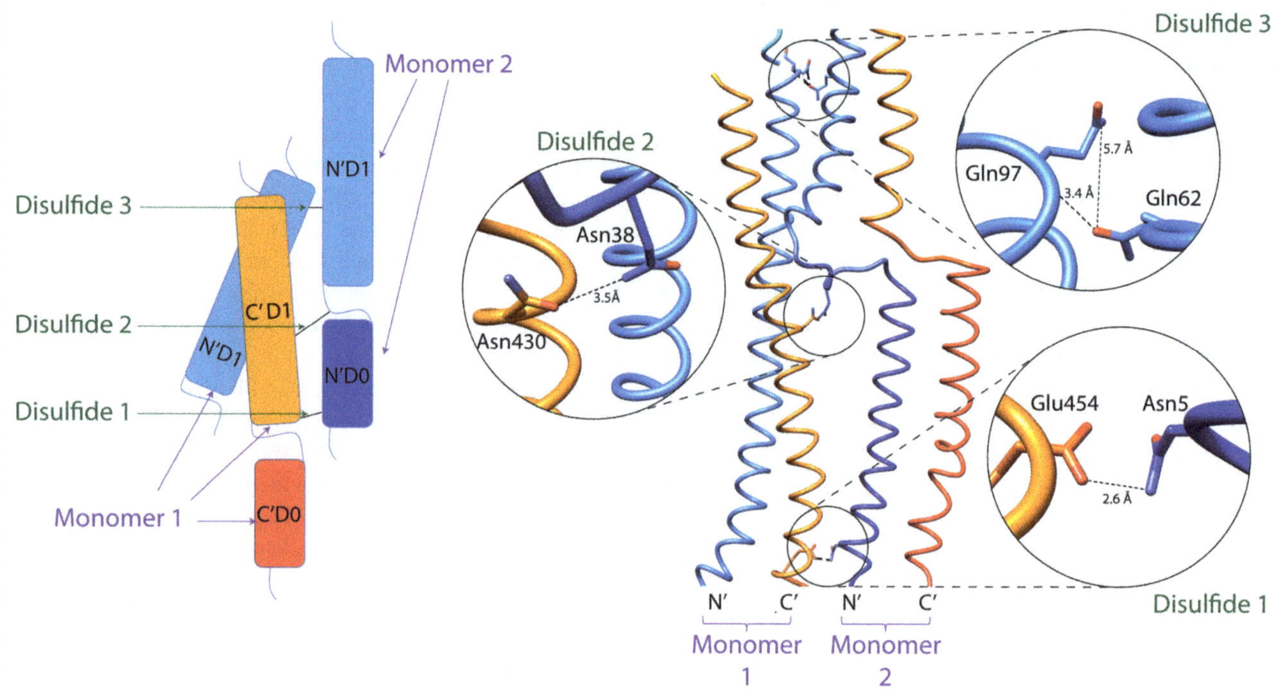


**Fig. 1** Introducing intermolecular disulfide bonds in FliC protofilaments. (Left) Depiction of the helical translations and proposed locations to introduce disulfide bonds between neighboring flagellin protofilaments. Domains N'D0 and N'D1 are colored dark and light blue respectively while C'D0 and C'D1 are orange and red, respectively. (Right) Molecular graphic representation of sites of interest within neighboring protofilament monomers. Insets show a zoomed and rotated view of the amino acid interactions with the specific distances shown: 1) Glu454[Oε1] – Asn5[Nδ1], 2) Asn430[Oδ1] – Asn38[Nδ1], and 3) Gln97[N] – Gln62[Oε1], Gln97[Nε2] – Gln62[Oε1]

domain with an intact virus protein domain, as we have reported previously [5].

**Cysteine replacements produced stabilized protofilaments with characteristic alignments**

Cysteine substitutions were performed in the three candidate pairs, generating FliC$^{N5C,E454C}$, FliC$^{N38C,N430C}$ and FliC$^{Q62C,Q97C}$ (Additional file 2: Figure S1), and each was expressed *in trans* under control of the native *fliC* promoter in the *S.* Typhimurium Δ*fliC* strain SPN313 [7]. To ensure production and also assess the structure of flagella containing the point mutations, SPN313 cells harboring the above plasmids were stained with an Alexa Fluor 488-labelled carboxylic acid TFP ester dye and imaged using confocal microscopy. As shown in Fig. 2a, imaging of the stained filaments showed that for each of the paired cysteine replacements, characteristic filament wavelengths and curvatures were generated. Measurement of flagellar helical pitch and radius allowed for calculation of twist and curvature associated with known and theoretical filament waveforms based on the model that each protofilament exists in either left- or right-handed conformation that differs in length [8–14]. The ratio of left- and right-handed protofilaments (L:R) determined the overall filament helical shape. Compared to the "normal" (L:R of 2:9) appearing wild-type flagella, the FliC$^{N5C,E454C}$ filament closely resembled a "coiled" (3:8) structure, while both FliC$^{N38C,N430C}$ and FliC$^{Q62C,Q97C}$ filaments resulted in "curly" (5:6) filaments (Fig. 2b), suggesting that either the cysteine substitution(s) or presence of disulfide bonds alter the composition of left- vs. right-handed protofilaments.

Stabilization of filaments in specific protofilament alignments might have consequences for bacterial mobility. With wild-type flagella, torsional stress due to flagellar rotation during bacterial motility is thought to be reflected in different protofilament alignments with corresponding filament wavelengths. A change in rotation of the filament causes a tumbling movement to enable changes in direction, and the change in torsional stress also causes changes in protofilament alignments. To determine whether the stabilization of the cysteine-substituted flagella in fixed protofilament alignments affected the function of the flagella, wild-type FliC, FliC$^{N5C,E454C}$, FliC$^{N38C,N430C}$ and FliC$^{Q62C,Q97C}$ expressed *in trans* in the non-motile *S.* Typhimurium SPN313 was examined for their ability to restore swimming motility on semi-solid media. Soft agar motility assays revealed that, despite the production of full length flagella, SPN313 expressing FliC$^{N38C,N430C}$ and FliC$^{Q62C,Q97C}$ remained completely non-motile while

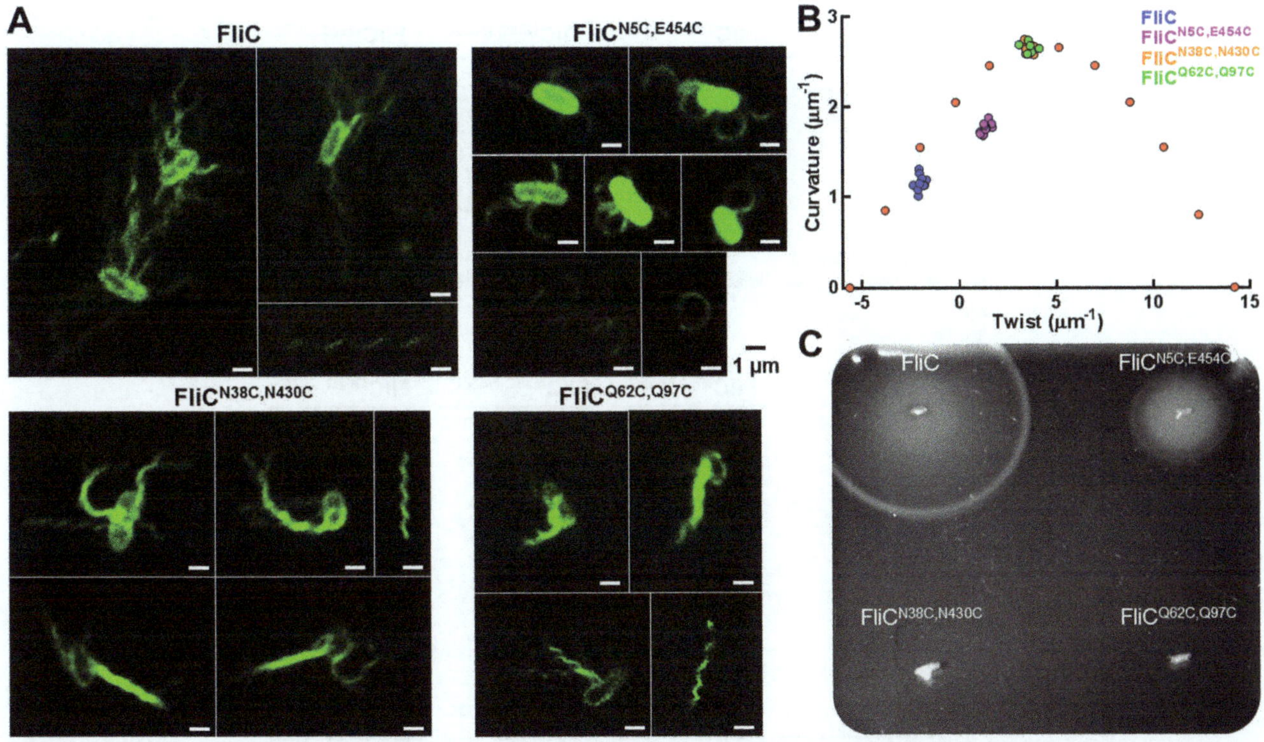


**Fig. 2** Assessing the structure and function of cysteine-substituted flagella. **a** Fluorescent staining of S. Typhimurium shows production and structural appearance of the respective flagella. Images included from three independent experiments. **b** Measurements of helical curvature ($\mu M^{-1}$) and twist ($\mu M^{-1}$) from each modified flagella compared to theoretical helical structures (red symbols). Multiple measurements were taken from both attached and detached filaments. Data ($n \geq 6$) from two independent experiments. **c** Soft-agar motility assay depicting the effect of cysteine substitutions on the swimming motility of S. Typhimurium. Image representative of three independent experiments

expression of FliC$^{N5C,E454C}$ restored approximately 50% of normal motility compared to wild-type FliC (Fig. 2c).

### Cysteine substitutions crosslink and stabilize FliC filaments

To assess protofilament crosslinking, extracted intact flagella from *S.* Typhimurium SPN313 expressing wild-type FliC, FliC$^{N5C,E454C}$, FliC$^{N38C,N430C}$ or FliC$^{Q62C,Q97C}$ were analyzed using SDS-PAGE under reducing and non-reducing conditions (Fig. 3a). Consistent with crosslinked multimers, each of the paired cysteine replacements generated flagella retaining high-order oligomeric structures while the wild-type flagella ran solely as the 52 kDa monomer. Importantly, the polymers were reduced to monomeric form with the addition of β-mercaptoethanol (β-ME), signifying that the FliC$^{N5C,E454C}$, FliC$^{N38C,N430C}$ and FliC$^{Q62C,Q97C}$ protofilaments were indeed crosslinked with disulfide bonds.

Traditional depolymerization of flagellar filaments is performed by lowering the pH with strong acids or incubating at temperatures 37 °C and above [15, 16]. To test the inherent thermal and acidic stability of the crosslinked filaments, extracted flagella were incubated in either 50 mM HCl or at 60 °C for 10 min. Native-PAGE analysis revealed that while wild-type FliC filaments were depolymerized following treatment with both HCl and 60 °C treatments, FliC$^{N5C,E454C}$, FliC$^{N38C,N430C}$ and FliC$^{Q62C,Q97C}$ substitutions retained a mixture of both full-length filaments (black arrow) and depolymerized monomers (red arrow), with FliC$^{Q62C,Q97C}$ completely unaffected by the treatments compared to untreated samples (Fig. 3b). Densitometry quantification of the respective filament and depolymerized protein bands from replicate Native-PAGE images is shown in Fig. 3c.

### Cysteine substitutions and bacterial viability

While the cysteine substitutions enabled disulfide crosslinks between monomers, the variable numbers of flagellin multimers observed using gel electrophoresis suggested that many of the available cysteines remained unconjugated. These cysteines might have not have been exposed to sufficiently oxidizing conditions to form disulfide bonds with neighboring cysteines, or they might simply not have been close enough to an available cysteine during filament formation. To examine the oxidizing and reducing capacity of SPN313 expressing wild-type FliC, FliC$^{N5C,E454C}$, FliC$^{N38C,N430C}$ and FliC$^{Q62C,Q97C}$ a broth microdilution

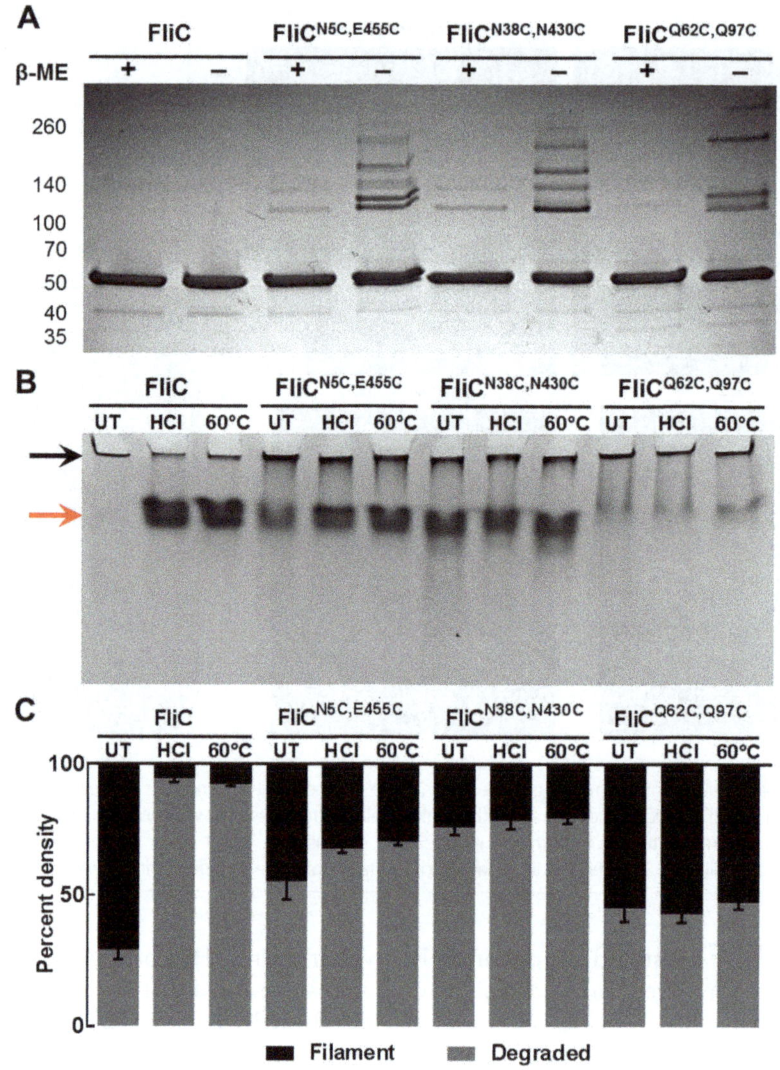


**Fig. 3** Crosslinking and stability of cysteine-substituted flagella. **a** Extracted filaments subjected to SDS-PAGE with and without β-mercaptoethanol (β-ME) reduction signifying the presence of disulfide bonds in all three cysteine-substituted flagella. Image representative of three independent experiments. **b** Thermal and acidic filament stability assessed by Native-PAGE following treatment with 50 mM HCl or heating at 60 °C for 10 min compared to untreated (UT). Image representative of two independent experiments. **c** Densitometry quantitation of replicate Native-PAGE gels as shown in **b**. Data included from two independent experiments with standard deviation shown

assay was performed using hydrogen peroxide (Fig. 4a) and β-mercaptoethanol (Fig. 4b) to assess their respective minimal inhibitory concentrations (MIC). Interestingly, when exposed to increasing concentrations of oxidizing (peroxide) or reducing (β-mercaptoethanol) conditions, the cysteine-substituted flagellins significantly enhanced bacterial viability. Thus, unbound cysteines in the flagella still appeared to have sufficient solvent exposure to provide capacity for hydrogen peroxide buffering in the surrounding environment. Alternatively, reduction of the crosslinked cysteines also allowed for enhanced survival in toxic conditions, together indicating that additional oxidation of the disulfide bonds is possible in downstream production and may provide enhanced filament stability.

### Crosslinked FliC filaments retain innate immune activation and epithelial invasion

The flagellin monomer is known to trigger innate immune responses via the Toll-like receptor TLR5 and the inflammasome NLRC4. The linear peptide sequence recognized by TLR5 is within the alpha helical domain and would be expected to be hidden when monomers are contained within the polymerized flagellar filament. Stabilization of the filament by disulfide cross-linking might alter the availability of the TLR5 recognition

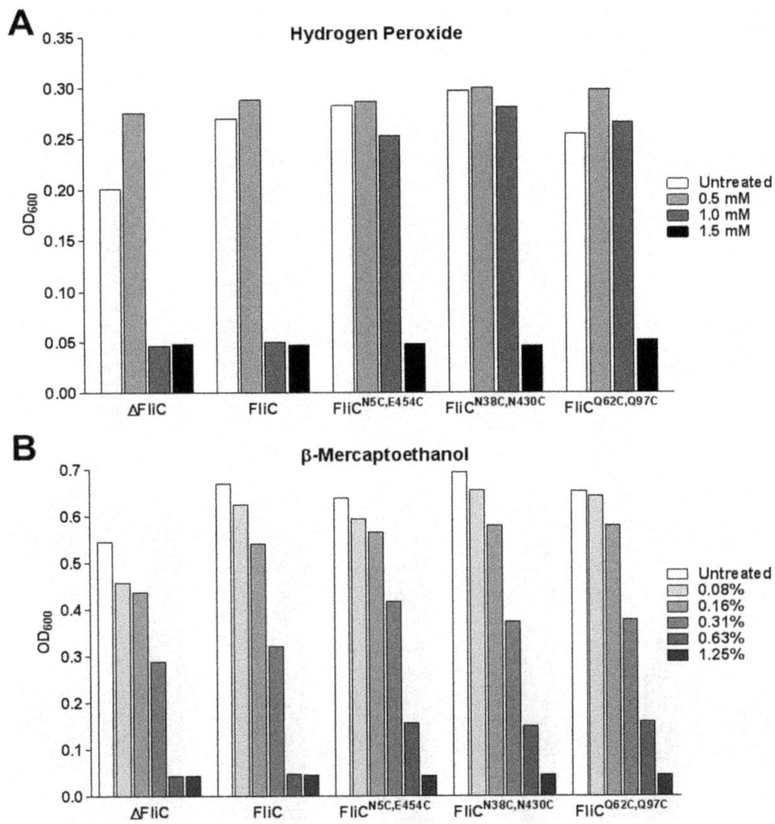


**Fig. 4** Oxidant and reductant buffering compacity of cysteine-substituted flagella. Growth of SPN313 in (**a**) hydrogen peroxide (**b**) β-mercaptoethanol using a broth microdilution method. Cultures harboring the indicated plasmids were incubated statically for 16 h at 35 °C, and $OD_{600}$ was measured to determine the MIC preventing growth. ΔFliC represents *S.* Typhimurium SPN313 harboring the empty parent vector. Data representative of two independent experiments

sequence, thereby affecting the potency in activating innate immune responses. To test the ability of the crosslinked FliC variants to induce an immune response, Caco-2 intestinal epithelial cells transfected with an NF-κB luciferase reporter vector were treated with extracted flagellar filaments from *S.* Typhimurium SPN313. As shown in Fig. 5a, compared to extract from the empty plasmid, unfiltered $FliC^{N5C,E454C}$, $FliC^{N38C,N430C}$ and $FliC^{Q62C,Q97C}$ induced NF-κB activity to a similar extent as WT FliC. To further asses the contributions of the individual fractions, flagella preparations were filtered to separate longer filaments from monomers and tested for relative potency in activating NF-κB responses. Interestingly, filaments retained by the filters proved be more potent stimulators of NF-κB (Fig. 5a).

To further examine the ability of the crosslinked flagella to function during in vivo-related processes, *S.* Typhimurium adherence and invasion assays were performed using Caco-2 epithelial cells (Fig. 5b). Compared to SPN313 containing the parent vector (ΔFliC), expression of all three crosslinked flagella variants enabled adherence and invasion of the epithelial cell layer equal to the expression wild-type FliC. Together, these data indicate that the crosslinked filaments retain their ability to function as normal flagella during host/immune processes.

### Long-term storage stability of crosslinked flagella at elevated temperatures

The $FliC^{Q62C,Q97C}$ mutation appeared to confer significant thermal stability and resistance to depolymerization in acidic pH solutions, so we assessed the long-term stability of $FliC^{Q62C,Q97C}$ filaments compared to wild-type FliC filaments in solution. Solutions of extracted flagella filaments were incubated for 7 days at 4 °C, 25 °C, 37 °C, and 42 °C, and filament survival was examined by Native-PAGE, including a depolymerized (DP) control sample. As shown in Fig. 6, a marked portion of WT FliC was depolymerized and degraded following incubation at 37 °C and 42 °C while that of $FliC^{Q62C,Q97C}$ remained remarkably intact.

## Discussion

In our previous studies using an engineered hybrid flagellin, we replaced the D3 central domain with the

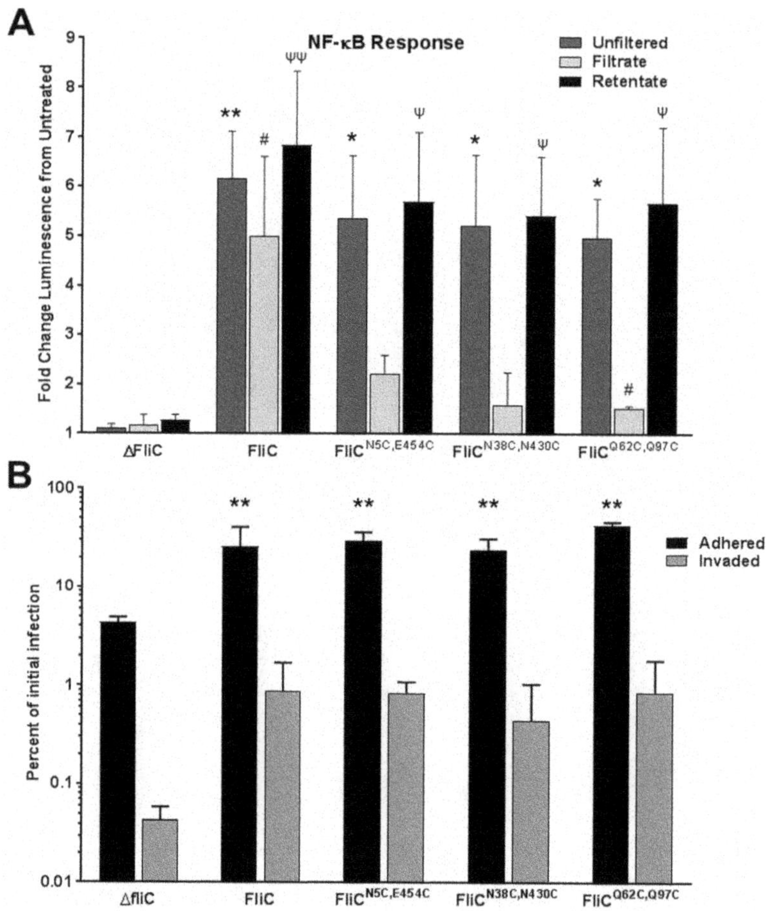


**Fig. 5** Immune response and functional effect of crosslinked flagella in Caco-2 cells. **a** Portions of extracted flagella were centrifuged though 100kdA filters to separate retentate filaments from filtrate monomers and, in addition to the original unfiltered extracts, 120 ng of each was added to Caco-2 epithelial monolayers harboring a NF-κB luciferase reporter vector. After 6 h, firefly and renilla luciferase were measured to assess the induction of NF-κB. Results include data from three independent experiments with standard error shown. Asterisk, hash, and psi symbols indicate statistically-significance difference comparing the unfiltered, filtrate, and retentate treatments, respectively, to ΔFliC. **b** Adherence and invasion assay performed with *S.* Typhimurium SPN313 cells harboring the empty vector (ΔFliC) or encoding the FliC derivatives were added to the apical side of Caco-2 cells. Invasion and adherence were assessed after 2 h and compared to the initial inoculum. Data from two independent experiments with standard deviation shown. Asterisk symbol indicates statistically-significance difference compared to ΔFliC. Single symbol; $p < 0.05$, double symbol; $p < 0.01$ as determined by 2-way ANOVA with Tukey's multiple comparisons test

coding sequence of the Dengue virus Env protein, and found that the monomer retained its ability to form polymeric filaments [5]. The filament proved to be more potent as an immunogen than the monomer, attributable in part to its potency as a polymeric antigen, but the protein might also have been protected from degradation by its assembly into the filaments. However, since flagella can depolymerize in aqueous solutions, we sought additional strategies to stabilize the filament structure that may provide resistance to depolymerization across a range of conditions including survival in vivo when administered as an engineered vaccine. The present study shows that a variety of amino acid substitutions based on computer modeling of the protofilament assembly can be used to produce filaments stabilized by disulfide cross links.

In these studies the D3 domain was intact, so the cysteine-replacement mutant proteins were normal size and transfection into FliC-negative *Salmonella* allowed production of flagella by the bacteria. As a result, we were able to gain additional insights into the structure and function of filaments produced by the cysteine replacement mutants. For example, replacement mutants were found to stabilize filaments into different protofilament alignments with distinct impacts on bacterial motility. Whether these conformations will have differential impacts on immune activation in vaccine formulations remains to be tested. In addition, the availability of thiol groups on exposed cysteine side chains appeared to provide microbial resistance to different redox conditions, and this may also have some impact on the stability of the filaments in vivo.

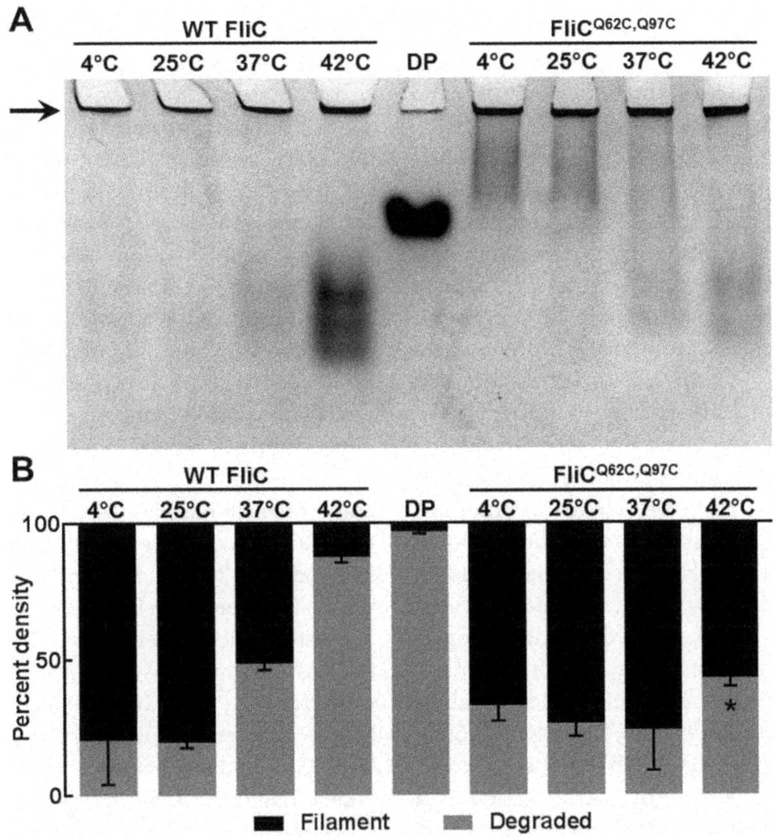


**Fig. 6** Long-term temperature stability of FliC$^{Q62C,Q97C}$ in solution. **a** Solutions of extracted WT and FliC$^{Q62C,Q97C}$ flagella filaments were incubated at 4 °C, 25 °C, 37 °C, and 42 °C for 7 days, and 3 ng was subjected to Native-PAGE to assess filament stability, depolymerization, and degradation. Image representative of two independent experiments. DP indicates an 80 °C depolymerized WT control sample. **b** Densitometry quantification of filaments (black arrow) and degraded / depolymerized protein. Data includes measurements from two independent experiments with standard deviation shown. Asterisk indicates statistically-significance ($p < 0.05$) difference compared to WT FliC as determined by 2-way ANOVA with Tukey's multiple comparisons test

The most important practical result from these studies is that computer modeling of the protofilament packing provided useful predictions on which amino acids could be substituted with cysteine to enable disulfide cross-linking of flagellin monomers. The resulting replacement mutants were all capable of forming flagella, and disulfide cross links were readily demonstrated in all cases, providing increased stability of the filaments. It is not clear what factors influence the percentage of cysteines able to form disulfide bonds. Preliminary studies indicated that treatment of filaments with peroxide did not increase crosslinking, so it may be that the protofilament conformations associated with each replacement mutant keep many of the available cysteines too far away from nearby cysteines to form disulfide bonds.

In sum, we have shown that an engineered version of bacterial flagellin can be produced that provides increased stability through the introduction of disulfide crosslinks while retaining the general structural features of the flagellar filaments including innate immune agonist activity. This scaffold will be a useful starting point in the design of engineered synthetic vaccines that provide protein stability, adjuvant activity, and polymeric antigen presentation to provide more potent activation of both T helper-dependent and -independent B lymphocyte responses.

## Conclusions

Computer modeling of the polymerized flagellin filament identified nearby amino acid residues in adjacent monomers. Substituting cysteine residues at paired adjacent sites allowed us to test whether monomers could form covalent disulfide bonds in the flagellin filament. Bacteria expressing these cysteine-substituted flagellin sequences produced flagellae with a high proportion of disulfide cross-linked flagellin multimers. The filaments showed stabilized helical conformations as well as high thermal stability in aqueous solution. These engineered filaments show promise as a stable scaffold for polymeric vaccines.

## Methods

### Computational modeling

The program UCSF Chimera [17] was utilized for visualization, intermolecular interaction analysis, and mutation design. An R-type flagellar filament structure, made of flagellin monomers (PDB Code 1UCU [10]) was used for modeling. Guided by structural and physicochemical analysis, potential sites for cross-linking through the introduction of disulfide bridges, and thus stabilization of monomer-monomer interfaces, were sought at the flexible elongated domains D0 and D1, while leaving D2 and antigenic domain D3 unrestrained. Mutations of asparagine, glutamine, and glutamic acid residues to cysteines were introduced at critical positions in D0 and D1 domains (Fig. 1). Given the largely polar nature of inter-monomer interfaces, mutations involved sites of side chain interactions that contribute to the stability of the dimers only through weak hydrogen bonds. The rationale was to replace these weak, and easily breakable, non-bonded interactions with strong covalent disulfide bridges, and thus strengthen the stability of the dimers. Asparagine, glutamine, and glutamic acid residues were selected in mutation pairs, E454(A)-N5(B), N430(A)-N38(B), and Q97(A)-Q62(B), because combinations of their side chain rotameric states can assume the distance required for the formation of an inter-monomer disulfide bridge, upon replacement by cysteine. Finally, mutations were chosen at locations that would not potentially compromise the integrity of the helical structures of D0 and D1. Finally, each chosen disulfide bridge contains at least one cysteine at a (typically dynamically fraying) helical terminus (D0 or D1), or at a (typically dynamically mobile) loop (D0-D1), to assure necessary flexibility for the steric accommodation of the disulfide bridge.

### Bacterial strains and cloning techniques

The *fliC* gene and promoter elements were amplified from the *Salmonella enterica* serovar Typhimurium IR715 ("*S.* Typhimurium" throughout the text) chromosome using oligonucleotides fliC-F and fliC-R (Additional file 1: Table S1) which encode restriction sites BamHI and EcoRI used for cloning into the pENTR3C vector (Invitrogen, USA), generating pFliC. Oligonucleotides encoding desired polymorphisms were used to divergently amplify the entire plasmid. Resulting PCR products were treated with DpnI (New England Biolabs, USA) to remove template plasmid prior to transformation into *E. coli* DH5α. Plasmid sequences were confirmed at the UC Riverside IIGB core facility. All pFliC variants were transformed into the *S.* Typhimurium SPN313 (IR715 Δ*fliC fljB*::Mu*d*Cm) [7].

### Collection of flagellar filaments

Overnight cultures of *S.* Typhimurium SPN313 harboring FliC-encoding plasmids were diluted 1:100 in LB broth (Fisher, USA) containing 28 mM glucose and 50 μg/mL kanamycin and incubated for 16 h at 30 °C with 125 RPM aeration. Cultures were then centrifuged and cells washed with PBS. Cell suspensions were vortexed for 5 min and sheared flagellin were separated from the cells by centrifugation. Supernatants containing intact flagella were precipitated with acetone or pelleted by centrifugation for 1 h at 48,400 x g, 4 °C and resolubilized in PBS. Where required, flagellar filaments were enriched by centrifugation through 100,000 MWCO filters (Millipore, USA) and, when necessary, filaments were depolymerized at 80 °C for 10 min.

### Luminescence reporter assay

Caco-2BBe cells were cultured similarly to standard methods [18]. The SRE or NF-κB luciferase reporter vector (Qiagen, USA) was transfected into $5 \times 10^4$ freshly harvested Caco-2 cells using Lipofectamine 3000 (Invitrogen). Following overnight incubation, transfected cells were treated for 6 h with 120 ng of flagellar protein. Firefly and renilla luminescence was detected using the Dual-Glo luciferase assay (Promega, USA). Three independent experiments were performed using multiple, independent flagellin collections.

### Flagella filament staining

Fluorescent imaging of *S.* Typhimurium flagella was performed similarly as previously described [19]. Briefly, late-logarithmic and early-stationary phase cultures of *S.* Typhimurium harboring FliC-encoding plasmids were washed with PBS and diluted to $OD_{600}$ 0.2 in 0.1 mL Alexa Fluor 488 5-Tetrafluorophenyl (TFP) Ester (Invitrogen, USA) prepared according to manufacturer recommendations. Cells were incubated for 1 h in the dark at room temperature with occasional mixing and washed with PBS prior to imaging. Images were acquired on a CSU-X-1 spinning-disk confocal imager (Yokogawa, Japan) attached to a Zeiss Axio Observer inverted microscope. Hardware, including the Prim 95B digital camera (Photometrics, USA) was controlled by Micro-Manager imaging software. Three independent experiments were performed. Images were further optimized using ZEN software [Zeiss, Germany] and flagella measurements taken using ImageJ/Fiji software.

### Gel electrophoresis and densitometry

Native- and SDS-PAGE was performed using precast 4–12% mini-PROTEAN TGX gels (Bio-Rad, USA) with appropriate manufacturer-recommended sample and running buffers. Equal amounts of flagellin protein was loaded as measured by BCA (Thermo, USA) and protein molecular weights were determined using the Spectra BR ladder (Thermo, USA). Densitometry analysis was performed using ImageJ with individual histogram peaks representing the filament and all depolymerized and/or degraded products quantified separately. Each fraction is presented as a percentage of total area under the curve. The number of independent experiments performed are described in the figure legends.

### Broth microdilutions for MIC quantification

Overnight cultures of *S.* Typhimurium SPN313 harboring FliC-encoding plasmids were diluted to $OD_{600} = 0.01$ in 0.1 mL of LB broth containing a range of hydrogen peroxide or β-mercaptoethanol concentrations within 96-well plates. Cultures were incubated statically for 16 h at 35 °C and absorbance at 600 nm was read to asses growth. Two independent experiments were performed.

### Adherence and invasion assay

After the Caco-2 monolayer reached full differentiation (21 days) the medium was exchanged to antibiotics-free DMEM for 24 h. Bacteria were cultured overnight in LB + $Kan^{50}$ at 35 °C with 150 RPM aeration then diluted 1:100 in fresh LB + $Kan^{50}$ and cultured the same way for 5 h. Bacteria were then pelleted at 5,000×g for 5 min, washed with PBS, and diluted to $OD_{600}$ 0.1 in DMEM (for MOI of 10). 500 μL of the bacterial suspension was added to the apical side of Caco-2 monolayers and incubated for 1 h at 37 °C. For invasion assays: after 1 h of incubation, cells were washed 2x with PBS and treated for 1 h with 50 μg/mL gentamicin in fresh DMEM + HEPES. For adherence assays: the 1 h gentamicin treatment was omitted. In all assays: 2 h post-infection, cells were washed 2x with PBS and lysed with 1% Triton X-100 for 10 min. Lysates were serial diluted in PBS and plated on LB agar to enumerate CFUs. Two independent experiments were performed.

### Statistics

When appropriate, statistical analyses were performed. Significant differences between experimental groups were determined with two-way analysis of variance (ANOVA) followed by Tukey's multiple-comparison test, within Prism 7 (GraphPad, La Jolla, CA). For all analyses, a $P$ value of less than 0.05 was considered statistically significant.

## Additional files

**Additional file 1: Table S1.** Oligonucleotides and plasmids. (DOCX 16 kb)

**Additional file 2: Figure S1.** *Salmonella enterica* Typhimurium fliC. (PDF 164 kb)

### Abbreviations

ANOVA: Analysis of variance; APC: Antigen presenting cell; BCA: Bicinchoninic acid; CFU: Colony-forming units; MIC: Minimal inhibitory concentration; NF-κB: nuclear factor kappa B; NLR: NOD-like receptor; NLRC4: NLR family CARD domain-containing protein 4; PAGE: Polyacrylamide gel electrophoresis; PAMP: Pathogen-associated molecular pattern; PBS: Phosphate-buffered saline; PRR: Pattern recognition receptors; SDS: Sodium dodecyl sulfate; TFP: Tetrafluorophenyl; TLR: Toll-like receptor; TLR5: Toll-like receptor 5; WT: Wild type; β-ME: Beta mercaptoethanol

### Acknowledgments

We are grateful to Dr. Andreas Bäumler (University of California, Davis) for sharing *S.* Typhimurium SPN313, and to Dr. Keiichi Namba (Osaka University) for sharing a filament structure of FliC and transformation operations to generate filament models from monomeric flagellin structures. We also thank Kailuh Menefee (University of California, Riverside) for assistance with MIC assays. Finally, we note the passing of Dr. Dimitrios Morikis on May 27, 2019.

### Authors' contributions

DM and DDL conceived the modeling and design studies, RRM performed the computer modeling studies, CMG designed and executed the studies on flagellin filament structure and filament stability, and MIC studies. All authors assisted in the writing, review and approval of the manuscript and figures.

### Funding

Research reported in this publication was supported by NIAID of the National Institutes of Health under award number R01AI063426 to DDL. The content is solely the responsibility of the authors and does not necessarily represent the official views of the National Institutes of Health. The funders had no role in the design of the study and collection, analysis and interpretation of data and in writing the manuscript.

### Ethics approval and consent to participate

Not applicable; this study did not involve human subjects, human data, human tissue, or any individual person's data (including individual details, images, or videos) in any form from any laboratory or person. This study also did not involve research in animals.

### Consent for publication

Not applicable; this study did not involve human subjects, human data, human tissue, or any individual person's data (including individual details, images, or videos) in any form from any laboratory or person. This study also did not involve research in animals.

### Competing interests

D. Morikis was Editor-in-Chief of BMC Biophysics. The authors declare that they have no competing interests to report.

### Author details

[1]Division of Biomedical Sciences, School of Medicine, University of California, Riverside, USA. [2]Department of Bioengineering, University of California, Riverside, USA.

### References

1. WHO. Assessing the programmatic suitability of vaccine candidates for WHO prequalification. Geneva: World Health Organization; 2012. Available at: http://apps.who.int/iris/bitstream/10665/76537/1/WHO_IVB_12.10_eng.pdf)
2. Smith KD, Andersen-Nissen E, Hayashi F, Strobe K, Bergman MA, Barrett SL, Cookson BT, Aderem A. Toll-like receptor 5 recognizes a conserved site on flagellin required for protofilament formation and bacterial motility. Nat Immunol. 2003;4(12):1247–53.
3. Coffman RL, Sher A, Seder RA. Vaccine adjuvants: putting innate immunity to work. Immunity. 2010;33(4):492–503.

4. Huleatt JW, Jacobs AR, Tang J, Desai P, Kopp EB, Huang Y, Song L, Nakaar V, Powell TJ. Vaccination with recombinant fusion proteins incorporating toll-like receptor ligands induces rapid cellular and humoral immunity. Vaccine. 2007;25(4):763–75.
5. Bennett KM, Gorham RD Jr, Gusti V, Trinh L, Morikis D, Lo DD. Hybrid flagellin as a T cell independent vaccine scaffold. BMC Biotechnol. 2015;15:71.
6. Maki-Yonekura S, Yonekura K, Namba K. Conformational change of flagellin for polymorphic supercoiling of the flagellar filament. Nat Struct Mol Biol. 2010;17(4):417–22.
7. Winter SE, Thiennimitr P, Nuccio SP, Haneda T, Winter MG, Wilson RP, Russell JM, Henry T, Tran QT, Lawhon SD, et al. Contribution of flagellin pattern recognition to intestinal inflammation during Salmonella enterica serotype typhimurium infection. Infect Immun. 2009;77(5):1904–16.
8. Calladine CR. Change of waveform in bacterial flagella: the role of mechanics at the molecular level. J Mol Biol. 1978;118(4):457–79.
9. Calladine CR. Construction of bacterial flagella. Nature. 1975;255(5504):121–4.
10. Yonekura K, Maki-Yonekura S, Namba K. Complete atomic model of the bacterial flagellar filament by electron cryomicroscopy. Nature. 2003; 424(6949):643–50.
11. Kamiya R, Asakura S. Helical transformations of Salmonella flagella in vitro. J Mol Biol. 1976;106(1):167–86.
12. Calladine CR, Luisi BF, Pratap JV. A "mechanistic" explanation of the multiple helical forms adopted by bacterial flagellar filaments. J Mol Biol. 2013;425(5):914–28.
13. Wang F, Burrage AM, Postel S, Clark RE, Orlova A, Sundberg EJ, Kearns DB, Egelman EH. A structural model of flagellar filament switching across multiple bacterial species. Nat Commun. 2017;8(1):960.
14. Hasegawa K, Yamashita I, Namba K. Quasi- and nonequivalence in the structure of bacterial flagellar filament. Biophys J. 1998;74(1):569–75.
15. Ibrahim GF, Fleet GH, Lyons MJ, Walker RA. Method for the isolation of highly purified Salmonella flagellins. J Clin Microbiol. 1985;22(6):1040–4.
16. Kajikawa A, Midorikawa E, Masuda K, Kondo K, Irisawa T, Igimi S, Okada S. Characterization of flagellins isolated from a highly motile strain of Lactobacillus agilis. BMC Microbiol. 2016;16:49.
17. Pettersen EF, Goddard TD, Huang CC, Couch GS, Greenblatt DM, Meng EC, Ferrin TE. UCSF chimera--a visualization system for exploratory research and analysis. J Comput Chem. 2004;25(13):1605–12.
18. Natoli M, Leoni BD, D'Agnano I, Zucco F, Felsani A. Good Caco-2 cell culture practices. Toxicol In Vitro. 2012;26(8):1243–6.
19. Turner L, Ryu WS, Berg HC. Real-time imaging of fluorescent flagellar filaments. J Bacteriol. 2000;182(10):2793–801.

# 8

# Development of a luciferase/luciferin cell proliferation (XenoLuc) assay for real-time measurements of Gfp-Luc2-modified cells in a co-culture system

Sin-Yeang Teow[1,2], Kitson Liew[1], Mohd Firdaus Che Mat[1], Marini Marzuki[1], Norazlin Abdul Aziz[1], Tai-Lin Chu[1], Munirah Ahmad[1] and Alan Soo-Beng Khoo[1*]

**Abstract**

**Background:** In vitro modelling of cancer cells is becoming more complex due to prevailing evidence of intimate interactions between cancer cells and their surrounding stroma. A co-culture system which consists of more than one cell type is physiologically more relevant and thus, could serve as a useful model for various biological studies. An assay that specifically detects the phenotypic changes of cancer cells in a multi-cellular system is lacking for nasopharyngeal carcinoma (NPC).

**Results:** Here, we describe a luciferase/luciferin (XenoLuc) assay that could specifically measure changes in the proliferation of cancer cells in the co-culture system using two modified NPC patient-derived tumour xenograft (PDTXs) cells: Xeno284-gfp-luc2 and XenoB110-gfp-luc2. Through this assay, we are able to show that the growth of NPC xenograft cells in both two-dimensional (2D) and three-dimensional (3D) models was enhanced when co-cultured with normal human dermal fibroblasts (NHDFs). In addition, potential applications of this assay in in vitro drug or inhibitor screening experiments are also illustrated.

**Conclusions:** XenoLuc assay is specific, sensitive, rapid and cost-effective for measuring the growth of luciferase-expressing cells in a co- or multiple-culture system. This assay may also be adapted for tumour microenvironment studies as well as drug screening experiments in more complex 3D co-culture systems.

**Keywords:** Luciferase assay, Luminescence, Nasopharyngeal carcinoma, Nasopharyngeal neoplasms, Patient-derived xenograft, Fibroblasts, Tumour microenvironment, Stromal interaction, Proliferation, Co-culture, Spheroids, Drug screening, Cancer immunology

## Background

Cell-based assays remain the leading platform for cancer research in preclinical settings. Majority of high-throughput compound screening programmes carried out by major pharmaceutical companies utilise cell-based assays to identify high-quality leads [1]. Such assays have the advantage over biochemical-based assays since these assays provide information on the cellular responses of a particular drug candidate. Moreover, cell-based assays could be reliably used to provide early indications of drug toxicity [1]. The recent advancement in cell culture technology has paved way for the development of in vitro cell-based assays that could be carried out in a more biologically relevant microenvironments. Such assays may include the incorporation of extracellular matrices (ECMs) and scaffolds as well as stromal cells in the presence of tumour cells [2, 3]. However, challenges remain in the design of these assays, particularly on measuring phenotypic changes of a specific cell population in complex multi-cell type systems. Notably, there is a scarcity of convenient cell-based assays that

* Correspondence: alankhoo@imr.gov.my; alankhoo.imr@gmail.com
[1]Molecular Pathology Unit, Cancer Research Centre, Institute for Medical Research, National Institutes of Health (NIH Complex), Ministry of Health Malaysia, Level 4, Block C7, No: 1, Jalan Setia Murni U13/52, Section U13, Setia Alam, 40170 Shah Alam, 50588 Kuala Lumpur, Selangor, Malaysia

could discriminate phenotypes of different cell types in a co-culture setting [4, 5]. With greater understanding of the microenvironment surrounding a tumour, it is increasingly important to employ appropriate in vitro assays that are able to address issues of tumour complexity and heterogeneity [2].

Cancer cells are often grown as monolayers, of which 2D (monolayer) culture models are commonly employed for in vitro drug testing. However, these 2D models are bound by several limitations that may potentially negate their usefulness. For example, there is a lack of 3D cell-cell and cell-ECM interactions in 2D models that are present in in vivo models [6]. These interactions generate signalling cues that are pivotal for numerous important cellular functions such as proliferation, differentiation and survival. Most of these functions are lost or have been compromised in 2D culture models. As a result, data generated from 2D in vitro drug testing could be misleading and non-predictive for in vivo responses [6]. The spatial arrangement of cancer cells within the 3D model, possibly together with the presence of other cell types and ECM components, mimic their natural microenvironment and hence, may reinstate the important signalling that are lost or compromised in 2D culture systems [7]. For these reasons, 3D culture models offer greater clinical and biological relevance than in vitro models and therefore, could bridge the gap between the monolayer culture and animal works.

Our group had developed two PDTXs, XenoB110 and Xeno284 from Malaysian NPC patients [8]. To facilitate the monitoring of in vivo tumour growth, these PDTXs were co-transduced with gfp-luc2 reporter genes (XenoB110-gfp-luc2 and Xeno284-gfp-luc2). The XenoLuc assay described in this paper is a luciferase-based assay that uses commercial D-Luciferin as the substrate for the luciferase in the transduced PDTXs to generate a luminescent signal that is proportional to cell numbers. We demonstrate that the XenoLuc assay is sensitive and could specifically measure real-time proliferation of PDTXs both in vitro and in vivo. As 3D models are gaining prominence as in vitro cell-based assays, we also demonstrate the robustness of XenoLuc assay in both mono- and multi-cellular spheroid cultures in addition to the conventional 2D monolayer model. This assay is able to measure the PDTXs growth enhancement that resulted from the addition of growth supplements as well as from the effect of co-culturing with other human cell types. In addition, the assay may also be used to evaluate the loss of cell viability, such as that induced by cisplatin treatment. Since the proliferation and viability of PDTX cells could be specifically detected in the complex 3D co-culture systems, XenoLuc assay may represent a cost-effective and a potentially useful cell-based assay for studies on tumour microenvironment, high-throughput compound screening and preclinical drug response prediction.

## Results

### Specificity and sensitivity of XenoLuc assay

The specificity of XenoLuc assay was assessed by determining if it specifically detected luminescent signals from luciferase-expressing cells compared to non-expressing cells. All experiments were carried out using freshly isolated xenograft cells harvested from three separate tumours from different mice and used for not more than two passages. Non-treated/non-cultured cells are used in assays as internal controls to avoid batch-to-batch variation which could arise from multiple causes. Based on Fig. 1a, luminescence was detected in *luc2*-transduced xenograft cells but not in parental xenograft cells and NHDFs, indicating the specificity of XenoLuc assay. Additional file 1: Figure S1 shows the linear correlation between luminescence signals versus various titrations of XenoB110-gfp-luc2 and Xeno284-gfp-luc2 cells measured at day-0 (30 min after cell seeding when the cells had settled at the bottom of wells, but had not adhered). The mouse cell-depleted xenograft cells exhibited higher luminescent signals compared to the non-depleted xenograft cells (Fig. 1a and b). To examine the assay sensitivity, two-fold serial dilutions of cells were plated and the readings were taken after 4 days. Figure 1b shows that cell seeding number had a positive linear correlation with luminescence in both 2D and 3D culture models, of which the lowest seeding density tested was 2500 cells/well in a 96-well plate. Overall in 2D and 3D models, XenoB110-gfp-luc2 cells exhibited higher signals (about 20-fold) than Xeno284-gfp-luc2 cells. This could be due to a higher expression of luciferase in XenoB110-gfp-luc2 cells.

We then asked whether XenoLuc assay could be used to measure cell proliferation. Figure 1c shows that the luminescent signals increased in a time-dependent manner whereby the maximum growth of both xenograft cells was recorded at Day 6 and Day 4 for 2D and 3D culture models, respectively. Following that, the growth of both xenograft cells either reached plateau phase or declined. In another experiment, we captured the images of XenoB110-gfp-luc2 cells (2D and 3D cultures) under FITC channel using IN-Cell high content cellular analyser, and the GFP images were subjected to fluorescence intensity analysis by the IN-Cell Developer software. Additional file 2: Figure S2A and S2B show that the fluorescence intensity of non-depleted and mouse cell-depleted xenograft cells in 2D culture correlated positively to the number of cells seeded. The signal intensity also increased following the increase in cell proliferation from day 2 to 4. Similar correlations (between signal intensity and cell seeding numbers) were seen when

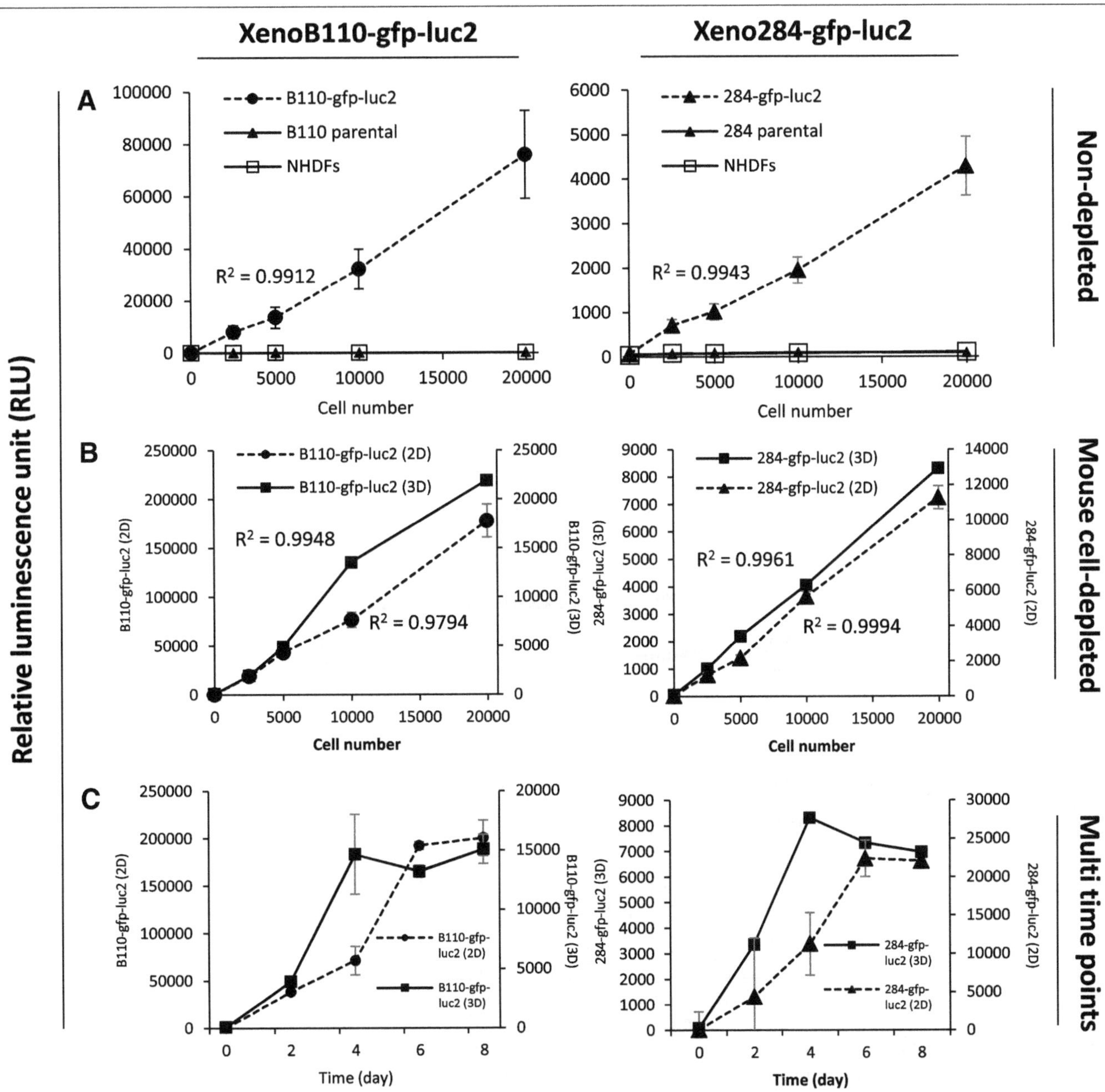


**Fig. 1** XenoLuc assay for in vitro cell proliferation measurement. **a** Measurement of luminescence in two-fold serially diluted non-depleted xenograft cells (parental and modified) and NHDF cells. These cells were seeded onto 96-well plates as 2D monolayer culture, and the luminescence was measured after 4 days. **b** Measurement of luminescence of mouse cell-depleted xenograft cells cultured in vitro in 2D and 3D models. **c** Proliferation rate of depleted xenografts cells as measured by XenoLuc assay at a 2-day interval. Results are represented by average of triplicate from 3 mice ± SD. Left panel: XenoB110-gfp-luc2; Right panel: Xeno284-gfp-luc2. $R^2$ shows the linear correlation of luminescence and cell number

XenoB110-gfp-luc2 cells were grown as 3D spheroids (Additional file 2: Figure S2C). However, in these 3D cultures, unlike luminescence signals, no obvious difference was found in fluorescence intensity between day-2 and day-4 readings.

### Characterization of XenoLuc assay

The principle of XenoLuc assay was illustrated in Fig. 2a. Light or luminescence is generated as the result of the enzymatic reaction between the D-Luciferin substrate and the intracellular luciferase in the presence of oxygen and adenosine triphosphate (ATP). In addition to the luminescence, other by-products are oxyluciferin, AMP, PPi (pyrophosphate), and carbon dioxide [9]. Figure 2b shows that the XenoLuc luminescent assay may be used either as endpoint (lytic) or non-destructive (non-lytic) assays. In line with the current trend of multiplexing cell-based assays, the non-lytic assay is preferable to the

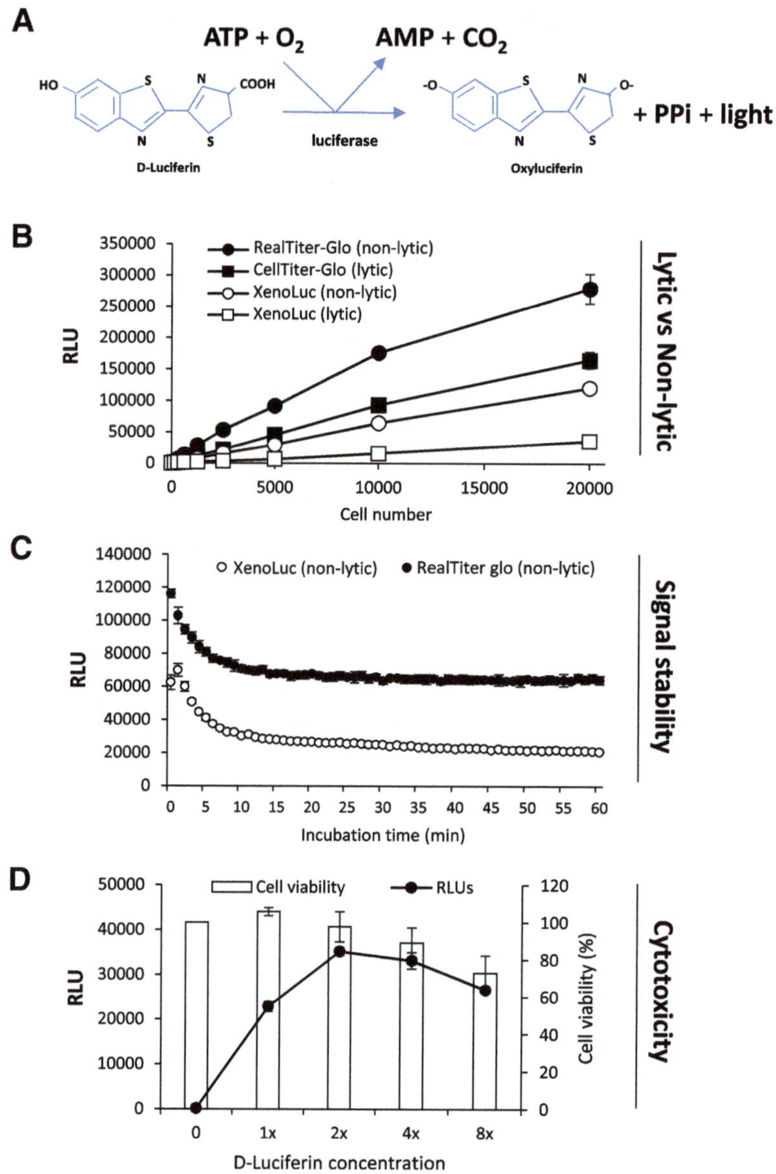


**Fig. 2** Characterization of XenoLuc assay. **a** Diagram showing the basic concept of luciferase/luciferin system. **b** Comparison of signal intensity detected by XenoLuc assay and two commercial assays: CellTiter-Glo (lytic) and RealTime-Glo (non-lytic). **c** Comparison of signal stability of XenoLuc and RealTime-Glo. **d** Cytotoxicity of D-Luciferin substrate after 24-h incubation as determined by MTS assay. Results are represented by average of triplicate from 3 mice ± SD

traditional endpoint assay because no cell lysis is required, and the same well may be multiplexed for various assays [10]. Comparatively, the luminescent signals detected by XenoLuc assay were relatively lower than that of commercial assays (Fig. 2b). RealTime-Glo and CellTiter-Glo employ exogenous luciferase and possibly more sensitive proprietary luciferin substrate along with other signal enhancers or co-factors such as $Mg^{2+}$, recombinant ATP and Coenzyme A (CoA) [11]. The non-lytic version of XenoLuc assay was found to give higher signals than the lytic version of the assay at various tested cell numbers (Fig. 2b). We then assessed the luminescent signal stability of the XenoLuc assay in comparison with RealTime-Glo by recording the signals for 60 min at a 1-min interval. Figure 2c shows that the signals in both assays initially dropped from the first reading, and eventually stabilized after 10 min of incubation. These signals were stable for up to 60 min. We then selected 10 min as the incubation time for luciferin substrate with the luc2-expressing cells for subsequent experiments. As shown in Fig. 2d, the increase of D-luciferin substrate concentration to 2X from 1X increased the luminescence signal by 1.5 folds. The increment in signal declined when D-luciferin substrate was used at higher concentrations of 4X and 8X. More importantly, the presence of luciferin at 2X for 24 h did not cause toxicity to XenoB110-gfp-luc2 cells (Fig. 2d). This is consistent with Tiffen et.al [12]. However, cell viability was reduced to 89 and 73%, respectively when incubated with 4X and 8X luciferin substrate (Fig. 2 d). This explains the reduction in luminescent signals at these concentrations. Similar trends were also observed in Xeno284-gfp-luc2 cells (data not shown).

### XenoLuc assay for anti-cancer drug screening

Figure 3 shows that cisplatin exhibited dose-dependent cytotoxicity against both xenograft cells. Of the two xenograft cells, the decrease in cell viability of Xeno284-gfp-luc2 cells were less possibly because the cells were more resistant to cisplatin treatment ($IC_{50}$ of 45 μM and 30 μM) as compared to XenoB110-gfp-luc2 cells ($IC_{50}$ of 2.5 μM and 6.25 μM) in 2D and 3D culture models. A parallel comparison revealed that the overall luminescence of Xeno284-gfp-luc2 cells was lower than XenoB110-gfp-luc2 cells, consistent with Fig. 1. Nevertheless, comparisons were made possible because of the inclusion of untreated cells derived from the same cell isolates and carried out within the same experimental batch to avoid batch-to-batch variability.

### Determination of xenograft cell growth in a co-culture system

As demonstrated by Fig. 2, XenoLuc assay specifically detected luc2-bearing tumour cells. Additionally, it

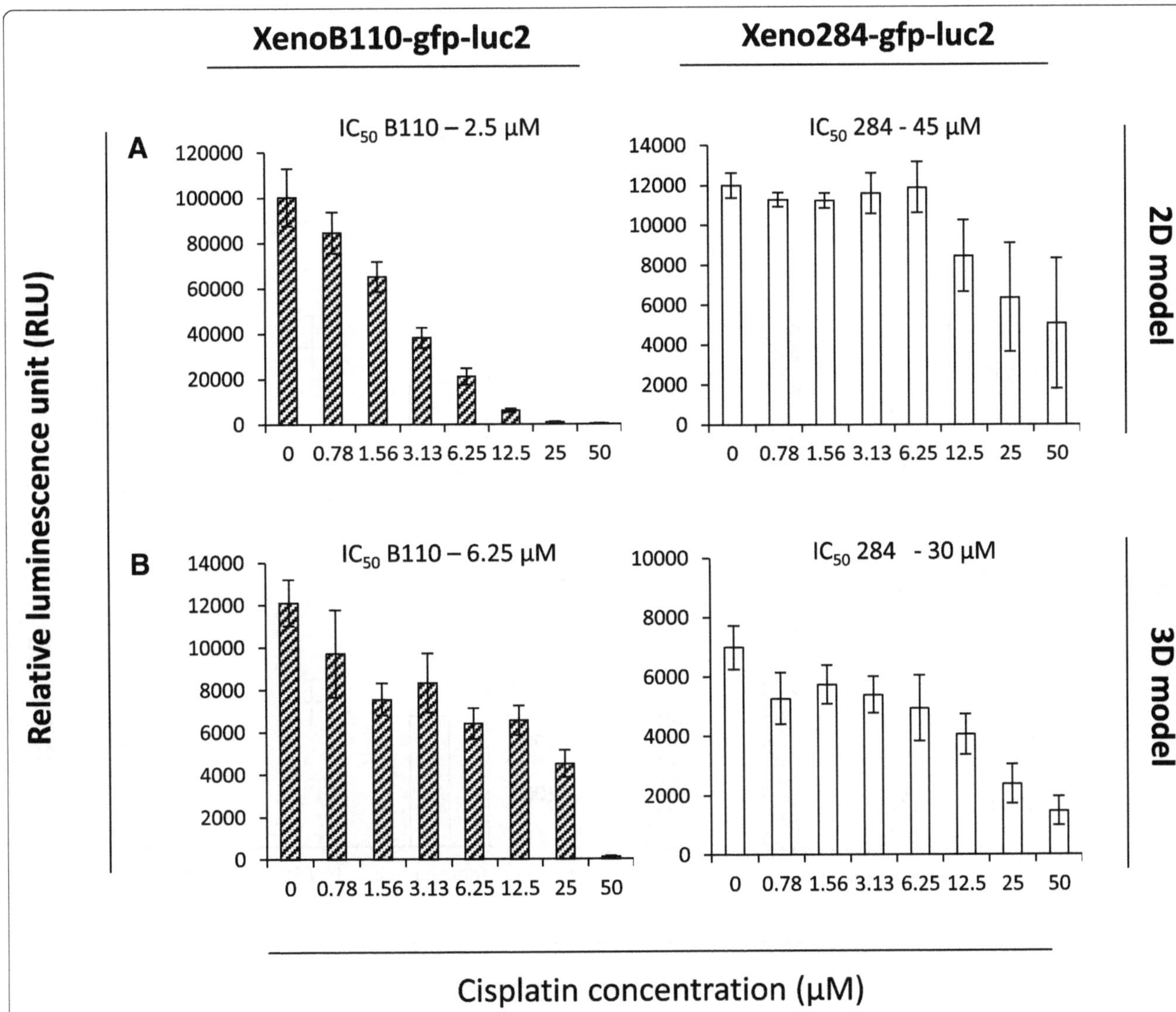


**Fig. 3** Inhibition of NPC xenograft cell viability by cisplatin as measured by XenoLuc assay. Xenograft cells ($1 \times 10^4$ cells/well) were seeded onto 96-well plates (***a***: 2D model; ***b***: 3D model) and treated with various concentrations of cisplatin as described in *Methods*. Luciferase activities were measured at Day 3. Results are represented by average of triplicate from 3 mice ± SD

would be beneficial if it could be used to measure the growth of tumour cells (either inhibition or enhancement) that were co-cultured with other cell types. To validate the use of XenoLuc assay in this application, we co-cultured the xenograft cells with either fibroblasts (NHDFs) (adherent cells) or PBMCs (suspension cells), and measured the luminescence signals to examine the growth inhibition or enhancement of xenograft cells. Figure 4a shows that the NHDFs enhanced xenograft cell growth in 2D culture model. However, only XenoB110-gfp-luc2 cells displayed a significant growth enhancement ($p < 0.05$). The growth enhancement in XenoB110-gfp-luc2 cells was also dependent on the number of seeded NHDFs (Additional file 3: Figure S3A). Co-culture of xenograft cells with PBMCs did not yield any significant difference in the growth as compared to that of monoculture (Fig. 4b and Additional file 3: Figure S3B). We then investigated whether this enhancing effect was reproducible in 3D culture model. Similarly, NHDFs significantly promoted XenoB110-gfp-luc2 cell growth in a seeding number-dependent manner ($p < 0.05$) (Fig. 4b and Additional file 3: Figure S3B). Figure 5 shows the microscopic analysis of GFP-labelled xenograft cells co-cultured with NHDFs and PBMCs individually in 3D models. Xenograft cell-NHDF co-cultures resulted in irregular and multiple spheroids compared to xenograft cell-PBMC co-cultures. Consistent with the luminescence data, the fluorescence intensity analysis showed that NHDFs promoted the growth of XenoB110-gfp-luc2 cells in both 2D and 3D cultures (Additional file 4: Figure S4).

## Discussion

In the present study, we developed a specific and sensitive luciferase-based assay known as XenoLuc to detect and measure real-time proliferation of luc2-transduced cells. The assay used freshly isolated xenograft cells which were passaged for not more than two times. This approach was established to circumvent the difficulty of

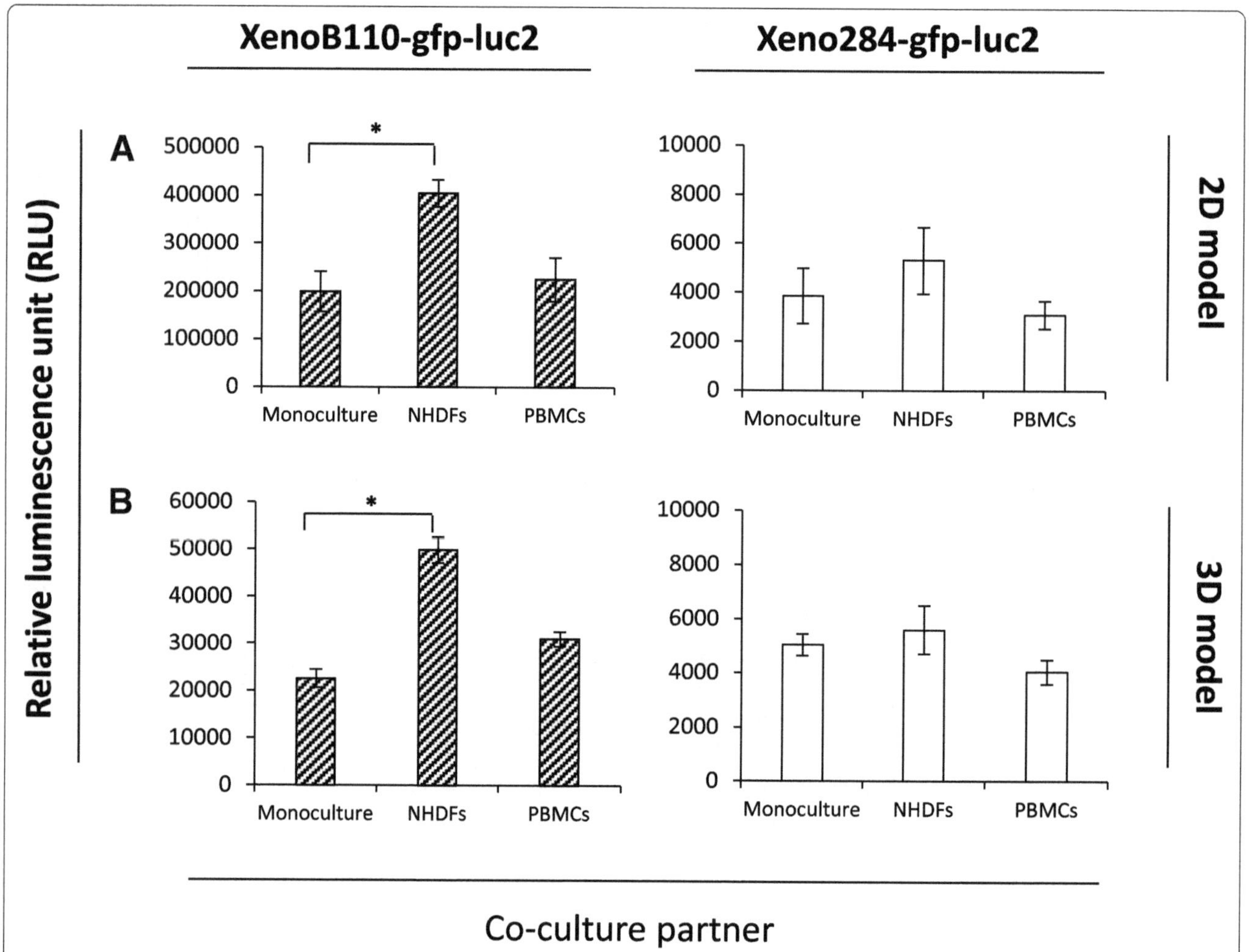


**Fig. 4** Measurement of luc2-modified xenograft cell growth in a co-culture system. Xenograft cells ($1 \times 10^4$ cells/well) were co-cultured with two different cell types as described in *Methods* in (**a**) 2D culture model; and (**b**) 3D culture model. Luminescence was measured at Day 4. Left panel: XenoB110-gfp-luc2; Right panel: Xeno284-gfp-luc2. Results are represented by average of triplicate from 3 mice ± SD. * $p < 0.05$

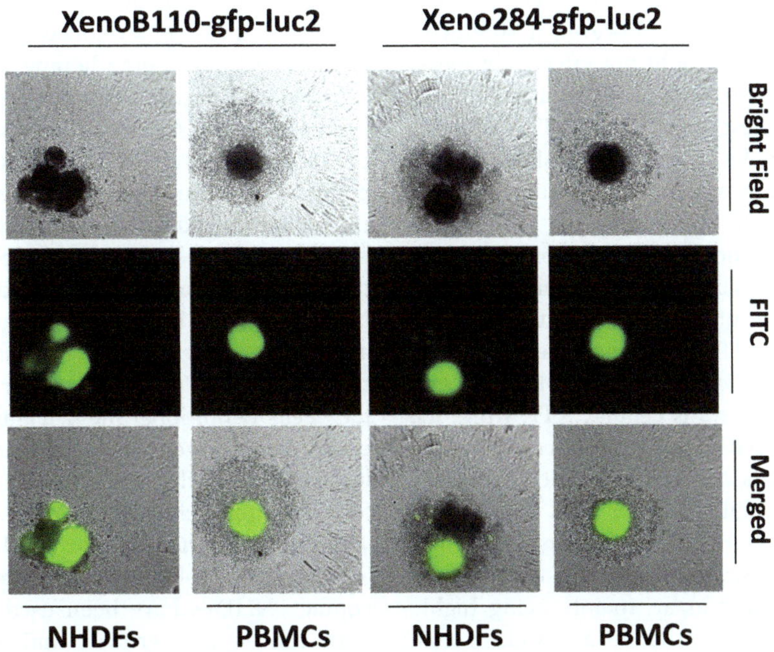


**Fig. 5** Microscopic examination of xenograft cells co-cultured with NHDFs and PBMCs in 3D culture model. 3D spheroids from xenograft cell-NHDF co-culture tended to form irregular shapes unlike those of xenograft cell-PBMC co-culture which were single, smooth and uniformed aggregates. (left panel: XenoB110-gfp-luc2; right panel: Xeno284-gfp-luc2)

establishing stable cell lines of NPC and allowed the use of xenografts which could not be propagated indefinitely as cell lines to be used for in vitro assays. This assay could be performed without lysing the cells, thereby allowing repeated and multiparametric measurements at multiple time-points in a single well (Fig. 1c). This assay could also capture the different growth patterns of culture models as shown in Fig. 1c. Both B110 and 284, the 3D culture models stopped growing at day 4, possibly due to the nutrient deprivation as basal RPMI1640 medium without any supplementation was used for culturing 3D spheroids. The luminescent signal was found to be stable up to 60 min, and D-Luciferin at working concentration was not toxic to the cells after 24-h exposure (Fig. 2c and d). We also showed that the assay is robust and could be applied on different cell model systems. Furthermore, it was reliably used for multiple applications, including drug/ inhibitor screening and growth measurement in a co-culture system (Figs. 3 and 4).

The luciferase/luciferin system has been adopted in various cell-based assays, and is widely used in biomedical research. This assay is sensitive, non-toxic, suitable for various applications and does not interfere with downstream applications [10, 13]. Conventional luciferase-based assays use exogenous or endogenous luciferase to generate bioluminescence upon the addition of an external substrate (luciferin). The amount of emitted light depends on the concentration of ATP, a widely accepted marker of viable cells that is also the limiting reactant in luciferase-based assays. Non-viable cells are unable to synthesize ATP and rapidly exhaust the existing ATPs through their endogenous ATPases. Hence, the amount of cellular ATP in viable cells determines the intensity of luminescent signal which is proportionate to the viable cell number [14, 15].

In addition to luminescence, GFP fluorescence of the transduced cells could also be detected, in contrast to the commercially available luminescence-based assays (Fig. 5 and Additional file 2: Figure S2C). The current assay utilizes dual *gfp* and *luc2* reporters to provide more cellular information. The endogenous luciferase (encoded by *luc2* reporter gene) in viable cells reacts chemically with an addition of luciferin to generate a luminescent signal, whereas the GFP signal (encoded by *gfp* reporter gene) provides a fluorescent visualization of transduced tumour cells. The 2-in-1 detection in XenoLuc assay helps researchers to differentiate cell types in a co-culture system. GFP signal intensity could also be determined using imaging software as a second measure of cell proliferation. It is noteworthy that traditional metabolic-based viability assays such as MTT, MTS and XTT do not discriminate the metabolic activity between cancer and stromal cells when they are cultured together. This results in an inability to measure accurately the viability of either cell population when the above metabolic assays are used in co-culture systems [16]. As luciferase and GFP expressions are confined within transduced cancer cells, measurement of luminescence and/or GFP fluorescence will accurately reflect cell proliferation changes in co-cultures. However, it should be noted that GFP fluorescence could not effectively measure the growth of 3D spheroid culture (Additional file 2: Figure S2C) unlike luminescence (Fig. 1c). This suggests that GFP fluorescence is less sensitive, possibly because the GFP signals do not penetrate well enough through the cells from within the 3D spheroids.

To demonstrate that XenoLuc assay is comparable, if not better than commercially available luminescent assays, we performed some of the experiments using our assay in parallel with CellTiter-Glo and RealTime-Glo kits from Promega, USA. Of note, the rate-limiting factor of both commercial assays is cellular ATP, while the supplied luciferin and luciferase are in excess. On the other hand, the limiting factor in XenoLuc could be the endogenous luciferase, cellular ATP, or both (Fig. 2a). By using these commercial kits as a benchmark, we showed that XenoLuc assay met several important criteria as a cell proliferation or viability assay, mainly being highly sensitive, rapid, non-toxic, quantitative, and yields stable signals (Fig. 2b–d). In addition, a comparison of the features among these assays was tabulated in Table 1. Despite having lower signal intensity, XenoLuc assay has several notable advantages over other luminescent assays which include having a lower cost per reaction but with its performance and robustness comparable to that of commercial assays. Thus, the XenoLuc assay appears to be the more cost-effective option for researchers. More importantly, XenoLuc assay may be used to measure specifically the proliferation of *gfp* and *luc2*-labeled cells of interest in multi-cellular systems such as co-cultures, unlike existing luminescent assays.

As a proof that XenoLuc assay could be a reliable tool for high-throughput screening for drug discovery, we tested the effect of cisplatin on the viability of our NPC xenograft cells. XenoLuc assay was able to show that although both transduced xenograft cells were sensitive to cisplatin, Xeno284-gfp-luc2 cells displayed some resistance to it as consistently observed in both 2D and 3D cultures. This observation could be attributed to the difference in disease type: Xeno-284 was established from recurrent NPC while XenoB110 was from untreated primary NPC. In the former, we speculate that the patient might have developed cisplatin resistance as a result of repetitive chemotherapy.

In the past decades, tumour cell monoculture has been used as a model for drug screening and the study of cancer biology. As increasing attention is drawn towards the complexity of tumour microenvironment, a co-culture model is believed to be more physiologically relevant for cancer research [2, 3]. The growth of XenoB110-gfp-luc2 cells was markedly enhanced when co-cultured with NHDFs (Fig. 4). In addition to luminescence data, it was also validated by the GFP fluorescence intensity analysis that the NHDFs promoted the growth of XenoB110-gfp-luc2 cells (Additional file 4: Figure S4). Unlike luminescence measurement that is in direct proportion to ATP levels [17], the fluorescence intensity analysis is not affected by the cellular ATP levels from the co-cultures. This growth enhancement was likely resulted from the activation of proliferative pathways upon direct cell-cell contact. NHDFs have been used as feeder cells to support cancer cell growth and to simulate epithelial-stromal cell interaction and paracrine signalling in an in vitro culture setting [18–21]. In addition, we also used PBMCs as the co-culture partner for xenograft cells. There were varying results in the growth of NPC xenograft cells when cultured in 2D and 3D models (Fig. 4). It could be due to the heterogeneity within PBMCs population as well as differential cell-cell contacts due to culture techniques.

## Conclusions

In conclusion, we have developed a specific, sensitive, real-time, non-toxic, and cost-effective luminescent assay to measure the proliferation of *luc2*-transduced cells. This robust assay known as XenoLuc assay could be used for multiple applications ranging from a simple experiment such as cell doubling time measurement, to complicated studies such as cell activation and proliferation or drug sensitivity/resistance screening in a co-culture system. This assay may serve as an important tool to specifically examine the role of a selected cell types in complex co-culture systems as well as for drug screening experiments.

**Table 1** Comparison of XenoLuc assay with commercial luminescent assays

| Assay | Lytic/ Non-lytic | Price per reaction[a] | Incubation time | Signal intensity | Signal stability | Linearity ($R^2$) |
|---|---|---|---|---|---|---|
| XenoLuc (In-house) | Lytic | RM 0.013 (USD 0.003) | 15 min | 36,133 | > 1 h | 0.9915 |
| CellTiter-Glo (Promega, USA) | Lytic | RM 0.20 (USD 0.05) | 1 h 15 min | 165,000 | > 1 h | 0.9963 |
| XenoLuc (In-house) | Non-lytic | RM 0.013 (USD 0.003) | 10 min | 120,960 | ≥ 1 h | 0.999 |
| RealTime-Glo (Promega, USA) | Non-lytic | RM 1.86 (USD 0.46) | 1 h 15 min | 279,400 | ≥ 1 h | 0.9835 |

[a]Price acquired from Malaysia's distributor

## Methods

### Plasmid constructs and reagents

Lentiviral constructs (pMDL, RSV-REV, and VSVg) and gfp-luc2 DNA transfer plasmid were kindly provided by Marco Herold (Walter and Eliza Hall Institute of Medical Research, Melbourne, Australia). RPMI 1640 (#31800–022), fetal bovine serum (FBS) (#10082–139), GlutaMAX supplement (#35050–061), 0.05% Trypsin-EDTA (#25300–120), Pen-Strep (#15140–122), Pen-Strep-Fungizone (#15240–062), B-27 supplement (#17504–001), Insulin-Transferrin-Selenium (ITS) (#41400–045), Fibroblast growth factor (FGF)-Basic (bFGF) (#PHG0261), and Epidermal growth factor (EGF) (#PHG0311) were obtained from Gibco, USA. Hydrocortisone (#H0135), Collagenase Type II (#C6885), DPBS (#D5652), HEPES (#H3375), and Sodium Bicarbonate (#S5761) were purchased from Sigma-Aldrich, USA. DNase I (#90083) and Lipofectamine 2000 (#11668–027) were purchased from Thermo Fisher Scientific, USA. Collagenase/Dispase (#11097113001) was obtained from Roche, USA. Rho kinase (ROCK) inhibitor (#SCM075) and Polybrene (#TR-1003-G) were obtained from Merck Millipore, USA. XenoLight D-Luciferin substrate (#122799) was purchased from PerkinElmer, USA and stored in small aliquots at − 20 °C in the dark. CellTiter 96 AQueous One Solution Cell Proliferation Assay (MTS) (#G3580), CellTiter-Glo Luminescent Cell Viability Assay (#G7571), and RealTime-Glo MT Cell Viability Assay (#G9711) were from Promega, USA. Cisplatin (#15663–27-1) was purchased from Acros Organics, USA. RBC lysis solution was purchased from Qiagen, USA. All reagents were dissolved, stored, and used according to the manufacturer's instruction.

### Cells and culture conditions

Primary normal human dermal fibroblasts (NHDFs) (#C-12300) were obtained from PromoCell, Germany and maintained using the medium and supplements supplied in the fibroblast growth medium kit (#C-23110) (PromoCell, Germany). HEK293T cells (#CRL-3216, ATCC, USA) were maintained in high glucose DMEM supplemented with 10% FBS (DMEM-10). Human peripheral blood mononuclear cells (PBMCs) were isolated from the blood samples collected from three donors via Ficoll-Paque (GE Healthcare, USA) centrifugation. All donors had agreed and signed the consent forms.

### Tissue dissociation and xenograft cell culture

Human NPC xenografts (XenoB110 and Xeno284) from NSG mice were processed as previously described by our group [8, 22]. All experiments were performed in accordance with the protocols approved by Animal Care and Use Committee (ACUC), Ministry of Health, Malaysia. Briefly, mice were humanely euthanized using carbon dioxide gas (flow rate of between 1.3 to 3.8 L/min) or cervical dislocation by trained personnel using protocols approved by the Animal Care and Use Committee. Harvested xenograft tumours were transferred to cold DPBS supplemented with 1X Pen-Strep-Fungizone, and processed immediately. Surrounding blood capillaries, fat, and/or necrotic tissue were removed and washed twice with DPBS with 1X Pen-Strep-Fungizone. The tumours were minced into 2 to 4-mm fragments and incubated with the appropriate dissociation solution (2 mg/mL Collagenase Type II and 200 U/mL DNase I for XenoB110; 1 mg/mL Collagenase/Dispase for Xeno284) on a belly dancer shaker with a constant agitation for 1–2 h at 37 °C, 5% $CO_2$. This was followed with an addition of complete growth media (RPMI-10, 1X Pen-Strep-Fungizone, 1X B-27 Supplement, 1X ITS, 10 μM ROCK Inhibitor, 10 ng/mL EGF, and 10 ng/mL bFGF), after which the entire suspension was filtered through 40 μm nylon mesh cell strainer (BD Falcon, USA). The released cells were centrifuged at 800 rpm for 5 min. RBC lysis solution was added onto the cells prior to centrifugation at 800 rpm for 5 min. The cells were then processed with the mouse cell depletion kit (MACS Miltenyi Biotec #130–104-694) following the manufacturer's instruction to enrich for human NPC xenograft cells. The enriched cells were cultured in the above mentioned complete growth medium for 2D monolayer culture, whereas basal RPMI-10 was used for 3D spheroid culture. The cells were detached using 0.05% Trypsin-EDTA at 70–80% confluence and were sub-cultured at 1:3 dilutions. To our best experience, xenograft cells could be maintained up to passage four (P4) in vitro in monolayer culture, and spheroids could be formed by medium exchange and using a commercial spheroid plate. To minimize biological variations, we only used freshly isolated xenograft cells which were passaged for up to 2 times for experiments. We used the terms xenograft cells for these cells in contrast to xenograft tumours in mice.

### Lentivirus production and cell transduction

$1 \times 10^6$ HEK293T cells were seeded on 10-cm culture dishes and incubated for 24 h. The gfp-luc2 lentiviral transfer vector with packaging and envelope plasmids (pMDL, RSV-REV and VSVg) were combined at a ratio of 4:2:1:1, respectively, and mixed with Lipofectamine 2000 for transfection according to the manufacturer's protocol. The viral supernatant was collected at 24, 48 and 72 h post-transfection. Cellular debris was pelleted by centrifugation at 1500 rpm for 5 min at 4 °C followed by filtration using PVDF Millex-HV filter, 0.45 μm (Millipore #SLHV033RS). The filtered lentivirus supernatant was then concentrated using Lenti-X Concentrator (Clontech #PT4421–2) according to the manufacturer's instruction. For cell transduction, the mouse cell-depleted xenograft cells were seeded on a 10-cm culture dish and transduced with the concentrated lentivirus at MOI 2.0 in the presence of 10 µg/mL Polybrene for 24 h. GFP positive cells were selected using a FACSAria SORP (BD Biosciences, USA).

### XenoLuc assay

XenoLuc assay was performed as follows for both 2D and 3D culture experiments. At each time point, 2X D-Luciferin substrate diluted in RPMI-10 was dispensed into each cell-containing well at 1:1 ratio. The plate was gently agitated and incubated at room temperature in the dark for 10 min to stabilize the luminescence. The luminescent signal of each plate was then read by EnVision multi-label plate reader (PerkinElmer) using the ultrasensitive mode. Following the completion of a time-point reading, the D-Luciferin containing medium was removed from the well. The cells were washed two times gently with 200 µL RPMI-10, and then replenished with 100 µL of fresh complete media until the subsequent reading. Alternatively, images of GFP-expressing xenograft cells from each well were captured using IN Cell Analyzer 2000, and the GFP fluorescence intensity was measured and compared using the IN Cell Developer software. For the lytic-based format of XenoLuc assay, the protocol described by Oba and co-workers was modified and adapted [23]. Briefly, PBS supplemented with 1% Triton X-100 and 1X protease inhibitor cocktail (Millipore #539134) was used as cell lysis buffer and the lysis was performed for 10 min. After lysis, the assay buffer containing 5 mM $MgCl_2$, 100 mM Tris-HCl (pH 7.8) and 2X D-Luciferin substrate was added to the well to generate the luminescence. The plate was immediately measured after 15-min incubation at room temperature in the dark.

### CellTiter-Glo and RealTime-Glo

CellTiter-Glo (end-point assay) and RealTime-Glo (continuous assay) were performed according to the manufacturer's instructions. To examine if the XenoLight D-Luciferin substrate reagent was toxic to the xenograft cells, cell viability assay (CellTiter 96 Aqueous One) was performed on xenograft cells which had been exposed to the substrate for 24 h.

### Applications of XenoLuc assay

We evaluated the suitability of the assay for inhibitor/drug screening and co-culture studies using both 2D and 3D culture models. For the drug screening study, cisplatin was used as the standard drug. Xenograft cells were similarly plated in 2D and 3D culture conditions as described above. Cells were then treated with various concentrations of cisplatin (0, 0.78, 1.56, 3.13, 6.25, 12.5, 25, and 50 µM) for 72 h followed by the XenoLuc assay. Prior to co-culturing, both NHDFs and PBMCs were gamma-irradiated separately at 35 Gy. For both 2D and 3D cultures, 10,000 xenograft cells were seeded simultaneously with the irradiated co-culture partner cells at 1:1 ratio into the same well. After an incubation of 4 days, XenoLuc assay was performed to measure the proliferation of xenograft cells in the presence of either irradiated NHDFs or PBMCs.

### Statistical analysis

For each experiment using xenograft cells, three tumours were harvested from three separate mice, and each resulting xenograft cells were seeded into three wells per isolate of xenograft cells for subsequent experiments. Each data point was represented by average of triplicate from 3 mice ± SD. GraphPad Prism 6 software (GraphPad Software Inc.) was used to calculate the $IC_{50}$ value of cisplatin against xenograft cells. $R^2$ value was also determined to show the linear association between two parameters in a linear curve. SPSS version 22.0 software (IBM SPSS, USA) was used for analysis of statistical significance. One-way analysis of variance with Bonferroni correction and Student's $t$ test were used to determine if the differences between two different groups are significant. The results presented are mean $\pm$ standard deviations. $p < 0.05$ was considered statistically significant.

## Additional files

**Additional file 1: Figure S1.** Relative luminescence units (RLUs) signals of non-adhering B110 xenografts cells. Luciferase activities were measured 30 min after cell seeding (day-0 reading). (A) XenoB110-gfp-luc2; (B) Xeno284-gfp-luc2. (PPTX 41 kb)

**Additional file 2: Figure S2.** Cell proliferation analysis of XenoB110-gfp-luc2 by GFP fluorescence intensity using IN-CELL Developer software. (A) Non-depleted xenograft cells; (B) Mouse cell-depleted xenograft cells; (C) 3D spheroids. The images inset shows an increasing size of spheroids captured at day 4 proportionate to cell seeding number. $R^2$ shows the linear correlation of luminescence and cell number. (PPTX 62 kb)

**Additional file 3: Figure S3.** Evaluation of xenograft cell growth by co-culture in 2D and 3D models. Xenograft cells ($1 \times 10^4$ cells/well) were co-cultured with various cell number of NHDFs and PBMCs in both (A) 2D; and (B) 3D model. Left panel: XenoB110-gfp-luc2; Right panel: Xeno284-gfp-luc2. (PPTX 44 kb)

**Additional file 4: Figure S4.** Analysis of XenoB110-gfp-luc2 cell growth in a co-culture system by GFP fluorescence intensity using IN-CELL Developer software. XenoB110-gfp-luc2 cells ($1 \times 10^4$ cells/well) were co-cultured with two different cell types as described in *Methods* in (A) 2D culture model; and (B) 3D culture model. GFP fluorescence intensity was determined at Day 4. (PPTX 38 kb)

#### Abbreviations

2D: Two-dimensional; 3D: Three-dimensional; ATP: adenosine triphosphate; EBER: Epstein-Barr virus-encoded small RNA; EBV: Epstein-Barr virus; ECM: extracellular matrix; EGF: epidermal growth factor; FACS: fluorescence-activated cell sorting; FGF: fibroblast growth factor; GFP: green fluorescence protein; H&E: Haematoxylin and Eosin; ITS: Insulin-Transferrin-Selenium; LCA: Leukocyte common antigen; luc: luciferase; NHDFs: normal human dermal fibroblasts; NPC: nasopharyngeal carcinoma; NSG: NOD *scid* gamma; PAN-CK: Cytokeratin Pan; PBMC: peripheral blood mononuclear cell; PBS: phosphate buffered saline; PDTX: patient-derived tumour xenograft; RBC: red blood cells; RFU: relative fluorescence unit; RLU: relative luminescence unit; ROCK: Rho kinase

**Acknowledgements**
The authors would like to thank the Director General of Health Malaysia for his permission to publish this study and the Director of the Institute for Medical Research for her support. We would like to acknowledge Specific Pathogen Free facility, Biobank and Molecular Pathology Unit, IMR staff for their technical assistance and support in this project as well as Dr. Susan LL Hoe for her critical appraisal of the manuscript.

**Authors' contributions**
SYT, KL, MFCM and CTL performed the experiments. SYT, KL, MFCM, NAA and MM analysed the data. SYT and KL drafted the manuscript. SYT, MA, and ASBK conceived the study, participated in its design and reviewed the manuscript. All authors read and approved the final manuscript.

**Funding**
This work was funded by the Ministry of Health Malaysia. The funder did not have any role in the design of the study, data collection, analysis, and interpretation of data or in writing the manuscript.

**Ethics approval**
All procedures and experimental protocols involving the use of NPC PDXs in this study were in accordance with the ethical standards of and approved by the Animal Care and Use Committee (ACUC), Ministry of Health, Malaysia [(ACUC/KKM/02 (03/2013), (05/2016), (16/2016) and (03/2017)]. The use of human blood specimens in this study was approved by the Medical Research and Ethics Committee (MREC), Ministry of Health, Malaysia [reference number: KKM/NIHSEC/P18-1062(10)]. Informed consent was obtained from all participating human subjects.

**Consent for publication**
Not applicable.

**Competing interests**
SYT, MFCM, MM, NAA, TLC, MA, ASBK are co-inventors of a patent filed related to the contents of this manuscript. (Khoo ASB, Norazlin AA, Teow SY, Mohd Firdaus CM, Marini M, Chu TL, Munirah A. Isolated nasopharyngeal carcinoma cells and derivatives prepared thereof. Malaysian Patent PI 2018702497 filed July 17, 2018 (Patent Pending).

**Author details**
[1]Molecular Pathology Unit, Cancer Research Centre, Institute for Medical Research, National Institutes of Health (NIH Complex), Ministry of Health Malaysia, Level 4, Block C7, No: 1, Jalan Setia Murni U13/52, Section U13, Setia Alam, 40170 Shah Alam, 50588 Kuala Lumpur, Selangor, Malaysia. [2]Present Address: Department of Medical Sciences, School of Healthcare and Medical Sciences, Sunway University, 47500 Bandar Sunway, Selangor Darul Ehsan, Malaysia.

**References**

1. Zang R, Li D, Tang IC, Wang J, Yang ST. Cell-based assays in high-throughput screening for drug discovery. Int J Biotechnol Wellness Ind. 2012;1:31–51.
2. Lovitt C, Shelper T, Avery V. Advanced cell culture techniques for cancer drug discovery. Biology. 2014;3(2):345–67.
3. Weigelt B, Ghajar C, Bissell M. The need for complex 3D culture models to unravel novel pathways and identify accurate biomarkers in breast cancer. Adv Drug Deliv Rev. 2014;69-70:42–51.
4. Zhang Y, Schedle A, Matejk M, Rausch-Fan X, Andrukhov O. The proliferation and differentiation of osteoblasts in co-culture with human umbilical vein endothelial cells: an improved analysis using fluorescence-activated cell sorting. Cell Mol Biol Lett. 2010;15(4):517–29.
5. Goers L, Freemont P, Polizzi K. Co-culture systems and technologies: taking synthetic biology to the next level. J R Soc Interface. 2014;11(96):20140065.
6. Edmondson R, Broglie J, Adcock A, Yang L. Three-dimensional cell culture systems and their applications in drug discovery and cell-based biosensors. Assay Drug Dev Technol. 2014;12(4):207–18.
7. Breslin S, O'Driscoll L. Three-dimensional cell culture: the missing link in drug discovery. Drug Discov Today. 2013;18(5–6):240–9.
8. Hoe SLL, Tan LP, Aziz NA, Liew K, Teow SY, Razak FRA, et al. CD24, CD44 and Epcam enrich for tumour-initiating cells in a newly established patient-derived xenograft of nasopharyngeal carcinoma. Sci Rep. 2017;7(1):12372.
9. Leitão JM, Esteves da Silva JC. Firefly luciferase inhibition. J Photochem Photobiol B. 2010;101(1):1–8.
10. Duellman SJ, Zhou W, Meisenheimer P, Vidugiris G, Cali JJ, Gautam P, et al. Bioluminescent, nonlytic, real-time cell viability assay and use in inhibitor screening. Assay Drug Dev Technol. 2015;13(8):456–65.
11. Fraga H. Firefly luminescence: a historical perspective and recent developments. Photochem Photobiol Sci. 2008;7(2):146–58.
12. Tiffen JC, Bailey CG, Ng C, Rasko JE, Holst J. Luciferase expression and bioluminescence does not affect tumor cell growth in vitro or in vivo. Mol Cancer. 2010;9:299.
13. Adams ST, Miller SC. Beyond D-luciferin: expanding the scope of bioluminescence imaging in vivo. Curr Opin Chem Biol. 2014;0:112–20.
14. Riss TL, Moravec RA, Niles AL, Duellman S, Benink HA, Worzella TJ, et al. Cell viability assays. 2013 may 1 [updated 2016 Jul 1]. In: Sittampalam GS, Coussens NP, Nelson H, et al., editors. Assay guidance manual. Bethesda: Eli Lilly & Company and the National Center for advancing translational Sciences; 2004.
15. Niles A, Moravec R, Riss T. In vitro viability and cytotoxicity testing and same-well multi-parametric combinations for high throughput screening. Curr Chem Genomics. 2009;3:33–41.
16. Yang N, Yan T, Zhu H, Liang X, Leiss L, Sakariassen P, et al. A co-culture model with brain tumor-specific bioluminescence demonstrates astrocyte-induced drug resistance in glioblastoma. J Transl Med. 2014;12:278.
17. Posimo JM, Unnithan AS, Gleixner AM, et al. Viability assays for cells in culture. J Vis Exp. 2014;83:e50645.
18. Friboulet L, Pioche-Durieu C, Rodriguez S, et al. Recurrent overexpression of c-IAP2 in EBV-associated nasopharyngeal carcinomas: critical role in resistance to toll-like receptor 3-mediated apoptosis. Neoplasia. 2008;10(11): 1183–94.
19. Vérillaud B, Gressette M, Morel Y, et al. Toll-like receptor 3 in Epstein-Barr virus-associated nasopharyngeal carcinomas: consistent expression and cytotoxic effects of its synthetic ligand poly(a:U) combined to a Smac-mimetic. Infect Agents Cancer. 2012;7:36.
20. Fang X, Sittadjody S, Gyabaah K, Opara E, Balaji K. Novel 3D co-culture model for epithelial-stromal cells interaction in prostate cancer. PLoS One. 2013;8(9):e75187.
21. Kayamori K, Katsube K, Sakamoto K, Ohyama Y, Hirai H, Yukimori A, et al. NOTCH3 is induced in cancer-associated fibroblasts and promotes angiogenesis in oral squamous cell carcinoma. PLoS One. 2016;11(4): e0154112.
22. Hoe SLL, Tan LP, Jamal J, Peh SC, Ng CC, Zhang WC, et al. Evaluation of stem-like side population cells in a recurrent nasopharyngeal carcinoma cell line. Cancer Cell Int. 2014;14:101.
23. Oba Y, Ojika M, Inouye S. Firefly luciferase is a bifunctional enzyme: ATP-dependent monooxygenase and a long chain fatty acyl-CoA synthetase. FEBS Lett. 2003;540(1):251–4.

9

# Characterization of tethered equine chorionic gonadotropin and its deglycosylated mutants by ovulation stimulation in mice

Kwan-Sik Min[1,2*], Jong-Ju Park[1], Munkhzaya Byambaragchaa[1] and Myung-Hwa Kang[3]

**Abstract**

**Background:** To directly assess the biological role of oligosaccharides in recombinant equine chorionic gonadotropin (rec-eCG) functioning, cDNA encoding the full-length eCGβ-subunit was fused with the mature protein part of the α-subunit, and we examined the expression levels of deglycosylated eCG mutants, the ovulation rate for deglycosylated mutants in C57BL/6 mice.

**Results:** The characterizations of heterodimeric and tethered mutants were studied following their respective secretions in culture medium, molecular weight and ovulation in vivo. Rec-eCG variants containing mutations at glycosylation sites at Asn82 of the α-subunit (eCGβ/αΔ82) and Asn13 of the β-subunit (eCGβΔ13/α) were not efficiently secreted into the culture medium from transfected cells. Western blot analysis revealed that the rec-eCGβ/α proteins have an approximate broad range of molecular weights of 40–46 kDa. Three rec-eCG mutants—a deglycosylated site at Asn56 of the α-subunit (eCGβ/αΔ56), a deletion of the C-terminal region of the β-subunit (eCGβ-D/α), and the double mutant (eCGβ-D/αΔ56)—turned out to have clearly lower (approximately 4–23 kDa) molecular weights. Protein N-glycosydase F (PNGase F) treatment markedly decreased the molecular weight to approximately 2–10 kDa. Normal oocytes were significantly more abundant in the natural eCG–treated group than in mutant rec-eCG–treated groups. In particular, numbers of nonfuntional oocytes were remarkably lower in all rec-eCG groups.

**Conclusions:** Our results indicate that the ovulation rates of oocytes are not affected by the deglycosylated rec-eCGβ/α mutant proteins. There are around 20% non-functional oocytes with natural eCG and only 2% with the rec-eCGs tested. These results provide insight into the molecular mechanisms underlying the production of rec-eCG hormones with excellent bioactivity in vivo.

**Keywords:** Rec-eCG, Glycosylated sites, Ovulation rate

## Background

Chorionic gonadotropin (CG) is a placental hormone that maintains the corpus luteum (CL) during pregnancy [1]. CG exist only in primates and equidaes and not in the other mammals. Gonadotropins are heterodimeric glycoprotein hormones, consisting of dissimilar α- and β-subunits that are noncovalently complexed [2]. The α-subunit of the glycoproteins is common for luteinizing hormone (LH), follicle-stimulating hormone (FSH), and thyroid-stimulating hormone (TSH) in a given species [3].

Equine CG (eCG) is a unique member of the gonadotropin family because it exhibits both LH- and FSH-like activities in non-equid species [4, 5]. The β-subunits of eCG and equine LH (eLH), being translated from the same gene, have an identical primary structure [6, 7]. Thus, eCG may be an ideal model for studying the structure-function

* Correspondence: ksmin@hknu.ac.kr
[1]Animal Biotechnology, Graduate School of Future Convergence Technology, Institute of Genetic Engineering, Hankyong National University, Ansung 17579, Korea
[2]Department of Animal Resource Science, Hankyong National University, Ansung 17579, Korea

relations of gonadotropins because it possesses properties of both its pituitary and placental counterparts [8, 9]. The difference between eCG and eLH lies in the structure of their carbohydrates, which are sialylated and sulfated in LH but only sialylated in CG [10, 11]. eCG is secreted from binucleate trophoblastic cells in endometrial cups, into maternal blood plasma during the first half of equine gestation. These cells get detached from the chorionic girdle of the conceptus between days 37 and 120 of pregnancy [12–14].

eCG administration has also been associated with an increase in ovulation rate [15], particularly in early-postpartum cows [16]. The standard dose of eCG required to promote single ovulation generally ranges between 200 and 1000 IU, whereas the dose necessary to induce superovulation is approximately 2500 IU [17]. Similarly, eCG administration to sheep has been demonstrated to significantly increase the level of proteins FSH receptor (FSHR) and gonadotropin-releasing hormone receptor (GnRHR) secreted by oocytes [18]. Thus, eCG enhances maturation and stimulates FSHR, LHR, and GnRHR expression. In some studies on the secretion and activity of recombinant eCG (rec-eCG) in mammalian cells, deletion of carboxy-terminal peptides (CTPs) from dimeric eCG induced a 50% decrease in the secretion of the truncated hormone as compared to the wildtype [19]. FSH activity strongly depends on amino acid residues (aa) 102–104 of the eCGβ-subunit [20], and correct folding and FSH activity are conferred by aa 104–109 of the eCGβ-subunit [9].

Some studies have revealed that glycosylation of α-subunit position 52 in human FSH (hFSH) [21], human chorionic gonadotropin (hCG) [22], and hTSH [23] is important for signal transduction, because cAMP or steroid formation is not stimulated, while the binding activity with FSHR was enhanced by 2- to 3-fold [24]. Thus, receptor binding and signal transduction are dissociable functions involving different sites on the FSH glycoprotein. Various studies suggest that CTPs of the deletion-containing (aa 115–145) hCGβ-subunit [25, 26], and deletion-containing (aa 122–145) hCGβ [27] are not important for receptor binding or in vitro signal transduction. Nontheless, the truncated form of hCGβ lacking aa 101–145 is the shortest form of the subunit known to retain biological activity [28]. In eLH, the deletion containing (aa 121–149) β-subunit is incapable of subunit association and receptor binding [29]. These results are consistent with findings of our previous research into the effects of deglycosylated rec-eCG mutants on estradiol and progesterone stimulation of rat granulosa cells and Leydig cells [5, 30].

To investigate the functional contribution of oligosaccharides in rec-eCGs and of carboxy-terminal extension (aa 114–149) of the eCGβ-subunit, we created a total of 20 expression vectors encoding 11 heterodimeric eCGs and nine tethered eCGs. We produced rec-eCGα/β and eCGβ/α proteins in CHO-K1 cells, characterized these proteins' biological activities in vivo. Our results indicate that the loss of glycosylation at site Asn82 of the α-subunit and at Asn13 of the β-subunit plays a pivotal role in secretion into the culture medium of CHO-K1 mammalian cells. Deglycosylated rec-eCGs showed a high oocyte ovulation rate in vivo.

## Results

### Quantities of heterodimeric rec-eCGα/β and tethered rec-eCGβ/α proteins

Oligosaccharide-directed mutagenesis was carried out examine the functional importance of oligosaccharides in eCG bioactivity. The eCGα-subunit contains two N-linked glycosylation sites at aa positions 56 and 82. The eCGβ-subunit contains one N-linked glycosylation site at aa position 13 and approximately 11 O-linked glycosylation sites in the C-terminal region. Thus, we constructed 20 expression vectors encoding 11 heterodimeric eCG mutants and nine tethered eCG mutants, respectively (Fig. 1).

The expression vectors were transiently transfected into cells and the culture supernatant was collected at 72 h after transfection. Stable colonies resistant to G418 were selected, and rec-eCGs secreted into the serum-free medium were collected and concentrated. The rec-eCGs were quantified by a pregnant mare serum gonadotropin (PMSG) enzyme-linked immunosorbent assay (PMSG ELISA). Although the rec-eCG protein was detected at a low concentration when it contained Asn13 in the eCGβ-subunit, the protein (deglycosylated mutant) was nearly undetectable when it contained a mutation of Asn82 in the eCGα-subunit and a mutation of Asn13 in the eCGβ-subunit (Fig. 2a, b). Expression levels of the heterodimeric rec-eCGs (α/β, αΔ56/β, α/β-D, and αΔ56/β-D) were 273 ± 38.1, 313 ± 53.5, 304 ± 47.1, and 294 ± 28.3 ng/mL, respectively. Expression levels of heterodimeric αΔ82/β and α/βΔ13 were 21 ± 2.5 and 80 ± 23.8 ng/mL, respectively. These levels were slightly lower for the tethered rec-eCGs. The secreted quantities of rec-eCG β/α, β/αΔ56, β-D/α, and β-D/αΔ56 were 136 ± 13.7, 255 ± 28.3, 154 ± 30.3, and 135 ± 8.8 ng/mL, respectively. Nevertheless, the secretion pattern was the same as that of dimeric eCG. These results indicate that a loss of glycosylation at site Asn82 of the α-subunit and at Asn13 of the β-subunit plays a pivotal role in secretion into the culture medium from CHO-K1 cells. On the other hand, deletion of the eCGβ-subunit CTP did not affect the secretion into the culture medium from CHO-K1 cells. We also compared the expression between mutants eCGβ/α and *myc* eCGβ/α; the latter contains a *myc*tag (Fig. 2c). The expression levles of these proteins were nearly the same.

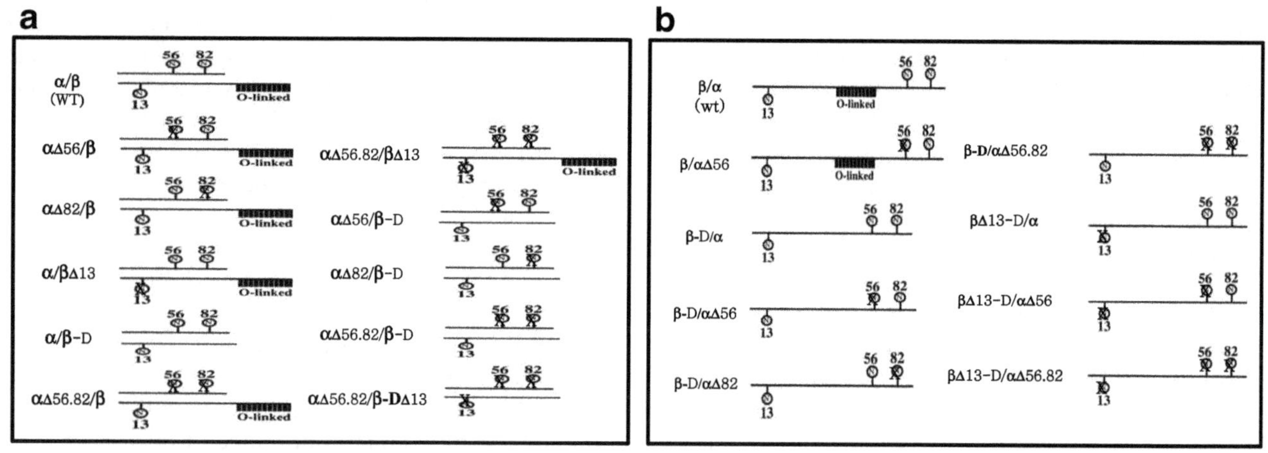


**Fig. 1** A schematic diagram of rec-eCGα/β mutants (**a**) and rec-eCGβ/α mutants (**b**). The wild-type protein and mutants with changed N- and O-linked oligosaccharide sites on eCG are shown. The Asn56 and 82 codons in the α-subunit were replaced by Gln or the CTP carrying O-linked oligosaccharides in the eCGβ-subunit was deleted by PCR mediated site-directed mutagenesis. The circle "N" denotes an N-linked oligosaccharide, "X" indicates the absence of an oligosaccharide, and "O-linked" represents an O-linked oligosaccharide. A total of 20 expression vectors (for 11 heterodimeric eCGs and nine tethered eCGs) were constructed (plasmids encoding heterodimeric eCGs were designated as pAB-eCGα/β, pAB-αΔ56/β, pAB-αΔ82/β, pAB-α/βΔ13, pAB-α/β-D, pAB-αΔ56,82/β, pAB-αΔ56/βΔ13, pAB-αΔ56/β-D, pAB-αΔ82/β-D, pAB-αΔ56.82/β-D, and pAB-αΔ56.82/βΔ13; plasmids encoding tethered eCGs were designated as pcDNA3-eCGβ/α, pcDNA3-β/αΔ56, pcDNA3-β-D/α, pcDNA3-β-D/αΔ56, pcDNA3-β-D/αΔ82, pcDNA3-β-D/αΔ56.82, pcDNA3-βΔ13-D/α, pcDNA3-βΔ13-D/αΔ56, and pcDNA3-βΔ13-D/αΔ56.82)

### Western blot analysis of rec-eCGβ/α glycosylation mutants

Next, we studied the tethered rec-eCG mutants to determine molecular weight and biological activity. By western blot analysis, we found an approximate range of molecular weight of 40–46 kDa for rec-eCGβ/α, as shown in Fig. 3a. Rec-eCGβ/αΔ56, without N-linked oligosaccharides at Asn56 of the α-subunit, was reduced to approximately 37–42 kDa, a decrease of ~ 3–4 kDa. Rec-eCGβ-D/α, with a deletiion of 35 aa including the O-glycosylated CTP of the β-subunit, remarkably diminished to approximately 27 kDa. The molecular weight of the double mutants (rec-eCGβ-D/αΔ56) decreased to ~ 23 kDa. After deglycosylation treatment with protein N-glycosydase F (PNGase F), the molecular weights of rec-eCGβ/α, β/αΔ56, β-D/α, and β-D/αΔ56 decreased to approximately 30–36, 33, 19, and 19 kDa, respectively (Fig. 3b). These results indicated that oligosaccharides were largely modified in rec-eCGs produced in CHO-K1 cells and in the mutant proteins that were deglycosylated in the N-linked and O-linked regions of eCGα- and β-subunits, confirming the loss of oligosaccharide chains.

### In vivo biological activity of rec-eCG mutants

To test the in vivo activity of natural eCG and of rec-eCG derivatives, we examined oocyte ovulation. The numbers of normal ovulated oocytes in the natural-eCG group were 22.6 ± 2.9 as presented in Table 1. By contrast, the other groups (β/α, β/αΔ56, β-Dα, and β-D/αΔ56), which were injected with rec-eCG proteins produced 13.5 ± 1.3, 14.8 ± 1.2, 14 ± 1.8, and 15 ± 0.6 oocytes, respectively. The number of normal oocytes after treatment with natural eCG was higher relative to the rec-eCG groups. Nonetheless, there were no differences among the rec-eCG groups. We also examined the nonfunctional oocytes which were detected in the natural-eCG–treated group (Additional file 1: Figure S1). Nearly the same numbers of nonfunctional oocytes were found in the rec-eCG–treated groups. These results meant that the rec-eCG proteins produced in this study had full activity and were capable of inducing ovulation.

## Discussion

We examined the biological roles of N-linked oligosaccharides and the O-linked CTP of eCG, as well as ovulation induction, in vivo by means of rec-eCG generated by site-directed mutagenesis. Another study [31] on rec-hCG suggest that mutagenesis of Asn at position 52, which is linked to an oligosaccharide, does not affect secretion of the mutant subunit. In contrast, the loss of the oligosaccharide at position 78 was found to cause the mutant subunit to be degraded quickly, and < 20% was secreted. The molecular weights of secreted wild-type hCG and mutant protein hCGαΔ52/β turned out to be 28 and 22 kDa, respectively. Furthermore, some researchers reported that the absence of both N-linked units (Asn13 and Asn30) in the hCGβ-subunit slows secretion by 2–2.4-fold [32]. As for rec-hTSH, wild-type TSH has a molecular weight of ~ 38 kD, whereas the TSH single-site mutant (TSHα1 or TSHα2) and TSH double mutant (TSHα1 + α2) are approximately 35 and 31 kDa, respectively [23]. Our results are consistent with those of other studies, suggesting that secretion into the

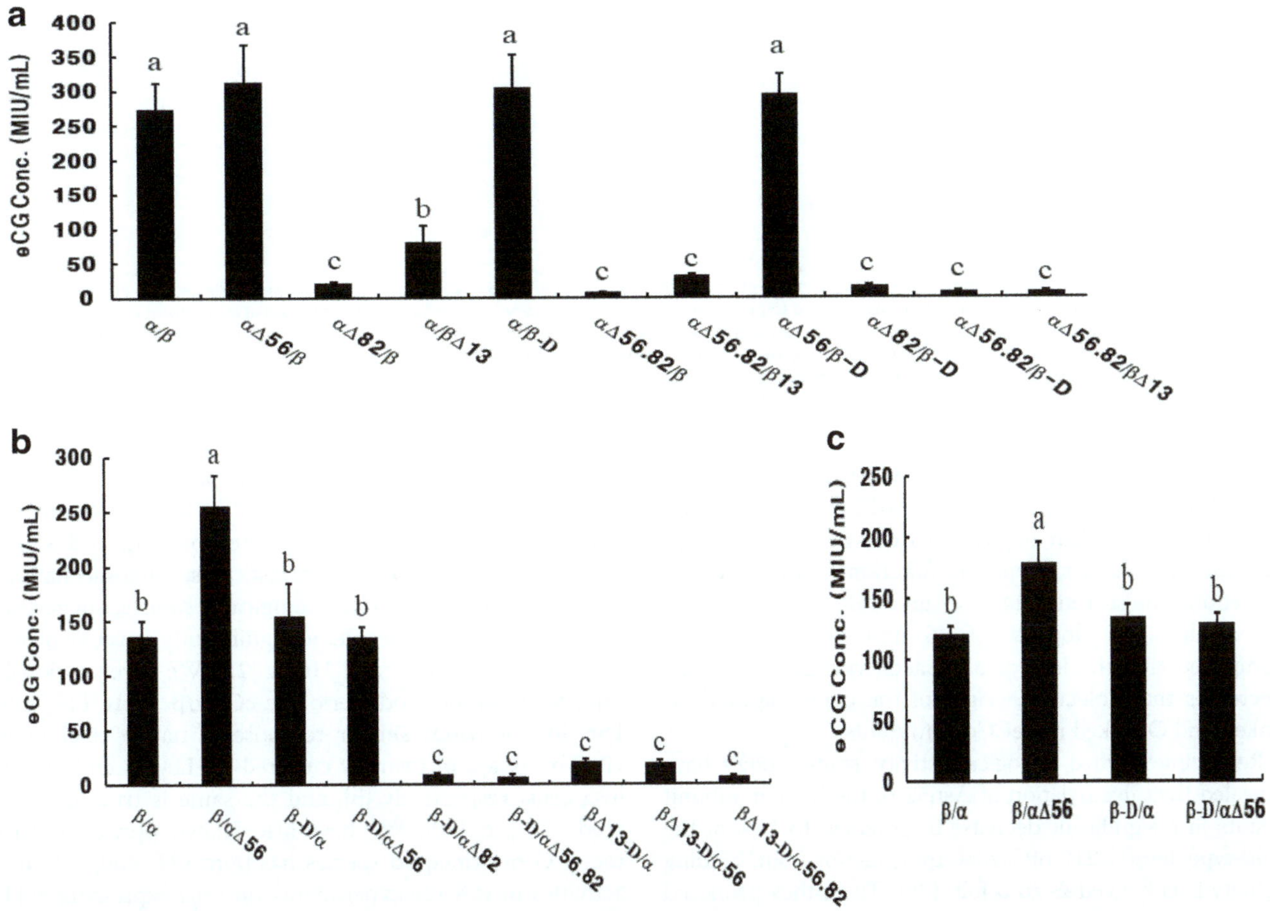


**Fig. 2** Quantitative analysis of rec-eCG mutants and myc-tagged rec-eCG mutants by ELISA after transient transfection into CHO-K1 cells. **a** Quantities of dimeric rec-eCG mutants were measured by an ELISA. **b** Tethered rec-eCG mutants were detected. **c** Quantities of *myc*-tagged rec-eCG mutants were analyzed by an ELISA. The *myc*-tag (Glu-Gln-Lys-Leu-Ile-Ser-Glu-Glu-Asp-Leu) was added between the first and second amino acid residue of the β-subunit of the mature eCG protein. Values with different superscripts are significantly different ($p < 0.05$)

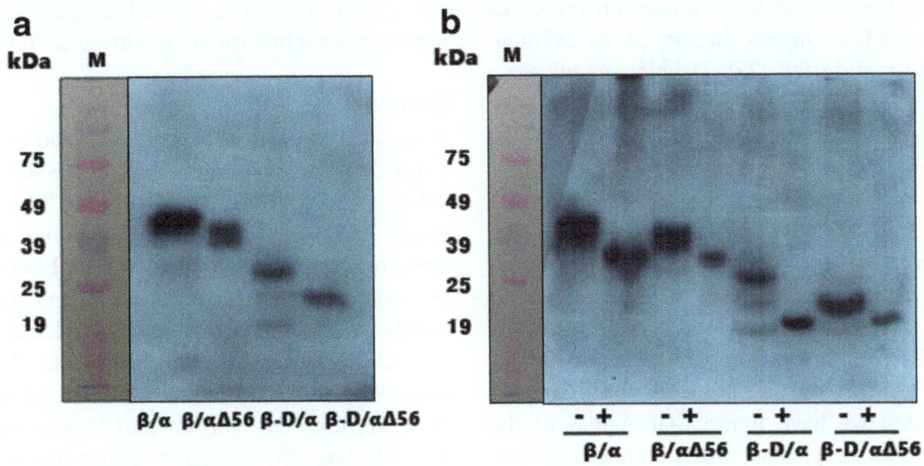


**Fig. 3** Western blot analysis of tethered rec-eCG mutants. **a** Rec-eCG mutants were expressed in CHO-K1 cells. The proteins in conditioned media were collected, separated by SDS-PAGE, and transferred to a blotting membrane. The proteins were detected with antibodies against the *myc*-tag. **b** Proteins for western blotting were also treated with N-GlycosidaseF. Lane 1: β/α, Lane 2: β/αΔ56, Lane 3: β-D/α, Lane 4: β-D/αΔ56,-: not treated, +: treated with N-Glycosidase-F

**Table 1** Superovulation of mouse oocytes between natural-eCG and tethered rec-eCGβ/α deglycosylated mutants

| | No. Oocytes | | |
|---|---|---|---|
| | Functional | Nonfunctional | Nonfunctional relative to total oocytes (%) |
| PMSG (*n* = 9) | 22.6 ± 2.9A | 4.78 ± 0.5a | 21.2 |
| rec-eCGβ/α (*n* = 6) | 13.5 ± 1.3B | 0.33 ± 0.2b | 2.4 |
| rec-eCGβ/αΔ56 (n = 6) | 14.8 ± 1.2B | 0b | 0 |
| rec-eCGβ-D/α (n = 6) | 14 .0 ± 1.8B | 0.33 ± 0.3b | 2.4 |
| rec-eCGβ-D/αΔ56 (n = 6) | 15.0 ± 0.6B | 0b | 0 |

The mice were superovulated by injection of 10 IU of PMSG or rec-eCGs and then 10 IU hCG after 48 h, and ovulated oocytes were collected into an oviduct ampulla after 13 h as described in Methods. 1 IU was assumed to be 100 ng according to the conversion factor of the suggested assay protocol corresponding PMSG by standard curve. The ovulated oocytes were determined under microscope as shown in Additional file 1: Figure S1. The percentage of nonfunctional relative to total oocytes was around 20% with natural eCG and only around 2% with rec-eCGs used

Values are expressed as mean ± SEM for at least three independent experiments

A-B, a-bValues with different superscripts are significantly different ($p < 0.05$)

culture medium is controlled by the specific glycosylation sites (Asn82 in the α-subunit and Asn13 in the β-subunit). Thus, the Asn-linked oligosaccharides of eCGα- and β-subunits may have site-specific functions with respect to secretion. The secreted single-chain eCG in COS-7 cells is detectable as a doublet of 46 and 44 kDa [9]. We confirmed that the loss of oligosaccharide chains greatly decreases the molecular weight of the deglycosylated N-linked and O-linked rec-eCGα/β mutants.

Regarding in vitro biological activity, some studies have revealed that glycosylation of Asn52 of the hFSHα-subunit results in a significant decrease in potency (to 26% of the wild-type level) [21] of signal transduction, but binding activity is enhanced 2- to 3-fold [24]. Thus, they proposed that receptor binding and signal transduction are dissociable functions involving different sites on the FSH glycoprotein. The α-subunit Asn52 oligosaccharide has a disproportionate role in signal transduction, and the amino acid sequence near β-subunit Asn24 functions in both binding and signal transduction [33]. In hCG, the Asn52 oligosaccharide has been found to be important for signal transduction; without this oligosaccharide, cAMP stimulation and steroid formation fail [22]. Deletion of oligosaccharide units from either site 1 (Asn52) or site 2 (Asn78) of the α-subunit increased the biological activity of the dimer in that study by approximately 30%. Nevertheless, carbohydrate unit at both sites (Asn52 and Asn78) of the hTSH α-subunit significantly reduces cAMP formation (by ~ 70%) and T3 secretion (by ~ 40%) as compared to wild-type hTSH [23]. Furthermore, the CTP of hCGβ does not interact directly with the choriogonadotropin hormone receptor (LH-CGR) complex and apparently does not influence the tertiary structure or folding of the β-subunit [22, 26, 31]. Besides, we have found that Asn56 of the eCGα-subunit does not enhance agonist stimulation internalization through rat LH/CGR (rLH/CGR) and rFSHR [34].

Various studies indicate that rec-eCG yield the formation and secretion of stable heterodimeric eCG in infected Sf9 cells [35] and COS-7 cells [36], and that rec-eCG has the same thermal stability as natural pituitary LH [37]. O-linked oligosaccharides of the CTP in the hCGβ-subunit play only a minor role in receptor binding and signal transduction in vitro. By contrast, these oligosaccharides are critical for the in vivo biological response, suggesting that truncated hCG results in significantly fewer ovulated oocytes at doses of 70 and 210 ng [22]. We have previously reported that heterodimeric rec-eCGα/β exert LH- and FSH-like activities similar to those of native eCG in in vitro bioassays of primary cultured rat Leydig and granulosa cells, respectively [5], and the same is true for tethered eCGβ/α [30]. We have also shown previously that rec-eCG of nonequid species has both LH- and FSH-like activities in in vitro experiments on cells expressing rLH/CGR and rFSHR [34] and in terms of receptor function [38]. Nevertheless, no studies have examined the ovulation rate in mice under the influence of rec-eCG derivatives. Here, we suggest that the ovulation rates in mice may not be affected by the deglycosylating mutations. Nevertheless, it is impossible to compare potencies by means of only one dose, 10 IU injection. Thus, biological activity of rec-eCG derivatives slightly differs in vitro and in vivo.

## Conclusions

Thus, secretion into the culture medium in rec-eCGβ/α is controlled by specific glycosylation sites: Asn82 of the α-subunit and Asn13 of the β-subunit. The Asn-linked oligosaccharides of eCGα- and β-subunits have site-specific functions with respect to secretion. Besides, we confirmed by removal of oligosaccharide chains that the molecular weight was decreased by deglycosylation of the N-linked and O-linked eCGα- and β-subunits. The nonfunctional oocytes were around 20% in the natural eCG-treated mice. However there are only around 2% with the rec-eCG derivatives tested. We also suggest that the differences between rec-eCG and natural-eCG could rather be due to the nature of the injected hormone. Therefore, rec-eCGβ/α derivatives have a good potential to produce functional oocytes with a much smaller number nonfunctional oocytes.

## Methods

### Materials

The oligonucleotides employed in this study were synthesized by Genotech (Daejon, Korea). The following reagents and materials were also used: restriction enzymes, polymerase chain reaction reagents, and the DNA ligation kit were purchased from Takara (Tokyo, Japan). CHO-K1 cells were obtained from the Japanese Cancer Research Resources Bank (Tokyo, Japan). Ham's F-12 medium, Opti-MEM I, serum-free CHO-S-SFM II, geneticin, and Lipofectamine 2000 were bought from Gibco BRL (Grand Island, NY, USA), and fetal bovine serum from Hyclone Laboratories (Logan, UT, USA). The QIAprep-Spin plasmid kit and RNeasy columns were acquired from QIAGEN, Inc. (Hilden, Germany), whereas the Pro-Prep™ protein extraction solution from Intron Biotechnology (Seoul, Korea). The Lumi-Light western blot kit was bought from Roche (Basel, Switzerland), and the pcDNA3 mammalian expression vector, PNGase F, and TRIzol reagent from Invitrogen (Carlsbad, CA, USA). The PMSG ELISA kit was purchased from DRG International, Inc. (Mountainside, NJ, USA), Centriplus Centrifugal Filter Devices from Amicon Bio separations (Merck, Billerica, MA, USA), and an anti-*myc* antibody from Santa Cruz Biotechnology (Dallas, Texas, USA). A peroxidase-conjugated anti-mouse IgG antibody was acquired from Bio-Rad (Hercules, CA, USA), whereas pregnant-mare serum gonadotropin (eCG) from Sigma-Aldrich Corp. (St. Louis, MO, USA), as were all other reagents. All experimental designs and procedures were in compliance with the approved Guidelines for Animal Experiments of Hankyong National University, Korea, and were approved by the Animal Care and Use Committee of Hankyong National University, Korea (Approval ID: 2015–8).

### Construction of heterodimeric eCGα/β and tethered eCGβ/α

The heterodimeric eCGα/β and its mutants were created using the pABWN vector as previously reported [5]. Additionally, cDNA encoding the tethered rec-eCGβ/α was inserted into the pcDNA3 vector and served as a template to construct mutants in which Asn (codon AAC) of the glycosylation site was substituted with Ala (codon CAG) or the CTP was deleted in the β-subunit as reported elsewhere [30]. The same method was used to add a *myc*-tag (Glu-Gln-Lys-Leu-Ile-Ser-Glu-Glu-Asp-Leu) between the first and second amino acid residues of the β-subunit of the mature eCG protein [38]. Site-directed mutagenesis was performed with three primers in a single-step PCR [5]. The primers employed in the experiment are summarized in Table 2). The first PCR was carried out with primers 2 and 3. DNA fragments were spliced and subjected to a second PCR involving primers 1 and 12 to generate tethered eCGs. The schematic diagrams of heterodimeric rec-eCGα/β and tethered rec-eCGβ/α mutants are depicted in Fig. 1.

We mutated Asn to Gln at each glycosylation site (Asn56, Asn82, Asn13, deletion of the β-subunit CTP region with one site and a combination). These fragments were digested with *Eco*RI and *Sal*I and ligated into eukaryotic expression vector pcDNA3. The plasmid DNAs were purified and sequenced in both directions by automated DNA sequencing to ensure that the correct mutations were introduced. A total of 20 expression vectors (11 heterodimeric eCGs and nine tethered eCGs) were constructed (plasmids expressing heterodimeric eCGs were designated as pAB-eCGα/β, pAB-αΔ56/β, pAB-αΔ82/β, pAB-α/βΔ13, pAB-α/β-D, pAB-αΔ56,82/β, pAB-αΔ56/βΔ13, pAB-αΔ56/β-D, pAB-αΔ82/β-D, pAB-αΔ56.82/β-D, and pAB-αΔ56.82/βΔ13; plasmids expressing tethered eCGs were designated as pcDNA3-eCGβ/α, pcDNA3-β/αΔ56, pcDNA3-β-D/α, pcDNA3-β-D/αΔ56, pcDNA3-β-D/αΔ82, pcDNA3-β-D/αΔ56.82, pcDNA3-βΔ13-D/α, pcDNA3-βΔ13-D/αΔ56, and pcDNA3-βΔ13-D/αΔ56.82) as previously reported [5, 30, 34].

### Cell culture and functional expression

Expression vectors were transfected into CHO-K1 cells by the liposome transfection method as described in ref. [30]. The transfected cells were cultured for 48 h in a serum-free medium (CHO-S-SFM-II) and then harvested and centrifuged at 100,000×*g* for 10 min. The supernatant was collected and stored at – 20 °C until the assay. Six to eight pools of stably transfected cells were selected by incubating the cells in a growth medium [Ham's F12 medium with 10% of fetal calf serum, penicillin (100 IU/mL), streptomycin (100 μg/mL), glutamine (2 mM), and G418 (800 μg/mL)] for 2–3 weeks after transfection, as previously reported [30]. The culture medium was centrifuged at 100,000×*g* for 10 min to remove cell debris. The supernatant was collected and concentrated in an Amicon Stirred cell concentrator and stored at – 20 °C until the assay.

### Quantification of rec-eCG proteins

The plasmids encoding dimeric eCG and single-chain eCG were transiently transfected into CHO-K1 cells, and the culture media were collected at 72 h after transfection, and rec-eCG was quantified with the PMSG ELISA kit (DRG Diagnostics). Briefly, the collected medium was centrifuged to remove cell debris at 5000 rpm for 3 min. The supernatant was collected and stored at – 20 °C until analysis. The PMSG standard and rec-eCG samples (100 μL) were dispensed into wells of a plate coated with a monoclonal antibody against a unique antigenic site on the eCG molecule. The plate was incubated for 60 min at room temperature without agitation. Then, we briskly shook out the contents of the wells and rinsed them three times with distilled water. After that, 100 μL of a anti-PMSG antibody conjugated to horseradish peroxidase was added into each

**Table 2** List of primers used for the construction of tethered eCG mutants

| | Primer name | Location | Primer Sequence |
|---|---|---|---|
| 1 | eCGβ/α β-subunit | - | 5′- T**GAATTC**ACCATGGAGACGGTCCAG -3′<br>***Eco R I* site** |
| 2 | *myc* eCGβ/α reverse | - | 5′- TCCATCAGGAAAAGAAGTCTTTATTGG -3′ |
| 3 | *myc* eCGβ/α forward | - | 5′- ATAAAGACTTCTTTTCCTTGATGGAGAG -3′ |
| 4 | eCGβ/α**Δ**56 reverse | α 56 | 5′- TGAGGTGATCTGCTTTGGGACCAACAT -3′ |
| 5 | eCGβ/α**Δ**56 forward | α 56 | 5′- ATGTTGGTCCCAAAGCAGATCACCTCA -3′ |
| 6 | eCGβ/α**Δ**82 reverse | α 82 | 5′- GGCCTCCTTCTCAGCGGCCAGAGTGGCGTTGAT -3′ |
| 7 | eCGβ/α**Δ**82 forward | α 82 | 5′- ATCAACGCCACTCTGGCCGCTGAGAAGGAGGCC -3′ |
| 8 | eCGβ**Δ**13/α reverse | β 13 | 5′- AGCAGCCAGAGTGGCGTTGATGGGCCGGCACAG -3′ |
| 9 | eCGβ**Δ**13/α forward | β 13 | 5′- CTGTGCCGGCCCATCAACGCCACTCTGGCTGCT -3′ |
| 10 | eCGβ-D/α reverse | β : 121-149 | 5′- TCCATCAGGAAAGGCCTGGGGGGCACAGGC -3′ |
| 11 | eCGβ-D/α forward | β : 121-149 | 5′- GCCCCCCAGGCCTTTCCTCATGGAGAGTTT -3′ |
| 12 | eCGβ/α α-subunit | - | 5′- CC**GTCGAC**TTTAAATCTTGTGGTGATAGCA -3′<br>***Sal I* site** |

well and incubated for 60 min at room temperature without agitation. The plate wells were rinsed five times, and residual water droplets were removed. A substrate solution (100 μL) was next added and incubated for 30 min at room temperature. Finally, 50 μL a stop solution was added to stop the enzymatic reaction. The absorbance of the product solution was read at 450 nm on a micro titer plate reader. The average absorbance of each standard was plotted against its corresponding concentration in a linear–log graph. We calculated the average absorbance of each sample to determine the corresponding PMSG value via simple interpolation by means of this standard curve. Finally, 1 IU was assumed to be 100 ng according to the conversion factor of the suggested assay protocol.

### Detection of rec-eCGs by western blotting and enzymatic digestion of N-linked oligosaccharides

For western blot analysis, samples of a concentrated medium were subjected to electrophores is under reducing conditions in sodium dodecyl sulfate (SDS) 12.5% polyacrylamide gels by the Laemmli method [39]. After SDS polyacrylamide gel electrophoresis (PAGE), the proteins were transferred to a polyvinylidene difluoride membrane (0.2 μm) at 100 V for 2 h in a Mini Trans-Blot Electrophoretic Transfer cell. After blotting, the membrane was blocked with a 1% blocking reagent for 1 h, and the monoclonal anti-*myc* antibody (1:5,000) was added for 2 h incubation. Next, the blot was washed to remove the unbound antibody, incubated with a secondary antibody (peroxidase-conjugated anti-mouse IgG antibody 37.5 μL/15 mL of the blocking solution) for 30 min, and washed. The membrane was then incubated for 5 min with 2 mL of the Lumi-Light substrate solution and placed with its protein side up on a plastic wrap. The membrane was covered with a second piece of the plastic wrap, and an X- ray film was exposed to the membrane for 1–10 min. The tethered rec-eCG protein was analyzed regarding removal of added glycans by *N*-glycosylation enzyme. To remove all N-linked glycans, rec-eCG was incubated for 24 h at 37 °C with PNGase F [2 μL of the enzyme (2.5 U/mL) per 30 μL of sample+ 8 μL of 5× reaction buffer]. The reaction was stopped by boiling for 10 min, and the samples were analyzed by SDS-PAGE and western blotting.

### In vivo biopotency

This property of rec-eCGβ/α proteins was evaluated by determining the ovulated-oocyte number. Oocytes were collected from 8-week-old B6D2F1 female mice. The mice were superovulated by injection of 10 IU of PMSG or rec-eCGs and then 10 IU hCG after 48 h, and ovulated oocytes were collected into an oviduct ampulla after 13 h. Cumulus cells were removed from the collected oocytes by means of 0.3% hyaluronidase. The oocytes were analyzed for several characteristics (e.g., cytoplasmic membrane, first polar body, and morphological features).

### Data and statistical analysis

Values are given as mean ± SEM. One-way ANOVA with Tukey's multiple-comparison test was conducted to compare the results between samples. In figures, superscripts indicate significant differences from a group ($p < 0.05$).

## Additional file

**Additional file 1: Figure S1.** Functional and nonfunctional oocytes. (PPT 289 kb)

### Abbreviations

CL: Corpus luteum; CTP: Carboxy-terminal peptides; FSH: Follicle-stimulating hormone; FSHR: FSH receptor; GnRHR: Gonadotropin releasing hormone receptor; LH: Luteinizing hormone; PMSG: Pregnant mare serum gonadotropin; rec-eCG: Recombinant equine chorionic gonadotropin; SDS-PAGE: Sodium dodecyl sulfate-polyacrylamide gel electrophoresis; TSH: Thyroid-stimulating hormone

### Acknowledgements

The authors thank Dr. HW Seong for a helpful discussion.

### Authors' contributions

JJP conducted the experiments. MB performed the ELISA analysis and interpretation of data. MHK and KSM wrote the main manuscript text and prepared the figures and supplementary file. All authors reviewed the final manuscript. All authors read and approved the final manuscript.

### Funding

This work was financially supported by Korean Research Foundation Program (2018007794), Republic of Korea. The funders role was providing funding for the study and all other aspects of the project (design, collection, analysis, and interpretation of data and writing of the manuscript) was executed by the authors.

### Ethics approval and consent to participate

Not applicable. All experimental designs and procedures were in compliance with the approved Guidelines for Animal Experiments of Hankyong National University, Korea and were approved by the Animal Care and Use Committee of Hankyong National University, Korea (Approval ID: 2015–8).

### Consent for publication

Not applicable.

### Competing interests

The authors declare that they have no competing interests.

### Author details

[1]Animal Biotechnology, Graduate School of Future Convergence Technology, Institute of Genetic Engineering, Hankyong National University, Ansung 17579, Korea. [2]Department of Animal Resource Science, Hankyong National University, Ansung 17579, Korea. [3]Department of Food Science and Nutrition, Hoseo University, Asan 31499, Korea.

### References

1. Talmadge K, Vamvakopoulos NC, Fiddes JC. Evolution of the gens for the b subunits of human chorionic gonadotropin and luteinizing hormone. Nature. 1984;307:37–40.
2. Pierce JG, Parsons TF. Glycoprotein hormones: structure and function. Annu Rev Biochem. 1981;50:465–95.
3. Murphy BD, Martinuk SD. Equine chorionic gonadotropin. Endocr Rev. 1991;12:27–44.
4. Chopineau M, Maurel MC, Combarnous Y, Durand P. Topography of equine chorionic gonadotropin epitopes relative to the luteinizing hormone and follicle-stimulating hormone receptor interaction sites. Mol Cell Endocrinol. 1993;92:229–39.
5. Min KS, Hattori N, Aikawa JI, Shiota K, Ogawa T. Site-directed mutagenesis of recombinant equine chorionic gonadotropin/luteinizing hormone: differential role of oligosaccharides in luteinizing hormone- and follicle-stimulating hormone-like activities. Endocr J. 1996;43:585–93.
6. Serman GB, Wolfe MW, Farmerie TA, Clay CM, Threadgill DS, Sharp DC, Nilson JH. A single gene encodes the β-subunit of equine luteinizing hormone and chorionic gonadotropin. Mol Endocrinol. 1992;6:951–9.
7. Min KS, Shinozaki M, Miyazawa K, Nishimura R, Sasaki N, Shiota K, Ogawa T. Nucleotide sequence of eCGα-subunit cDNA and its expression in the equine placenta. J Reprod Dev. 1994;40:301–5.
8. Guillou F, Combarnous Y. Purification of equine gonadotropins and comparative study of their acid-dissociation and receptor-binding specificity. Biochim Biophys Acta. 1983;755:229–36.
9. Galet C, Guillou F, Foulon-Gauze F, Combarnous Y, Chopineau M. The β104-109 sequence is essential for the secretion of correctly folded single-chain βα horse LH/CG and for its activity. J Endocrinol. 2009;203:167–74.
10. Smith PL, Bousfield GR, Kumar S, Fiete D, Baenziger JU. Equine lutropin and chorionic gonadotropin bear oligosaccharides terminating with $SO_4$-4-GalNAc and Siaα2,3Gal, respectively. J Biol Chem. 1993;268:795–802.
11. Matsui T, Mizuochi T, Titani K, Okinaga T, Hoshi M, Bousfield GR, Sugino H, Ward DN. Structural analysis of N-linked oligosaccharides of equine chorionic gonadotropin and lutropin β-subunits. Biochemistry. 1994;33:14039–48.
12. Boeta M, Zarco L. Luteogenic and luteotropic effects of eCG during pregnancy in the mare. Anim Reprod Sci. 2012;130:57–62.
13. Flores-Flores G, Velazquez-Canton E, Boeta M, Zarco L. Luteoprotective role of equine chorionic gonadotropin (eCG) during pregnancy in the mare. Reprod Dom Anim. 2014;49:420–6.
14. Conley AJ. Review of the reproductive endocrinology of the pregnant and parturient mare. Theriogenology. 2016;86:355–65.
15. Garcia-Ispierto I, Lopez-Helguera I, Martino A, Lopez-Gatius F. Reproductive performance of anoestrous high-producing dairy cows improved by adding equine chorionic gonadotrophin to a progesterone-based oestrous synchronizing protocol. Reprod in Dom Anim. 2012;47:752–8.
16. Rostami B, Niasari-Naslaji A, Vojgani M, Nikjou D, Amanlou H, Gerami A. Effect of eCG on early resumption of ovarian activity in postpartum dairy cows. Anim Reprod Sci. 2011;128:100–6.
17. Rensis FD, Lopez-Gatius F. Use of equine chorionic gonadotropin to control reproduction of the dairy cow: a review. Reprod Dom Anim. 2014;49:177–82.
18. Wei SC, Gong ZD, Zhao HW, Liang HQ, Lai LJ, Deng YY. Equine chorionic gonadotropin influence on sheep oocyte in vitro maturation, apoptosis, and follicle-stimulating hormone receptor and luteinizing hormone receptor expression. Gene Mol Res. 2016;15:1–13.
19. Galet C, Menck L, Bourhis C, Chopineau M, Le-Griec G, Perrin A, Magallon T, Attal J, Viglietta C, Houdebine LM, Guillou F. Expression of a single βα chain protein of equine LH/CG in milk of transgenic rabbits and its biological activity. Mol Cell Endocrinol. 2000;174:31–40.
20. Chopineau M, Martinat N, Galet C, Guillou F, Combarnous Y. β-Subunit 102~104 residues are crucial to confer FSH activity to equine LH/CG but are not sufficient to confer FSH activity to human CG. J Endocrinol. 2001;169:55–63.
21. Bishop LA, Robertson DM, Cahir N, Schofield PR. Specific roles for the asparagine-linked carbohydrate residues of recombinant human follicle stimulating hormone in receptor binding and signal transduction. Mol Endocrinol. 1994;8:722–31.
22. Matzuk MM, Keene JL, Boime I. Site specificity of the chorionic gonadotropin N-linked oligosaccharides in signal transduction. J Biol Chem. 1989;264:2409–14.
23. Fares FA, Gruener N, Kraie MZ. The role of the asparagine-linked olgosaccharides of the α-subunit in human thyrotropin activity. Endocrinology. 1996;137:555–60.
24. Valove FM, Finch C, Anasti JN, Froehlich J, Flack MR. Receptor binding and signal transduction are dissociable functions requires different sites on follicle-stimulating hormone. Endocrinology. 1994;135:2657–61.
25. Chen W, Bahl OP. Recombinant carbohydrate variant of human choriogonadotropin β-subunit (hCGβ) descarboxyl terminus (115-145). J Biol Chem. 1991;266:6246–51.
26. Matzuk MM, Hsueh AJW, Lapolt P, Tsafriri A, Keene JL, Boime I. Biological role of the carboxyl-terminal extension of human chorionic gonadotropin β-subunit. Endocrinology. 1990;126:376–83.
27. El-Deiry S, Kaetzel D, Kennedy G, Nilson J, Puett D. Site-directed mutagenesis of the human chorionic gonadotropin beta-subunit: bioactivity of a heterologous hormone, bovine alpha-human des-(122-145) beta. Mol Endocrinol. 1989;3:1523–8.
28. Chen W, Puett D. Delineation via site-directed mutagenesis of the carboxyl-terminal region of human choriogonadotropin β required for subunit assembly and biological activity. J Biol Chem. 1991;266:6904–8.

29. Bousfield GR, Liu WK, Ward DN. Effects of removal of carboxyl-terminal extension from equine luteinizing hormone (LH) β-subunit on LH and follicle-stimulating hormone receptor-binding activities and LH steroidogenic activity in rat testicular Leydig cells. Endocrinology. 1989; 124:379–87.
30. Min KS, Hiyama T, Seong HH, Hattori N, Tanaka S, Shiota K. Biological activities of tethered equine chorionic gonadotropin (eCG) and its deglycosylated mutants. J Reprod Dev. 2004;50:297–304.
31. Matzuk MM, Boime I. The role of the asparagine-linked oligosaccharides of the α subunit in the secretion and assembly of human chorionic gonadotropin. J Cell Bio. 1988;106:1049–59.
32. Matzuk MM, Boime I. Site-specific mutagenesis defines the intracellular role of the asparagine-linked oligosaccharides of chorionic gonadotropin β subunit. J Biol Chem. 1988;26:17106–11.
33. Flack MR, Froehlich J, Bennet AP, Anasti J, Nisula BC. Site-directed mutagenesis defines the individual roles of the glycosylation sites on follicle-stimulation hormone. J Biol Chem. 1994;269:14015–20.
34. Park JJ, Seong HK, Kim JS, Byambarachaa M, Kang MH, Min KS. Internalization of rat FSH and LH/CG receptors by rec-eCG in CHO-K1 cells. Dev Reprod. 2017;21:111–20.
35. Legardinier S, Klett D, Poirier JC, Combarnous Y, Cahoreau C. Mammalian-like nonsialyl complex-type N-glycosylation of equine gonadotropins in mimic™ insect cells. Glycobiology. 2005;15:776–90.
36. Chopineau M, Martinat N, Trospoux C, Marichatou H, Combarnous Y, Stewart F, Guillou F. Expression of horse and donkey LH in COS-7 cells: evidence for low FSH activity in donkey LH compared with horse LH. J Endocrinol. 1997;152:371–7.
37. Legardinier S, Poirier JC, Klett D, Combarnous Y, Cahoreau C. Stability and biological activities of heterodimeric and single-chain equine LH/chorionic goandotropin variants. J Mole Endocrinol. 2008;40:185–98.
38. Min KS, Liu X, Fabritz JE, Jaquette J, Abell AN, Ascoli M. Mutations that induce constitutive activation and mutations that impair signal transduction modulate the basal and/or agonist-stimulated internalization of the lutropin/choriogonadotropin receptor. J Biol Chem. 1998;273:34911–9.
39. Laemmli UK. Cleavage of structural proteins during the assembly of the head of bacteriophage T4. Nature. 1970;227:680–5.

# 10

# FlexiBAC: a versatile, open-source baculovirus vector system for protein expression, secretion, and proteolytic processing

Régis P. Lemaitre[1], Aliona Bogdanova[1], Barbara Borgonovo[1], Jeffrey B. Woodruff[2*] and David N. Drechsel[1,3*]

**Abstract**

**Background:** Baculovirus-mediated expression in insect cells is a powerful approach for protein production. However, many existing methods are time-consuming, offer limited options for protein tagging, and are unsuitable for secreted proteins requiring proteolytic maturation, such as TGF-β family growth factors.

**Results:** To overcome the limitations of traditional baculovirus expression systems, we engineered "FlexiBAC". This system allows recombinant baculovirus formation inside insect cells and reduces the time between initial cloning and protein production to 13 days. FlexiBAC includes 143 shuttle vectors that append combinations of purification tags, fluorescent markers, proteolytic cleavage sites, trafficking signals, and chemical conjugation tags to the termini of the target protein. This system also overexpresses recombinant furin convertase to allow efficient proteolytic processing of secreted proteins. We demonstrate that FlexiBAC can be used to produce high levels of mature, active forms of TGF-β family growth factors, such as Activin A, as well as other proteins that are typically difficult to reconstitute, such as proteins rich in coiled-coil, low complexity, and disordered domains.

**Conclusions:** FlexiBAC is a protein expression system for production of both cytosolic proteins and secreted proteins that require proteolytic maturation. The design of FlexiBAC and its expansive complementary shuttle vector system reduces cloning steps and simplifies baculovirus production.

**Keywords:** Baculovirus, Intrinsically disordered proteins, Protein expression, TGF-β

## Background

Protein production using baculovirus-infected insect cells combines biological safety with high-level expression of functional proteins. Eukaryotic proteins produced recombinantly in this system generally adopt a native, folded conformation and contain the appropriate post-translational modifications. The large baculovirus genome (ranging between 80 and 180 kbp, depending on the species) can accommodate large insertions, making it ideal for expressing large heteromeric protein complexes. Moreover, insect cells can grow in serum-free media, which greatly facilitates purification of secreted proteins from the conditioned media.

Despite these advantages, recombinant protein production using baculovirus is time-consuming. The time between initial cloning and protein expression lasted 3 to 4 weeks in the earliest baculovirus expression vector systems that relied on plaque selection of recombinants into the polyhedron locus after homologous recombination in insect cells [1, 2]. To make baculovirus production simpler and faster, more recent protocols have facilitated production of recombinant baculovirus genomes using site-specific transposition in *E. coli* (Bac-to-Bac®, Thermo Fisher Scientific) or homologous recombination in insect cells (flashBAC™, Oxford Expression Technologies). Both systems employ the

* Correspondence: jeffrey.woodruff@utsouthwestern.edu; david.drechsel@imp.ac.at
[2]Department of Cell Biology, Dept. of Biophysics, UT Southwestern Medical Center, Dallas, TX 75390, USA
[1]Max Planck Institute of Molecular Cell Biology and Genetics, Pfotenhauerstrasse 108, 01307 Dresden, Germany

polyhedrin promoter to drive high-level, late expression of the target protein(s) of interest. While the former system is more popular, the latter system is faster, as it reduces the window between initial cloning and protein expression to ~ 14 days. One major drawback of these commercial systems is that they only offer limited options for tagging the protein of interest: 6xHis, GST, and S-tag options are available, but fluorophores or tags for conjugating chemical probes are not. Purification of difficult proteins would be greatly facilitated by a baculovirus vector system that offers a larger combination of purification and solubility tags. Furthermore, as reconstituted systems are analyzed more and more via light microscopy, there is a pressing need for more fluorophore tagging options.

Most baculovirus expression systems are also not suited for high-level secretion of recombinant proteins. The secretory pathway in insect cells has limited capacity and cannot handle the efflux of recombinant proteins during baculovirus infection [3]. During late infection, baculovirus produce cathepsin and chitinase, two highly-expressed enzymes which are secreted by insect cells. In the wild, these enzymes are essential for liquefaction of the larval host, but they play no role in baculovirus replication in cultured cells. It was proposed that eliminating these proteins would reduce the viral protein load passing through the secretory pathway and would free up room for recombinant target proteins. Indeed, deletion of the genes encoding cathepsin and chitinase dramatically improved secretion of recombinant target proteins; unexpectedly, these mutations also reduced degradation of both secreted and non-secreted target proteins [3, 4]. Thus, knocking out genes encoding cathepsin and chitinase should generally improve the efficiency and versatility of baculovirus expression systems.

Another limitation of baculoviral expression systems is that they are not equipped to handle high-volume post-translational processing of secreted proteins. Many secreted proteins require proteolytic trimming to reach a fully active, mature form. One prominent example is the transforming growth factor beta (TGF-β) member Activin A, which must be cleaved by furin convertases in the Golgi apparatus to become an active signaling ligand [5]. Baculovirus-mediated expression of Activin A largely resulted in the secretion of an inactive pro-form peptide [6]. Co-infection of baculoviruses expressing furin and Activin A greatly increased secretion of mature Activin A, indicating that furin activity is limited due to insufficient endogenous expression levels [7]. Engineering a single baculovirus to express furin in addition to the target gene of interest would increase yields of active TGF-β and other secreted peptides that require furin processing, obviating the need for co-infection.

To overcome these limitations, we designed an open-source baculovirus expression system called FlexiBAC. Our system limits the time between initial cloning and protein isolation to 13 days and incorporates rescue of a doubly defective bacmid to ensure that all infective viruses contain the recombinant target gene. FlexiBAC includes a compatible, modular shuttle vector set (143 vectors) that appends combinations of various purification tags, fluorescent markers, proteolytic cleavage sites, trafficking signals, and chemical conjugation tags to the user's target protein. To allow high yields of secreted proteins, the FlexiBAC viral genome includes deletions of genes encoding cathepsin and chitinase. We also present a specialized version of FlexiBAC that improves the maturation of secreted proteins requiring furin proteolysis. We demonstrate that FlexiBAC is suited for the expression of proteins deemed difficult to reconstitute, including mature TGF-β family growth factors, intrinsically disordered proteins, as well as proteins containing numerous coiled-coil domains.

## Results

### Overview of the FlexiBAC system

To streamline the production of viral stocks, we engineered a bacmid encoding a replication-defective baculovirus, which we term "DefBac" (Fig. 1). This bacmid is designed to recombine with one of 143 compatible shuttle vectors, called pOCC vectors, where the target gene of interest (GOI) is inserted (see Additional files 1 and 2). Transfection of *Spodoptera frugiperda* (Sf9) insect cells with linearized DefBac and a pOCC vector permits their recombination and produces infectious baculovirus. The GOI is then expressed late in the lytic cycle using the polydedrin promoter. Viruses are produced and escape the Sf9 cells only if recombination between pOCC and DefBac was successful. One round of viral amplification is typically required before the virus is used to express the target protein. The general workflow is shown in Fig. 1d; a detailed, step-by-step protocol is included in the supplement (see Additional file 3). In the following sections, we describe construction of DefBac and the pOCC shuttle vectors.

### A doubly defective viral backbone ensures production of complete recombinant viruses containing the gene of interest

One of the earliest, and still very popular, systems used to generate recombinant baculovirus for protein expression (Bac-to-Bac®, Thermo Fisher Scientific) employs site-specific transposition of a baculovirus genome, propagated as a bacmid, with a cloning plasmid that contains the gene of interest. Transposition occurs in bacteria, which can take ~ 1 week to select and isolate positive clones. To streamline recombinant virus preparation, we

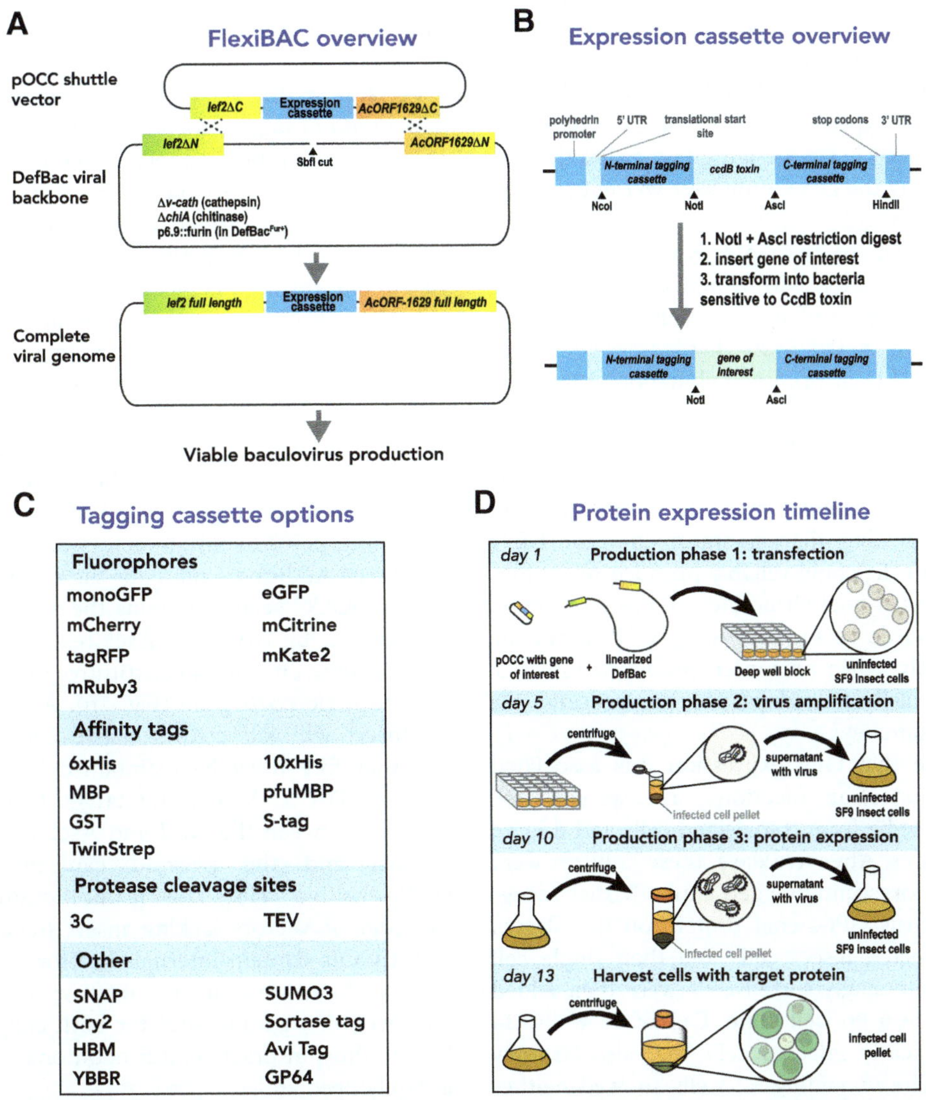


**Fig. 1** Overview of the FlexiBAC baculovirus expression system. **a** Recombination between a SbfI-linearized defective viral backbone ("DefBac") and a shuttle vector ("pOCC", which contains the target gene of interest) creates a viral genome capable of producing infectious virus. Recombination occurs between complementary *lef2* and *AcORF1629* truncations located on the pOCC vector and the DefBac viral backbone. Recombination generates full-length versions of *lef2* and *AcORF1629* genes, which are needed for baculovirus production. No virus is produced without proper recombination, thus eliminating the need for post-production screening for recombinant virus. The DefBac viral backbone also includes deletions in the genes encoding cathepsin and chitinase. A second version of DefBac, called DefBac$^{Fur+}$, expresses the convertase furin. **b** Each pOCC shuttle vector contains a modular expression cassette that allows insertion and swapping of the gene of interest, N-terminal tags, and C-terminal tags using classic cloning techniques (restriction sites are shown with arrowheads). A gene encoding the ccdB toxin selects against plasmids lacking the gene of interest. **c** pOCC shuttle plasmids encode a variety of tags that can be appended to the target protein of interest. Tags are easily combined and swapped to create customizable N-terminal or C-terminal fusions. Descriptions of each tag and the available combinations (143) are shown in Table 1 (see Additional file 1) and available upon request. The most commonly used plasmids (52) are shown in Table 2 (see Additional file 2) and are available at www.addgene.org. **d** Timeline for production of recombinant baculovirus using the FlexiBAC system. On day 1, the user transfects Sf9 insect cells with pOCC shuttle vector (with target gene of interest) and linearized DefBac DNA. Within some insect cells, pOCC and DefBac will recombine and this event generates an infectious baculovirus that propagates throughout the culture. On day 5, the user collects the conditioned media (which contains released baculovirus) and uses it to infect fresh Sf9 insect cells. On day 10, the user uses the conditioned media again to infect fresh Sf9 cells. On day 13, the user harvests the infected Sf9 cells containing the target protein of interest (represented by green, GFP+ cells). For secreted protein targets the conditioned medium is collected after this phase

modified the *Autographa californica* baculovirus bacmid (bMON14272) by deleting the 3′-ends of two essential genes: *lef-2* and *AcOrf-1629*. These genes flank the polyhedrin locus and are required for baculovirus production in cultured insect cells [8–11]. Cells transfected with this defective bacmid never appeared infected with baculovirus, and the cells continued to grow and divide at a rate similar to non-transfected controls (data not shown). In contrast, insect cells transfected with the unmodified parental bacmid produced infected cells 72 h post transfection, as expected. We refer to this doubly defective bacmid as H092. The design ensures that the parental bacmid lacking the gene of interest will not produce functional baculovirus.

We designed H092 so that infectivity is restored by site-specific recombination with a pOCC shuttle vector, which contains the gene of interest flanked by 3′-segments of *lef-2* and *AcOrf-1629* that complement the truncations in H092 (Fig. 1a). Recombination at the *lef-2* and *AcOrf-1629* loci restores functionality of these genes, thus producing an infective, recombinant baculovirus genome that expresses the gene of interest (Fig. 1a). Sf9 insect cells efficiently promote homologous recombination, suggesting that co-transfection of H092 and pOCC in Sf9 cells could reliably produce recombinant bacmid and functional virus. Indeed, co-transfection of H092 and a shuttle vector expressing GFP, under the control of the polyhedrin promoter, produced infected, GFP-positive Sf9 cells four days post-transfection (see Fig. 2 for more examples).

To demonstrate that GFP expression was associated with the production of an infectious virus, we collected the conditioned media from transfected cells and diluted it into fresh cultures. After five days, these cultures were incubated with fluorescently-tagged antibody that recognizes the baculovirus GP64 coat protein on the plasma membrane of infected insect cells and then single cell FACS sorted. These singly sorted cells were then added to fresh cultures in a 96-well plate. Out of 96 wells, 34 wells contained clearly infected cells that also strongly expressed GFP. The remaining 62 wells showed neither signs of infection nor GFP expression. These results show that rescue was both precise and efficient and that only recombinant virus expressing the transgene supplied by the shuttle vector are replication-competent. This design ensures that the parental bacmid lacking the gene of interest will not produce functional baculovirus.

### Deletion of two non-essential baculovirus genes improves protein expression

Proteins required for baculoviral host liquefaction are not necessary for infection in cell culture and can often interfere with target protein expression. Two examples include cathepsin (*v-cath*) and chitinase (*chiA*), which are abundantly secreted late in the lytic cycle. Kaba et al. (2004) demonstrated that deletion of these genes improved the integrity and yield of a secreted target protein, p67. To improve our system, we eliminated these genes in H092. The resulting bacmid, called DefBac, produced infective baculovirus when co-tranfected with a pOCC shuttle vector and enhanced protein expression of the TGF-β member Activin A (Fig. 2a; see below).

### A versatile shuttle vector set enables combinatorial appending of tags to the target protein

It is often desirable to append different tags to the target protein to improve expression levels, solubility, purity and visualization. To facilitate this optimization process, we developed a compatible set of 143 shuttle plasmids, called Optimized Classic Cloning (pOCC) vectors, designed to recombine with DefBac and express the user's GOI (Fig. 1b). Using this system, one GOI can be easily inserted into numerous vectors to produce target proteins with various combinations of cleavable N- and C-terminal tags (Fig. 1c). The Defbac viral genome and the most popular 52 pOCC shuttle vectors are available at www.addgene.org (see Additional file 2). The remaining pOCC shuttle vectors are available upon request (see Additional file 1 for the complete collection).

Each pOCC vector contains the same cloning cassette demarcated by NotI and AscI restriction enzyme sites. This cassette encodes a constitutive promoter (gb3) driving the toxic *ccdB* gene (Fig. 1b). A pOCC vector with no insert will kill ccdB-sensitive bacteria (e.g. DH5α) and therefore must be maintained in a resistant strain such as DB3.1. When the target GOI is successfully inserted between the NotI and AscI sites, the *ccdB* gene is lost and the plasmid can be maintained in ccdB-sensitive cells. This feature dramatically reduces isolation of vectors lacking insert (data not shown). If target GOIs contain internal sites for NotI and AscI enzymes, their removal by silent mutation is required. However, recognition sites for both enzymes are 8 bp in length; thus, in practice, this additional mutagenesis step is rarely needed.

On the 5′ and 3′ ends of the cloning cassette are modules encoding purification tags, fluorescent probes, conjugating tags, trafficking peptides, and cleavable linkers (Fig. 1c; see Additional file 1 for a full list and description of tags). The translational reading frame is adjusted so that the NotI and AscI cloning sites encode amino acids with small side chains often found in natural peptide linkers. Upstream of these sequences lie a polyhedrin promoter and a translational start site, while stop codons and transcriptional terminator elements lie downstream (Fig. 1b). Once a GOI is synthesized with compatible NotI and AscI ends, it can be inserted or subcloned into pOCC vectors to produce in-frame fusion proteins.

### An enhanced version of DefBac that highly expresses a pro-protein convertase

Many secreted proteins require pro-protein convertases to generate the functional, mature form. We hypothesized

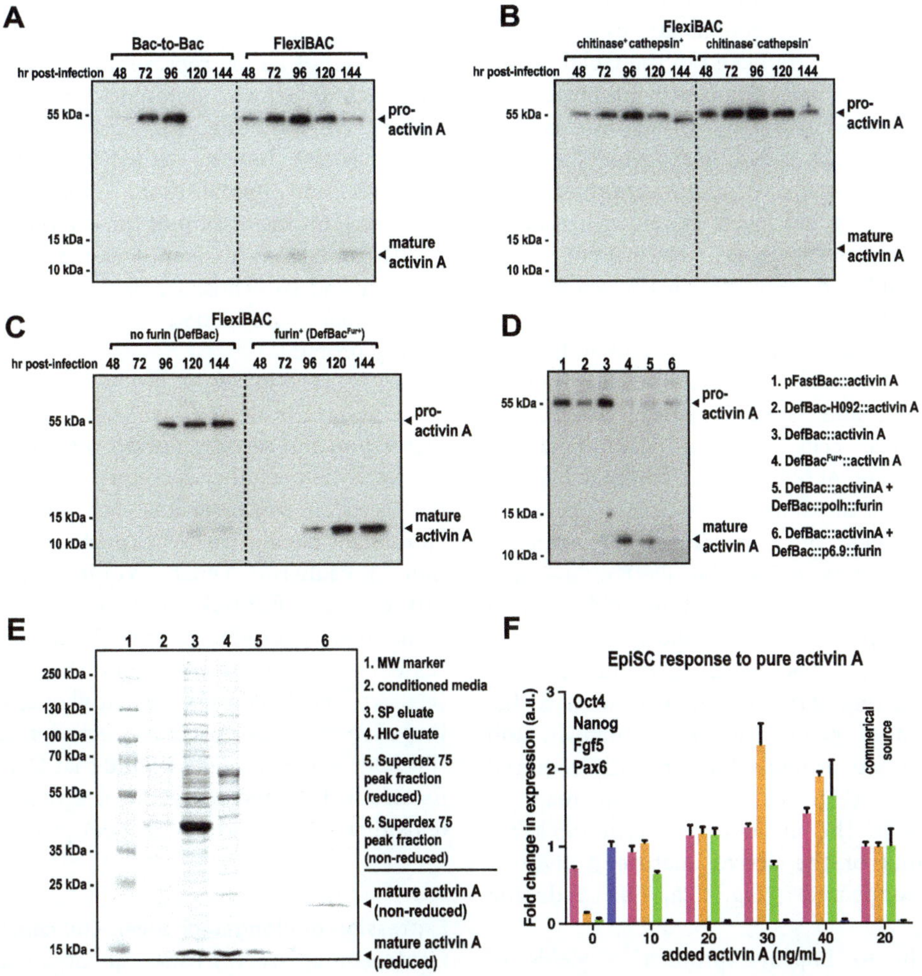


**Fig. 2** Efficient production of mature TGF-β member Activin A using FlexiBAC. **a** Sf9 insect cells were infected with recombinant baculovirus generated using the commercial Bac-to-Bac system (ThermoFisher Scientific) or the FlexiBAC system. Cell supernatants were collected at the indicated times after infection, resolved on 4–20% gradient SDS-PAGE gels under reducing conditions, then analyzed by western blot with anti-Activin A IgG. Bands corresponding to pro-Activin A (uncleaved) and mature Activin A (cleaved) are indicated. **b** Activin A secretion from insect cells infected with baculovirus generated using the precursor backbone for the FlexiBAC system (called "DefBac-H092", which expresses viral chitinase and cathepsin) or the current backbone (called "DefBac", which does not express viral chitinase and cathepsin). **c** Activin A maturation and secretion from insect cells infected with DefBac-derived baculovirus vs. DefBac$^{Fur+}$-derived baculovirus. In addition to expressing the target protein of interest, DefBac$^{Fur+}$ expresses furin convertase, which converts pro-Activin A to its mature form. **d** Insect cells were either singly infected or co-infected with the following virus: pFastBac::Activin A (lane 1), DefBac-H092::Activin A (lane 2), DefBac::Activin A (lane 3), DefBac$^{Fur+}$::Activin A (lane 4), DefBac::Activin A and DefBac::polh::Furin (lane 5), DefBac::Activin A and DefBac::p6.9::Furin (lane6). Each virus was added to insect cells at MOI = 0.2 except for DefBac::p6.9::Furin (MOI = 2). The conditioned media was analyzed by western blot. Samples taken from peak Activin A expression times are shown (96 hpi for conditions 1–5, 72 hpi for condition 6). **e** Multi-step purification of mature Activin A from conditioned media from insect cells infected with DefBac$^{Fur+}$::Activin A. Samples were resolved on an SDS-PAGE gel and stained with coomassie blue. Samples 1–5 were reduced with DTT prior to loading. **f** To assess activity, purified mature Activin A was added to epiblast derived stem cells (EpiSCs) cultured on fibronectin. Fold changes in gene expression, normalized to beta-actin, were determined over 11 days by quantitative polymerase chain reaction (qPCR) for the pluripotency makers Oct4, Nanog, Fgf5, and for the lineage marker, Pax6 (mean +/– S.D.; $n = 3$ separate experiments). Activin A from a commercial source was used as a control (grey box)

that high expression of such target proteins might overwhelm the endogenous proteolytic processing enzymes in the Golgi apparatus. To elevate furin protein levels, we replaced the conotoxin gene in DefBac with full-length cDNA encoding murine furin driven by the p6.9 promoter. This promoter directs transcription earlier during the lytic cycle compared to the strong polyhedrin (polh) promoter driving target gene expression [12]. Thus, increased levels of furin are present before the target is translated. We call this parental bacmid "DefBac$^{Fur+}$".

### Expression and secretion of mature TGF-β family proteins

Transforming growth factor Beta (TGF-β) family proteins are secreted proteins involved in inter-cellular communication during development and tissue homeostasis. After translation, these proteins are imported into the ER in a pro-form, then trafficked to the Golgi apparatus, where they undergo maturation through proteolytic cleavage by furin convertases. The mature form is then secreted to the extracellular space by exocytosis. Recombinant TGF-β proteins are difficult to express at high levels in baculovirus-infected insect cells due to limited capacity of the secretory pathway. To overcome this limitation, we employed our enhanced version of DefBac, called DefBac$^{Fur+}$, which expresses furin.

We focused on production of recombinant Activin A, a secreted TGF-β protein involved in inflammation, neural development, and hematopoiesis [13, 14]. We co-transfected insect cells with a pOCC plasmid (pOCC-RL) containing Activin A and parental bacmid, then collected conditioned media containing secreted protein for 6 days post infection (dpi). We achieved higher Activin A yields using the DefBac parental bacmid compared to a conventional bacmid used in commercially-available systems (bMONH092, the bacmid encoding the baculvirus genome in the Bac-to-Bac system) (Fig. 2a). This improvement is due to the elimination of genes encoding cathepsin and chitinase, two secreted proteins that are only required for host liquefaction (Fig. 2b). However, the majority of Activin A was in the inactive pro-form (55 kDa); only trace amounts of the active, mature Activin A form (13 kDa) appeared (Fig. 2a, b; reducing conditions).

We next sought to improve production yields of mature Activin A. To become fully active, TGF-β -family proteins require C-terminal cleavage by furin convertases that reside in the early Golgi apparatus and recognize the multi-basic motif RX(K/R)R [15]. Western blot analysis revealed that DefBac$^{Fur+}$ greatly improved the production of mature Activin A and its secretion into the extracellular medium, with levels peaking 120 hpi (Fig. 2c). We detected only trace amounts of the pro-form of Activin A in the DefBac$^{Fur+}$ sample (Fig. 2c). DefBac$^{Fur+}$ also produced more mature Activin A compared to cultures co-infected with virus expressing Activin A and virus expressing furin (Fig. 2d).

Next, we purified this mature Activin A and determined its activity. We isolated Activin A from the conditioned medium by cation exchange chromatography (SP), followed by hydrophobic interaction chromatography (HIC) and size exclusion chromatography (Superdex 75). From 1 L of conditioned medium, 0.5 mg of pure Activin A was obtained (Fig. 2e). Addition of purified Activin A (from 10 to 40 ng/mL) to epiblast-derived stem cells (epiSC) suppressed expression of the differentiation marker Pax6 and either maintained or potently promoted the expression of pluripotency markers such FGF5, Nanog, and Oct4, indicating that Activin A was active (Fig. 2f).

The TGF-β family represents a broad class of proteins that include growth, differentiation, and morphogenetic factors. In addition to Activin A, two prominent examples are human growth/differentiation factor 15 (GDF15) and human bone morphogenetic protein 4 (BMP4) [16]. Expression of these targets using DefBac$^{Fur+}$ greatly improved their processing and secretion levels (Fig. 3a and b). Thus, DefBac$^{Fur+}$ is generally suited for recombinant production of active forms of secreted TGF-β proteins.

#### Expression and secretion of processed E-cadherin

To test whether other secretory protein families similarly benefit from co-expression with furin, we generated recombinant baculovirus to express the ectodomain of canine E-cadherin, which requires maturation of the pro-form by a furin-like convertase. When the E-cadherin ectodomain is produced from DefBac virus, some unprocessed, pro-form is secreted into the conditioned medium (Fig. 3c, lane 1) However, when we used DefBac$^{Fur+}$, only the mature, fully-processed ectodomain was secreted (Fig. 3c, lane 2). We conclude that DefBac$^{Fur+}$ is suitable for production and secretion of a range of secreted proteins that require furin-dependent processing.

#### Expression of elongated, coiled-coil centrosome proteins

Centrosomes are micron-scale organelles that nucleate and organize microtubules needed for mitotic spindle assembly. Centrosomes comprise barrel-shaped centrioles surrounded by a less structured mass of protein called pericentriolar material (PCM). Mechanistic understanding of how PCM assembles and nucleates microtubules has long been hampered by lack of in vitro reconstitution approaches. This is attributed to the PCM scaffold being composed of elongated, high-molecular weight proteins that contain coiled-coil domains, which have been notoriously difficult to reconstitute in their full length.

The main PCM scaffold protein in the nematode *C. elegans*, SPD-5, contains 9 predicted coiled-coil domains, has a molecular weight of 135 kDa, and a hydrodynamic radius of ~ 8 nm [17, 18]. Compared to the Bac-to-Bac system (ThermoFisher Scientific), use of the FlexiBAC system improved the consistency and yield of full-length SPD-5 (Fig. 3d). We then used FlexiBAC to express and purify 12 additional full-length PCM proteins (Fig. 3e), which were sufficient to assemble minimal PCM and nucleate microtubule asters in vitro [17]. Thus, the FlexiBAC system is suited for expressing elongated coiled-coil proteins,

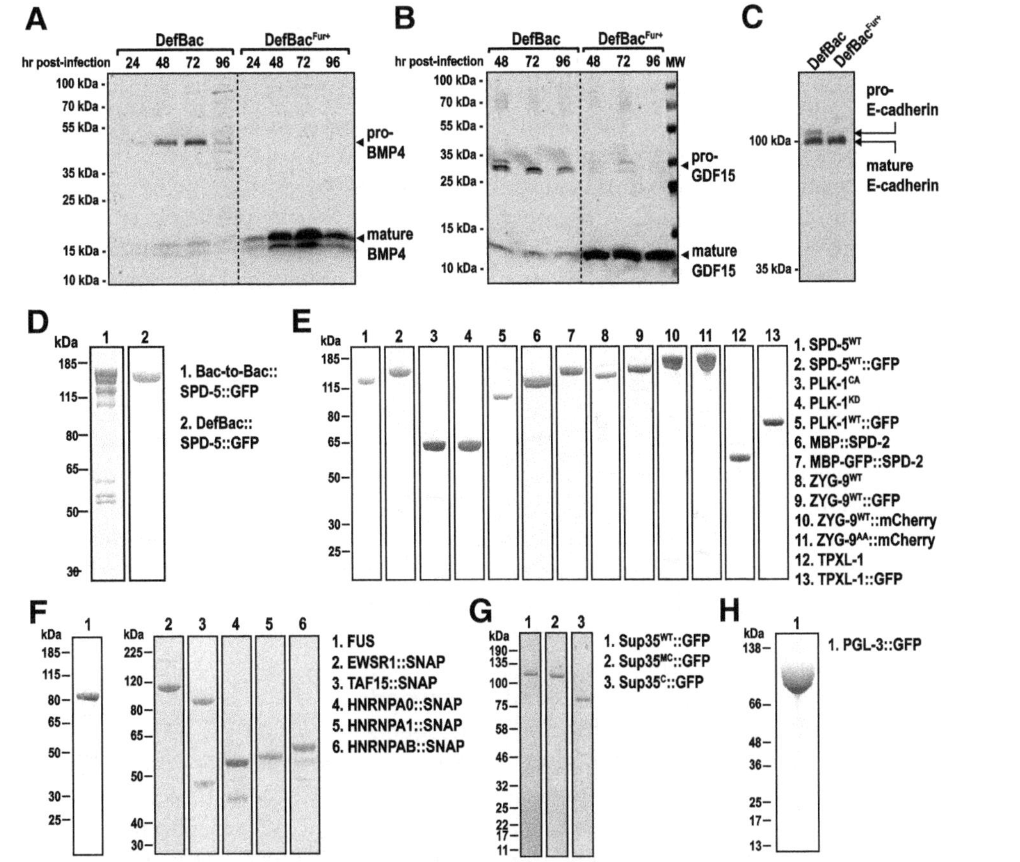


**Fig. 3** Production of additional recombinant proteins using FlexiBAC. **a** Human Bone morphogenetic protein 4 (BMP4) was expressed in insect cells using the two different FlexiBAC viral backbones that differ in their expression of recombinant furin: DefBac and DefBac$^{Fur+}$. BMP4 levels in the conditioned media were analyzed by western blot using an anti-BMP4 antibody. **b** Human Growth/differentiation factor 15 (GDF15) was expressed in insect cells using DefBac and DefBac$^{Fur+}$. GDF15 levels in the conditioned media were analyzed by western blot using an anti-GDF-15 antibody. **c** The ecto-domain of canine E-cadherin was expressed in insect cells using DefBac and DefBac$^{Fur+}$. E-cadherin levels in the conditioned media were analyzed by western blot using an anti-E-cadherin antibody. **d** Coomassie-stained gels depicting 6xHis-tagged SPD-5::GFP expressed with the Bac-to-Bac system (ThermoFisher Scientific) or with DefBac. In both lanes, SPD-5::GFP was purified from insect cell lysate using Ni-NTA affinity chromatography, then eluted with 250 mM imidazole. **e** All proteins in (B-E) were expressed in SF9 insect cells using the FlexiBAC system. Coomassie-stained gels depicting purified *C. elegans* centrosome proteins. For more information, see [17, 27]. **f** Coomassie-stained gels of purified human proteins that contain long stretches of intrinsically-disordered regions. For more information, see [22, 25]. **g** Coomassie-stained gels depicting purified yeast prion-like proteins. MC and C represent middle and C-terminal domains of Sup35. For more information, see [24]. **h** Coomassie-stained gel of purified PGL-3, a protein that constitutes P granules in *C. elegans* embryos. For more information, see [23]

as well as globular kinases and microtubule-stabilizing enzymes.

### Expression of intrinsically-disordered proteins that self-assemble

Approximately 20% of the proteome in eukaryotes comprises protein sequences that lack globular structure, termed "intrinsically-disordered" proteins (IDPs) [19]. Numerous non-membrane-bound organelles—such as nucleoli, stress granules, and germ granules—assemble through the coalescence of IDPs through multivalent interactions with themselves, other IDPs, or RNAs [20, 21]. Historically, IDPs have been difficult to reconstitute, as they are prone to aggregation and end up in inclusion bodies.

The FlexiBAC system was successful in expressing full-length IDPs that form the basis of stress granules (in human cells and yeast) and germ granules (in *C. elegans*) (Fig. 3f-h). Target proteins were isolated from infected cells after multi-step purifications in sufficient yields (1–100 mg/L culture) to carry out extensive biochemical reconstitution assays. Under specific buffer conditions, each of these proteins self-assembles into micron-scale condensates [22–25]. We conclude that FlexiBAC is suitable for expression of disordered proteins that exhibit a strong tendency to self-assemble.

## Discussion

Baculovirus expression has long been used for generating high yields of eukaryotic proteins that are properly folded and have native post-translational modifications. To make baculovirus expression simpler, more reliable, and more flexible, we generated a new system called FlexiBAC. We have demonstrated that FlexiBAC can be used to express diverse targets, including secreted proteins that require proteolytic processing in the Golgi apparatus, as well as intrinsically disordered and coiled-coil proteins.

FlexiBAC exhibits a number of benefits. First, the time from initial cloning to protein production can be as little as 13 days. Second, FlexiBAC is versatile and amenable for screening, as we designed a compatible shuttle vector set (143 vectors) harboring numerous tags used for purification, visualization, trafficking, and labeling with chemical probes. The target gene of interest is easily swapped between the plasmids using classic restriction enzyme cloning techniques. Third, FlexiBAC is suitable for expression of secreted proteins and is ideal for proteins requiring processing by the Golgi-resident furin convertase. Using a viral backbone that overexpresses furin, called DefBac$^{Fur+}$, we produced high yields of Activin A, a TGF-β family protein. We expect our system to be useful for expressing mature versions of other targets requiring processing by a pro-protein convertase. Finally, the FlexiBAC system is open-source and freely available through www.addgene.org.

FlexiBAC has been useful for expressing difficult proteins in numerous studies [17, 22–30]. We envision additional modifications to specifically address other classes of protein targets. For example, with proteins found in heteromeric complexes, expression of individual subunits often is unsuccessful but is greatly improved by stoichiometric co-expression of all the subunits. We are currently adapting the FlexiBAC system to co-express multiple genes for stoichiometric expression of all subunits [31]. In addition, FlexiBAC could also be expanded to express additional factors necessary for native post-translational modifications such as other pro-protein convertases, kinases, phosphatases, and glycosylation enzymes. Use of the DefBac$^{Fur+}$ bacmid might also improve maturation of recombinant proteins secreted by baculovirus-infected insect larvae. Larval expression produces very high levels of secreted recombinant proteins and is more economical compared to expression in cultured insect cells [32, 33]. Thus, use of FlexiBAC might be applied for large-scale production of growth factors or nanobodies.

The speed, simplicity, and flexibility of the FlexiBAC system make it an ideal starting platform for expression and screening of diverse types of recombinant proteins. Given that FlexiBAC is open-source, we invite improvements that will expand its applicability and facilitate production of a broader range of proteins.

## Conclusions

FlexiBAC is an open-source baculovirus expression system that is freely available at www.addgene.org. The FlexiBAC system includes a shuttle vector set that can append 143 combinations of tags to the user's protein of interest. Finally, FlexiBAC is engineered to permit high level expression of secreted proteins that require furin-dependent proteolytic maturation. By increasing protein tagging possibilities, allowing bacmid formation within insect cells, and providing selection against improper baculoviruses, FlexiBAC simplifies baculovirus-mediated expression of cytosolic proteins and secreted proteins that require proteolytic processing.

## Methods

### Reagents

Restriction enzymes were purchased from New England Biolabs (Ipswich, MA, USA). *E. coli* DH10Bac containing bMON14272 [34] was purchased from Thermo Fisher Scientific (Waltham, MA, USA). *E. coli* GB2005 and pSC101-BAD-gba-Cm [35] were kindly provided by Francis Stewart (Tech. Univ. Dresden). Colicin E1 was expressed and purified as a His6-tagged fusion from pETMM11_colE1_imm, an expression construct derived from pT7-7_ColE1/Imm that was kindly provided by William A. Cramer (Purdue Univ.). pOET1, pBAC4X-1, the human activin A gene in pCMV6-AC (SC319183) were purchased from Oxford Expression Technologies (Oxford, UK), EMD Millipore (Billerica, MA, USA), and OriGene (Rockville, MD, USA) respectively. DNA constructs and maps were designed using Gene Construction Kit 3.5 from Textco Biosoftware, Inc. (West Lebanon, NH, USA). pSVL fur-mur (D1352) containing murine furin was a kind gift from L. Robertson (Univ. of Oxford) [36]. Sf-9ET cells developed by Hopkins and Esposito (NCI, Bethesda, MD) [37] for baculovirus titering were kindly provided by Sabine Suppmann (MPI-Biochemie, München). Sf9 ESF insect cells and ESF 921 Insect Cell Culture Medium, Protein-Free, were purchased from Expression Systems (Davis, CA, USA). Escort™ IV transfection reagent was purchased from Sigma-Aldrich (St. Louis, MO, USA). Oligonucleotides were purchased from Metabion (Planegg, Oberbayern, Germany). Infected insect cells were sorted using the BD FACS Aria cell sorter from BD Biosciences (San Jose, CA, USA).

### Generation of the double defective baculovirus genome (DefBac) using recombineering

We used the selection/counterselection system employing TolC, an *E. coli* outer membrane protein, in combination with the λ redgam recombination system to generate seamless insertions or deletions in both a bacmid encoding the baculovirus genome as well as the *E. coli* genome [38]. This method generates rapid and "scar free" modifications and avoids the use of classic antibiotic drug selections

Prior to carrying out recombineering on the baculovirus bacmid, the genomic copy of TolC was deleted from the host *E. coli* strain, GB2005. This deletion was made by inducing Lambda Red recombination enzymes using pSC101-BAD-gba-Cm followed by transformation with an oligo (B22I3; all oligonucleotides sequences are listed in Table S1 in Additional file 1), homologous to genomic sequences flanking the TolC gene. Correct clones were identified as having resistance to the toxin Colicin E1. This deletion strain, named RL001, was sensitive to sodium dodecyl sulfate (SDS) and could host seamless TolC selection/counterselection recombineering of the bacmid encoding the baculovirus genome (bMON14272).

To generate baculovirus bacmid that could not self-replicate, we truncated two genes, *lef2* and Ac-*Orf1629*, which flank the *polyhedrin* locus using Lambda Red recombineering in RL001. For each deletion, the C-terminus of the encoded protein was replaced with a cassette composed of the TolC gene flanked with 50 base pair homology arms for precise integration into the target DNA. Subsequently, the TolC gene was removed by counterselection after recombination of the bacmid with an oligonucleotide composed of a direct fusion of those same homology arms. The strain harboring the bacmid with these two disabled baculovirus genes was named H092.

TolC recombineering was then used to delete *v-cathepsin* and *v-chitinase* genes from the H092 bacmid to generate H099, which is the final baculoviral genome we termed "DefBac". More detailed information about intermediate steps of bacmid production can be found in the supplement.

### Generation of the pOCC shuttle vectors

We modified the vector pOET1 (Oxford Expression Technologies) following the Uracil-Specific Excision Reagent (USER) cloning technique (New England Biolabs). Briefly, the vector backbone of EcoRI linearized pOET1 was amplified by PCR using primers B28B9 and B28C1. Primers B22F6 and B22F7 were then used to amplify the polyhedrin promoter/MCS/polyA terminator cassette of EcoRV linearized pFastBacM11. The two PCR products were then treated with the USER enzyme and transformed into DH5α cells to generate the desired construct, pOEM1.

To make the first pOCC baculovirus shuttle vector, pOCC5, a PCR fragment containing the *ccdB* gene under the constitutive *gb3* promoter was amplified from the vector pNCP2 [39] with primers to generate NcoI and HindIII ends. This fragment was cut with NcoI and HindIII and then sub-cloned into pOEM1. We then inserted sequences for both N- or C-terminal tagging (see Fig. 1). In the course of this a work a diverse collection of baculovirus shuttle vectors was generated in this manner (see Additional file 1). All the pOCC shuttle vectors and defective bacmids described herein are available either at www.addgene.org or upon request. Detailed information about construction of intermediate plasmids and viral backbones can be found in the supplement.

### Antibodies

Goat anti-Activin (#AF338) and Goat anti-BMP4 (#AF757) antibodies were purchased from R&D Systems (Minneapolis, MN, USA). Rabbit anti-Furin (#PA1–062) antibody was purchased from Thermo Fisher Scientific (Waltham, MA, USA). PE labeled Mouse anti-GP64 antibody (#sc-65,498 PE) was purchased from Santa Cruz Biotechnology (Dallas, TX, USA). Mouse Penta-His antibody (#34660) was purchased from Qiagen (Hilden, Germany). Rabbit anti-GDF15 (Mic-1/D2A, #8479; Rabbit mAb) was purchased from Cell Signaling.

### Cell culture

Sf9-ESF insect cells (Expression systems, Davis, CA, USA) were cultured at 27 °C in suspension in glass culture flasks in ESF 921 serum-free medium (Expression Systems, Davis, CA, USA).

### Western blot analysis

After infections, the supernatant samples are resolved into reducing 4–20% gradient SDS–PAGE gels. Cell lysates are prepared with PBS buffer supplemented with a cocktail of protease inhibitors. Separated proteins are then transferred onto nitrocellulose membranes, blocked for 2 h, and probed overnight with Goat anti-Activin IgG (1:5000) (R&D systems, Minneapolis, MN), Rabbit anti-Furin IgG (1:2000) (Thermo Scientific) or Goat anti-BMP4 (1:5000) (R&D systems). Immunoreactive bands are revealed using ECL detection.

### Activin a activity assay

Epiblast derived stem cells (EpiSCs) were cultured on fibronectin in N2B27 medium with 12 ng/ml FGF2 and the respective amount of ActA for 11 days over 4 passages. Fold changes in gene expression, by quantitative polymerase chain reaction (qPCR), were determined for cells for the pluripotency makers Oct4, Nanog, Fgf5, and for the lineage marker, Pax6 and normalized to levels of beta-actin. Results are presented as the mean of triplicate determinations with error bars showing the standard deviation.

## Additional files

**Additional file 1: Table S1.** Inventory of available pOCC shuttle vectors. The user's gene of interest can be inserted into 143 available vectors using NotI and AscI restriction cloning (XLSX 14 kb)

**Additional file 2: Table S2.** Inventory of pOCC shuttle vectors available at www.addgene.org. The 52 most popular shuttle vectors were deposited to Addgene (XLSX 11 kb)

**Additional file 3:** Supplementary Materials and Methods. A detailed protocol for using FlexiBAC and more information on plasmid and bacmid construction. (DOCX 39 kb)

### Abbreviations

dpi: Days post infection; EpiSCs: Epiblast derived stem cells; FACS: Fluorescence activated cell sorting; GFP: Green fluorescent protein; GOI: Gene of interest; IDP: Intrinsically disordered protein; MOI: Multiplicity of infection; Sf9: *Spodoptera frugiperda* cell line; TGF-β: Transforming growth factor Beta

### Acknowledgements

We thank the support of the Max Planck Society and the Protein Expression Purification and Characterization (PEPC) Facility at the MPI-CBG. We gratefully acknowledge Katrin Neumann (Tech. Univ. Dresden) for assaying the activity of Activin A in epiSCs.

### Funding

J.B.W. is supported by the Cancer Prevention Research Institute of Texas (CPRIT grant RR170063) and the Endowed Scholars program at UT Southwestern. D.N.D was supported by the Max Planck Society. The funders played no role in the design or execution of this study.

### Authors' contributions

RPL and DD performed the experiments. DD, RPL, AB, BB created the plasmids and viral backbone. JBW optimized expression and purification protocols. DD conceived of the project. DD and JBW wrote the manuscript. All authors have read and approved the manuscript.

### Ethics approval and consent to participate

Not applicable.

### Consent for publication

Not applicable.

### Competing interests

The authors declare no competing interests.

### Author details

[1]Max Planck Institute of Molecular Cell Biology and Genetics, Pfotenhauerstrasse 108, 01307 Dresden, Germany. [2]Department of Cell Biology, Dept. of Biophysics, UT Southwestern Medical Center, Dallas, TX 75390, USA. [3]Research Institute of Molecular Pathology (IMP), Campus-Vienna-Biocenter 1, 1030 Vienna, Austria.

### References

1. Jarvis DL. Baculovirus-insect cell expression systems. Methods Enzymol. 2009;463:191–222.
2. van Oers MM, Pijlman GP, Vlak JM. Thirty years of baculovirus-insect cell protein expression: from dark horse to mainstream technology. J Gen Virol. 2015;96(Pt 1):6–23.
3. Kaba SA, Salcedo AM, Wafula PO, Vlak JM, van Oers MM. Development of a chitinase and v-cathepsin negative bacmid for improved integrity of secreted recombinant proteins. J Virol Methods. 2004;122(1):113–8.
4. Hitchman RB, Possee RD, Crombie AT, Chambers A, Ho K, Siaterli E, et al. Genetic modification of a baculovirus vector for increased expression in insect cells. Cell Biol Toxicol. 2010;26(1):57–68.
5. Leduc R, Molloy SS, Thorne BA, Thomas G. Activation of human furin precursor processing endoprotease occurs by an intramolecular autoproteolytic cleavage. J Biol Chem. 1992;267(20):14304–8.
6. Cronin CN, Thompson DA, Martin F. Expression of bovine activin-a and inhibin-a in recombinant baculovirus-infected Spodoptera frugiperda sf 21 insect cells. Int J Biochem Cell Biol. 1998;30(10):1129–45.
7. Laprise MH, Grondin F, Dubois CM. Enhanced TGFbeta1 maturation in high five cells coinfected with recombinant baculovirus encoding the convertase furin/pace: improved technology for the production of recombinant proproteins in insect cells. Biotechnol Bioeng. 1998;58(1):85–91.
8. Kitts PA, Possee RD. A method for producing recombinant baculovirus expression vectors at high frequency. BioTechniques. 1993;14(5):810–7.
9. Je YH, Chang JH, Choi JY, Roh JY, Jin BR, O'reilly DR, et al. A defective viral genome maintained in Escherichia coli for the generation of baculovirus expression vectors. Biotechnology letters. 2001;8:575–82.
10. Zhao Y, Chapman DA, Jones IM. Improving baculovirus recombination. Nucleic Acids Res. 2003;31(2):E6.
11. Wu CP, Huang YJ, Wang JY, Wu YL, Lo HR, Wang JC, et al. Autographa californica multiple Nucleopolyhedrovirus LEF-2 is a capsid protein required for amplification but not initiation of viral DNA replication. J Virol. 2010;84(10):5015–24.
12. Hill-Perkins MS, Possee RD. A baculovirus expression vector derived from the basic protein promoter of Autographa californica nuclear polyhedrosis virus. J Gen Virol. 1990;71 (Pt 4:971–6.
13. Vale W, Rivier J, Vaughan J, McClintock R, Corrigan A, Woo W, et al. Purification and characterization of an FSH releasing protein from porcine ovarian follicular fluid. Nature. 1986;321(6072):776–9.
14. Chen YG, Wang Q, Lin SL, Chang CD, Chuang J, Ying SY. Activin signaling and its role in regulation of cell proliferation, apoptosis, and carcinogenesis. Exp Biol Med (Maywood). 2006;231(5):534–44.
15. Thomas G. Furin at the cutting edge: from protein traffic to embryogenesis and disease. Nat Rev Mol Cell Biol. 2002;3(10):753–66.
16. Wakefield LM, Hill CS. Beyond TGFbeta: roles of other TGFbeta superfamily members in cancer. Nat Rev Cancer. 2013;13(5):328–41.
17. Woodruff JB, Ferreira Gomes B, Widlund PO, Mahamid J, Honigmann A, Hyman AA. The centrosome is a selective condensate that nucleates microtubules by concentrating tubulin. Cell. 2017;169(6):1066–77 e10.
18. Hamill DR, Severson AF, Carter JC, Bowerman B. Centrosome maturation and mitotic spindle assembly in C. Elegans require SPD-5, a protein with multiple coiled-coil domains. Dev Cell. 2002;3(5):673–84.
19. Peng Z, Yan J, Fan X, Mizianty MJ, Xue B, Wang K, et al. Exceptionally abundant exceptions: comprehensive characterization of intrinsic disorder in all domains of life. Cell Mol Life Sci. 2015;72(1):137–51.
20. Banani SF, Lee HO, Hyman AA, Rosen MK. Biomolecular condensates: organizers of cellular biochemistry. Nat Rev Mol Cell Biol. 2017;18(5):285–98.
21. Hyman AA, Brangwynne CP. Beyond stereospecificity: liquids and mesoscale organization of cytoplasm. Dev Cell. 2011;21(1):14–6.
22. Patel A, Lee HO, Jawerth L, Maharana S, Jahnel M, Hein MY, et al. A liquid-to-solid phase transition of the ALS protein FUS accelerated by disease mutation. Cell. 2015;162(5):1066–77.
23. Saha S, Weber CA, Nousch M, Adame-Arana O, Hoege C, Hein MY, et al. Polar positioning of phase-separated liquid compartments in cells regulated by an mRNA competition mechanism. Cell. 2016;166(6):1572–84 e16.
24. Franzmann TM, Jahnel M, Pozniakovsky A, Mahamid J, Holehouse AS, Nuske E, et al. Phase separation of a yeast prion protein promotes cellular fitness. Science. 2018;359(6371).

25. Wang J, Choi JM, Holehouse AS, Lee HO, Zhang X, Jahnel M, et al. A molecular grammar governing the driving forces for phase separation of prion-like RNA binding proteins. Cell. 2018.
26. Woodruff JB, Hyman AA. Method: in vitro analysis of pericentriolar material assembly. Methods Cell Biol. 2015;129:369–82.
27. Woodruff JB, Wueseke O, Viscardi V, Mahamid J, Ochoa SD, Bunkenborg J, et al. CentrosomesRegulated assembly of a supramolecular centrosome scaffold in vitro. Science. 2015;348(6236):808–12.
28. Boke E, Ruer M, Wuhr M, Coughlin M, Lemaitre R, Gygi SP, et al. Amyloid-like self-assembly of a cellular compartment. Cell. 2016;166(3):637–50.
29. Murray DH, Jahnel M, Lauer J, Avellaneda MJ, Brouilly N, Cezanne A, et al. An endosomal tether undergoes an entropic collapse to bring vesicles together. Nature. 2016;537(7618):107–11.
30. Hernandez-Vega A, Braun M, Scharrel L, Jahnel M, Wegmann S, Hyman BT, et al. Local nucleation of microtubule bundles through tubulin concentration into a condensed tau phase. Cell Rep. 2017;20(10):2304–12.
31. Weissmann F, Petzold G, VanderLinden R, Brown NG, Lampert F, et al. biGBac enables rapid gene assembly for the expression of large multisubunit protein complexes. Proc Natl Acad Sci U S A. 2016;113(19):E2564–9.
32. Gomez-Sebastian S, Nunez MC, Garaicoechea L, Alvarado C, Mozgovoj M, Lasa R, et al. Rotavirus A-specific single-domain antibodies produced in baculovirus-infected insect larvae are protective in vivo. BMC Biotechnol. 2012;12:59.
33. Dolinska MB, Wingfield PT, Sergeev YV. Purification of recombinant human Tyrosinase from insect larvae infected with the Baculovirus vector. Curr Protoc Protein Sci. 2017;89(1):6–15.
34. Luckow VA, Lee SC, Barry GF, Olins PO. Efficient generation of infectious recombinant baculoviruses by site-specific transposon-mediated insertion of foreign genes into a baculovirus genome propagated in Escherichia coli. J Virol. 1993;67(8):4566–79.
35. Wang J, Sarov M, Rientjes J, Fu J, Hollak H, Kranz H, et al. An improved recombineering approach by adding RecA to lambda red recombination. Mol Biotechnol. 2006;32(1):43–53.
36. Constam DB, Robertson EJ. Regulation of bone morphogenetic protein activity by pro domains and proprotein convertases. J Cell Biol. 1999;144(1):139–49.
37. Hopkins R, Esposito D. A rapid method for titrating baculovirus stocks using the sf-9 easy titer cell line. BioTechniques. 2009;47(3):785–8.
38. DeVito JA. Recombineering with tolC as a selectable/counter-selectable marker: remodeling the rRNA operons of Escherichia coli. Nucleic Acids Res. 2008;36(1):e4.
39. D'Urzo N, Martinelli M, Nenci C, Brettoni C, Telford JL, Maione D. High-level intracellular expression of heterologous proteins in Brevibacillus choshinensis SP3 under the control of a xylose inducible promoter. Microb Cell Factories. 2013;12:12.

11

# MicroRNA-218 promotes early chondrogenesis of mesenchymal stem cells and inhibits later chondrocyte maturation

Song Chen[1†], Zhenyu Xu[2†], Jiahua Shao[1†], Peiliang Fu[1*] and Haishan Wu[1*]

**Abstract**

**Background:** MicroRNAs (miRNAs) reportedly participate in the mesenchymal stem cell (MSC) chondrogenic differentiation regulation. We objected to examine how miR-218 regulate chondrogenic differentiation of synovium-derived MSCs (SDSCs) and the maturation of RCJ3.1C5.1 chondrocytes. SDSCs were sourced from the knee joint synovium of New Zealand white rabbits, and their multilineage differentiation potentials were examined. The level of miR-218 was measured during SDSC chondrogenic differentiation, together with determination of SDSCs chondrogenic markers and RCJ3.1C5.1 chondrocytes maturation markers expression level after transfection of miR-218 mimics/inhibitor.

**Results:** miR-218 expression was notably upregulated in early chondrogenesis but mostly ceased during the maturation phases of chondrogenic differentiation in SDSCs. The transfection of miR-218 mimics notably enhanced SDSCs chondrocytes differentiation, as evidenced by augmented expressions of chondrogenic markers (SOX9, COL2A1, ACAN, GAG, and COMP) in terms of mRNA and protein level, and the inhibition of miR-218 yielded opposite resutls. Additionally, miR-218 overexpression substantially suppressed the expression of osteogenic markers (ALP, BSP, COL1A1, OCN and OPN) during the early stage of chondrogenesis while increasing that of chondrogenic markers (SOX9, COL2A1, ACAN, GAG and COMP). However, miR-218 mimics notably suppressed maturation markers (CMP, COL10A1, MMP-13 and VEGF) expression in RCJ3.1C5.18 chondrocytes, and the miR-218 inhibitor promoted the expression of these maturation markers. We proposed miR-218 plays a regulatory role on 15-hydroxyprostaglandin dehydrogenase (HPGD), which plays a key role in chondrogenic differentiation, and this finding indicates that miR-218 directly regulates HPGD expression in SDSCs.

**Conclusion:** Our study suggests that miR-218 contributes to early chondrogenesis while suppressing later chondrocyte maturation. The miR-218-HPGD pathway offers us a perspective into how SDSCs differentiate into chondrogenic cells.

**Keywords:** Articular cartilage, microRNA-218, Mesenchymal stem cells, Chondrogenesis, Chondrocyte maturation

## Background

Articular cartilage (AC) is an avascular tissue, and thus is characterised with restricted ability to repair itself after trauma or degenerative disease [1]. Various methods are attempted to treat AC injuries, including abrasion under arthroscope, transplantations with mature chondrocytes and mesenchymal stem cells (MSCs), or grafted with tissues like periosteum. Autologous chondrocyte transplantation (ACT), which delivers autologous chondrocytes into the area of defect [2], has been proposed by most clinicians as an feasible approach for cartilage repair, though randomized trials have reported little clinical benefit with this procedure compared with subchondral bone microfracture [3]. An apparent drawback with ACT is it involved morbidity to the donor-site and therefore resultant osteoarthritic changes in the donor joint [4], emphasizing the need for alternative sources of chondrocyte cells. Adult stem cells have been proposed as an excellent cell candidate for cartilage repair due to their proliferation and multilineage differentiation capacity. Especially,

* Correspondence: fupeiliang@163.com; drisland@163.com
[†]Song Chen, Zhenyu Xu and Jiahua Shao contributed equally to this work.
[1]Department of Arthroplasty Surgery, Changzheng Hospital, Second Military Medical University, Shanghai 200003, China

synovium-derived MSCs (SDSCs) are unique because of their tissue specificity for cartilage regeneration [5], and these cells have been shown to have great potential in cartilage regeneration research.

MicroRNAs (miRNAs) is a class of small, noncoding, single-stranded RNAs that regulate at posttranscriptional level through either mRNA translation inhibition or mRNA degradation promotion. There are many reports on critical roles miRNAs play in biology, such as differentiation, development, proliferation, and tumorigenesis. Moreover, several miRNAs have been identified to modulate cartilage differentiation and degradation [6–8]. For example, miR-365 targets histone deacetylase 4 and enhances chondrogenesis [9], and miR-145 inhibits early chondrogenesis by targeting Sox9 [10]. Recent reports have found that miR-218 targets the transcription factor Runx2 to control development pathway in osteogenic lineage [11]. It also activates WNT signaling, which play a role in the MSCs chondrogenic differentiation [12, 13]. However, the paticipation of miR-218 in SDSC chondrogenic differentiation and maturation stays unclear.

As an important regulator, prostaglandin E2 (PGE2) significantly promotes chondrogenic markers (SOX9, ACAN, COMP and COL2A1) expression in the early stage of chondrogenic differentiation and inhibits chondrocyte maturation in the late stage. HPGD is a known effecter that regulates PGE2 by converting it to its biologically inactive metabolite. Taken together, Runx2 and HPGD (dehydrogenase of PGE2) play key roles in chondrogenic differentiation and chondrocyte maturation during chondrogenic differentiation.

## Methods

### Isolation and culture of SDSCs

Synovial tissue was harvested from the knees of New Zealand White rabbits which were obtained from the Animal Center of Second Military Medical University. Surgeries were approved by Animal Experimentation Ethics Committee of Second Military Medical University. The rabbits were anaesthetized through the intravenous injection of 0.6% pentobarbital sodium; the injection dosages were 22–26 mg/kg, and the injection rates were 10–20 ml/5–10 min. At a respiratory rate of 46–68/min, the heart rate was 180–210/min, and the corneal reflex was weakened, indicating that the rabbit was under anesthesia. After the isolation of synovial tissue, all the rabbits were alive and fed at the Animal Center of Second Military Medical University. The synovial tissue was finely minced and digested with 0.1% trypsin (Roche DiACANostics, Mannheim, Germany) at 37 °C for 30 min and then with 0.1% collagenase D (Roche) for 2 h. Cells were collected after filtering and plated at clonal density in 60-$cm^2$ culture dishes (Nalge Nunc International, Rochester, N.Y., USA) in complete medium (α-minimum essential medium containing 10% fetal bovine serum, 100 U/mL penicillin, 100 mg/mL streptomycin, and 0.25 mg/mL fungizone [Invitrogen, Carlsbad, CA]) and incubated at 37 °C with 5% humidified $CO_2$. Non-adherent cells were removed when the medium was changed every two days. SDSCs were harvested when adherent cells reached ~ 80% confluence.

### Detection of SDSC surface markers

A flow cytometry analysis was performed to confirm isolated SDSCs surface markers as previously described [14]. Anti-rabbit CD44-PE, CD90-FITC (Abcam, Cambridge, MA, USA), CD105-FITC, CD14-FITC, CD34-FITC, and CD45-FITC (eBioscience, San Diego, CA, USA) were employed as monoclonal antibodies. Flow cytometry was conducted with a FACSCalibur cytometer (BD Biosciences), and data analyses were performed using BD FAC Suite software (BD Biosciences).

### Adipogenic, osteogenic and SDSCs chondrogenic differentiation

SDSCs at passage 3 were collected and seeded in a six-well plate at a density of $1 \times 10^5$/well. When they reached ~ 80% confluence, the SDSCs were incubated in either adipogenic differentiation medium, osteogenic differentiation medium, or chondrogenic differentiation medium supplemented with 10 ng/mL transforming growth factor beta 3 (TGF-β3). All three types of differentiation media were purchased from BD Biosciences (Bedford, MA, USA), and the culture medium was changed every 3 days. Staining for adipogenic, osteogenic, or chondrogenic differentiation was performed with fresh Oil Red O solution (Sigma-Aldrich, MO, USA) for 2 h, 0.1% Alizarin Red (Sigma-Aldrich, MO, USA) solution for 40 min, or 1% Alcian Blue solution (Sigma-Aldrich) for 30 min at room temperature, respectively.

### Lentivirus-mediated cell transfection

Recombinant lentivirus miR-218-mimics and miR-218-inhibitor were purchased from GeneChem (Shanghai, China). For the lentivirus infection, differentiated SDSCs and RCJ3.1C5.18 chondrocytes were seeded in a 12-well plate and cultured overnight to 70% confluence according to the manufacturers' instructions. The cells were then transfected with different lentiviruses for 24 h.

### Reverse transcription quantitative polymerase chain reaction (RT-qPCR)

Total RNA was extracted using TRIzol reagent (Life Technologies Inc., Grand Island, NY) and RNeasy Mini Kit (Qiagen, Valencia, CA) according to the manufacturer's instructions. Approximately 1 μg of mRNA was used for reverse transcriptase with a High-Capacity cDNA Archive Kit (Applied Biosystems) at 37 °C for 2 h. For miRNA detection, reverse transcription was performed using miRNA-specific stem-loop primers, and real-time PCR was performed using SYBR green Master Mix. Chondrogenic marker genes *SOX9, COL2A1, ACAN, GAG* and *COMP,* osteogenic marker genes *ALP, BSP, Col1a1, OCN* and *OPN,* and maturation markers *CMP, Col10a1, MMP-13,* and *VEGF* were customized by Applied Biosystems as part of Custom TaqMan Gene Expression Assays (Table 1). β-Actin and U6 were employed as internal controls to determine mRNA and miRNA expression levels. The RT-PCR conditions were as follows: an initial 10-min incubation at 95 °C, 40 cycles of 95 °C for 10 s, 60 °C for 20 s and 72 °C for 30 s, and 5 min at 4 °C. Relative quantification analysis was conducted using the $2^{-\Delta\Delta CT}$ method. Each sample was analyzed in triplicate, and all experiments were carried out three times independently.

### Western blot analysis

For total protein extraction, cells were lysed in RIPA buffer (GuHCl, Sigma-Aldrich) supplemented with phenylmethanesulfonyl fluoride (PMSF) and centrifuged at 12,000×g and 4 °C for 10 min. The supernatant was collected, and total protein was quantified using a BCA Protein Assay Kit (Sigma-Aldrich) according to the manufacturer's protocols. Equal amounts (40 μg) of protein samples were separated by 10% sodium dodecyl sulfate-polyacrylamide gel electrophoresis (SDS-PAGE) and transferred onto polyvinylidene fluoride (PVDF) membranes (Millipore, Billerica, MA, USA). After blocking with 5% nonfat milk in TBST (100 mM Tris-HCl, 0.9% NaCl, 1% Tween 20, pH 7.5) for 1 h, membranes were individually incubated with primary monoclonal antibodies against chondrogenic markers SOX9, COL2A1, ACAN, GAG and COMP, osteogenic markers ALP, BPS, COL1A1, OCN and OPN, and maturation markers CMP, COL10A1, MMP-13 and VEGF in 1% nonfat milk in TBST for 3 h at room temperature. Membranes were then incubated with a secondary horseradish peroxidase (HRP)-conjugated goat anti-rabbit antibody (dilution factor of 1:5000) for 40 min at room temperature, followed by exposure in SuperSignal West Femto Maximum Sensitivity Substrate using CL-XPosure Film.

### Immunocytochemistry

A total of $2.5 \times 10^5$ passage-3 SDSCs were seeded into 15-mL centrifuge tubes and cultured in 0.5 mL of chondrogenic differentiation medium for 21 days. The pellets ($n = 2$) were fixed in 4% paraformaldehyde at 4 °C overnight, followed by dehydration in a gradient ethanol series, clearing with xylene, and embedding in paraffin blocks. Sections with a thickness of approximately 5 mm were stained with Alcian Blue (Sigma; counterstained with Fast Red) and Safranin O (Sigma; counterstained with hematoxylin) for sulfated glycosaminoglycans (GAGs). For immunohistochemistry analysis, sections were probed with primary antibodies against collagen II (II-II6B3; DSHB, Iowa City, IA, USA), followed by the secondary antibody biotinylated horse anti-rabbit IgG

**Table 1** Sequences of primers used for real-time PCR analysis

| Name | Forward primer (5'-3') | Reverse primer (5'-3') |
|---|---|---|
| Chondrogenic markers | | |
| Sox9 | TGGCAAGGCTGACCTGAAG | GCTCAGCTCGCCGATGT |
| Col2a1 | TCCTGGCCTCGTGGGT | GGGATCCGGGAGAGCCA |
| Aggrecan | GCCACTGTTACCGCCACTT | CACTGGCTCTCTGCATCCA |
| COMP(cartilage oligomeric matrix protein) | ACTGCCTGCGTTCTAGTGC | CGCCGCATTAGTCTCCTGAA |
| Osteogenic markers | | |
| ALP | CCCTTCACTGCCATCCTGTAC | CCATGGAGACGTTCTCTCTCTCA |
| BSP (bone sialoprotein) | ATGCCTGCCTTGTACCACGAGC | TCCATCGAAGAATCAAAGCAGAG |
| Col1a1 | CCGTGCCCTGCCAGATC | CAGTTCTTGATTTCGTCGCAGATC |
| OCN(Osteocalcin) | GGCAGGGAAGTCAGGGTAG | CCCGTGGTTTCCTGGTC |
| OPN(Osteopontin) | GTACCCTGATGCTACAGACG | TTCATAACTGTCCTTCCCAC |
| Mature markers | | |
| CMP(cartilage matrix protein) | AAGAACGACGACCAAAAGGAC | CATCCCCTATACCATCGCCA |
| Col10a1 | GGCCCGGCAGGTCATC | TGGGATGCCTTTTGGTCCTT |
| MMP-13 | AGTTTGGCCATTCCTTAGGTCTTG | GGCTTTTGCCAGTGTAGGTATAGAT |
| VEGF | GGCTGGCAACATAACAGAGAA | CCCCACATCTATACACACCTCC |
| β-actin | CTGAACCCTAAGGCCAACCG | GTCACGCACGATTTCCCTCTC |

(Vector, Burlingame, CA, USA). Immunoactivity was detected using Vectastain ABC reagent (Vector) with 3,3′-diaminobenzidine (DAB) as a substrate. Hematoxylin (Vector) served as a counterstain.

### Luciferase reporter assay

pMiR-REPORT-mut-HPGD 3'UTR constructs containing the HPGD 3′ untranslated region (UTR) with three point mutations in the seed sequence were produced using a QuikChange II Site-Directed Mutagenesis kit (Stratagene). HEK293 cells were seeded in a 96-well plate for 24 h and then cotransfected with pMIR-REPORT-HPGD 3'UTR plasmid or pMIR-REPORT-mut-HPGD 3'UTR plasmid, internal control pRL-TK-Renilla-luciferase plasmid and the indicated RNAs. In contrast, the pMIR-REPORT-HPGD 3'UTR plasmid or pMIR-REPORT-mut-HPGD 3'UTR plasmid and pRL-TK plasmid were transfected into SDSCs in 24-well plates. After the induction of chondrogenic differentiation for 14 days, luciferase activities were determined using a Dual Luciferase Reporter Assay Kit (Promega). All transfection experiments were conducted in triplicate and repeated three times independently.

### Statistical analysis

The experimental values are presented as the means ± standard deviations (SDs). The statistical analyses were conducted with GraphPad Prism Software Version 6.0 (GraphPad Software Inc., La Jolla, CA, USA). Significant differences between groups were evaluated by one-way ANOVA. $P<0.05$ was considered statistically significant.

## Results

### Identificaiton and determination of multilineage differentiation capacity of SDSCs

To confirm whether our cultured SDSCs exhibit the characteristics of MSCs, passage-3 cells were analyzed by flow cytometry. Rabbit SDSCs were positive for CD44, CD90, and CD105 but negative for CD14, CD34 and CD45 (Fig. 1). To further identify the multilineage

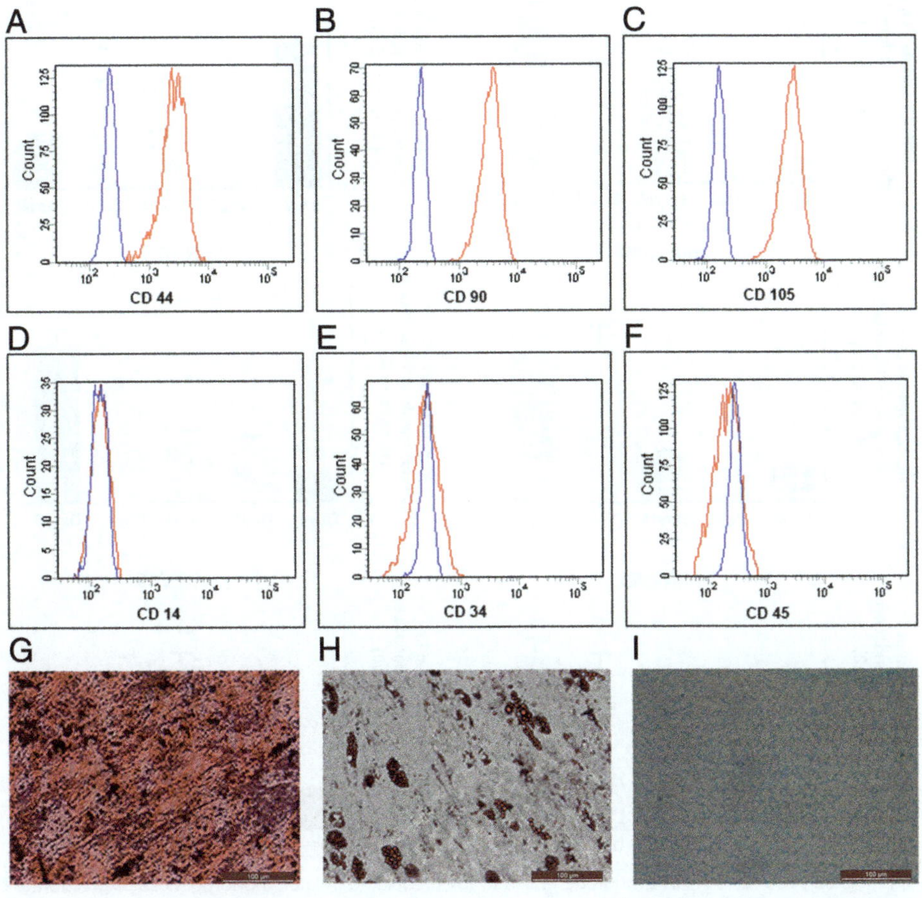


**Fig. 1** Characterization of MSC properties (**a-f**) and multilineage differentiation capacity (**g-i**) of SDSCs. Flow cytometry was used to characterize MSC surface phenotypic markers CD44 (**a**), CD90 (**b**), CD105 (**c**), CD14 (**d**), CD34 (**e**), and CD45 (**f**) in SDSCs. Adipogenic, osteogenic and chondrogenic differentiation were assessed using Oil Red O staining (**g**), alkaline phosphatase positivity (**h**) and Alcian Blue staining (**i**), respectively. Bar 100 μm

differentiation potentials of SDSCs, adipogenic, osteogenic or chondrogenic differentiation was induced in the third-passage cells. Following induction for 21 days, SDSCs were positive for Oil Red O staining, alkaline phosphatase or Alcian Blue staining (Fig. 1).

### miR-218 is upregulated in early chondrogenesis but downregulated in late chondrogenesis in SDSCs

Expression of 6 miRNAs implicated in chondrogenesis was evaluated by RT-qPCR at 0, 7, 14, 21 and 28 days of chondrogenic differentiation [6–10]. Our results showed that expression of chondrogenesis-related miRNAs miR-140, miR-199a, miR-365, and miR-675 increased gradually during the entire chondrogenesis process, whereas miR-145 levels gradually decreased. Of note, miR-218 increased significantly in a time-dependent manner during early chondrogenesis and peaked at 14 days, though expression of miR-218 was downregulated in late stages of SDSC chondrogenesis (Fig. 2).

### miR-218 enhances SDSCs chondrogenic differentiation during the early stage

Twenty-one days after induction of chondrogenic differentiation, immunohistochemistry staining was employed to detect type II collagen (Col II); Alcian Blue and Safranin O were used to stain sulfated glycosaminoglycan (GAG) and aggrecan (ACAN). The mRNA and protein levels of SOX9, COL2A1, ACAN and COMP in SDSCs were measured at 0, 7, 14 or 21 days of chondrogenic differentiation. As shown in Fig. 3a, the size of the cell pellet increased with miR-218 mimic transfection but decreased following miR-218 inhibitor transfection. Additionally, RT-qPCR results showed that miR-218 mimic transfection led to a significant increase in chondrogenic marker mRNA (Fig. 3b) and protein expression

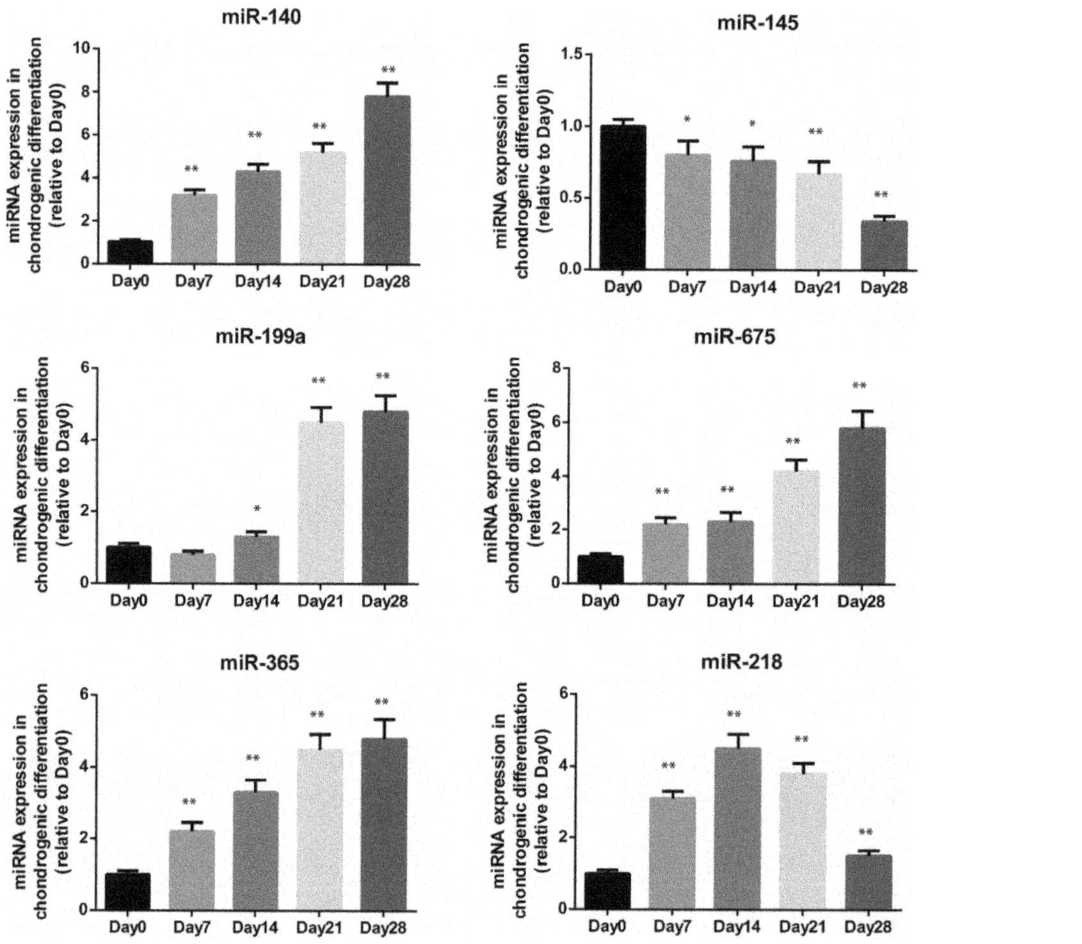


**Fig. 2** miR-218 was upregulated in SDSCs early in chondrogenesis but downregulated in later chondrogenesis. Expression of 6 miRNAs, miR-140, miR-199a, miR-365, miR-675, miR-145 and miR-218, was determined using quantitative real-time PCR on days 0, 7, 14, 21 and 28 of chondrogenic differentiation. The expression levels of miRNA were normalized to those of U6. The data are presented as the means ± SDs ($n = 3$). *$P < 0.05$ vs. Day 0, **$P < 00.01$ vs. Day 0

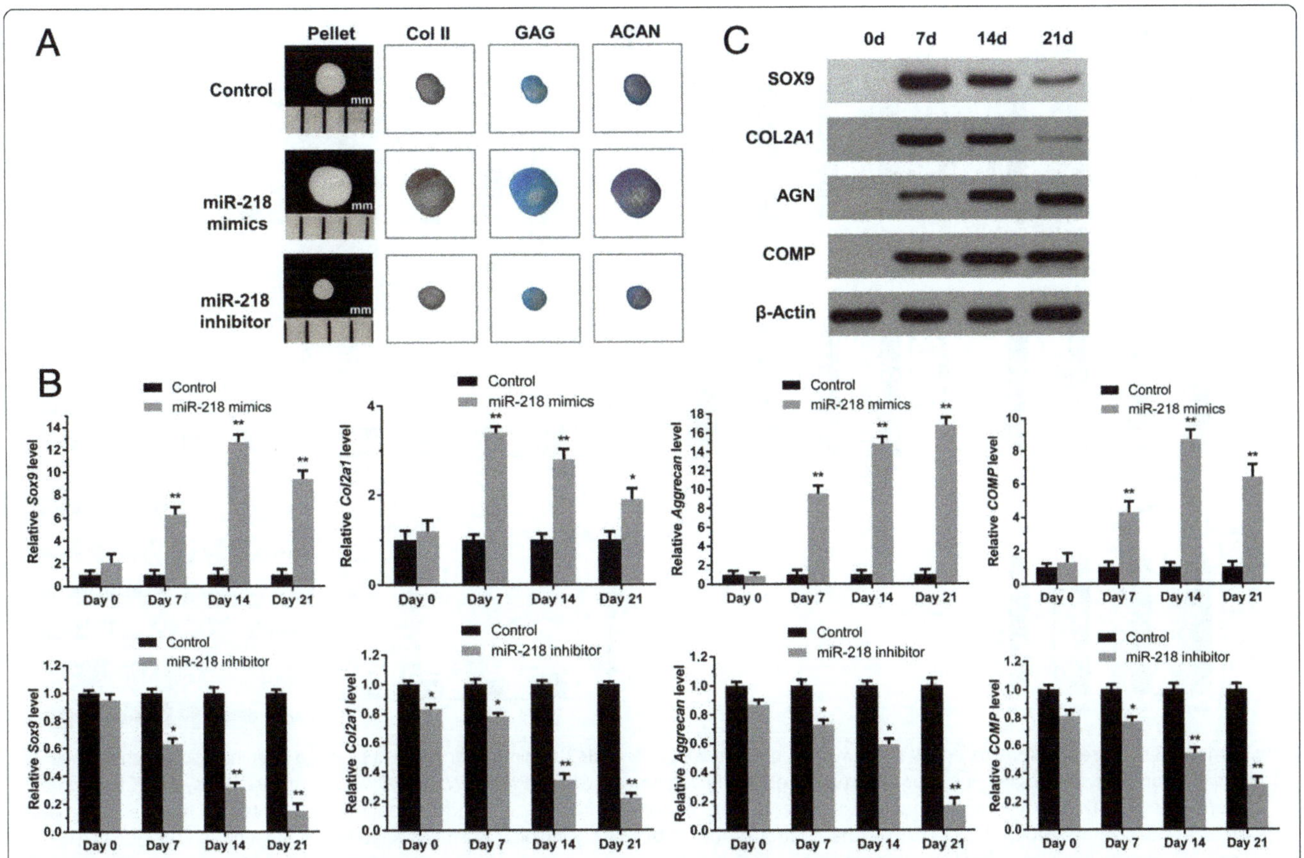


**Fig. 3** miR-218 promotes SDSC chondrogenesis. SDSCs were transfected with either miR-218 mimics or miR-218 inhibitor. After induction of chondrogenic differentiation for 21 days, (**a**) immunohistochemistry was used to detect Col II, and Alcian Blue and Safranin O were utilized to stain sulfated GAG or ACAN, respectively. **b** RT-PCR was used to measure expression of chondrogenic marker genes, including *SOX9, COL2A1, Aggrecan*, and *COMP*, in miR-218-overexpressing and miR-218-inhibited SDSCs. Three independent experiments were performed in triplicate, and the data are presented as the means ± SDs. * $P < 0.05$ vs. the control group, ** $P < 0.01$ vs. the control group. **c** miR-218 mimics were transfected into SDSCs, and SOX9, COL2A1, Aggrecan and COMP protein levels were assessed by western blot analysis

levels (Fig. 3c), whereas these chondrogenic markers were markedly downregulated in miR-218 inhibitor-transfected SDSCs.

### miR-218 represses early osteogenesis of SDSCs

Twenty-one days after induction of chondrogenic differentiation, the mRNA and protein levels of osteogenic markers ALP, BSP, Col1a1, OCN, and OPN were examined at days 0, 7, 14 and 21. As shown in Fig. 4a, RT-qPCR results showed that miR-218 mimic transfection led to a significant decrease in osteogenic marker mRNA expression levels; in contrast, these osteogenic markers were markedly upregulated in miR-218 inhibitor-transfected SDSCs, and western blotting results were consistent with the results of RT-qPCR (Fig. 4b).

### miR-218 inhibits chondrocyte maturation

Twenty-eight days after induction of chondrogenic differentiation, the mRNA and protein levels of the maturation markers CMP, COL10A1, MMP-13, and VEGF were examined at days 0, 21 and 28. RT-qPCR revealed that miR-218 mimic transfection led to a significant decrease in mRNA expression of maturation marker, whereas levels of these markers were markedly increased in miR-218 inhibitor-transfected C5.18 chondrocytes (Fig. 5a). Western blotting results were consistent with the RT-qPCR results (Fig. 5b).

### miR-218 directly regulates HPGD expression in SDSCs

The TargetScan bioinformatics tool was used to search for miRNAs targeting the HPGD 3'UTR, and only miR-218-5p and miR-26-5p were predicted as candidate miRNAs. Moreover, miR-218-5p was predicted to target two conserved regions of HPGD. Luciferase reporters carrying either the wild-type 3'UTR of HPGD or a mutant 3'UTR with three point mutations in the target site (as a negative control) were engineered to investigate whether HPGD is directly targeted by miR-218 (Fig. 6a) and were then cotransfected with miR-218 mimics into HEK293 cells. Compared with the negative control, the

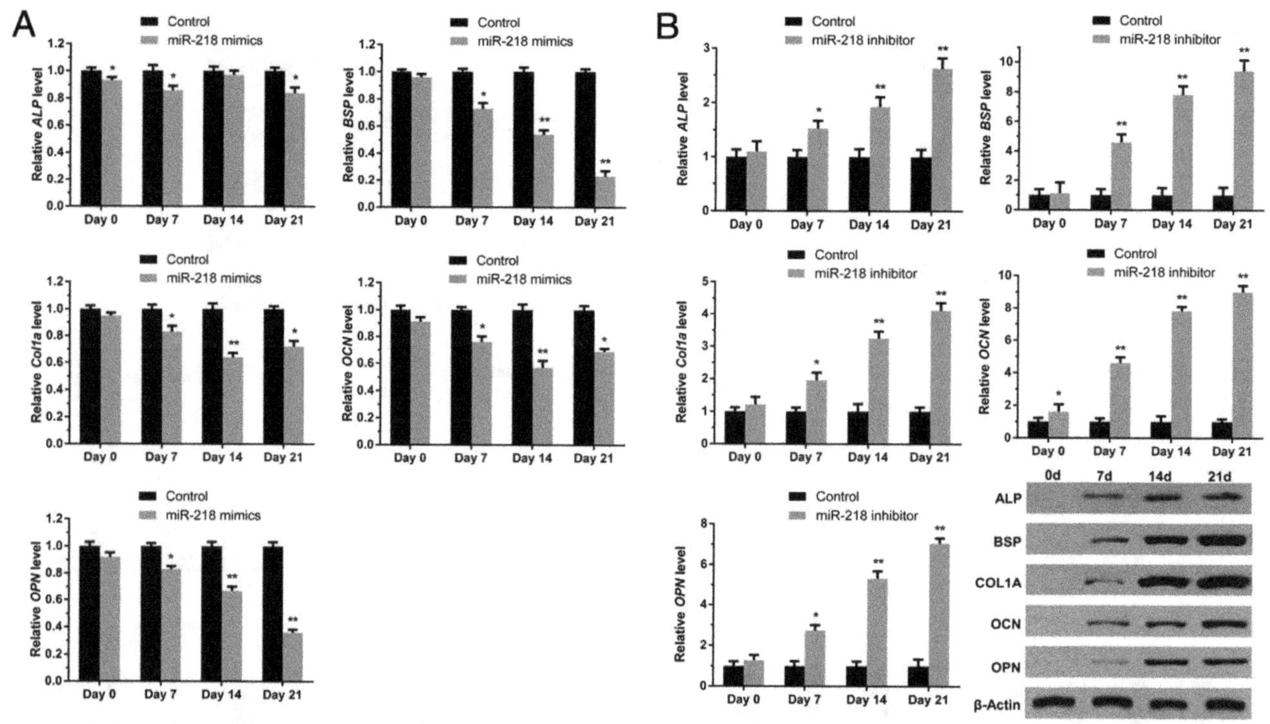

**Fig. 4** miR-218 represses SDSC osteogenesis. SDSCs were transfected with either miR-218 mimics or miR-218 inhibitor. **a** After induction of chondrogenic differentiation for 21 days, expression of osteogenic marker genes, including *ALP, BPS, Col1a, OCN*, and *OPN*, was measured by RT-PCR in miR-218-overexpressing SDSCs. Three independent experiments were performed in triplicate, and the data are presented as the means ± SDs. * $P < 0.05$ vs. the control group, ** $P < 0.01$ vs. the control group. P (**b**) miR-218 inhibitor was transfected into SDSCs, and ALP, BPS, COL1A1, OCN and OPN protein levels were measured by western blot analysis. * $P < 0.05$ vs. the control group, ** $P < 0.01$ vs. the control group

luciferase activity of the wild-type HPGD reporter were significantly reduced by miR-218 mimics. miR-199 was also cotransfected with reporters as a negative RNA control. In contrast, the mimics of miR-218 did not repress the expression of mutant reporters (Fig. 6b). To examine the role of endogenous miR-218 in inhibiting the HPGD 3'UTR reporter in chondrogenic differentiation, SDSCs were cotransfected with the wild-type 3'UTR luciferase reporter and the luciferase reporter with the mutant UTR as a negative control at 14 days after chondrogenic differentiation, and significant inhibition of the wild-type HPGD 3'UTR luciferase reporter activity was observed in the differentiated compared with undifferentiated SDSCs (Fig. 6c). Taken together, these results show that miR-218 directly regulates HPGD by targeting its 3'UTR and that miR-218 might play a prominent role during chondrogenic differentiation.

## Discussion

Because miRNAs are central to the regulations of self-renewal and differentiation of stem cells [15, 16], stem cells represent novel targets for regulating their behavior and differentiation. In this study, we have examined how miR-218 regulates SDSCs chondrogenic differentiation and RCJ3.1C5.1 chondrocytes maturation. miR-218 was found to be dramatically upregulated in SDSC early chondrogenesis but was downregulated in late stages. Moreover, miR-218 mimics strongly promoted the differentiation of SDSCs into chondrocytes. In contrast, miR-218 inhibition suppressed SDSC differentiation into chondrocytes. In addition, the level of osteogenic marker expression was found to significantly decrease in the early stage of chondrogenesis, and miR-218 overexpression might negatively regulate this process. Chondrogenic markers were dramatically increased by overexpression of miR-218. Furthermore, transfection of miR-218 mimics markedly suppressed expression of maturation markers in RCJ3.1C5.18 chondrocytes, but miR-218 inhibitor notably promoted expression of these maturation markers. miR-218 might directly regulate HPGD expression in SDSCs. These findings indicate that miR-218 possibly regulates SDSC chondrogenesis via the miR-218-HPGD pathway.

Articular cartilage is a unique tissue characterized by low cellular density and a lack of vascular components. The major composition of articular cartilage is proteoglycans and extracellular collagens and it has a very limited self-healing capacity in pathological conditions such as injury and degeneration [1]. Moreover, treatment options for cartilage trauma have dramatic differences in

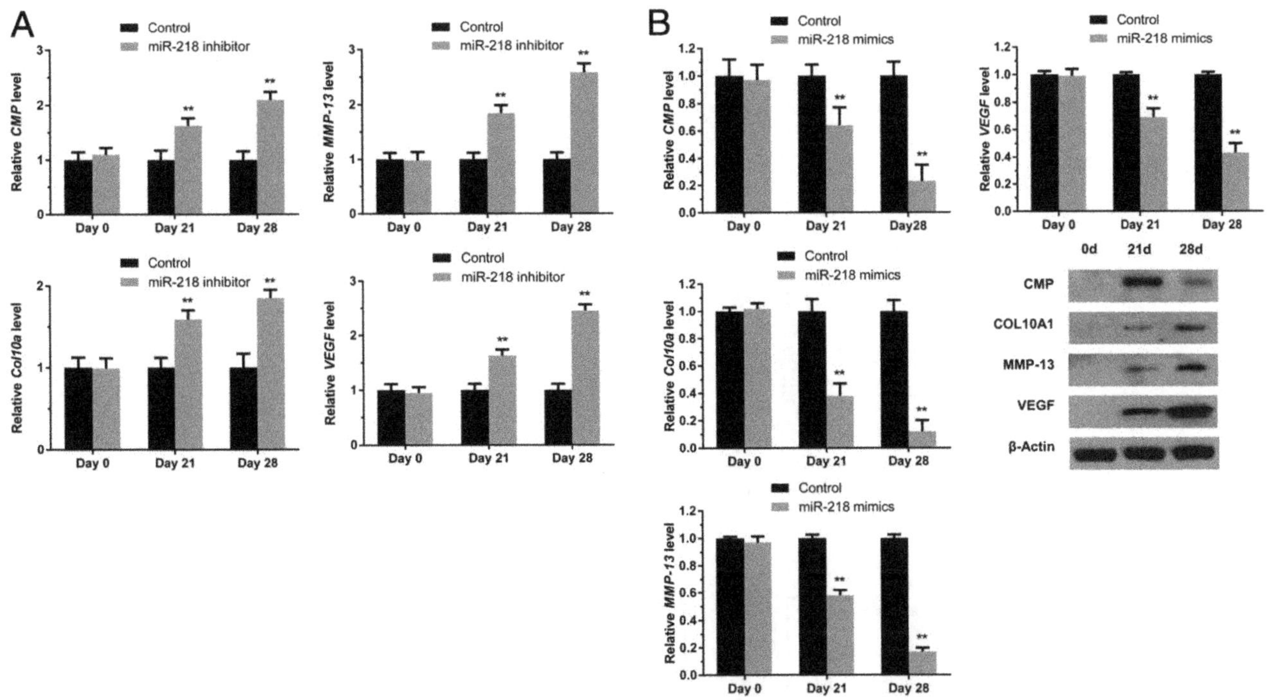


**Fig. 5** miR-218 inhibits chondrocyte maturation. C5.18 chondrocyte cells were transfected with either miR-218 mimics or miR-218 inhibitor. **a** After induction of chondrogenic differentiation for 28 days, expression of mature marker genes, including *CMP, Col10a1, MMP-13* and *VEGF*, was measured by RT-PCR in miR-218-overexpressing C5.18 chondrocyte cells. Three independent experiments were performed in triplicate, and the data are presented as the means ± SDs. ** $P < 0.01$ vs. the control group. **b** miR-218 inhibitor was transfected into C5.18 chondrocyte cells, and CMP, COL10A1, MMP-13 and VEGF protein levels were measured by western blot analysis. ** $P < 0.01$ vs. the control group

terms of effectiveness. Although early clinical studies indicated the effectiveness of ACT [17], randomized studies have reported little value of this ACT procedure in terms of effectiveness when compared with inducing subchondral bone microfracture [3]. ACT also involves donor-site morbidity and joint osteoarthritic alterations [4]. Therefore, alternative cell sources are required. MSCs have been recommended as candidates because they are capable of differentiating into chondrocytes with biological functions. Additionally, the similarity between differentiated MSCs and host cells render them ideal for cartilage replacement therapy [18]. Accordingly, transplantation of MSCs possessing chronic differentiation capacity might be an effective treatment strategy for articular cartilage injury. However, despite successful reports, MSC transplantation has also been associated with drawbacks; for example, transplantation of MSCs may lead to heterogeneous undifferentiated and differentiated populations [19]. Thus, new strategies are needed to enhance dedicated chondrocytes differentiation from MSCs.

There are abundant experimental findings pointing at the critical regulatory role that miRNAs play in MSC chondrogenic differentiation. By suppression of certain genes, several miRNAs have been shown to regulate chondrocytes differentiation from MSCs. For instance, miR-140 [20], miR-23b [21], miR-455 [20], and miR-335 [22] were found to boost chondrogenic differentiation by suppressing *ADAMTS5*, *PRKACB*, *Smad2/3*, and *ROCK1*. On the other hand, miR-574 [23], miR-29a [24], and miR-495 [25] were found to downregulate chondrogenesis by inhibiting *RXRα*, *FOXO3A*, and *SOX9*. WNT signaling plays an important role in the MSCs chondrogenic differentiation [13], and miR-218 activates WNT signaling pathway by targeting its negative regulatory factors [12]. Based on these results, we hypothesized that miR-218 might positively regulate MSC chondrogenic differentiation as well as maturation of RCJ3.1C5.1 chondrocytes. Our findings clarified that miR-218 acts as a promoter during early chondrogenic differentiation of SDSCs and as a silencer during later chondrocyte maturation.

Chondrogenesis has been divided into two phases: early chondrogenesis and hypertrophy. The former stage is featured by upregulated *COL2A1* and *SOX9*, and the latter stage is evidenced by upregulated *Col10a1* and *Runx2* [26]. Both phases are inter-regulated via interactions among several signaling pathways [27], and they antagonize each other [28–31]. Runx2, a key regulatory gene in osteogenic differentiation, mediates many osteogenic-related genes. However, Runx2 suppresses MSCs chondrogenic differentiation in the early stage and promotes later chondrocyte maturation [32]. PGE2

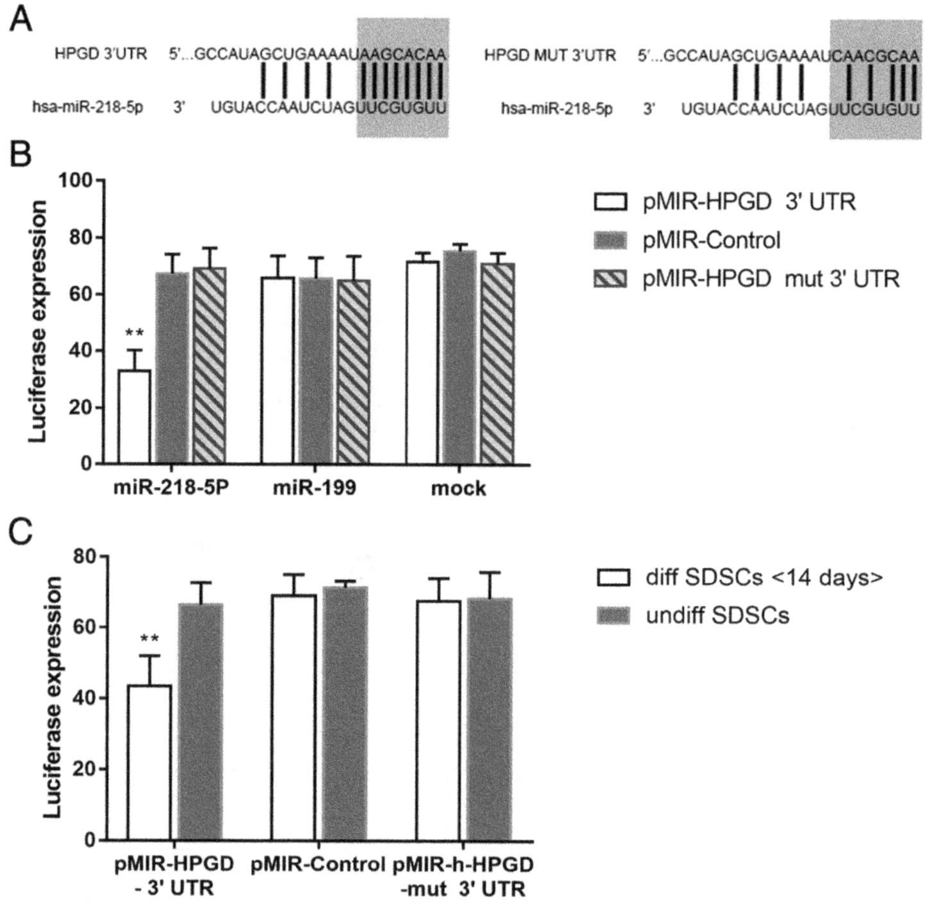


**Fig. 6** miR-218 directly regulates HPGD expression in SDSCs. The predicted pairing of miRNAs and their target regions in the wild-type HPGD 3'UTR or the mutant (mut) HPGD 3'UTR are shown (**a**). HEK293 cells were cotransfected with the firefly luciferase-expressing vector pMIR-REPORT containing the wild-type HPGD 3'UTR, the mut HPGD 3'UTR or the control insert as well as the internal control Renilla-luciferase-expressing vector pRL-TK and the indicated RNAs. miR-199 was cotransfected with reporters as a negative RNA control. Luciferase activities were measured after 48 h. The data were normalized by dividing firefly luciferase activity with that of Renilla luciferase; **, $P < 0.01$ (**b**). Luciferase activities were measured in SDSCs cotransfected with the pMIR-REPORT plasmid containing wild-type HPGD 3'UTR, mut HPGD 3'UTR or control insert and the internal control pRL-TK. All samples were assayed in duplicate ($n = 3$). **, $P < 0.01$ (**c**)

may promote cell proliferation and increase *SOX9*, *COL2A1*, and *aggrecan* expression in the early phase of MSCs chondrogenic differentiation while restraining later chondrocyte maturation. These findings reveal that Runx2 and PGE2 may play critical roles in early chondrogenic differentiation and later chondrocyte maturation.

In conclusion, our results showed significant upregulation of miR-218 early in SDSC chondrogenic differentiation and downregulation later during chondrocyte maturation. miR-218 overexpression enhances expression of chondrogenic markers, promoting early SDSC chondrogenic differentiation and suppressing later chondrocyte maturation. Moreover, miR-218 may regulate SDSC chondrogenesis via the miR-218-HPGD pathway. Therefore, miR-218 mimics may constitute a therapeutic strategy when applying SDSC-based therapy for the treatment of cartilage-related disorders.

## Conclusion

Our study suggests that miR-218 contributes to early chondrogenesis while suppressing later chondrocyte maturation. The miR-218-HPGD pathway offers us a perspective into how SDSCs differentiate into chondrogenic cells and constitute a therapeutic strategy when applying SDSC-based therapy for the treatment of cartilage-related disorders.

### Abbreviations

AC: Articular cartilage; ALP: Alkaline phosphatase; COMP: Cartilage oligomeric matrix protein; GAGs: Glycosaminoglycans; HPGD: 15-Hydroxyprostaglandin dehydrogenase; MMP-13: Metal matrix proteinase 13; MSC: Mesenchymal stem cell.; OCN: Osteocalcin; OPN: Osteopontin; PBS: Phosphate-buffered saline; PGE2: Prostaglandin E2; Runx2: Runt-related transcription factor 2; SDs: Standard deviations; SDSCs: Synovium-derived MSCs; SOX9: SRY-related gene 9; TGF-β3: Transforming growth factor beta 3; VEGF: Vascular endothelial growth factor

### Acknowledgements

Not applicable.

**Funding**
This project was partially supported by the Natural Science Foundation of Shanghai City, China (Grant No. 15ZR1414000, to PLF), and the Natural Science Foundation of China (Grant No. 81601889, to SC).

**Author disclosure statement**
No competing financial interests exist.

**Authors' contributions**
Conceived and designed the experiments: SC, ZYX, JHS, PLF, HSW. Performed the experiments: SC, ZYX, JHS. Analyzed the data: SC, ZYX, JHS. Contributed reagents/materials/analysis tools: HSW. Wrote the paper: SC, ZYX, JHS. All the authors reviewed the results and approved the final version of the manuscript.

**Ethics approval**
The experimental protocols were approved by the Animal Care and Protection Committee of Second Military Medical University.

**Consent for publication**
Not applicable.

**Competing interests**
The authors declare that they have no competing interests.

**Author details**
[1]Department of Arthroplasty Surgery, Changzheng Hospital, Second Military Medical University, Shanghai 200003, China. [2]State Key Laboratory for Diagnosis and Treatment of Infectious Diseases, and Collaborative Innovation Center for Diagnosis and Treatment of Infectious Diseases, School of Medicine, Zhejiang University, Hangzhou 310003, China.

**References**

1. Redman SN, Oldfield SF, Archer CW. Current strategies for articular cartilage repair. European cells & materials. 2005;9:23–32 discussion 23-32.
2. Peterson L, Minas T, Brittberg M, Nilsson A, Sjogren-Jansson E, Lindahl A. Two- to 9-year outcome after autologous chondrocyte transplantation of the knee. Clin Orthop Relat Res. 2000;374:212–34.
3. Knutsen G, Engebretsen L, Ludvigsen TC, Drogset JO, Grontvedt T, Solheim E, Strand T, Roberts S, Isaksen V, Johansen O. Autologous chondrocyte implantation compared with microfracture in the knee. A randomized trial. J Bone Joint Surg Am. 2004;86-A(3):455–64.
4. Lee CR, Grodzinsky AJ, Hsu HP, Martin SD, Spector M. Effects of harvest and selected cartilage repair procedures on the physical and biochemical properties of articular cartilage in the canine knee. J Orthop Res. 2000;18(5): 790–9.
5. Jones BA, Pei M. Synovium-derived stem cells: a tissue-specific stem cell for cartilage engineering and regeneration. Tissue Eng B Rev. 2012;18(4):301–11.
6. Kobayashi T, Lu J, Cobb BS, Rodda SJ, McMahon AP, Schipani E, Merkenschlager M, Kronenberg HM. Dicer-dependent pathways regulate chondrocyte proliferation and differentiation. Proc Natl Acad Sci U S A. 2008;105(6): 1949–54.
7. Dong S, Yang B, Guo H, Kang F. MicroRNAs regulate osteogenesis and chondrogenesis. Biochem Biophys Res Commun. 2012;418(4):587–91.
8. Hong E, Reddi AH. MicroRNAs in chondrogenesis, articular cartilage, and osteoarthritis: implications for tissue engineering. Tissue Eng B Rev. 2012; 18(6):445–53.
9. Guan YJ, Yang X, Wei L, Chen Q. MiR-365: a mechanosensitive microRNA stimulates chondrocyte differentiation through targeting histone deacetylase 4. FASEB J. 2011;25(12):4457–66.
10. Yang B, Guo H, Zhang Y, Chen L, Ying D, Dong S. MicroRNA-145 regulates chondrogenic differentiation of mesenchymal stem cells by targeting Sox9. PLoS One. 2011;6(7):e21679.
11. Zhang Y, Xie RL, Croce CM, Stein JL, Lian JB, van Wijnen AJ, Stein GS. A program of microRNAs controls osteogenic lineage progression by targeting transcription factor Runx2. Proc Natl Acad Sci U S A. 2011; 108(24):9863–8.
12. Hassan MQ, Maeda Y, Taipaleenmaki H, Zhang W, Jafferji M, Gordon JA, Li Z, Croce CM, van Wijnen AJ, Stein JL, et al. miR-218 directs a Wnt signaling circuit to promote differentiation of osteoblasts and osteomimicry of metastatic cancer cells. J Biol Chem. 2012;287(50):42084–92.
13. Akiyama H, Lyons JP, Mori-Akiyama Y, Yang X, Zhang R, Zhang Z, Deng JM, Taketo MM, Nakamura T, Behringer RR, et al. Interactions between Sox9 and beta-catenin control chondrocyte differentiation. Genes Dev. 2004;18(9): 1072–87.
14. Pei M, He F, Vunjak-Novakovic G. Synovium-derived stem cell-based chondrogenesis. Differentiation; research in biological diversity. 2008; 76(10):1044–56.
15. Gangaraju VK, Lin H. MicroRNAs: key regulators of stem cells. Nat Rev Mol Cell Biol. 2009;10(2):116–25.
16. Lakshmipathy U, Hart RP. Concise review: MicroRNA expression in multipotent mesenchymal stromal cells. Stem Cells. 2008;26(2):356–63.
17. Brittberg M, Lindahl A, Nilsson A, Ohlsson C, Isaksson O, Peterson L. Treatment of deep cartilage defects in the knee with autologous chondrocyte transplantation. N Engl J Med. 1994;331(14):889–95.
18. Oreffo RO, Cooper C, Mason C, Clements M. Mesenchymal stem cells: lineage, plasticity, and skeletal therapeutic potential. Stem cell reviews. 2005;1(2):169–78.
19. Ham O, Lee CY, Kim R, Lee J, Oh S, Lee MY, Kim J, Hwang KC, Maeng LS, Chang W. Therapeutic potential of differentiated mesenchymal stem cells for treatment of osteoarthritis. Int J Mol Sci. 2015;16(7):14961–78.
20. Swingler TE, Wheeler G, Carmont V, Elliott HR, Barter MJ, Abu-Elmagd M, Donell ST, Boot-Handford RP, Hajihosseini MK, Munsterberg A, et al. The expression and function of microRNAs in chondrogenesis and osteoarthritis. Arthritis Rheum. 2012;64(6):1909–19.
21. Ham O, Song BW, Lee SY, Choi E, Cha MJ, Lee CY, Park JH, Kim IK, Chang W, Lim S, et al. The role of microRNA-23b in the differentiation of MSC into chondrocyte by targeting protein kinase a signaling. Biomaterials. 2012; 33(18):4500–7.
22. Lin X, Wu L, Zhang Z, Yang R, Guan Q, Hou X, Wu Q. MiR-335-5p promotes chondrogenesis in mouse mesenchymal stem cells and is regulated through two positive feedback loops. J Bone Miner Res. 2014;29(7):1575–85.
23. Guerit D, Philipot D, Chuchana P, Toupet K, Brondello JM, Mathieu M, Jorgensen C, Noel D. Sox9-regulated miRNA-574-3p inhibits chondrogenic differentiation of mesenchymal stem cells. PLoS One. 2013;8(4):e62582.
24. Guerit D, Brondello JM, Chuchana P, Philipot D, Toupet K, Bony C, Jorgensen C, Noel D. FOXO3A regulation by miRNA-29a controls chondrogenic differentiation of mesenchymal stem cells and cartilage formation. Stem Cells Dev. 2014;23(11): 1195–205.
25. Lee S, Yoon DS, Paik S, Lee KM, Jang Y, Lee JW. microRNA-495 inhibits chondrogenic differentiation in human mesenchymal stem cells by targeting Sox9. Stem Cells Dev. 2014;23(15):1798–808.
26. Hou C, Yang Z, Kang Y, Zhang Z, Fu M, He A, Zhang Z, Liao W. MiR-193b regulates early chondrogenesis by inhibiting the TGF-beta2 signaling pathway. FEBS Lett. 2015;589(9):1040–7.
27. Shimizu H, Yokoyama S, Asahara H. Growth and differentiation of the developing limb bud from the perspective of chondrogenesis. Develop Growth Differ. 2007;49(6):449–54.
28. Cheng A, Genever PG. SOX9 determines RUNX2 transactivity by directing intracellular degradation. J Bone Miner Res. 2010;25(12):2680–9.
29. Yamashita S, Andoh M, Ueno-Kudoh H, Sato T, Miyaki S, Asahara H. Sox9 directly promotes Bapx1 gene expression to repress Runx2 in chondrocytes. Exp Cell Res. 2009;315(13):2231–40.
30. Chen H, Ghori-Javed FY, Rashid H, Adhami MD, Serra R, Gutierrez SE, Javed A. Runx2 regulates endochondral ossification through control of chondrocyte proliferation and differentiation. J Bone Miner Res. 2014;29(12):2653–65.
31. Ulrich C, Rolauffs B, Abele H, Bonin M, Nieselt K, Hart ML, Aicher WK. Low osteogenic differentiation potential of placenta-derived mesenchymal stromal cells correlates with low expression of the transcription factors Runx2 and Twist2. Stem Cells Dev. 2013;22(21):2859–72.
32. Komori T. Regulation of bone development and extracellular matrix protein genes by RUNX2. Cell Tissue Res. 2010;339(1):189–95.

# 12

# Rapid high throughput template preparation (rHTTP) method: a novel cost effective method of direct PCR for a wide range of plants

Prassan Choudhary, Sudipta Das, Hillol Chakdar*, Arjun Singh, Sanjay Kumar Goswami and Anil Kumar Saxena

**Abstract**

**Background:** Conventional plant DNA isolation methods are complex, time consuming and require technical expertise. These limitations were overcome using the DNA isolation kits which, however significantly add to the research costs. Hence the present study was aimed to develop a high throughput, rapid and inexpensive method of PCR ready DNA template preparation from plant materials.

**Methods:** Concentration of SDS in lysis buffer, amount of starting material, period and temperature for lysis were optimized for obtaining PCR ready templates from plant materials. The method was tested using RAPD and ITS specific primers for different plant species like rice, wheat, mustard, pea, soybean, pigeonpea, tomato, maize, march lilly, bougainvillea, Indian blanket flower, nerium, petunia, purple pirouette petunia, moses-in-the-cradle, golden cane palm, duranta, periwinkle, chrysanthemum and two xerophytes viz. *Dipterygium glaucum* and *Crotaleria burhia*. SSR markers RM18398 and RM26108 showed successful amplification in rice varieties Improved Pusa Basmati 1 and KS Dev 12. The effectiveness of the method was tested using fresh as well as 1 year old tissues. The storability of the lysate was also tested.

**Results:** In this report, we developed a novel method called rapid high throughput template preparation (rHTTP) method to prepare PCR ready DNA templates. Most striking feature of this technique is that it can be done anywhere where water can be boiled by any means. Using rHTTP method, PCR ready templates can be prepared in just 10 min. Robust and reproducible amplification for all the test plants were recorded with RAPD, plant ITS primers and SSR markers following this method. rHTTP methods works well for both fresh as well as old plant tissues. The lysates had a shelf life of 1 month when stored at 4 °C and 3 days when stored at room temperature.

**Conclusions:** rHTTP method has several advantages over the other protocols like ease of execution, no requirement of tissue grinding/liquid nitrogen/hazardous chemicals and above all, equally effective for both fresh and old samples. Using this method, costs per prep comes down ~ 10–50 times as compared to most commercial kits. This method can be used for on-field experiments like molecular diagnostics, varietal identification etc.

**Keywords:** Plant- DNA- high throughput- PCR-SSR

* Correspondence: hillol.chakdar@gmail.com
ICAR-National Bureau of Agriculturally Important Microorganisms (NBAIM), Mau, Uttar Pradesh 275103, India

## Background

DNA extraction is one of the most important and critical steps of a molecular biology laboratory. The conventional methods of DNA isolation involved multiple steps to rupture the cells and purify nucleic acids from the complex mixture of macromolecules using hazardous chemicals like phenol, chloroform etc. Such methods were also considerably lengthy and impeding the speed of work when DNA had to be isolated from a large number of samples. In the early 1990s, efforts were being made to simplify the cumbersome methods of obtaining good quality plant DNA which could be useful for downstream processing [1–4]. Rapid DNA fingerprinting techniques were also developed necessitating the scientists to work towards reducing the time and effort required for DNA isolation [4]. Varma et al. (2007) reviewed the various aspects of plant DNA extraction complexities and concluded that the chemical heterogeneity and diversity of plant systems rendered it very difficult for a single method to be employed for isolating pure, restrictable plant genomic DNA [5–11]. In 1993, Wang et al. reported a rapid NaOH based method for DNA template preparation for direct PCR [12]. The methodology was reported to be unsuitable for amplification of products more than 600 bp and consequently, Steiner et al. developed a new buffer (ROSE buffer) composition for DNA extraction [13]. Gradually, a number of modifications of the NaOH based methods have been reported by a number of researchers [14–17]. Apart from NaOH, sucrose has also been used for rapid extraction of DNA from plant tissues [18]. Few "touch and go" methods of direct PCR have also been reported [18, 19]. Research in this regard got a flip with the development of commercial kits which mostly rely on spin columns that capture the nucleic acids from cleared homogenates. However, the use of commercial kits attracts a considerable financial burden for research works. Despite the continuous development of a number of rapid DNA isolation methodologies since last 25 years, their use has been very limited. Satya et al. (2013) emphasized that reliability, wide application, storability of the extracts or even sometimes simple mental bias might have been the reasons behind their poor popularity among the scientific community.

On comparison of few reported DNA template preparation methods, it was realized that most of the so called "rapid methods" still require liquid nitrogen or lyophilized tissue, essentially grinding of the tissue and use of instruments like centrifuge or vortex. These operations reduce the sample processing efficiency as many of these had to be done individually or separately for each sample. Hence, these so called "rapid methods" seemed not to be very rapid when we need to process hundreds of samples together. However, such methodologies have potential to be used for large scale genotyping, DNA barcoding, molecular breeding and even in molecular diagnostics. In a recent report, Wang et al. (2016) conclusively demonstrated the use of rapid DNA template preparation in loop-mediated isothermal amplification (LAMP) for field detection of transgenic lines of rice in China.

Therefore, there was a need to develop a really rapid, simple method which can be used to yield PCR ready templates from a wide range of plant species and simultaneously cheaper than the commercially available kits. In this study, we report a novel, rapid and inexpensive method to prepare PCR ready template from a variety of plants and showed its usefulness in molecular biology experiments.

## Results

### Development and optimization of the method

The results on optimization of SDS concentration in lysis buffer revealed that highest DNA yield was achieved when 1% SDS was used (Table 1). The concentration of SDS did not specifically influence the purity of DNA. For further optimization studies, 1% SDS was used for lysis. Among the two different amounts of tissue used for lysis, the results on DNA yield was significantly better when two leaf discs (~ 3.2 mg) were used as a start material for lysis (Table 1). Among the two time periods (10 or 20 min) used for lysis,10 min was finally selected as the DNA yield was higher as compared to 20 min of lysis (Table 1). Hence, the optimized protocol for DNA template preparation included lysis of two discs (~ 3.2 mg) of leaf tissue in 100 μl of 1% SDS and heated for 10 min at 99 °C in a thermal cycler.

The success of the DNA template was judged by PCR amplification using RAPD primers (OPB06 and OPB07). An aliquot of 2.5 μl of crude lysate, 10X and 20 X diluted lysate were used as a template for two series of amplification reactions with or without BSA in PCR reactions for two different reaction volumes viz. 20 μl and 40 μl. No amplification was achieved with the crude lysate in presence or absence of BSA (Fig. 1). Likewise the bacterial DNA dissolved in 0.1% SDS did not yield any amplification product. Both 10X and 20X diluted lysate yield amplified products with no significant differences in the amplification profile both in presence or absence of BSA (Fig. 1). Therefore, 20X diluted lysate was used for further validation of our protocol for DNA template preparation followed by PCR amplification with no addition of BSA in the reaction mixture.

### Effectiveness of the optimized method

The usefulness of this rHTTP method was validated for 11 different rice varieties and DNA template obtained resulted in robust RAPD profile upon PCR amplification

**Table 1** DNA quality and yield in lysates prepared with different concentrations of SDS, different amount of tissue and different time duration

| Parameters | | DNA Purity ($A_{260}/A_{280}$) | DNA Concentration (ng/μl) [a] |
|---|---|---|---|
| SDS conc. (%) | 0.5 | 1.03 | 451.67 ± 8.82 |
| | 1.0 | 1.01 | 768.33 ± 6.01 |
| | 1.5 | 1.03 | 376.67 ± 4.41 |
| | 2.0 | 1.04 | 252.17 ± 1.48 |
| | 2.5 | 1.03 | 438.33 ± 14.24 |
| | 3.0 | 1.04 | 261.17 ± 21.17 |
| | 3.5 | 1.05 | 218.33 ± 3.33 |
| | 4.0 | 1.06 | 260.17 ± 3.06 |
| | 4.5 | 1.05 | 326.67 ± 1.67 |
| | 5.0 | 1.04 | 446.67 ± 14.81 |
| Amount of tissue | One disc (~ 1.6 mg) | 1.06 | 420.00 ± 10.08 |
| | Two disc(~ 3.2 mg) | 1.04 | 760.00 ± 6.75 |
| Time durations | 10 min | 1.08 | 876.67 ± 37.12 |
| | 20 min | 1.05 | 808.33 ± 6.01 |

[a]DNA concentration is represented as Mean ± SE

(Additional file 1: Fig. S1). The leaf lysates of wheat, pigeonpea, mustard, soybean, pea, tomato, maize, march lilly, bougainvillea, Indian blanket flower, nerium, petunia, purple pirouette petunia, moses-in-the-cradle, golden cane palm, duranta, periwinkle, chrysanthemum, *Dipterygium glaucum* and *Crotaleria burhia* yielded a ~700 bp amplicon with universal plant specific primers for internal transcribed spacer (ITS) region (Fig. 2 a-d). The rHTTP method was equally effective for old tissues as lysis of 2 months old rice leaves of four different varieties yielded a DNA template that gave strong amplification using RAPD primer OPB07 (Additional file 2: Fig. S2). Figure 3 showed

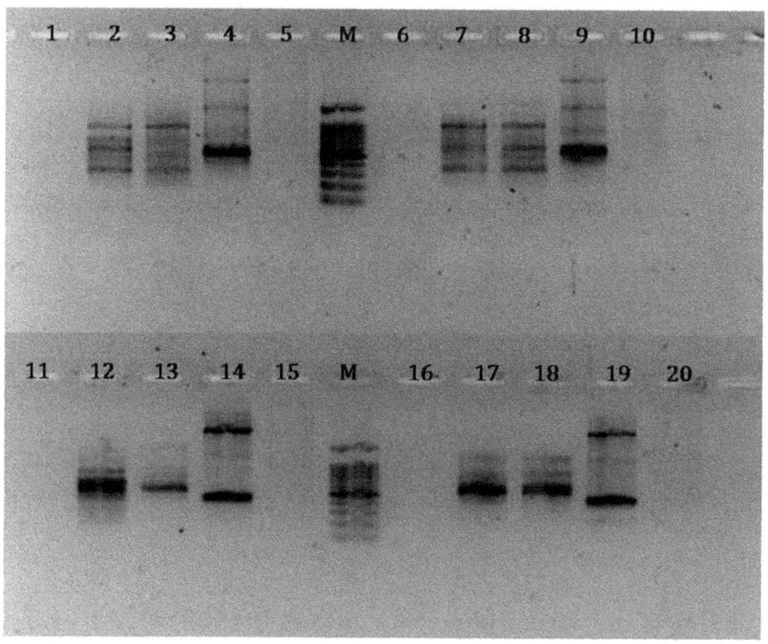


**Fig. 1** OPB6 (lane 1 to 10) and OPB7 (lane 11 to 20) primers showed good amplification results using10X and 20X dilutions of the crude template DNA lysate. Lane 1:Crude lysate (Cr) -without 1% BSA ($BSA^-$); Lane 2: 10 times diluted (10X) lysate-$BSA^-$; Lane 3: 20times diluted (20X) lysate-$BSA^-$; Lane 4:Bacterial DNA (Bac)-$BSA^-$; Lane 5: Bac- with 0.1% SDS-$BSA^-$; M: 100 bp marker (Promega); Lane 6: Cr-with 1% BSA ($BSA^+$); Lane 7: 10X lysate -$BSA^+$; Lane 8: 20X lysate-$BSA^+$; Lane 9: Bac-$BSA^+$; Lane 10: Bac- with 0.1% SDS-$BSA^+$; Lane 11: Cr-$BSA^-$; Lane 12: 10X lysate-$BSA^-$; Lane13:20X lysate-$BSA^-$; Lane 14: Bac-$BSA^-$; Lane 15: Bac- with 0.1% SDS-$BSA^-$; M: 100 bp marker (Promega); Lane 16: Cr-$BSA^+$; Lane 17: 10X lysate-$BSA^+$; Lane 18: 20X lysate-$BSA^+$; Lane 19: Bac-$BSA^+$; Lane 20: Bac- with 0.1% SDS- $BSA^+$

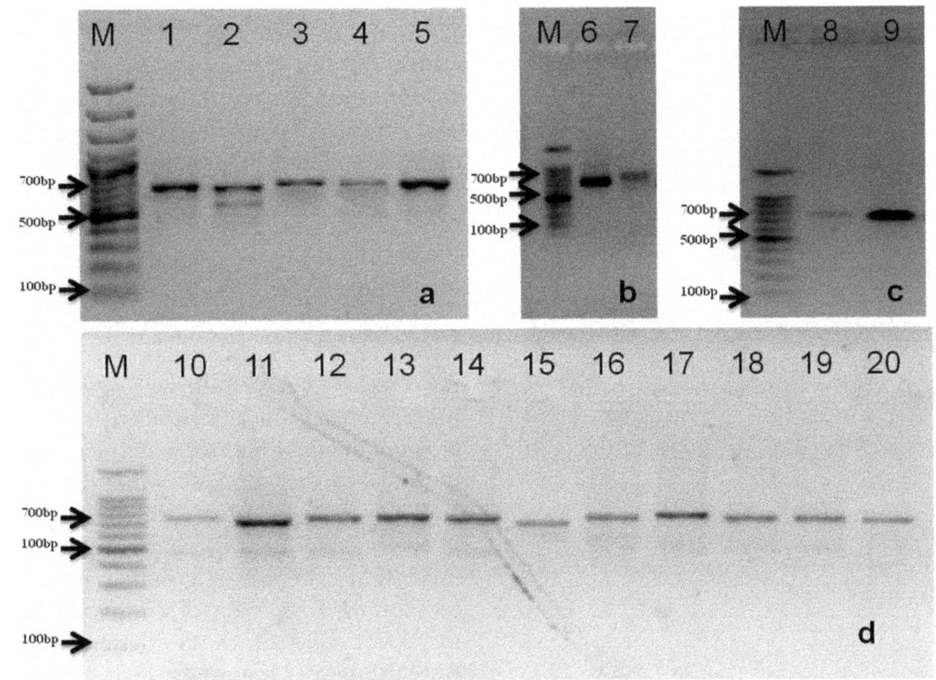


**Fig. 2** **a** Universal plant ITS primers depicted excellent resolution with the plant species tested. M: 100 bp marker (Promega); Lane 1: Mustard; Lane 2: Soybean; Lane 3: Pigeonpea; Lane 4: Wheat; Lane 5: Pea. **b** M: 100 bp marker (Promega); Lane 6: Tomato; Lane 7: Maize. **c** 100 bp marker (Promega); Lane 8: *Dipterygium glaucum*; Lane 9: *Crotaleria burhia*. **d** M: 100 bp marker (Promega); Lane 10:March Lilly; Lane 11: Bougainvillea; Lane 12: Indian blanket flower; Lane 13: Nerium; Lane 14: Petunia; Lane 15: Purple pirouette petunia; Lane 16: Moses-in-the-cradle; Lane 17: Golden cane palm; Lane 18: Duranta; Lane 19: Periwinkle; Lane 20: Chrysanthemum

the results of PCR using templates stored at − 20 °C for 30 days. DNA templates prepared following rHTTP method showed intense bands of ITS regions of the tested crop plants. This indicated that the DNA template preparation using rHTTP method was stable over 1 month. In another experiment, the lysates were kept at room temperature (~ 20–22 °C) and checked for PCR amplification after 3 days. Additional file 3: Fig. S3 showed that even after storage of 3 days at room temperature, enough DNA remained in the lysate for PCR amplification.

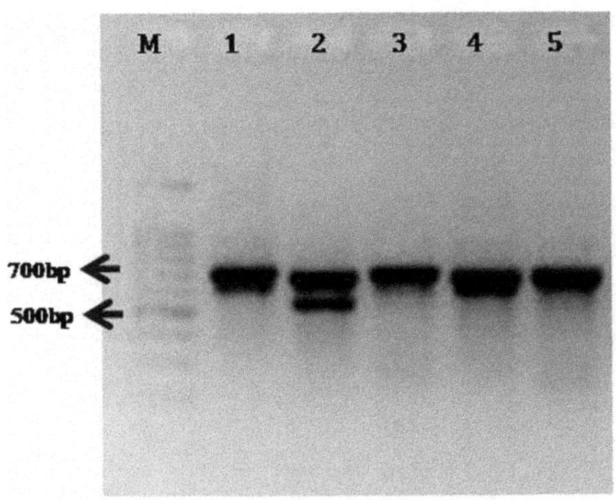


**Fig. 3** ITS amplification using 30 days old template DNA of the different plant species stored at − 20 °C. M: 100 bp marker (Promega); Lane 1: Mustard; Lane 2: Soybean; Lane 3: Pigeonpea; Lane 4: Wheat; Lane 5: Pea

The validated rHTTP protocol yielded a strong ITS amplicon even when boiling water (100–102 °C) was used for 10 and 20 min instead of heating the lysis mixture in thermal cycler. Strong ITS amplification was observed for both the time periods (Fig. 4).

SSR markers RM18398 (305 bp) and RM26108 (195 bp) showed good amplification with template prepared by rHTTP method as shown in Fig. 5a-b.

Comparative analysis of RAPD profiles of two rice varieties was done and amplification of ITS region following the different protocols detailed in methods revealed positive results with excellent resolution with newly developed rHTTP method and protocol described by Satya et al. (Figs. 6 and 7). Protocols developed by Steiner et al. (1995) and Wang et al. (2016) failed to yield any amplification product or RAPD profiles. In case of the method described by Wang et al. (2016), very faint amplicon resulted in case of rice variety HUR 917 (Fig. 7).

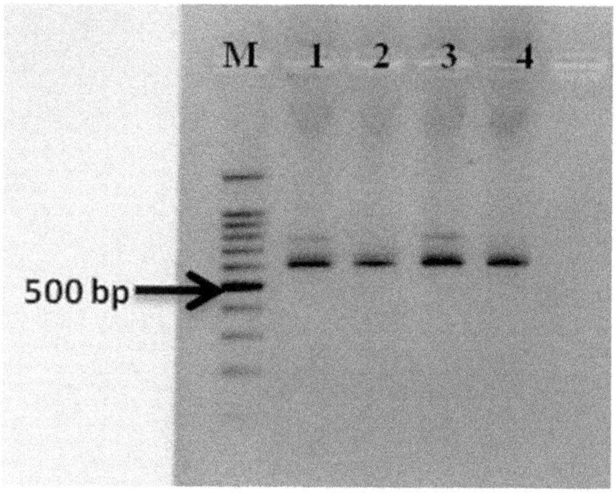


**Fig. 4** Amplification of ITS region of rice (varieties Improved Pusa Basmati 1 and KS Dev 12) using templates prepared in boiling water for 10 and 20 min. M: 100 bp marker (Promega); Lane 1: Improved Pusa Basmati 1_10 min; Lane 2: KS Dev 12_10 min; Lane 3: Improved Pusa Basmati 1_20 min; Lane 4: KS Dev 12_20min

## Discussion

In the present study, we have developed a simple, rapid, inexpensive protocol to prepare DNA template for PCR assays. In the rHTTP method development, the two most important factors were the concentration of SDS and the dilution factor. It is a well-known fact that SDS, an anionic detergent is inhibitory for Taq Polymerases [20]. In the present study also it was observed that bacterial DNA containing a final concentration of 0.0125% SDS in PCR reactions did not produce any amplification but amplification was obtained in bacterial DNA devoid of SDS. SDS has been listed among the chemicals having low concern regarding its environmental risks according to the United States Environmental Protection Agency and other relevant studies [21]. No amplification in case of crude extract from rHTTP lysates could be attributed to high concentration of both SDS (0.125%) and DNA (768.3 ng/μl). However, 10X and 20X diluted lysates for which the final concentration of SDS were 0.0062 and 0.0031% respectively showed strong amplifications. Despite no variation in amplification patterns between 10X and 20X diluted lysates, the latter was chosen to completely rule out any chance of presence of PCR inhibitors. Earlier also, Satya et al. (2013) emphasized that the dilution factor was the most important factor for success of such direct PCR methods [15].

Although, available literature showed a number of reports [12–15, 18] for rapid DNA template preparation for direct PCR assay, their use is still very limited. It has been reported that reliability, reproducibility of such techniques, shelf life of the extracts, quality and quantity of the DNA extracted etc. might have affected the utility and popularity of such techniques [15]. However, most of the techniques still require a good amount of instrumentation and tissue grinding which is probably the most monotonous act of any molecular biology laboratory. In many of these methods, liquid nitrogen or lyophilization is required which further add to the costs involved in these techniques. The rHTTP method developed in this study requires almost no technical expertise to execute and it requires no tissue grinding, vortexing or centrifugation. One most striking feature of this method is that it can be used outside any laboratory and without any sophisticated equipment. The only thing required is a means to boil water. Hence, the rHTTP method can be used for on-field experiments like molecular diagnostics, varietal identification, plant genotyping etc. Wang et al. (2016) reported a NaOH based modified DNA extraction protocol which they used for identification of transgenic lines in China [14]. However, the method reported by Wang et al. (2016) did not produce good results when we compared it to our method along with few other methods. The rHTTP method in

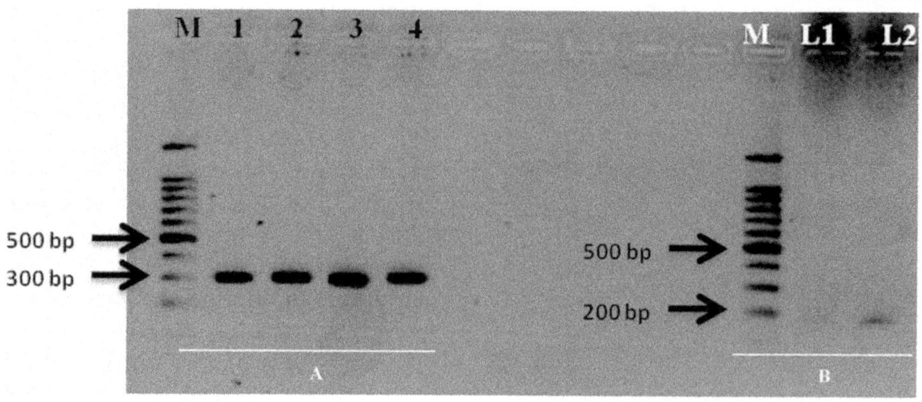


**Fig. 5** Amplification of SSR markers RM18398 (Panel A) and RM26108 (Panel B) showed successful amplification in rice varieties. Panel A: M: 100 bp marker (Promega); Lane 1: Pusa basmati 1121; Lane 2: Improved Pusa Basmati 1; Lane 3: KS Dev 12; Lane 4: MTU 5204. Panel B: M: 100 bp marker (Promega); L1: Improved Pusa Basmati 1; L2: KS Dev 12

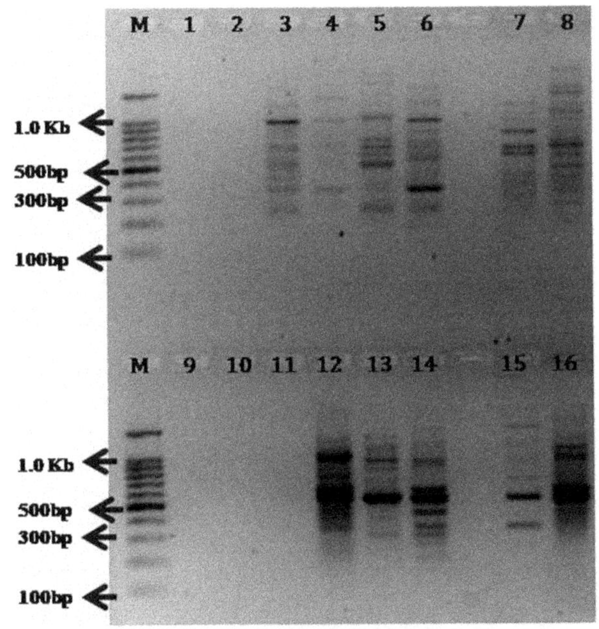


**Fig. 6** Comparative analysis of RAPD profiles of rHTTP and methods reported by Wang et al. (2016), Wang et al. (1993) and Satya et al. (2013). TKM13 and HUR917 varieties of rice were used for this PCR assay. Lane 2 to 9: Primer OPB06; Lane 10 to 16: Primer OPB07. M: 100 bp marker (Promega); Lane 1: TKM 13_Wang 2016; Lane 2: HUR 917_Wang 2016; Lane 3: TKM 13_Wang 1993; Lane 4: HUR 917_Wang 1993; Lane 5: TKM 13_Satya 2013; Lane 6: HUR 917_Satya 2013; Lane 7: TKM 13_rHTTP method; Lane 8: HUR 917_rHTTP method; M: 100 bp marker (Promega); Lane 9: TKM 13_Wang 2016; Lane 10: HUR 917_Wang 2016; Lane 11: TKM 13_Wang 1993; Lane 12: HUR 917_Wang 1993; Lane 13: TKM 13_Satya 2013; Lane 14: HUR 917_Satya 2013; Lane 15: TKM 13_rHTTP method; Lane 16: HUR 917_rHTTP method

conjunction with isothermal amplification technology can be a revolution in case of molecular diagnostics.

rHTTP method is high throughput method which can bring down the average sample processing time to few seconds. For example, if we want to process 100 samples at a time, the time required will be 10 min plus the time required for dispensing the lysis buffer and time required for making dilutions. However, for the same 100 samples, the other methods as well as any standard kit will take hours as the samples have to be ground or vortexed or centrifuged individually or in a series of very limited numbers. We have made a comparison with five other methods of rapid DNA template preparation and few commercially available kits for plant DNA extraction (Tables 2 and 3) with special reference to time, cost and instrumentation. Thermo Fisher Scientific Phire Plant Direct PCR Kit recommends the use of young leaves and essentially requires Phire Hot Start II DNA Polymerase which significantly increases the cost per sample and the use of commonly available Taq polymerases is not applicable in this case. Its dilution protocol involves tissue grinding as well as spin centrifuge which increases time and effort. Among the commercial kits available globally, only the QuickExtract™ DNA Plant DNA Extraction Solution marketed by Lucigen, USA is comparable to rHTTP method in terms of time and requirement of instrumentation. However, the sample needs to be heated at two different temperatures in case of this kit which in turn necessitates the use of a thermal cycler or a heating device with instant temperature controller. On the other hand, rHTTP method can be executed using boiling water which eliminates the temperature controlling step altogether. Nonetheless, the cost for template preparation per sample for this kit is almost 10 times higher as compared to rHTTP method. In majority of the commercially available kits, the cost per preparation ranges between $ 3–4.5 (INR 200–300) while the same is approximately $ 0.06 (INR 4.5; Additional file 4: Table S3) per preparation i.e., almost 50–75 times lesser for rHTTP method (Table 3). However, we believe that the rHTTP method is more suitable for PCR based studies where the quality of the DNA can be sacrificed to

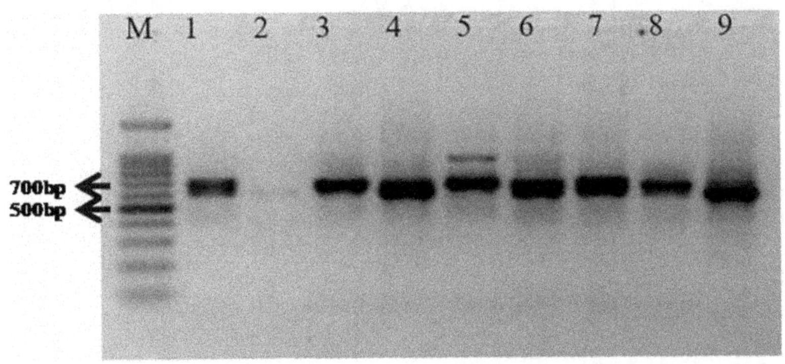


**Fig. 7** Comparative amplification of rice ITS using universal primer. Rice varieties Rajendra Sweta, HUR917, and TKM13 were used for the assay. M: 100 bp marker (Promega); Lane 1: TKM 13_Wang 2016; Lane 2: HUR 917_Wang 2016; Lane 3: TKM 13_Wang 1993; Lane 4: HUR 917_Wang 1993; Lane 5: TKM 13_Satya 2013; Lane 6: HUR 917_Satya 2013; Lane 7: TKM 13_rHTTP method; Lane 8: HUR 917_rHTTP method; Lane 9: Rajendra sweta_rHTTP method

**Table 2** Comparison of rHTTP method with other reported methods of quick plant DNA extraction

| | Rapid One Step Extraction (ROSE) method (Steiner et al., 1995) | Sucrose Prep Method (Berendzen et al. 2005) | NaOH Method (Wang et al., 1993) | NaOH Method (Satya et al. 2013) | NaOH Method (Wang et al., 2015) | rHTTP method (Present study) |
|---|---|---|---|---|---|---|
| Type and form of plant material | Lyophilized plant tissue | Fresh tissue | Young tissue | Fresh tissue | Fresh tissue | Fresh/old tissue |
| Liquid nitrogen | No | Yes/Ice | No | No | No | No |
| Extraction/lysis buffer | Tris-HCl, EDTA, Sodium Sarkosyl sulfate, PVPP | Tris-HCl, NaCl, Sucrose | NaOH, Tris | NaOH, Tris-EDTA | NaOH | SDS |
| Instrumentation | Lyophilizer, Shaker, hybridization Oven/ Water bath | Water Bath/Thermal cycler, Centifuge, grinding assembly/ Mortar pestle | Grinding assembly/ Mortar-pestle | Grinding assembly/ Mortar-pestle, Centrifuge | Grinding assembly/ Mortar-pestle/ centrifuge | Thermal cycler/hot plate/any means to boil water |
| Duration to make templates for PCR (as per original publication) | ~ 45 min after lyophilization | ~ 15 min | Should not be more than 10 min | 10–12 min | 5 min | 10 min |
| Actual time required to process five samples at a time | ND | 59 min (11.8 min per sample) *For larger number of samples time will increase proportionately due to grinding | 31 min (6.2 min per sample) *For larger number of samples time will increase proportionately due to grinding | 65 min (13 min per sample) *For larger number of samples time will increase proportionately due to grinding | 29 min (5.8 min per sample) *For larger number of samples time will increase proportionately due to grinding | 12.5 min (2.5 min per sample) *For larger number of samples, the time will remain almost same as there is no grinding or maceration. The time will be 10 min plus time required for pipetting 100 µl lysis buffer |

**Table 3** Comparison of rHTTP method with few commercially available kits

| Kits/methods | Time required per sample | Instrumentation | Cost |
|---|---|---|---|
| DNeasy Plant Mini Kit (Qiagen) | ~ 60 min | Grinding assembly/ Mortar-pestle, Centrifuge, vortex (6 centrifugation steps) | INR 282/– per prep |
| Synergy™ 2.0 (OPS Diagnostics) | 43 min | Centrifuge, vortex (5 centrifugation steps) | INR 153/– per prep |
| Plant DNA isolation reagent (Takara) | 30 min. | Centrifuge, vortex (3 centrifugation steps) | INR 138/– per prep |
| Phire Plant Direct PCR Kit (Thermo Fisher Scientific) | – | Thermal cycler (sample lysis and PCR done in a single tube). Crushing and spinning required in Dilution protocol | INR 51/– per prep |
| QuickExtract™ Plant DNA extraction solution (Lucigen) | 8 min | Heating device with temperature controller (it requires incubation at two different temperatures) | INR 40/– per prep |
| rHTTP method | 10 min | Thermal cycler/ heating device/ any means to boil water | INR 4.5/– per prep |

certain extent to save a considerable amount of time and cost. For identification, diagnostics and genotyping methods using PCR will find the major applications of the present method. The purity of the template DNA prepared using rHTTP method suggests that it may not be suitable for restriction analyses or genome sequencings. The DNA templates prepared with rHTTP method has an excellent storability. At − 20 °C, templates can be stored for 1 month while at room temperature it could be stored for at least 3 days.

rHTTP method can be used for a variety of plant species belonging to Poaceae, Fabaceae, and Brassicaceae which encompass majority of the agriculturally important crops and most of the marker assisted breeding programs are targeted to these crops. Both random markers like RAPD and targeted amplification of Internal Transcribed Spacer (ITS) were obtained for all the plant species used. In order to further validate the applicability of the method, SSR markers were amplified in rice following the rHTTP method. This method can be very helpful in molecular breeding programs, plant genotyping where huge number of varieties or lines need to be screened using a large number of PCR based markers.

## Conclusions

The rHTTP holds a number of advantages over most of the above mentioned protocols- (i) No requirement of ice, liquid nitrogen or lyophilization, (ii) No tissue grinding, (iii) No centrifugation, (iv) No harmful chemicals, (v) This is high throughput method. Unlike, other methods, average sample processing time goes down with increasing number of samples (vi) Equally efficient for fresh as well as old tissue, (vii) It can work for a wide range of plant species, (viii) Requires almost no technical expertise, (ix) A single prep can be used for more than 500 PCR reactions (40 μl), (ix) The lysates have a month long shelf life when stored at − 20 °C, and (x) Cost per preparation is very low as compared to any other direct PCR method or commercially available kit. Hence, the rHTTP method described here is simpler, cheaper and more rapid as compared to other direct PCR systems reported till date and will be very useful in plant biotechnology projects having versatile PCR based applications.

## Methods

### Plant samples

A total of 21 different plant species belonging to 12 different families viz. Amaryllidaceae, Nyctaginaceae, Asteraceae, Apocynaceae, Solanaceae, Commelinaceae, Arecaceae, Fabaceae, Capparaceae, Verbenas, Poaceae, and Brassicaceae collected from different parts of India were used in the study (Refer Additional file 4: Table S1 for details of samples). All samples were collected in polypropylene bags and stored at 4 °C prior to use.

### Protocol development and optimization

A linear gradient of 0.5 to 5% (w/v with an interval of 0.5%) molecular biology grade Sodium Dodecyl Sulfate (HiMedia, India) prepared in MiliQ water was tested as lysis buffer. Final concentration of SDS was selected on the basis of DNA quantity and quality in the lysate prepared with two leaf discs. Initially, the lysis was done at 99 °C for 20 min. Varying amount of tissue was used for optimizing the amount of start sample as it affects the quantity and quality of DNA template to be prepared. For this, variable number of leaf discs (~ 0.5 cm, diameter) viz. one disc (~ 1.6 mg) and two discs (~ 3.2 mg) were used for lysis as mentioned above. After optimizing the concentration of SDS and amount of tissue to be lysed, the time for lysis was further optimized using two time periods for lysis viz. 10 and 20 min at 99 °C.

For protocol optimization, leaves of the rice variety HUR917 were used. For lysis, the leaf discs were placed in a 200 μl PCR tubes (Tarsons, India) and 100 μl of lysis buffer was added to it and heated in a thermal cycler at 99 °C (PeqStar, VWR International Ltd., UK). All lysis and spectrophotometric observations were carried out in five replicates.

**Initial assessment of DNA template quality and quantity**

DNA concentration and purity ($A_{260}/A_{280}$) was determined using ScanDrop spectrophotopmeter (Analytik Jena AG, Germany).

**PCR assay**

For optimization of PCR amplification, crude lysate, 10X, and 20X dilutions of the lysates were used as template. Dilutions were prepared in MiliQ water. GoTaq® Green Mastermix (Promega, USA) was used throughout the experiments. An aliquot of 2.5 µl of the template was used in two reaction volumes viz. 20 and 40 µl. For 40 µl reaction volume, 10X dilutions as well as 20X dilutions were used and for 20 µl reaction volume, only crude extract was used. With an assumption that lysates will have a significant amount of PCR inhibitors, all PCR reactions were made in two sets, one with Bovine Serum Albumin (Promega, USA) as a PCR enhancer with a final concentration of 0.01 µg/ml and another without BSA. For comparison, bacterial genomic DNA suspended in 0.1% SDS and TE buffer was also used for RAPD in a 20 µl reaction volume.

**Primers and PCR conditions**

For optimization of initial PCR conditions, RAPD primers listed in Additional file 4: Table S2 were used. The following conditions were used-Initial denaturation: 94 °C for 45 s, ten cycles of (denaturation: 94 °C for 45 s, primer annealing: 45 °C for 60 s, primer extension: 72 °C for 2 min), 30 cycles of (denaturation: 94 °C for 45 s, primer annealing: 45 °C for 45 s, primer extension: 72 °C for 2 min), final primer extension: 72 °C for 10 min. For comparison with other methods, both RAPD primers and universal plant internal transcribed spacer (ITS) specific primers were used (Additional file 4: Table S2) [22]. For amplification of plant specific ITS region, following PCR conditions were used- Initial denaturation: 94 °C for 4 min, 34 cycles of (denaturation: 94 °C for 30 s, primer annealing: 58 °C for 40 s, primer extension: 72 °C for 60 s), final primer extension: 72 °C for 10 min.

**Effectiveness of the optimized method**

To check the effectiveness of the optimized method, it was tested on 11 different rice varieties, wheat, pigeonpea, mustard, soybean, pea, tomato, maize, march lilly, bougainvillea, Indian blanket flower, nerium, petunia, purple pirouette petunia, moses-in-the-cradle, golden cane palm, duranta, periwinkle, chrysanthemum, *Dipterygium glaucum* and *Crotaleria burhia*. SSR markers specific for rice viz. RM18398 and RM26108 (Additional file 4: Table S2) were used for amplification with rHTTP templates [23]. Further the effectiveness of the protocol was tested with the leaves of 2 months old leaf tissues of rice varieties PR108, PR111 and two other local varieties collected from Punjab, India. The integrity of the template DNA prepared using optimized protocol during storage was studied by keeping the lysates at − 20 °C or at room temperature. The lysates were stored at − 20 °C for 30 days and PCR amplification was checked at an interval of 10 days. Likewise the lysates were kept at room temperature for a week and checked through PCR amplification at an interval of 3 days.

**Testing of the method using boiling water**

After optimization of all the parameters viz. SDS concentration, amount of tissue and time, the protocol was further tested for its execution using boiling water instead of thermal cycler so that it can be performed even outside the laboratory. For this, water was boiled on hot plate (Impact Icon Instruments Company, India) and temperature of the water was monitored with a digital thermometer (Frontier Multi- Thermometer, Model: ST-9283, India). The PCR tubes containing leaf discs and lysis buffer were put into a beaker containing boiling water. Lysis was carried out for two time periods viz. Ten and 20 min as the surface temperature of the boiling water ranged from (100–102 °C).

**Comparison with other similar methods**

Four quick plant DNA extraction protocols previously reported by- (i) Wang et al. (1993), (ii) Wang et al.(2016), (iii) Satya et al. (2013), and (iv) Steiner et al. (1995) were compared with the optimized protocol in terms of amplification profiles using RAPD and ITS specific primers. A comparison in terms of time, cost and instrumentation was also made with few popular commercial kits for plant DNA isolation.

**Additional file 1: Figure S1.** RAPD profile of 11 different varieties of rice using OPB06 primer. M: 100 bp marker (Promega); Lane 1: TKM 13; Lane 2: Rajendra Sweta; Lane 3: HUR 105; Lane 4: Improved Pusa Basmati 1; Lane 5: GM 96; Lane 6: GM 99; Lane 7: GM 137; Lane 8: GM 113; Lane 9: Pusa Basmati 1121; Lane 10: PR 115; Lane 11: PR111.

**Additional file 2: Figure S2.** RAPD profile (with OPB07) of four different rice using 2 months old leaf tissue. Lane 1: PR111; Lane 2: PR108; Lane 3: Local variety 1; Lane 4: Local variety 2.

**Additional file 3: Figure S3.** Gel photograph showing the results of ITS amplification using 3 days old template DNA of rice and wheat stored at room temperature. M: 100 bp marker (Promega); Lane 1: Wheat (var. HD2967); Lane 2: Rice (var. HUR917)

**Additional file 4: Table S1.** Details of plant samples used in the study. **Table S2.** Primers (RAPD, ITS, and SSR) used in the study. **Table S3.** Calculation of cost per preparation for rHTTP method.

**Abbreviations**

BSA: Bovine Serum Albumin; CRIJAF: Central Research Institute for Jute and Allied Fibers; DNA: Deoxyribonucleic acid; ICAR: Indian Council of Agricultural Research; IISS: Indian Institute of Seed Science; ITS: Internal Transcribed Spacer; LAMP: Loop Mediated Isothermal Amplification; NBAIM: National

Bureau of Agriculturally Important Microorganisms; PCR: Polymerase Chain Reaction; RAPD: Random Amplified Polymorphic DNA; rHTTP: Rapid High Throughput Template Preparation; ROSE: Rapid One-step Extraction; SDS: Sodium Dodecyl Sulphate; SE: Standard Error; TE: Tris EDTA

**Acknowledgements**
Authors gratefully acknowledge the financial assistance received under the project entitled "Development of gene-chip for detection of major fungal plant pathogens" from ICAR Network Project on Application of Microorganisms in Agriculture and Allied Sector (AMAAS). Authors are thankful to Dr. Soham Ray, Scientist, ICAR-CRIJAF, Barrackpore, India for his inputs during the experiments using SSR markers.

**Authors' contributions**
HC conceptualized and designed the experiments. PC and SD executed the experiments. PC conducted additional experiments during the revision of the manuscript. PC with the help of SKS compiled the results and prepared the first draft of the manuscript. HC and AS revised the first draft of the manuscript. AKS edited and finalized the manuscript. All the authors have read and approved the manuscript in its current form.

**Funding**
This study was supported by financial assistance received from ICAR funded Network Project on Application of Microorganisms in Agriculture and Allied Sector (AMAAS). The funding agency did not play any role in the design or execution of the study including collection, analysis, and interpretation of data or in writing the manuscript.

**Ethics approval and consent to participate**
Not applicable.

**Consent for publication**
Not applicable.

**Competing interests**
The authors declare that they have no competing interests.

**References**

1. Langridge U, Schwall M, Langridge P. Squashes of plant tissue as substrate for PCR. Nucleic Acids Res. 1991;19:6954.
2. Edwards K, Johnstone C, Thompson C. A simple and rapid method for the preparation of plant genomic DNA for PCR analysis. Nucleic Acids Res. 1991; 19:1349 http://www.ncbi.nlm.nih.gov/pmc/articles/PMC333874/.
3. Doyle JJ, Doyle JL. A rapid DNA isolation procedure for small quantities of fresh leaf tissue. Phytochem Bull. 1987;19:11–5. https://doi.org/10.2307/4119796.
4. Williams JGK, Kubelik AR, Livak KJ, Rafalski JA, Tingey SV. DNA polymorphisms amplified by arbitrary primers are useful as genetic markers. Nucleic Acids Res. 1990;18:6531–5.
5. Aljanabi SM, Martinez I. Universal and rapid salt-extraction of high quality genomic DNA for PCR-based techniques. Nucleic Acids Res. 1997;25:4692–3.
6. Stewart CN. Rapid DNA Extraction from Plants. In: Micheli M.R., Bova R. (eds) Fingerprinting Methods Based on Arbitrarily Primed PCR. Springer Lab Manuals. Berlin, Heidelberg: Springer; 1997. p. 25–8.
7. Sharma R, Mahla HR, Mohapatra T, Bhargava SC, Sharma MM. Isolating plant genomic DNA without liquid nitrogen. Plant Mol Biol Report. 2003;21:43–50.
8. Ahmad SM, Ganaie MM, Qazi PH, Verma V, Basir SF, Qazi GN. Rapid DNA isolation protocol for angiospermic plants. Bulg J Plant physiol. 2004;30:25–33.
9. Manen J-F, Sinitsyna O, Aeschbach L, Markov AV, Sinitsyn A. A fully automatable enzymatic method for DNA extraction from plant tissues. BMC Plant Biol. 2005;5:23.
10. Varma A, Padh H, Shrivastava N. Plant genomic DNA isolation: an art or a science. Biotechnol J. 2007;2:386–92.
11. Allen GC, Flores-Vergara MA, Krasynanski S, Kumar S, Thompson WF. A modified protocol for rapid DNA isolation from plant tissues using cetyltrimethylammonium bromide. Nat Protoc. 2006;1:2320–5. https://doi.org/10.1038/nprot.2006.384.
12. Wang H, Qi M, Cutler AJ. A simple method of preparing plant samples for PCR. Nucleic Acids Res. 1993;21:4153–4.
13. Steiner JJ, Polemba CJ, Fjellstrom RG, Elliott LF. A rapid one-tube genomic DNA extraction process for PCR and RAPD analyses. Nucleic Acids Res. 1995; 23:2569–70.
14. Wang L, Wang R, Yu Y, Zhang F, Wang X, Ying Y, et al. A powerless on-the-spot detection protocol for transgenic crops within 30 min, from leaf sampling up to results. Anal Bioanal Chem. 2016;408:657–62.
15. Satya P, Mitra S, Ray DP, Mahapatra BS, Karan M, Jana S, et al. Rapid and inexpensive NaOH based direct PCR for amplification of nuclear and organelle DNA from ramie (*Boehmeria nivea*), a bast fibre crop containing complex polysaccharides. Ind Crop Prod. 2013;50:532–6.
16. Zhang L, Wang B, Pan L, Peng J. Recycling isolation of plant DNA, a novel method. J Genet Genomics. 2013;40(1):45–54.
17. Zhang L, Dai L, Gou J, Peng J. An effective protocol to solve the problem in genomic DNA isolation of tung tree. J Plant Biochem Biotechnol. 2013;22(4): 492–97.
18. Berendzen K, Searle I, Ravenscroft D, Koncz C, Batschauer A, Coupland G, et al. A rapid and versatile combined DNA/RNA extraction protocol and its application to the analysis of a novel DNA marker set polymorphic between *Arabidopsis thaliana* ecotypes Col-0 and Landsberg erecta. Plant Methods. 2005;1:4.
19. Li Y, Zhao H, Yan X, Li M, Chen P, Zhang S. A universal method for direct PCR amplification of plant tissues. Anal Methods. 2017;9:1800–5.
20. Schrader C, Schielke A, Ellerbroek L, Johne R. PCR inhibitors - occurrence, properties and removal. J Appl Microbiol. 2012;113:1014–26.
21. Bondi CAM, Marks JL, Wroblewski LB, Raatikainen HS, Lenox SR, Gebhardt KE. Human and environmental toxicity of sodium lauryl sulfate (SLS): evidence for safe use in household cleaning products. Environ Health Insights. 2015;9:EHI-S31765.
22. Cheng T, Xu C, Lei L, Li C, Zhang Y, Zhou S. Barcoding the kingdom Plantae: new PCR primers for ITS regions of plants with improved universality and specificity. Mol Ecol Resour. 2016;16:138–49.
23. Ray S, Bose L K, Ray J, Ngangkham U, Katara JL, Samantaray S, et al. Development and validation of cross-transferable and polymorphic DNA markers for detecting alien genome introgression in Oryza sativa from Oryza brachyantha. Mol Genet Genomics. 2016;291(4):1783–94.

13

# Enhancing methane production using anaerobic co-digestion of waste activated sludge with combined fruit waste and cheese whey

Seyed Mostafa Hallaji[1*], Mohammad Kuroshkarim[1] and Seyede Parvin Moussavi[2]

**Abstract**

**Background:** Recently, it has been indicated that anaerobic co-digestion of waste activated sludge with other waste streams at wastewater treatment plants is a promising strategy for enhancing methane production and materials recovery. The enhanced methane production can be used as a renewable source of energy in wastewater treatment plants. It can also reduce the amount of greenhouse gas emission in landfilling of the waste streams.

**Results:** According to the results obtained in this study, anaerobic co-digestion of waste activated sludge with mixed fruit waste and cheese whey improves methane production and the quality of digested sludge in comparison to the anaerobic digestion of waste activated sludge individually. It was indicated that carbon/nitrogen ratio (C/N) in the mixture of waste activated sludge, fruit waste and cheese whey improved considerably, leading to better anaerobic organisms' activity during digestion. With assessing the activity of protease and cellulase, as the main enzymes hydrolyzing organic matter in anaerobic digestion, it was indicated that co-digestion of waste activated sludge with mixed fruit waste and cheese whey enhances the activity of these enzymes by 22 and 9% respectively. At the end of digestion, the amount of cumulative methane production significantly increased by 31% in the reactor with 85% waste activated sludge and 15% mixed fruit waste and cheese whey, compared to the reactor with 100% waste activated sludge. In addition, chemical oxygen demand (COD) and volatile solid (VS) in digested sludge was improved respectively by 9 and 7% when mixed fruit waste and cheese whey was used.

**Conclusions:** This study revealed that mixed fruit waste and cheese whey is potentially applicable to anaerobic digestion of waste activated sludge, as fruit waste and cheese whey have high C/N ratio that enhance low C/N in waste activated sludge and provide a better diet for anaerobic organisms. This is of significant importance because not only could higher amount of renewable energy be generated from the enhanced methane production in wastewater treatment plants, but also capital costs of the companies whose waste streams are being transported to wastewater treatments plants could be reduced considerably.

**Keywords:** Anaerobic co-digestion, Waste activated sludge, Fruit waste, Cheese whey, Methane production, enzymatic activity

* Correspondence: mostafa.hallaji69@ut.ac.ir
[1]School of Environment, College of Engineering, University of Tehran, Tehran, Iran
Full list of author information is available at the end of the article

## Background

Nowadays, the importance of wastewater treatment plants in protecting water streams from pollution is widely recognized. Activated sludge process is a conventional method being widely used in wastewater treatment plants throughout the world due to its high efficiency in removing organic matter and simplicity of its application [1]. However, in this process, a large amount of waste activated sludge is produced, which increases sludge treatment costs in wastewater treatment plants. This is important because sludge treatment unit is the most cost-intensive section in wastewater treatment plants, accounting for around 60% of total operation costs [2, 3].Therefore, recent studies have been focused on enhancing digestion of sewage sludge, whereby sludge treatment costs would be reduced [4–9].

Anaerobic digestion is considered as a promising method for stabilizing sewage waste activated sludge, which not only dispenses with aeration facilities and its related capital and operation costs, but it also produces methane as a renewable energy resource in wastewater treatment plants [10–13]. In addition, anaerobic digestion harnesses methane that results in lower greenhouse gas emission when digested sludge is used in farms, forests, and landfills. This is of great importance because around 40% of total greenhouse gas emission in wastewater treatment plants is produced from sludge treatment sections [14–17]. However, an issue attributing to anaerobic digestion of waste activated sludge is its poor biochemical methane potential and slow fermentation process [11, 18–20].

Anaerobic co-digestion of sewage sludge with other waste streams is an environmentally friendly and economically attractive method to overcome low biochemical methane potential of sewage sludge [13, 21–28]. Hublin et al. [21] revealed that anaerobic co-digestion of cheese whey and cow manure reduces $CO_2$ and $CH_4$ emissions by 3.5 ktonnes $CO_2$/year and 5.7 ktonnes $CO_2$-eq/year respectively. They, also, indicated that a medium-scale biogas plant is profitable for 12 and 15 year effectuation periods. Sewage waste activated sludge has a C/N ratio of around 10, but the suitable ratio for anaerobic digestion of organic matter is around 20 to 30 [29]. Therefore, the low C/N in waste activated sludge results in imbalanced diet for anaerobic organisms and low biodegradability of organic matter [30]. C/N ratio in waste activated sludge is low because it mainly consists of microbial cells leading to high nitrogen content [31]. High organic loading rate (OLR) can also cause low C/N ratio due to ammonia accumulation that produces inhibitory effects on anaerobic organisms' activity [32, 33]. Consequently, anaerobic co-digestion of waste activated sludge with other waste streams, which have higher carbon content, can provide a balanced diet for anaerobic organisms' activity. Hosseini Koupaie et al. [34] revealed that anaerobic co-digestion of fruit-juice waste and municipal sludge reduces total net costs up to 37%. Anh et al. [35] investigated the effect of anaerobic digestion of sewage sludge with carbon-rich organic waste on microbial community. They showed that adding higher proportion of carbon-rich organic waste reduces diversity of microbial community, but increases biogas production. Fitamo et al. [36] demonstrated that co-digestion of food waste, grass clippings, garden waste with mixed municipal sludge enhances methane production by 35–48% and decreases hydraulic retention time (HRT) to 15 days. Maragkaki et al. [23] revealed that anaerobic co-digestion of sewage primary sludge (95%) with mixed olive mill and cheese whey wastewater (5%) in continuous reactors can enhance methane yield and biogas production around 127% and improve organic removal accordingly.

In this study, we hypothesized that combined cheese whey and fruit waste could enhance C/N ratio in waste activated sludge significantly, leading to better anaerobic microorganisms' performance, meanwhile provide simultaneous digestion of these waste streams, which reduces related costs and environmental impacts in comparison to individual digestion of the waste streams. This is the first investigation focusing on anaerobic co-digestion of mixed cheese whey and fruit waste with sewage waste activated sludge. Substrate characterizations in different proportions of waste activated sludge with mixed fruit waste and cheese whey were measured before conducting the anaerobic co-digestion process. During anaerobic co-digestion, key hydrolytic enzymes activity, biogas and methane production, volatile solid (VS) and chemical oxygen demand (COD) were measured. In this study, different ratios of waste activated sludge and mixed fruit waste and cheese whey were used in biochemical methane potential tests, so as to investigate the feasibility of enhancing methane production and quality of the digested waste streams in the anaerobic co-digestion.

## Results

### Substrates and inoculum

In this study waste activated sludge and inoculum were collected from the south wastewater treatment of Tehran, the biggest wastewater treatment plant of the Middle East. This plant treats wastewater of 4,200,000 people living in central and southern parts of Tehran, Iran. In this study, the waste activated sludge was collected from belt thickener and the inoculum was collected from six anaerobic digesters being operated in mesophilic condition at this plant. The different characterizations of the collected sludge such as total chemical oxygen demand (TCOD), soluble chemical oxygen demand (SCOD), volatile solid (VS), volatile suspended solid (VSS), total solid (TS), total suspended solid (TSS), alkalinity, pH and C/N were measured as soon as the sludge was conveyed to the university's laboratory (Table 1).

**Table 1** Characterizations of the substrates and inoculum

| Characterizations | Waste activated sludge | Fruit waste | Cheese whey | inoculum |
|---|---|---|---|---|
| TCOD (g/l) | 52.10 ± 0.70 | 32.51 ± 0.71 | 92.45 ± 0.82 | 38.21 ± 0.02 |
| SCOD (g/l) | 4.02 ± 0.50 | 12.45 ± 0.30 | 53.65 ± 0.91 | 3.550 ± 0.078 |
| VS (g/l) | 33.12 ± 0.52 | 20.52 ± 0.52 | 70.12 ± 0.81 | 25.15 ± 0.031 |
| VSS (g/l) | 30.01 ± 0.36 | 11.64 ± 0.12 | 32.10 ± 0.40 | 23.11 ± 0.09 |
| TS (g/l) | 41.25 ± 0.62 | 23.12 ± 0.48 | 84.34 ± 0.79 | 32.34 ± 0.012 |
| TSS (g/l) | 36.8 ± 0.56 | 14.43 ± 0.21 | 39.66 ± 0.57 | 28.45 ± 0.12 |
| pH | 6.54 ± 0 | 5.71 ± 0 | 5.88 ± 0 | 7.72 ± 0 |
| C/N | 10 ± 0.2 | 19 ± 0.4 | 35 ± 0.5 | 7 ± 0.3 |
| Alkalinity (mg CaCO3/l) | 3050 ± 25 | 1100 ± 25 | 2550 ± 44 | 3710 ± 52 |

Cheese whey used in this study was collected from a local cheese-producing factory, working with traditional cheese producing technologies in the same region. Fruit waste was also collected from fruit stores near the university's laboratory. After shredding and homogenizing, the fruit waste was diluted with distilled water to reduce its dry solid (DS) to around 2 g/L, so as to set overall DS of the mixtures (substrate and inoculum) at 3 to 9%, which is suggested by Metcalf and Tchobanoglou [37] as the best DS in anaerobic digesters for uniform nutrient and heat distribution, and proper mixers' performance. The fruit waste consisted of homogenized apple, tomato, carrot, orange and potato with equal proportion (based on weight). Characterizations of the cheese whey and fruit waste were measured as soon as they was conveyed to the university's laboratory (Table 1).

For assessing the influence of adding mixed fruit waste and cheese whey to anaerobic digestion of waste activated sludge, four ratios of waste activated sludge with mixed cheese whey and fruit waste were considered; A mixture with 100% waste activated sludge was considered as a control, the three other ratios were 95, 90 and 85% waste activated sludge with 5, 10 and 15% mixed fruit waste and cheese whey respectively. Characterizations of the mixtures with different ratios were measured as soon as the waste streams were completely mixed with each other (Table 2).

### Key hydrolytic enzymes activity

Protease and cellulase play an important role in decomposition of organic matter and provision of more readily biodegradable organic matter for anaerobic organisms. Cellulase catalyzes hydrolyze of polysaccharides to monoses and protease degrades proteins to amino acids [12]. In this study, the activity of protease and cellulase, as two key hydrolytic enzymes, were measured in the biochemical methane potential reactors with and without mixed fruit waste and cheese whey (Table 3). According to the data obtained, both protease and cellulase activities increased significantly ($p < 0.05$) in the bioreactors containing mixed cheese whey and fruit waste in comparison to the control bioreactors. The highest increase in the activity of protease and cellulase was observed in the reactor containing 85% waste activated sludge and 15% mixed fruit and cheese whey that accounted respectively for 22 and 10% in comparison to the control reactor. Followed by that, the reactor containing 90% waste activated sludge and 10% cheese whey had the second highest protease and cellulase activity with 17 and 6% higher than the control.

### Daily biogas and methane production

Daily biogas production from different bioreactors was measured during anaerobic digestion. The amount of daily biogas production enhanced in all of the bioreactors containing mixed fruit waste and cheese whey in comparison to the control bioreactor (Fig. 1). The maximum daily biogas was achieved in the bioreactor containing 85% waste activated sludge and 15% mixed cheese whey and fruit waste with 350 mL in the third day of digestion (Additional file 1). Analogously, cumulative biogas production, which can be obtained by calculating the area below each chart, was the highest in this bioreactor with 2970 mL at the end of digestion process (30th day). This was 33% higher than the cumulative biogas produced from the control bioreactor with 2225 mL at the end of digestion process.

Methane production from the bioreactors was regularly measured during the anaerobic digestion. According to the data obtained, the amount of methane production from the bioreactors containing cheese whey and fruit waste enhanced considerably in comparison to the control bioreactor (Fig. 2-a). For normalizing the data, the produced methane (mL) was divided by the added volatile solids (gram) to each bioreactors. The highest amount of methane production was achieved in the bioreactor containing 15% mixed fruit waste and cheese whey with 85% waste activated sludge that accounted for 384.1 mL $CH_4$/g VS $_{added}$ (Additional file 1). This is 31% higher than that of the control bioreactor with 292.76 mL $CH_4$/g VS $_{added}$ and 5% higher than that of the bioreactor containing 90% waste

**Table 2** Characterizations of mixed substrates

| Waste activated sludge: mixed fruit and cheese whey | 100:0 | 95:5 | 90:10 | 85:15 |
|---|---|---|---|---|
| COD (g/l) | 52.1 ± 0.06 | 53.12 ± 0.06 | 54.78 ± 0.06 | 55.31 ± 0.06 |
| SCOD (g/l) | 4.020 ± 0.08 | 6.23 ± 0.11 | 7.12 ± 0.09 | 8.51 ± 0.21 |
| VS (g/l) | 33.12 ± 0.02 | 34.41 ± 0.82 | 34.91 ± 0.92 | 35.33 ± 0.82 |
| VSS (g/l) | 30.01 ± 0.06 | 28.91 ± 0.52 | 29.45 ± 0.61 | 30.27 ± 0.4 |
| TS (g/l) | 41.25 ± 0.02 | 42.68 ± 0.55 | 44.11 ± 0.71 | 44.72 ± 0.52 |
| TSS (g/l) | 36.8 ± 0.06 | 35.96 ± 0.72 | 35.21 ± 0.81 | 34.87 ± 0.83 |
| pH | 6.54 ± 0 | 6.51 ± 0 | 6.49 ± 0 | 6.46 ± 0 |
| C/N | 10 ± 0.2 | 13.5 ± 0.3 | 15.12 ± 0.25 | 16.33 ± 0.52 |
| Alkalinity (mg CaCO3/l) | 3050 ± 25 | 2651 ± 0.43 | 2570 ± 56 | 2450 ± 63 |

activated sludge and 10% mixed cheese whey and fruit waste (with 370.77 $CH_4$/ g VS $_{added}$).

The amount of methane content during the anaerobic digestion fluctuated between 53 and 65% (Fig. 2-b). Assessing the mean value of methane content from different bioreactors, it was indicated that the mean value in the bioreactors containing mixed cheese whey and fruit waste during digestion period increased by 3%, reaching to 59% from 56%.

### Removal of organic matter

The amount of COD and VS in digested sludge play an important role in applicability of the sludge to agricultural lands and forests as they should meet the local environmental criteria. In addition, VS is an important characteristic affecting transportation costs of sludge [38] and it represents the metabolic status of microbial community in anaerobic system [39]. Therefore, it is imperative to reduce the costs of sludge management in wastewater treatment plants that accounts for around 60% of total operation costs of wastewater treatment plants [2, 20]. In this study, the amount of VS and COD were regularly measured during the anaerobic digestion process. It was demonstrated that the removal efficiency of COD and VS enhanced in the bioreactors containing mixed cheese whey and fruit waste (Fig. 3-a and -b respectively). The highest improvement achieved in the bioreactor containing 15% mixed cheese whey and fruit waste with 85% waste activated sludge, in which COD and VS removal enhanced by 9 and 7% in comparison to the control.

**Table 3** Protease and cellulase activity in different proportions of mixed cheese whey and fruit waste

| Waste activated sludge: mixed fruit and cheese whey (%) | 100/0 | 95/5 | 90/10 | 85/15 |
|---|---|---|---|---|
| Protease (mm) | 21.0 ± 0.3 | 23.5 ±0.7 | 24.6 ± 0.6 | 25.6 ± 0.8 |
| Cellulase (mm) | 26.5 ± 0.4 | 27.5 ± 0.6 | 28.1 ± 0.4 | 29.3 ± 0.8 |

### Discussion

The enhancement of enzymes' activity in the reactors contained cheese whey and fruit waste could be mainly attributed to the balanced C/N ratio, which provides higher amount of nutrients for microbial activity. Analogously, Feng et al. [30] revealed that the activity of protease, as a protein degrading enzyme, increases by 25% when waste activated sludge (with $C/N = 7.1$) is anaerobically co-digested with rice ($C/N = 20$), corroborating the effectiveness of balanced C/N in improving the activity of key hydrolytic enzymes.

The enhancement of biogas was marked from the first day of digestion process, indicating the positive effect of hydrolytic key enzymes activity in providing more readily biodegradable organic matter for better microbial activity. Higher soluble chemical oxygen demand in these reactors in comparison to the control provided higher amount of readily biodegradable organic matter for anaerobic organisms that may lead to increase of biogas production. However, in previous studies it was indicated that extremely high SCOD may lead to accumulation of volatile fatty acids, making a delay in biogas production or reducing pH of bioreactors [4, 40, 41].

The methane production enhanced as the proportion of mixed cheese whey and fruit waste increased. This is probably attributed to more balanced nutrients available to microbial community, positive synergism effect, and higher hydrolytic enzymes activity (Table 3). The enhancement in the bioreactors containing mixed cheese whey and fruit waste was significant ($p < 0.05$) in comparison to the control, corroborating the effectiveness of anaerobic co-digestion of waste activated sludge with mixed cheese whey and fruit waste in enhancing methane production. The enhanced methane is of paramount importance because not only does it enhance renewable energy generation in wastewater treatment plants, but it also reduces methane emission to the atmosphere, as a major greenhouse gas emission [17]. However, the population means of methane content was not significantly different in the bioreactors containing 5

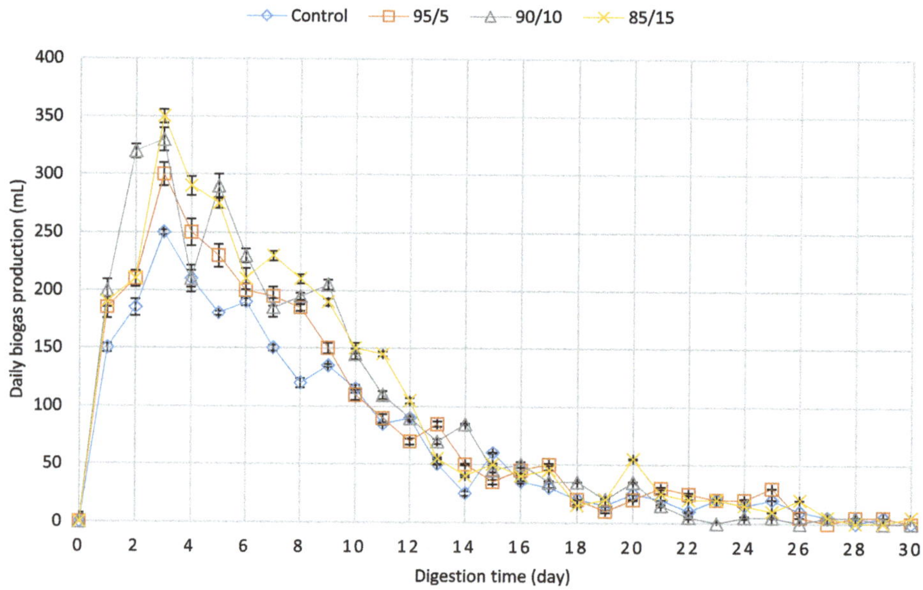


**Fig. 1** Daily biogas production from the bioreactors during digestion time. Error bars represent standard errors obtained from triplicate measurements

and 10% mixed cheese whey and fruit waste ($p = 0.41 > 0.05$), but that of the bioreactor contained 15% cheese whey and fruit waste was of significant difference ($p = 0.01 < 0.05$). This confirms that mixed cheese whey and fruit waste in higher proportion can be effective in improving methane content of the biogas. In comparison to this study, higher methane content (67%) was achieved in Maragkaki et al. [23] study, in which sewage sludge was co-digested with mixed cheese whey, food waste and olive mill wastewater. Also, Habiba et al. [42] assessed anaerobic co-digestion of waste activated sludge with fruit and vegetable waste, in which methane content fluctuated around 60% in all of the bioreactors. The difference in results can be attributed to wastes specifications and operation conditions employed in these studies, compared to the current study. In this study, methane production from the bioreactors containing mixed cheese whey and fruit waste enhanced considerably from the first day of digestion, indicating lower lag time in these bioreactors in comparison to the control. Analogous trend was observed in Zou et al. [22] study, in which lower lag time (0.2 day) was achieved in anaerobic co-digestion of residual sludge and lignocellulosic wastes. Also, Callaghan et al. [43] revealed that cattle slurry with fruit waste can enhance methane production to 450 mL $CH_4$/g VS $_{added}$ and adding chicken manure to this mixture declines methane production and VS reduction due to the accumulation of free ammonia.

The improved removal of organic matter can be linked to higher biogas and methane production, which lead to higher degradation of organic matter during anaerobic co-digestion of waste streams. The improved VS and COD are of great significance because not only does it reduce the transportation costs of digested waste streams, but it also paves the way for having integrated, sustainable system for reusing digested wastes in agricultural lands and forests. Compared to the current research, the amount of VS reduction was higher in Habiba et al. [42] study, in which waste activated sludge with fruit and vegetable waste was anaerobically digested in sequencing batch reactors.

In spite of the significant results this study represents, applying new methods to sludge treatment section of wastewater treatment plants entails inspecting long-term effect of the anaerobic co-digestion on microbial community and assessing economic viability of the proposed methods. Therefore, in prospective research, the effect of the mixed waste streams on microbial behaviour, economic viability, and environmental impacts of the proposed method should be put into perspective.

## Conclusions

This study investigated the viability of enhancing anaerobic digestion of waste activated sludge using mixed fruit waste and cheese whey. It was revealed that anaerobic co-digestion of waste activated sludge with mixed fruit and cheese whey could stimulate hydrolytic enzymes activity and providing better nutrients balance for microbial communities, whereby not only methane production increases, but also organic matter removal enhances at the end of digestion process. The increased methane production is of paramount importance because higher amount of renewable energy could be produced from anaerobic digestion. It also reduces methane emission from sludge treatment units with harnessing

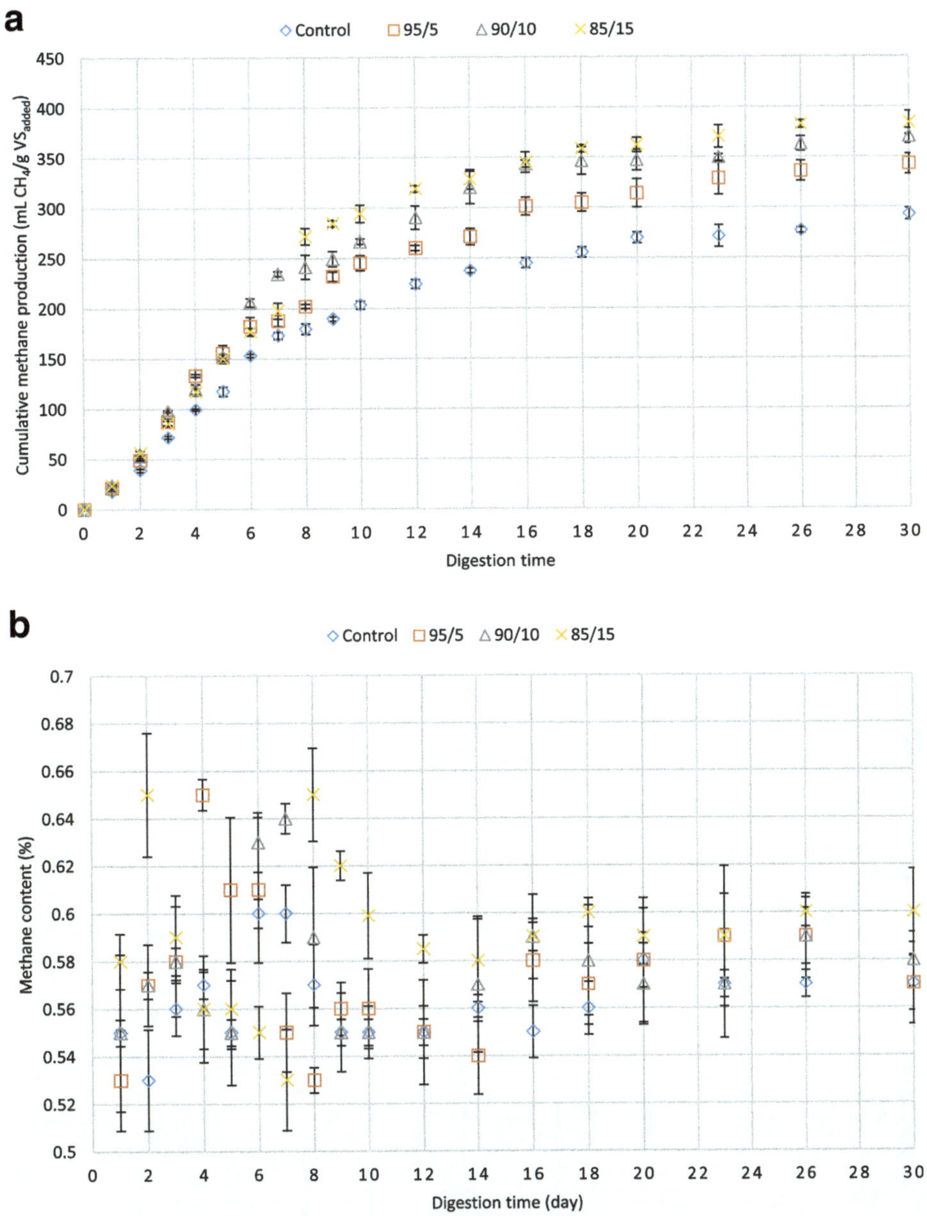


**Fig. 2 a** Cumulative methane production from the bioreactors during digestion time. Error bars represent standard error obtained from triplicate tests. **b** Methane content during anaerobic digestion. Error bars represent standard error obtained from triplicate tests

methane, as a major greenhouse gas, and converting it to renewable energy. The enhanced removal of organic matter paves the way for having more integrated system via heightening quality of the digested waste streams, which increases applicability of the digested wastes to agricultural lands and reduces transportation costs of the wastes in wastewater treatment plants.

## Methods

### Analytical methods

TS, VS, TSS, VSS, COD, SCOD and alkalinity were measured according to standard methods for the examination of water and wastewater [44]. To separate solid particles from liquid phase, samples were centrifuged at 10000 rpm for 30 min, then the supernatant was filtered through 0.45 μm pore size glass fiber filters, using Buchner funnel and vacuum pump. C/N ratio was measured in a commercial laboratory. Total nitrogen (TN) was measured according to persulfate digestion method using Hach spectrophotometer. Total carbon (TC) was measured by catalytic oxidation on a total organic carbon (TOC) Euroglace analyser.

Protease and cellulase activities were measured according to well agar diffusion method [45, 46], in which

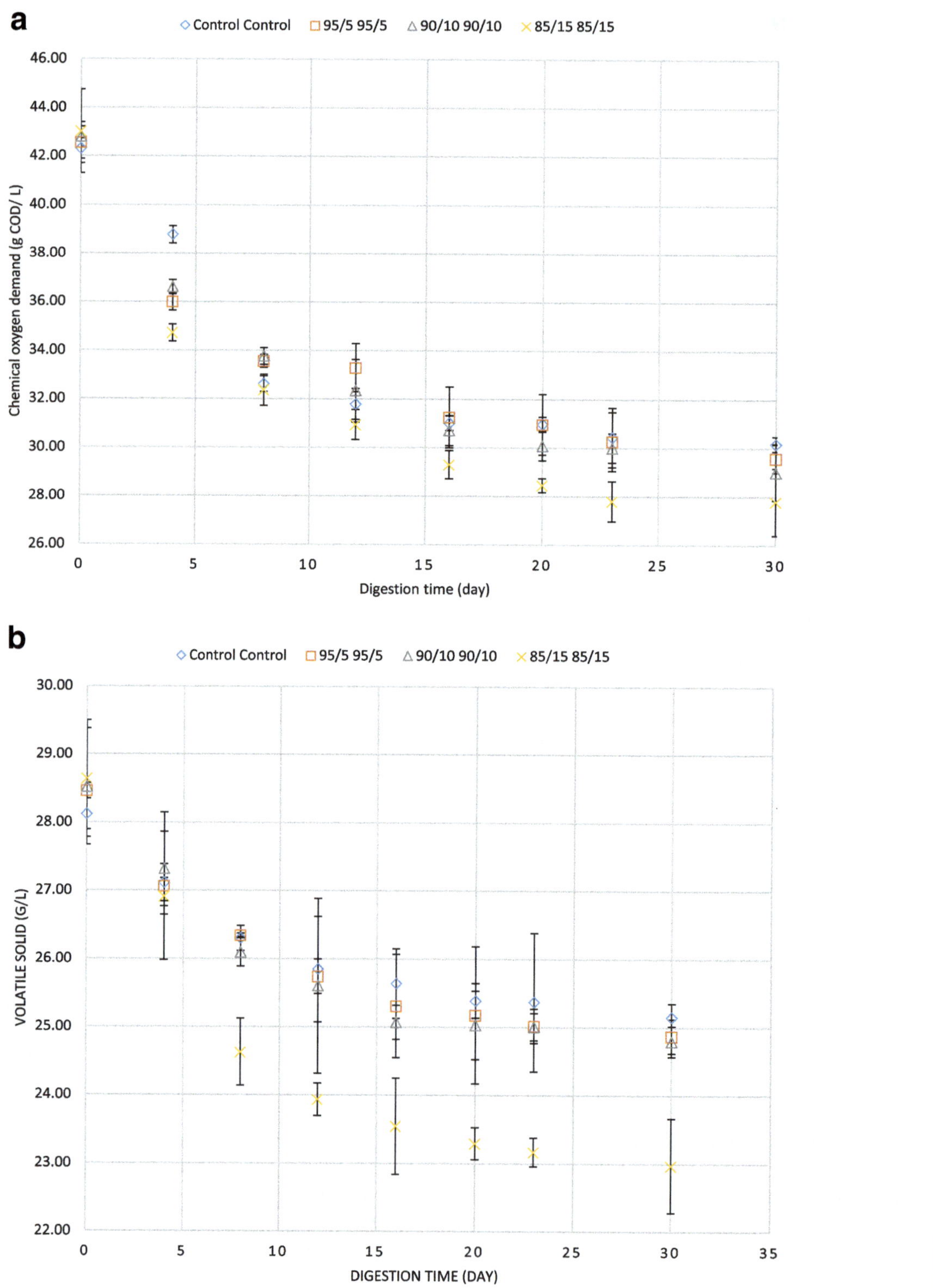


**Fig. 3 a** Chemical oxygen demand from the bioreactors during digestion time. Error bars represent standard error obtained from triplicate tests. **b** Volatile solid from the bioreactors during digestion time. Error bars represent standard error obtained from triplicate tests carboxymethyl cellulase agar (CMCA) and plate count skim milk agar (PSMA) were used as growth medium for protease and cellulose enzymes. Methanogens affect protease and cellulase activity by consuming hydrolyzed organic matter, so for removing the errors caused by methanogens in measuring enzymatic activity, sludge samples were heated at 102 °C for 30 min prior to well agar diffusion tests and the methanogens were removed from the samples [47, 48]. After removing methanogens, the reactors were flushed with $N_2$ gas. Then, they were sealed and put into water bath (36 ± 1 °C) for 72 h. Then the samples were used in well agar diffusion method.

Biogas and methane production were measured regularly during the digestion process. For measuring biogas production, liquid displacement method was employed, in which liquid barrier was 100% saturated with NaCl and acidified with $H_2SO_4$ (pH = 2), so as to reduce dissolution of $CO_2$ and $CH_4$ into the liquid barrier [49]. Methane production was measured with gas chromatography (GC), using thermal conductivity detector (TCD), and helium as carrier gas. The biogas was collected in Tedlar gas bags prior to use in GC tests.

### Biochemical methane potential tests

Biochemical methane potential tests were carried out in 1000 mL bioreactors with working volume of 500 mL (Fig. 4). Fruit waste and cheese whey were mixed (50%:50% *V*/V) prior to mix with waste activated sludge. Then, the mixed fruit waste and cheese whey was combined with waste activated sludge as substrate for the inoculums. The ratio between the inoculum and substrate (I/S) was considered as 2 in this study for better activity of anaerobic organisms [50]. A control reactor without mixed fruit waste and cheese whey was considered for comparing different bioreactors. The pH of substrates was set at 7 and their temperature was increased to 37 °C prior to mix with the inoculums, so as to prevent pH and temperature shock to anaerobic organisms. After the substrates and inoculums were mixed, they were flushed with $N_2$ gas for 2 min (1 L/min) and precisely sealed for creating strict anaerobic condition. The bioreactors were kept in warm water bath, heated by automatic heaters at 36 ± 1 °C during the anaerobic digestion. For adequate distribution of nutrients to various microbial communities and making uniform temperature in different parts of each bioreactor, they were constantly blended with magnetic stirrers at 100 rpm. The measurements were performed for 30 days, when no biogas production was detected from the bioreactors.

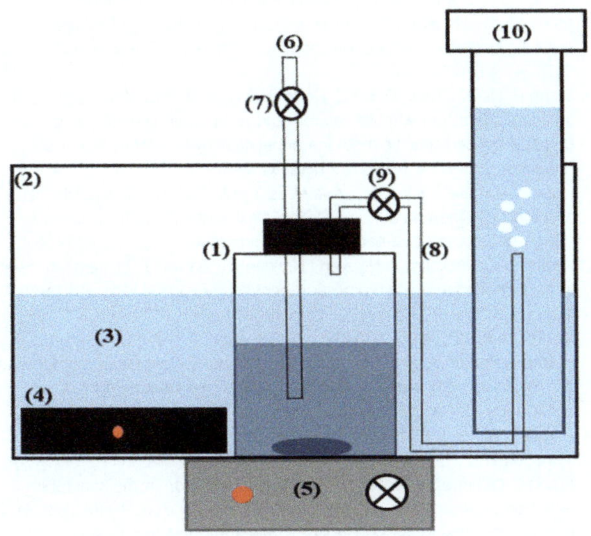


**Fig. 4** Schematic of experimental system. 1) BMP reactor, 2) aquarium 3) saturated and acidified water, 4) automatic heater, 5) magnetic stirrer, 6) sampling pipe, 7) control valve, 8) biogas collecting pipe, 9) biogas control valve, 10) graduated cylinder

### Data analyses

The significance of difference between the studied parameters was specified with one-way factor Analysis of Variance ANOVA and the significance level was considered as $p < 0.05$. Data analyses and graph processing were conducted with Microsoft Excel software (2010).

## Additional file

**Additional file 1: Table S1.** Daily biogas production during digestion process (average of triplicate tests). **Table S2.** Cumulative methane production during digestion process (average of triplicate tests). **Figure S1.** Gas chromatography samples. (DOCX 50 kb)

#### Abbreviations

CMCA: Carboxymethyl cellulase agar; COD: Chemical oxygen demand; GC: Gas chromatography; PSMA: Plate count skim milk agar; SCOD: Soluble chemical oxygen demand; TC: Total carbon; TCD: Thermal conductivity detector; TN: Total nitrogen; TOC: Total organic carbon; TS: Total solid; TSS: Total suspended solid; VS: Volatile solid; VSS: Volatile suspended solid

#### Acknowledgments

The authors acknowledge the University of Tehran for providing materials and experimental facilities and Tehran Sewerage Company for providing the required samples and data for this study.

#### Funding

The authors would appreciate Tehran Sewerage Company for financially assisting this study, sending the sludge samples to the university laboratory, and providing some general information about the south wastewater treatment of Tehran.

#### Authors' contributions

SMH reviewed literatures and chose the subject. SMH, MK, and SPM designed and provided the BMP systems. SMH and KM conducted experiments. SMH and SPM wrote the paper. SMH proofed data and text. All authors read and approved the final manuscript.

#### Ethics approval and consent to participate

Not applicable.

**Consent for publication**
Not applicable.

**Competing interests**
The authors declare that they have no competing interests.

**Author details**
[1]School of Environment, College of Engineering, University of Tehran, Tehran, Iran. [2]Environmental Health Research Center, International Branch of Shahid Sadoughi University of Medical Sciences and Health Services, Yazd, Iran.

**References**
1. Nazari L, Yuan Z, Santoro D, Sarathy S, Ho D, Batstone D, et al. Low-temperature thermal pre-treatment of municipal wastewater sludge: process optimization and effects on solubilization and anaerobic degradation. Water Res. 2017;113:111–23.
2. Neyens E, Baeyens J, Dewil R, De Heyder B. Advanced sludge treatment affects extracellular polymeric substances to improve activated sludge dewatering. J Hazard Mater. 2004;106:83–92.
3. Appels L, Degrève J, Van der Bruggen B, Van Impe J, Dewil R. Influence of low temperature thermal pre-treatment on sludge solubilisation, heavy metal release and anaerobic digestion. Bioresour Technol. 2010;101(15): 5743–8.
4. Hallaji SM, Torabian A, Aminzadeh B, Zahedi S, Eshtiaghi N. Improvement of anaerobic digestion of sewage mixed sludge using free nitrous acid and Fenton pre-treatment. Biotechnol Biofuels. 2018;11(1):233.
5. Zahedi S, Icaran P, Yuan Z, Pijuan M. Enhancing sludge biodegradability through free nitrous acid pre-treatment at low exposure time. Chem Eng J. 2017;321:139–45.
6. Wei W, Wang Q, Zhang L, Laloo A, Duan H, Batstone DJ, et al. Free nitrous acid pre-treatment of waste activated sludge enhances volatile solids destruction and improves sludge dewaterability in continuous anaerobic digestion. Water Res. 2018;130:13–9.
7. Svensson K, Kjørlaug O, Higgins MJ, Linjordet R, Horn SJ. Post-anaerobic digestion thermal hydrolysis of sewage sludge and food waste: effect on methane yields, dewaterability and solids reduction. Water Res. 2018;132: 158–66.
8. Peng L, Appels L, Su H. Combining microwave irradiation with sodium citrate addition improves the pre-treatment on anaerobic digestion of excess sewage sludge. J Environ Manag. 2018;213:271–8.
9. Pei H, Hu W, Liu Q. Effect of protease and cellulase on the characteristic of activated sludge. J Hazard Mater. 2010;178(1–3):397–403.
10. Mata-Alvarez J, Mace S, Llabres P. Anaerobic digestion of organic solid wastes. An overview of research achievements and perspectives. Bioresour Technol. 2000;74(1):3–16.
11. Tiehm A, Nickel K, Zellhorn M, Neis U, Tiehm A. Ultrasonic waste activated sludge disintegration for improving anaerobic stabilization. Water Res. 2001; 35(8):2003–9.
12. Feng Y, Zhang Y, Quan X, Chen S. Enhanced anaerobic digestion of waste activated sludge digestion by the addition of zero valent iron. Water Res. 2014;52:242–50.
13. Lin Y, Wang D, Liang J, Li G. Mesophilic anaerobic co-digestion of pulp and paper sludge and food waste for methane production in a fed-batch basis. Environ Technol. 2012;33(23):2627–33.
14. McCarty PL, Bae J, Kim J. Domestic wastewater treatment as a net energy producer-can this be achieved? Environ Sci Technol. 2011;45(17):7100–6.
15. Ni JQ, Naveau H, Nyns EJ. Biogas: exploitation of a renewable energy in Latin America. Renew Energy. 1993;3(6–7):763–79.
16. Xie S, Lawlor PG, Frost JP, Hu Z, Zhan X. Effect of pig manure to grass silage ratio on methane production in batch anaerobic co-digestion of concentrated pig manure and grass silage. Bioresour Technol. 2011; 102(10):5728–33.
17. Pilli S, More TT, Yan S, Tyagi RD, Surampalli RY. Fenton pre-treatment of secondary sludge to enhance anaerobic digestion: energy balance and greenhouse gas emissions. Chem Eng J. 2016;283:285–92.
18. Dai X, Li X, Zhang D, Chen Y, Dai L. Simultaneous enhancement of methane production and methane content in biogas from waste activated sludge and perennial ryegrass anaerobic co-digestion: the effects of pH and C/N ratio. Bioresour Technol. 2016;216:323–30.
19. Sanscartier D, MacLean HL, Saville B. Electricity production from anaerobic digestion of household organic waste in Ontario: techno-economic and GHG emission analyses. Environ Sci Technol. 2012;46(2):1233–42.
20. Appels L, Baeyens J, Degrève J, Dewil R. Principles and potential of the anaerobic digestion of waste-activated sludge. Prog Energy Combust Sci. 2008;34(6):755–81.
21. Hublin A, Schneider DR, Džodan J. Utilization of biogas produced by anaerobic digestion of agro-industrial waste: energy, economic and environmental effects. Waste Manag Res. 2014;32(7):626–33.
22. Zou H, Chen Y, Shi J, Zhao T, Yu Q, Yu S, et al. Mesophilic anaerobic co-digestion of residual sludge with different lignocellulosic wastes in the batch digester. Bioresour Technol. 2018;268:371–81.
23. Maragkaki AE, Vasileiadis I, Fountoulakis M, Kyriakou A, Lasaridi K, Manios T. Improving biogas production from anaerobic co-digestion of sewage sludge with a thermal dried mixture of food waste, cheese whey and olive mill wastewater. Waste Manag. 2018;71:644–51.
24. Stroot PG, McMahon KD, Mackie RI, Raskin L. Anaerobic codigestion of municipal solid waste and biosolids under various mixing conditions - II: microbial population dynamics. Water Res. 2001;35(7):1804–16.
25. Heo NH, Park SC, Kang H. Effects of mixture ratio and hydraulic retention time on single-stage anaerobic co-digestion of food waste and waste activated sludge. J Environ Sci Heal Part A. 2004;39(7):1739–56.
26. Kim S-H, Han S-K, Shin H-S. Feasibility of biohydrogen production by anaerobic co-digestion of food waste and sewage sludge. Int J Hydrog Energy. 2004;29(15):1607–16.
27. Rizk MC, Bergamasco R, Tavares CRG. Anaerobic co-digestion of fruit and vegetable waste and sewage sludge. Int J Chem React Eng. 2007;5:CP6–U148.
28. Estevez MM, Linjordet R, Morken J. Effects of steam explosion and co-digestion in the methane production from Salix by mesophilic batch assays. Bioresour Technol. 2012;104:749–56.
29. Parkin GF, Owen WF. Fundamentals of anaerobic digestion of wastewater sludges. J Environ Eng. 1986;112(5):867–920.
30. Feng L, Chen Y, Zheng X. Enhancement of waste activated sludge protein conversion and volatile fatty acids accumulation during waste activated sludge anaerobic fermentation by carbohydrate substrate addition: the effect of pH. Environ Sci Technol. 2009;43(12):4373–80.
31. Pilli S, Yan S, Tyagi RD, Surampalli RY. Overview of Fenton pre-treatment of sludge aiming to enhance anaerobic digestion. Rev Environ Sci Biotechnol. 2015;14(3):453–72.
32. Dai X, Hu C, Zhang D, Dai L, Duan N. Impact of a high ammonia-ammonium-pH system on methane-producing archaea and sulfate-reducing bacteria in mesophilic anaerobic digestion. Bioresour Technol. 2017;245:598–605.
33. Wang G, Dai X, Zhang D, He Q, Dong B, Li N, et al. Two-phase high solid anaerobic digestion with dewatered sludge: improved volatile solid degradation and specific methane generation by temperature and pH regulation. Bioresour Technol. 2018;259:253–8.
34. Hosseini Koupaie E, Barrantes Leiva M, Eskicioglu C, Dutil C. Mesophilic batch anaerobic co-digestion of fruit-juice industrial waste and municipal waste sludge: process and cost-benefit analysis. Bioresour Technol. 2014;152:66–73.
35. Nguyen AQ, Wickham R, Nguyen LN, Phan HV, Galway B, Bustamante H, et al. Impact of anaerobic co-digestion between sewage sludge and carbon-rich organic waste on microbial community resilience. Environ Sci Water Res Technol. 2018;4(12):1956–65.
36. Fitamo T, Boldrin A, Boe K, Angelidaki I, Scheutz C. Co-digestion of food and garden waste with mixed sludge from wastewater treatment in continuously stirred tank reactors. Bioresour Technol. 2016;206:245–54.
37. Tchobanoglous G, Burton FL, Stensel HD. Wastewater engineering. Management. 1991;7:1–4.
38. Arnaiz C, Gutierrez JC, Lebrato J. Biomass stabilization in the anaerobic digestion of wastewater sludges. Bioresour Technol. 2006;97(10):1179–84.
39. Elango D, Pulikesi M, Baskaralingam P, Ramamurthi V, Sivanesan S. Production of biogas from municipal solid waste with domestic sewage. J Hazard Mater. 2007;141(1):301–4.

40. Zahedi S. Energy efficiency: importance of indigenous microorganisms contained in the municipal solid wastes. Waste Manag. 2018;78:763–9.
41. Chen Y, Cheng JJ, Creamer KS. Inhibition of anaerobic digestion process: a review. Bioresour Technol. 2008;99:4044–64.
42. Habiba L, Hassib B, Moktar H. Improvement of activated sludge stabilisation and filterability during anaerobic digestion by fruit and vegetable waste addition. Bioresour Technol. 2009;100(4):1555–60.
43. Callaghan FJ, Wase DAJ, Thayanithy K, Forster CF. Continuous co-digestion of cattle slurry with fruit and vegetable wastes and chicken manure. Biomass Bioenergy. 2002;22(1):71–7.
44. APHA. Standard Methods. For the examination of water and wastewater. 22 nd. 2012;
45. Lorian V. Antibiotics in laboratory medicine: Lippincott Williams & Wilkins; 2005.
46. Singh Rathore S, Narendhirakannan RT. Journal Of Advanced Scientific Research Screening Of Cellulase Producing Microorganisms From Lake Area Containing Water Hyacinth For Enzymatic Hydrolysis Of Cellulose. J Adv Sci Res J Adv Sci Res. 2014;5(53):23–30.
47. Oh S-E, Van Ginkel S, Logan BE. The relative effectiveness of pH control and heat treatment for enhancing biohydrogen gas production. Environ Sci Technol. 2003;37(22):5186–90.
48. Basu SK, Oleszkiewicz JA, Sparling R. Effect of sulfidogenic and methanogenic inhibitors on reductive dehalogenation of 2-chlorophenol. Environ Technol. 2005;26(12):1383–91.
49. Walker M, Zhang Y, Heaven S, Banks C. Potential errors in the quantitative evaluation of biogas production in anaerobic digestion processes. Bioresour Technol. 2009;100(24):6339–46.
50. Boulanger A, Pinet E, Bouix M, Bouchez T, Mansour AA. Effect of inoculum to substrate ratio (I/S) on municipal solid waste anaerobic degradation kinetics and potential. Waste Manag. 2012;32(12):2258–65.

# 14

# Fast gene disruption in *Trichoderma reesei* using in vitro assembled Cas9/gRNA complex

Zhenzhen Hao and Xiaoyun Su*

**Abstract**

**Background:** CRISPR/Cas9 has wide application potentials in a variety of biological species including *Trichoderma reesei*, a filamentous fungus workhorse for cellulase production. However, expression of Cas9 heterologously in the host cell could be time-consuming and sometimes even troublesome.

**Results:** We tested two gene disruption methods in *T. reesei* using CRISPR/Cas9 in this study. The intracellularly expressed Cas9 led to unexpected off-target gene disruption in *T. reesei* QM9414, favoring inserting 9- or 12-bp at 70- and 100-bp downstream of the targeted *ura5*. An alternative method was, therefore, established by assembling Cas9 and gRNA in vitro, followed by transformation of the ribonucleoprotein complex with a plasmid containing the *pyr4* marker gene into *T. reesei* TU-6. When the gRNA targeting *cbh1* was used, eight among the twenty seven transformants were found to lose the ability to express CBH1, indicative of successful *cbh1* disruption through genome editing. Large DNA fragments including the co-transformed plasmid, chromosomal genes, or a mixture of these nucleotides, were inserted in the disrupted *cbh1* locus.

**Conclusions:** Direct transformation of Cas9/gRNA complex into the cell is a fast means to disrupt a gene in *T. reesei* and may find wide applications in strain improvement and functional genomics study.

**Keywords:** CRISPR/Cas9, *Trichoderma reesei*, Gene disruption, Ribonucleoprotein

## Background

*Trichoderma reesei* is a well-known filamentous fungus workhorse for production of cellulase and other industrially important proteins [1]. Gene disruption is a critical technique for functional genomics study of *T. reesei*. From a practical perspective, deleting key regulatory genes that repress cellulase expression has proved to be very useful in strain improvement [2]. In addition, removal of major cellulase and protease genes is an essential step towards creating a platform strain for heterologous protein production [3]. Therefore, a method is needed to efficiently disrupt the genes of interest in *T. reesei*.

In *T. reesei*, gene disruption is routinely achieved by gene replacement through homology-based DNA recombination (HDR) [4]. However, one should note that the efficiency of HDR in *T. reesei* is notoriously low. Knocking out *ku70* or *mus53* involved in non-homologous end-joining (NHEJ) can improve the HDR rate [5], but time is needed to obtain a recipient strain with abated NHEJ function. The CRISPR/Cas9 system (with CRISPR standing for *c*lustered *r*egularly *i*nterspaced *s*hort *p*alindromic *r*epeats) is the bacterial type-II adaptive immune system [6], which can serve as an alternative means for gene disruption in *T. reesei*. In this system, the gRNA recognizes specific target DNA sequence, forms DNA/RNA duplex at the locus of recognition, and guides the endonuclease Cas9 to cleave the DNA. The CRISPR/Cas9 technology has been widely used for genome editing in mammalian cells, plants, and microbes. Genome disruption using CRISPR/Cas9 has been reported for a variety of filamentous fungi [7], which commonly requires expression of *cas9* in vivo. In contrast, transformation of the in vitro assembled Cas9 and gRNA complex provides a fast means of genome disruption [8], which however has not been reported in *T. reesei*.

* Correspondence: suxiaoyun@caas.cn
Key Laboratory for Feed Biotechnology of the Ministry of Agriculture, Feed Research Institute, Chinese Academy of Agricultural Sciences, No. 12 South Zhongguancun Street, Beijing 100081, China

In the present study, it was first found that the intracellularly expressed Cas9 led to unexpected off-target gene disruption in *T. reesei* QM9414. An alternative means, i.e. direct transformation of the in vitro assembled Cas9/gRNA, was evaluated for its efficacy in gene disruption using the cellobiohydrolase I (*cbh1*) gene as a model. The sequences of the disrupted gene were analyzed, identifying insertion of large chromosomal DNAs or plasmid fragments.

## Results

### Intracellularly expressed Cas9 led to unexpected off-target genome editing

The *cas9* gene has been expressed constitutively or induced successfully in *T. reesei*, in both cases which can be used in successful genome editing in QM6a and RUT-C30 [7]. The codon-optimized *cas9* gene [7] was synthesized, ligated downstream of the strong, constitutive *pdc1* promoter, and transformed into *T. reesei* QM9414. Nine positive transformants were obtained, among which strain C5 had the highest *cas9* transcript level (determined by RT-qPCR, data not shown) and was therefore selected for subsequent gene disruption.

The *ura5* gene has been used as the target when the CRISPR/Cas9 technology was established in *T. reesei* [7]. The in vitro transcribed gRNA targeting the same *ura5* locus was transformed into C5. Nineteen 5-fluoroorotic acid (5-FOA) resistant transformants were obtained. While no mutation was found in the expected editing locus for all these transformants, seven had an unexpected insertion of 12-bp at the 100-bp downstream of the expected editing locus (Fig. 1, Additional file 1: Figure S1). One (QM9414-C5/T18) had an insertion of 9-bp, while another (QM9414-C5/T13) had a deletion of this 9-bp fragment, at the 70-bp downstream of the expected editing locus. Interestingly, the inserted short DNA fragments were a direct repeat of the nucleotides downstream of the insertion site. Moreover, two direct repeats (7- and 11-bp, respectively) were found in both sites (Fig. 1), suggestive of biased off-target editing.

### Designing gRNA for *cbh1* gene disruption

Direct transformation of Cas9/gRNA complex may be an alternative for genome editing. Indeed, this technique has already been successfully used for gene disruption in various filamentous fungi including *Mucor circinelloides* [9], *Aspergillus fumigatus* [10], and *Fusarium oxysporum* [11]. For those filamentous fungi that are difficult to manipulate, the transformantion of Cas9/gRNA complex provides a means to speed up genome editing. With this advantage, nevertheless, there was no report of using such a technique in *T. reesei*. CBH1 is the major cellulase secreted by *T. reesei*, accounting for above 50–60% in the fermentation broth [12]. Disruption of *cbh1* will result in loss of the major band on SDS-PAGE gel, facilitating identification of the strains with successful genome editing. Therefore, for ease of verification, *cbh1* was selected as the target gene. Three different gRNAs on *cbh1* were designed and synthesized by in vitro transcription. Cas9 and a certain gRNA was incubated with the *Bcu*I-linearized pT3cbh1 plasmid (4717-bp) containing the *cbh1* gene. gRNA-3 was the best one to guide Cas9-mediated digestion of pT3cbh1 into two DNA fragments with expected sizes (Fig. 2).

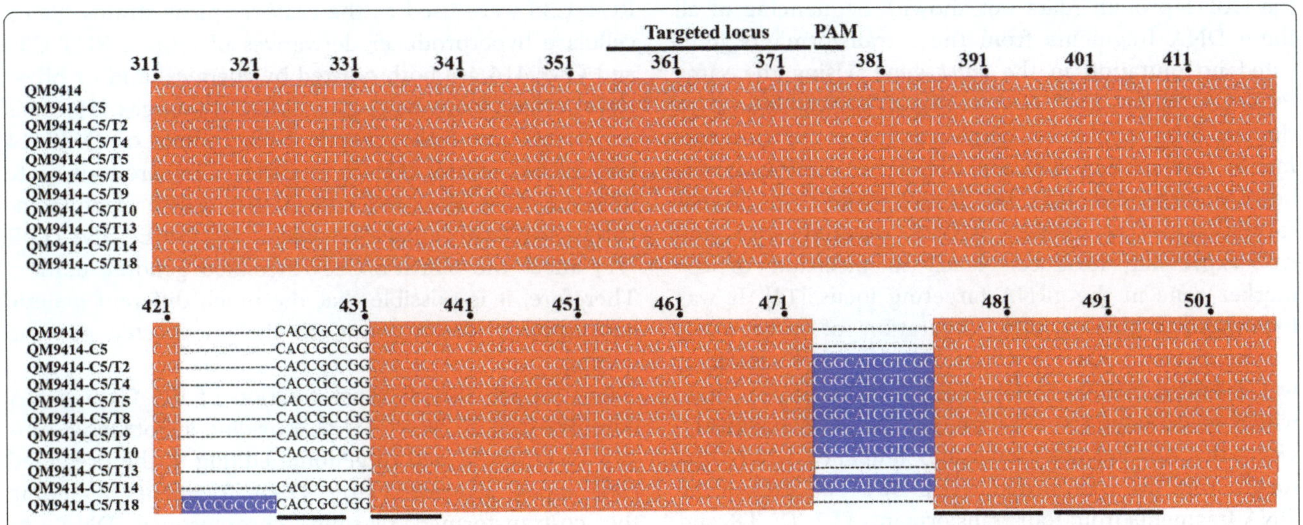


**Fig. 1** Intracellularly expressed Cas9 led to unexpected insertion and deletion in *ura5*. The inserted oligonucleotides were labeled in blue and the direct repeats were underlined

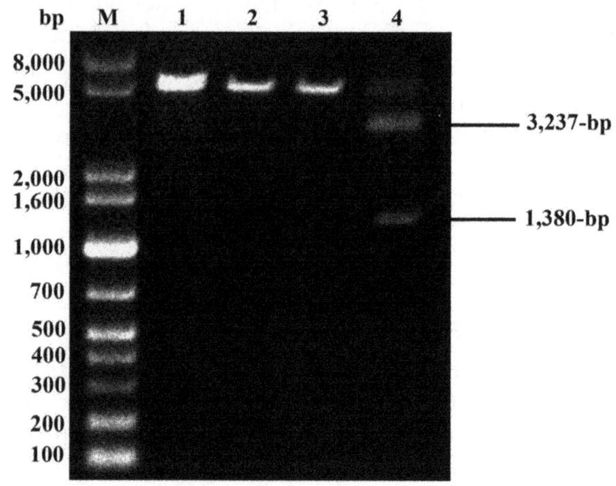


**Fig. 2** In vitro digestion of *cbh1* by Cas9 and gRNAs. M: DNA molecular mass marker; lane 1: *BcuI*-linearized plasmid; lane 2–4: *BcuI*-linearized plasmid incubated with Cas9 and gRNA1 (lane 2), gRNA2 (lane 3), and gRNA3 (lane 4)

### Disruption of *cbh1* involved insertion of large DNA fragments

Twenty-seven transformants were obtained in two transformations of Cas9/gRNA with pSKpyr4 in TU-6. Through SDS-PAGE analysis, it was suggested that eight among the 27 transformants could have the *cbh1* gene disrupted since the band corresponding to CBH1 disappeared (Fig. 3A). Interestingly, in these strains, expression of other proteins apparently increased (Fig. 3A), similar to improved secretion of a heterologous lipase in a *cbh1*-silenced *T. reesei* strain [13]. Using the primers specific for *cbh1* (Additional file 1: Table S1), a DNA fragment of 1676-bp was amplified from the parent strain TU-6 (Fig. 3B), as well as from the nineteen transformants still secreting the CBH1 protein (data not shown). Sequencing of all these DNA fragments from these transformants indicated no mutation in the *cbh1* gene. Using the same PCR condition, no DNA fragment corresponding to the 1676-bp could be amplified from the transformants (except T1) that did not secret CBH1. It has been pointed out that in *Nodulisporium* sp., transformation of a linear marker plasmid into the *cas9*-expressing host cell leads to insertion of the marker gene in the gRNA targeting locus [14]. It was hypothesized that similar integration of large DNA fragments could take place. Therefore, the PCR condition was modified, mainly by lengthening the extension time, which would allow amplification of larger DNA fragments. Using this method, we successfully amplified DNA fragments from five transformants (Fig. 3B). The sizes of the DNA fragments from four transformants (T2, T7, T8, and T9) were from 5- to 8-kb, while that for one transformant (T1) was 1.9-kb.

Further sequence analyses indicated that there were gene editing events at the expected *cbh1* locus. Deletion of 91~705-bp in *cbh1* was observed in T2, T7, and T9 (Fig. 3D). The inserted fragments varied in length and sequence by containing: i) the transformed plasmid (T2, T7, and T8); ii) a chromosome fragment (T1); or iii) mixed chromosome and plasmid sequences (T9, Fig. 3C&D). Although T2 and T7 had different deletions, they had the same inserts. It seemed that the pSKpyr4 DNA fragments starting from nucleotide 23 were favored in these gene rearrangement events (Fig. 3D). The underlying reason for this is not known. All five had another fragment ("other insert" in Fig. 3C&D) which was not from pSKpyr4 or *T. reesei* chromosome but from *E. coli* chromosome (T1) or an unidentified plasmid(s) (T2, T7, T8, and T9), which was likely contaminants from pSKpyr4 preparation.

## Discussion

The intracellularly expressed Cas9 successfully disrupts the *ura5* gene in the *T. reesei* QM6a and RUT-C30 strains by creating small deletions [7]. However, disruption of the same gene was quite different in QM9414 in this study since no such small deletions were observed in the 5-FOA-resistant transformants. In contrast, insertion (or less frequently, deletion) of small DNA fragments was found downstream of the genomic locus expected for editing. These mutations resulted in 3–4 amino acids duplication (or 3 amino acids deletion) in the *ura5*-encoded orotate phosphoribosyl transferase, which might interfere with the normal function of this enzyme, thereby enabling the transfromants to resist the toxicity of 5-FOA. Since the sequence of gRNA was the same as that used in [7], the off-target genome editing was unexpected. In the previous report [7], QM6a and RUT-C30 were used as the *cas9*-recipient strains. Being cellulase hyperproducer derivatives of QM6a, RUT-C30 and QM9414 are both created by chemical and/or physical mutations but belong to different lineages [15]. It is conceivable that the genetic backgrounds of QM9414 and the two strains (QM6a and RUT-C30) are thus quite different. It is also well-known that cellular factors involved in DNA damage response [16], NHEJ, and HDR [17] affect the CRISPR/Cas9-mediated genome editing. Therefore, it is possible that the much different genetic backgrounds may account for the unexpected genome editing.

In contrast, direct transformation of the Cas9-gRNA complex into *T. reesei* TU-6, a uridine auxotrophic mutant of QM9414, led to genome editing in the predicted *cbh1* locus. However, large DNA fragments including the co-transformed plasmid, chromosomal DNA, or their hybrid form, were found to be inserted in the targeted *cbh1* gene. Insertion of large DNA fragments was

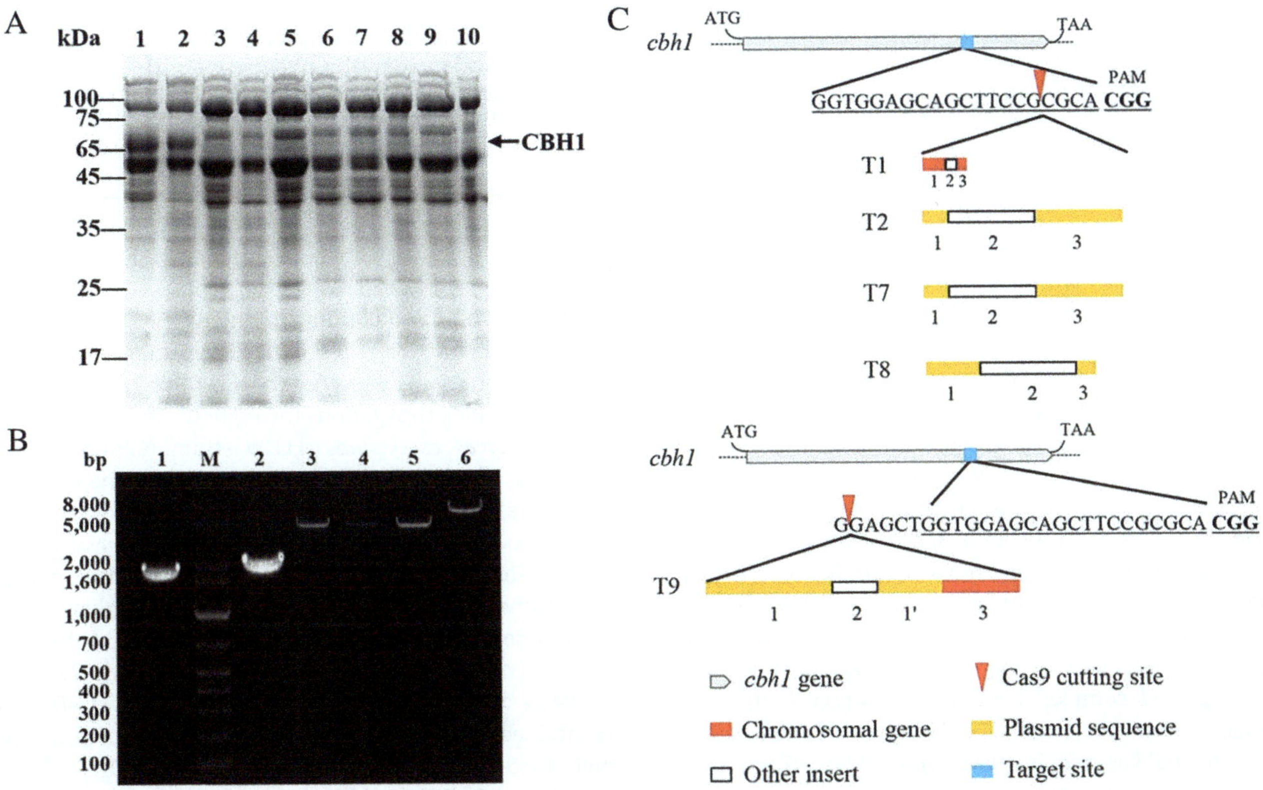


D

| Transformants | Deletion | Insert segment number | Insert segment | Chromosome of *T. reesei* |
|---|---|---|---|---|
| T1 | None | 1 | Gene ID: 122824 (1072 - 1192) | Chromosome VI |
| | | 2 | Other insert | ----- |
| | | 3 | Gene ID: 65223 (447 - 507) | Chromosome VI |
| T2 | 676-bp (*cbh1*: 687-1362) | 1 | pSKpyr4 (345 - 23) | ----- |
| | | 2 | Other insert | |
| | | 3 | pSKpyr4 (5752 - 4648) | |
| T7 | 705-bp (*cbh1*: 657-1362) | 1 | pSKpyr4 (345 - 23) | ----- |
| | | 2 | Other insert | |
| | | 3 | pSKpyr4 (5752 - 4648) | |
| T8 | None | 1 | pSKpyr4 (5149 - 5752) | ----- |
| | | 2 | Other insert | |
| | | 3 | pSKpyr4 (23 - 101) | |
| T9 | 91-bp (*cbh1*: 1293-1383) | 1 | pSKpyr4 (3584 - 23) | ----- |
| | | 2 | Other insert | |
| | | 1' | pSKpyr4 (4343 - 3697) | |
| | | 3 | Gene ID: 75230 (-261 - 640) | Chromosome VII |

**Fig. 3** (See legend on next page.)

(See figure on previous page.)
**Fig. 3** *cbh1* disruption in *T. reesei* by direct transformation of Cas9/gRNA complex and a plasmid containing the *pyr4* selection marker. A: SDS-PAGE analysis of the fermentation supernatants of the transformants. M: protein molecular mass marker, lanes 1–2: TU-6; lanes 3–11: transformants (T1, T2, T3, T4, T6, T7, T8, and T9, respectively) that did not express CBH1; B: agarose gel electrophoresis of the PCR products amplifying the *cbh1* locus from the transformants that did not express CBH1. M: DNA molecular mass marker; lane 1: TU-6; lane 2–6: the transformants T1, T2, T7, T8, and T9, respectively; C: Schematic diagram showing the inserted DNA fragments in the edited *cbh1* locus; D: Characteristics of inserted or deleted DNA fragments. The numbers for the *T. reesei* genes were counted from the start codon

not observed in *Mucor circinelloide* [9] and *Fusarium oxysporum* [11] when direct transformation of Cas9/gRNA was employed. For these two fungi, only the Cas9/gRNA complex, but none exogenous DNA, was used in the transformation. In *Nodulisporium* sp., similar insertion of large DNA fragment was also observed. However, in that case an exogenous plasmid containing the selection marker was co-transformed with the in vitro transcribed gRNA into a recipient strain with in vivo expressed Cas9. Note, however, unlike the discovery reported herein, only the exogenous plasmid sequence was found in the inserted fragments in *Nodulisporium* sp. The promiscuity of the inserted fragments in *T. reesei* suggested complex DNA-repair events in the target locus.

With the success in *cbh1* disruption, direct transformation of Cas9/gRNA complex provides a fast means for disruption of genes of interest in *T. reesei*. This method may also be used for HDR-mediated gene replacement. In a preliminary experiment we have observed gene replacement at the *cel3c* locus, albeit at a low frequency (5 out of 143 transformants), by co-transformation of Cas9/gRNA (targeting *cel3c* [18]) with a donor DNA in *T. reesei* (data not shown). Gene replacement efficiency can be improved by optimizing the amounts, nature and ratio of Cas9 and gRNA or using a strain with dysfunctional NHEJ DNA repair pathway.

## Conclusions

The intracellularly expressed Cas9 led to biased off-target editing in *ura5*, which might be explained by the different strains used in our study and by other researchers. Direct transformation of Cas9/gRNA complex allowed disruption of *cbh1* in *T. reesei* with large DNA inserts found in the edited locus. Compared with the previous report, our method avoids the time and labor needed for introducing the *cas9* gene into *T. reesei* and the uncertainty that may occur with its expression. In our hands, both in vivo expressed Cas9 and in vitro assembled Cas9/gRNA led to insertion of small or large DNA fragments during genome editing. This might be explained by the robust NHEJ process, which is the dominant DNA repair mechanism in *T. reesei*.

## Methods

### Strains, plasmids and culture conditions

The *Escherichia coli* Trans 1-T1 strain (Transgen, Beijing, China) was used as the host for plasmid construction and propagation throughout the study. The *Saccharomyces cerevisiae* AH109 strain (Clontech, San Francisco, CA) was used for construction of the *cas9*-expressing plasmid. The pPdc1-Cas9 plasmid for expressing the codon-optimized *Streptococcus pyogenes cas9* [7] was obtained by assembling *cas9* between the strong, constitutive *pdc1* promoter and *pdc1* terminator via DNA assembler using pRS424 (New England Biolabs, Beverly, MA) as a backbone plasmid [19]. The *T. reesei* QM9414 and its uridine auxotrophic strain TU-6 were maintained in our lab. *T. reesei* was grown in the minimal medium (MM, containing $(NH_4)_2SO_4$, 5.0 g/L; $KH_2PO_4$, 15 g/L; $MgSO_4$, 0.6 g/L;$CaCl_2$, 0.6 g/L; $FeSO_4{\cdot}7H_2O$, 0.005 g/L; $MnSO_4{\cdot}H_2O$, 0.0016 g/L; $ZnSO_4{\cdot}7H_2O$, 0.0014 g/L; $CoCl_2$, 0.002 g/L) supplemented with a certain kind of carbohydrate (2% glucose for mycelial growth and 2% Avicel cellulose for cellulase induction) as the sole carbon source. For sporulation, *T. reesei* was grown on potato dextrose agar (PDA) plates at 28 °C.

### Designing gRNA targeting *chh1*

The gene *cbh1* encoding the major cellulase cellobiohydrolase I (CBH1) was amplified from the genomic DNA of TU-6 by PCR using the primer pair Pcbh1F/R (Additional file 1: Table S1). It was then ligated into pEASY-T3 (Transgen, Beijing, China) to obtain pT3cbh1. Three gRNA sequences for *cbh1* were designed using the online E-CRISPR design server (http://www.e-crisp.org/E-CRISP/). The corresponding encoding DNAs were synthesized in the form of single-stranded, complementary oligonucleotides (Additional file 1: Table S1). The T7 promoter (5'-TAATACGACTCACTATAGG-3') was designed upstream of the gRNAs. Equal amounts (1 μg) of two complementary oligonucleotides were mixed, boiled for 10 min, and cooled down at room temperature for annealing. Using the annealed double-stranded DNAs as the template, gRNAs were obtained by in vitro transcription with a MEGAshortscript T7 Transcription Kit (Invitrogen, Carlsbad, CA) and the RNAs were purified by using the RNA Purification Kit (Tiangen, Beijing, China). To determine the ability to guide digestion, 1 μg of gRNA, 2 μg of recombinant Cas9 (GenScript, Nanjing, China), and 200 ng of the *Bcu*I-linearized pT3cbh1 plasmid were mixed and incubated at 37 °C for 2 h. At the end of reaction, the samples were taken out for agarose gel electrophoresis analysis.

### Transformation of *T. reesei*

For plasmid transformation, pPdc1-Cas9 and pRLMex30 (carrying the hygromycin B phosphotransferase gene) were co-introduced into the *T. reesei* QM9414 strain by polyethylene glycol (PEG)-mediated chemical transformation [20]. The transformants were selected on MM-glucose plates containing 150 μg/ml of hygromycin. For transformation of gRNA targeting *ura5*, 6.5 μg of in vitro transcribed gRNA were similarly introduced into *T. reesei* through the same PEG-mediated protoplast transformation method. MM-glucose supplemented with 5-fluoroorotic acid (5-FOA, 3 mg/ml) and 10 mM uridine was used to select the transformants. Expression of wild-type, functional *ura5* (encoding orotate phosploribosyl transferase) will convert 5-FOA into toxic 5′-fluorouridine monophosphate. For Cas9/gRNA transformation, 12 μg of the Cas9 protein and 6.5 μg of gRNA were mixed and incubated at 37 °C for 30 min for in vitro assembly into the Cas9/gRNA RNP complex. The complex was then co-transformed with 6 μg of the pSKpyr4 plasmid, which contains the *pyr4* marker gene expression cassette [13], into the *T. reesei* TU-6 protoplasts.

### Analyzing *cbh1* disruption in the transformants

The spores ($1 \times 10^8$) of TU-6 and its transformants were individually inoculated into 100 ml of liquid MM-glucose (2%) and cultured at 28 °C with shaking for 48 h. Mycelia were collected by filtering through a 200-mesh sifter, washed with MM, and transferred into the MM-Avicel cellulose (2%) medium to induce cellulase expression. The fermentation broth on the eighth day after cellulose induction was analyzed by SDS-PAGE. The genomic DNAs of TU-6 and its transformants were extracted from the harvested mycelia and were used as templates for PCR amplification of the *cbh1* locus. The PCR conditions were 94 °C for 1 min for initial denaturation, then 24 cycles including 94 °C for 25 s, 61 °C for 30 s (temperature reduced by 0.2 °C per cycle), and 70 °C for 5 min.

### Gene replacement of *cel3c*

The Cas9 (12 μg) was in vitro assembled with 6.5 μg of gRNA targeting *cel3c* (GenBank accession number: AY281375.1) by incubation at 37 °C for 30 min. The donor DNA was consisted of two homology arms (1.0 kb each) flanking the *cel3c* locus. The homology arms were intercepted by an expression cassette for *Melanocarpus albomyces* laccase (GenBank accession number: CAE00180, under the control of *cbh1* promoter and terminator) and the *pyr4* marker gene. The Cas9/gRNA complex and 3 μg of the donor DNA were co-transformed into *T. reesei* protoplasts and the transformants were transferred into MM-lactose plate containing 0.4 mM ABTS and 0.1 mM $CuSO_4$. Laccase oxidizes ABTS and blue halos form around the laccase-expressing colonies. Laccase-positive transformants were grown on PDA plates for sporulation. Replacement of *cel3c* in these transformants was investigated using diagnostic PCR amplifying the fragments traversing across the chromosome and the *cel3c* editing locus.

## Additional file

**Additional file 1: Table S1.** Primers used in this study. **Figure S1.** Sequencing the *ura5* gene of QM9414, C5 expressing Cas9 intracellularly, and the C5-derived transformants. (DOCX 1220 kb)

#### Abbreviations

5-FOA: 5-fluoroorotic acid; CBH1: cellobiohydrolase I; CRISPR/Cas9: Clustered Regularly Interspaced Short Palindromic Repeats/CRISPR associated protein 9; HDR: homology-based DNA recombination; MM: minimal medium; NHEJ: non-homologous end-joining; PDA: potato dextrose agar; PEG: polyethylene glycol; RNP: ribonucleoprotein

#### Acknowledgements

We thank Prof. Lina Qin for scientific discussion.

#### Funding

This research was supported by the National Key R&D Program of China (2016YFD0501409-02), the National Natural Science Foundation of China Grant 31672458, the Fundamental Research Funds for Central Non-profit Scientific Institution (1610382016004), and the Elite Youth Program of Chinese Academy of Agricultural Sciences.

#### Authors' contributions

XS designed the study; ZH carried out the experiments. Both authors analyzed the data, wrote the manuscript, and approved the final manuscript.

#### Ethics approval

Ethics approval was not needed for the study.

#### Consent for publication

Not applicable.

#### Competing interests

The authors declare that they have no competing interests.

**References**


1. Cherry JR, Fidantsef AL. Directed evolution of industrial enzymes: an update. Curr Opin Biotechnol. 2003;14(4):438–43.
2. Ilmen M, Saloheimo A, Onnela ML, Penttila ME. Regulation of cellulase gene expression in the filamentous fungus *Trichoderma reesei*. Appl Environ Microbiol. 1997;63(4):1298–306.
3. Suominen PL, Mantyla AL, Karhunen T, Hakola S, Nevalainen H. High frequency one-step gene replacement in *Trichoderma reesei*. II. Effects of deletions of individual cellulase genes. Mol Gen Genet. 1993;241(5–6):523–30.
4. Seiboth B, Hakola S, Mach RL, Suominen PL, Kubicek CP. Role of four major cellulases in triggering of cellulase gene expression by cellulose in *Trichoderma reesei*. J Bacteriol. 1997;179(17):5318–20.
5. Landowski CP, Huuskonen A, Wahl R, Westerholm-Parvinen A, Kanerva A, Hanninen AL, Salovuori N, Penttila M, Natunen J, Ostermeier C, et al. Enabling low cost biopharmaceuticals: a systematic approach to delete proteases from a well-known protein production host *Trichoderma reesei*. PLoS One. 2015;10(8):e0134723.
6. Sternberg SH, Redding S, Jinek M, Greene EC, Doudna JA. DNA interrogation by the CRISPR RNA-guided endonuclease Cas9. Nature. 2014; 507(7490):62–7.
7. Liu R, Chen L, Jiang YP, Zhou ZH, Zou G. Efficient genome editing in filamentous fungus *Trichoderma reesei* using the CRISPR/Cas9 system. Cell Discov. 2015;1.
8. Matsu-Ura T, Baek M, Kwon J, Hong C. Efficient gene editing in *Neurospora crassa* with CRISPR technology. Fungal Biol Biotechnol. 2015;2:4.
9. Nagy G, Szebenyi C, Csernetics A, Vaz AG, Toth EJ, Vagvolgyi C, Papp T. Development of a plasmid free CRISPR-Cas9 system for the genetic modification of *Mucor circinelloides*. Sci Rep. 2017;7(1):16800.
10. Al Abdallah Q, Ge W, Fortwendel JR. A simple and universal system for gene manipulation in *Aspergillus fumigatus*: *in vitro*-assembled Cas9-guide RNA ribonucleoproteins coupled with microhomology repair templates. mSphere. 2017;2(6).
11. Wang Q, Cobine PA, Coleman JJ. Efficient genome editing in *Fusarium oxysporum* based on CRISPR/Cas9 ribonucleoprotein complexes. Fungal Genet Biol. 2018;117:21–9.
12. Su X, Schmitz G, Zhang M, Mackie RI, Cann IK. Heterologous gene expression in filamentous fungi. Adv Appl Microbiol. 2012;81:1–61.
13. Qin LN, Cai FR, Dong XR, Huang ZB, Tao Y, Huang JZ, Dong ZY. Improved production of heterologous lipase in *Trichoderma reesei* by RNAi mediated gene silencing of an endogenic highly expressed gene. Bioresour Technol. 2012;109:116–22.
14. Zheng YM, Lin FL, Gao H, Zou G, Zhang JW, Wang GQ, Chen GD, Zhou ZH, Yao XS, Hu D. Development of a versatile and conventional technique for gene disruption in filamentous fungi based on CRISPR-Cas9 technology. Sci Rep. 2017;7(1):9250.
15. Seidl V, Gamauf C, Druzhinina IS, Seiboth B, Hartl L, Kubicek CP. The *Hypocrea jecorina* (*Trichoderma reesei*) hypercellulolytic mutant RUT C30 lacks a 85 kb (29 gene-encoding) region of the wild-type genome. BMC Genomics. 2008;9:327.
16. Ihry RJ, Worringer KA, Salick MR, Frias E, Ho D, Theriault K, Kommineni S, Chen J, Sondey M, Ye C, et al. p53 inhibits CRISPR-Cas9 engineering in human pluripotent stem cells. Nat Med. 2018;24(7):939–46.
17. Chu VT, Weber T, Wefers B, Wurst W, Sander S, Rajewsky K, Kuhn R. Increasing the efficiency of homology-directed repair for CRISPR-Cas9-induced precise gene editing in mammalian cells. Nat Biotechnol. 2015; 33(5):543–8.
18. Qin L, Jiang X, Dong Z, Huang J, Chen X. Identification of two integration sites in favor of transgene expression in *Trichoderma reesei*. Biotechnol Biofuels. 2018;11:142.
19. Shao ZY, Zhao H, Zhao HM. DNA assembler, an *in vivo* genetic method for rapid construction of biochemical pathways. Nucleic Acids Res. 2009:37(2).
20. Penttila M, Nevalainen H, Ratto M, Salminen E, Knowles J. A versatile transformation system for the cellulolytic filamentous fungus *Trichoderma reesei*. Gene. 1987;61(2):155–64.

# 15

# Purification and biochemical characterization of a novel thermostable protease from the oyster mushroom *Pleurotus sajor-caju* strain CTM10057 with industrial interest

Maroua Omrane Benmrad[1], Sondes Mechri[1], Nadia Zaraî Jaouadi[1,2], Mouna Ben Elhoul[1,2], Hatem Rekik[1,2], Sami Sayadi[3], Samir Bejar[1,2], Nabil Kechaou[4] and Bassem Jaouadi[1,2*]

**Abstract**

**Background:** Proteases are hydrolytic enzymes that catalyze peptide linkage cleavage reactions at the level of proteins and peptides with different degrees of specificity. This group draws the attention of industry. More than one protease in three is a serine protease. Classically, they are active at neutral to alkaline pH. The serine proteases are researched for industrial uses, especially detergents. They are the most commercially available enzyme group in the world market. Overall, fungi produced extracellular proteases, easily separated from mycelium by filtration.

**Results:** A new basidiomycete fungus CTM10057, a hyperproducer of a novel protease (10,500 U/mL), was identified as *Pleurotus sajor-caju* (oyster mushroom). The enzyme, called SPPS, was purified to homogeneity by heat-treatment (80 °C for 20 min) followed by ammonium sulfate precipitation (35–55%)-dialysis, then UNO Q-6 FPLC ion-exchange chromatography and finally HPLC-ZORBAX PSM 300 HPSEC gel filtration chromatography, and submitted to biochemical characterization assays. The molecular mass was estimated to be 65 kDa by sodium dodecyl sulfate polyacrylamide gel electrophoresis (SDS-PAGE), Native-PAGE, casein-zymography, and size exclusion by HPLC. A high homology with mushroom proteases was displayed by the first 26 amino-acid residues of the $NH_2$-terminal aminoacid sequence. Phenylmethanesulfonyl fluoride (PMSF) and diiodopropyl fluorophosphates (DFP) strongly inhibit SPPS, revealing that it is a member of the serine-proteases family. The pH and temperature optima were 9.5 and 70 °C, respectively. Interestingly, SPPS possesses the most elevated hydrolysis level and catalytic efficiency in comparison with SPTC, Flavourzyme® 500 L, and Thermolysin type X proteases. More remarkably, a high tolerance towards organic solvent tolerance was exhibited by SPPS, together with considerable detergent stability compared to the commercial proteases Thermolysin type X and Flavourzyme® 500 L, respectively.

**Conclusions:** This proves the excellent proprieties characterizing SPPS, making it a potential candidate for industrial applications especially detergent formulations.

**Keywords:** Protease, Oyster mushroom, *Pleurotus sajor-caju*, Organic solvent, Peptide synthesis, Detergent formulations

* Correspondence: bassem.jaouadi@cbs.rnrt.tn; biotechecozym@gmail.com
[1]Laboratory of Microbial Biotechnology and Engineering Enzymes (LMBEE), Centre of Biotechnology of Sfax (CBS), University of Sfax, Road of Sidi Mansour Km 6, P.O. Box 1177, 3018 Sfax, Tunisia
[2]Biotech ECOZYM Start-up, Business Incubator, Centre of Biotechnology of Sfax (CBS), University of Sfax, Road of Sidi Mansour Km 6, P.O. Box 1177, 3018 Sfax, Tunisia

## Background

Enzymes are natural catalysts, which are increasingly more required in various industrial fields and bioprocesses. Today, researchers focus on detecting new enzymes endowed with extremely satisfactory properties for commercial applications [1, 2]. Using enzymes in industrial process has a large number of advantages, namely their specificity and lower pressures, i.e. lower cost. Besides, they can be preserved for a long time. They are also biodegradable, therefore causing less environmental pollution. Global demand for industrial enzymes is expected to rise up to almost 7.1 billion dollars in 2018, www.FreedoniaGroup.com (Cleveland, USA). Compared to plant and animal enzymes, microbial enzymes are the most attractive ones to industrialists owing to their ease of production [3]. In fact, the microbial proteases production is beneficial thanks to their distinctive features such as short generation time, easy manipulation of microorganisms genes, and large availability in nature [1]. Besides, their preferred properties in biotechnology are widely approved. Proteases isolated from plant and animal do not receive industrial demand save for some uses. Around three quarts of industrial biocatalysts are hydrolyses [4, 5], 65% of them are proteases of the total industrial enzyme market [6, 7]. The business applications of proteases are quickly growing thanks to the current technological advances. These hydrolyzes possess have original properties and substrate specificities [8]. They have been well-studied, being considered as the most attractive group [9]. The majority of the total sales (89%) of protease are used in detergent formulation [10]. Historically, proteases were the first to be used in laundry detergents [2, 11, 12], about two-third share of this industry [13, 14]. Thanks to these enzymes, detergent industries have made considerable progress [11, 13]. It is noteworthy that protease is a hydrolytic biocatalyst which catalyzes the interruption of protein polymers [9, 12, 15, 16]. It contributes significantly in the metabolism of almost all organisms [4]. Besides, bio-additives improving the efficiency and the performance of laundry detergents are receiving increasing demands. In fact, a large number of researches are focusing on the isolation and purification of alkaline proteases for detergence purpose. Although industrial applications resort to bacterial proteases, the process of cell elimination is expensive, thus their use is restricted [17]. It is noteworthy that bacteria produce the majority of proteases of microbial origin. In fact, *Bacillus* genus produces neutral and alkaline proteases. The excellent properties of basidiomycetes' proteases pave the way for further study. Although the 3D structural and functional proprieties of basidiomycetes proteases occurred more than 30 years ago in scientific research, the variety and complexity of xylotrophic basidiomycetes enzymes action has resulted in interesting studies. In this respect, our researches are focused on fungi protease thanks to their easy cell separation. The study of micromycetes proteases has recently shed lights on the specific characteristics of enzymes [5, 18]. Moreover, basidiomycota are a secret treasure proteolytic variety [19]. Several species of strains including fungi (viz. *Aspergillus flavus*, *Aspergillus clavatus*, *Aspergillus melleu*, *Aspergillus niger*, *Fusarium graminarum*, *Chrysosporium keratinophilum*, *Penicillium griseofulvin*, *Penicillium chrysogenum*, *Scedosporium apiosermum*, and *Trametes cingulata*) and bacteria (viz. *Bacillus licheniformis*, *Bacillus amyloliquefaciens*, *Bacillus pumilus*, *Bacillus safensis*, *Bacillus firmus*, *Bacillus alcalophilus*, *Bacillus proteolyticus*, *Bacillus subtilis*, and *Bacillus thuringiensis*) are described as foodstuffs proteases [6, 12, 20, 21]. However, the fungi protease origin, *Penicillium* and *Aspergillus* are widely studied, as a greater species and can produce different proteases types, such as *Aspergillus oryzae*. In fact, mushroom species have been proven to possess various biological effects. For centuries, mushrooms have been used as food and food materials, as well as being topics of study in many fields, namely medicine and cosmetics. Oyster mushrooms, particularly, represent a type of specialty mushroom that is much prized around the world and most popularly in the United States. Indeed, mushrooms are a source of protein, vitamins C and B, and minerals, and mainly endowed with a very low fat content [22]. Many bioactive molecules, including protease, have been isolated from superior fungi. More than twenty proteases have been isolated from fungi species [23]. Moreover, the study of basidiomycota's proteases is still limited. Thus, the study of new *Pleurotus* proteases is a biotechnology area that necessitates to be discovered. The *Pleurotus* species are extremely valued in cooking thanks to their refined flavor and they have also been examined in different bioactive molecules such as antitumor, hypocholesterolemic, anti-inflammatory antiviral, antioxidant, antibiotic, antidiabetic, antitumor, immunomodulatory, antihyperlipidemic, and hepatoprotective compounds etc. [1]. Numerous researches have demonstrated the fibrinolytic activity of edible mushrooms, including *Flammulina velutipes*, *Pleurotus ostreatus*, *Grifola frondosa*, *Tricholoma saponaceum*, and *Armillaria mellea* [23]. The genus *Pleurotus* is known for laccase production. This is linked to the ligninolytic metabolism with the presence of proteases. In addition to laccase production, this genus is known for the hydrolytic enzymes production by *Pleurotus* spp. that have been designated, resulting in these properties of this genus for future research [1].

The study of saprophytic basidiomycetes proteases has led to the isolation of different classes of proteases [19]. As white rot fungi, the genus *Pleurotus* has ligninolytic

[24] and proteolytic [1, 5, 19] enzymes. A few studies have investigated the proteolytic enzymes isolated from *Pleurotus* spp. as reported. This microorganism is a white rot fungi and the second crucial group of fungi cultivated in life, with more than 30 species [1]. It is the most produced edible mushroom, withstanding any temperature. It was found in tropical and subtropical regions and can grow artificially with no difficulties. This genus can be cultivated at any substrate and temperature. In fact, its easy cultivation is performed by means of cheap nutriments. Besides, it is characterized by excellent adaptation [5, 25], and therefore, a high yield of production.

In this study, a novel alkaline and detergent-stable protease, called SPPS, from the fungus *Pleurotus sajor-caju* strain CTM10057, was purified to homogeneity and its enzymatic characteristics were examined. Moreover, the enzymatic performance evaluation of the SPPS was compared with other used proteases. This present work provides valuable information on biochemical characteristics of SPPS from the oyster mushroom.

## Methods

### Substrates, chemicals, and used comparative proteases

All the chemicals and substrates used in this study were reagent position, unless specified otherwise. General reagents were obtained from commercial suppliers. The SPTC enzyme is a fungi protease from *Trametes cingulata* CTM10101 [26]. The Flavourzyme® 500 L is a commercial fungal proteolytic enzyme from *Aspergillus oryzae* supplied by Novozymes A/S (Bagsvaerd. Denmark). The Thermolysin type X is a commercial protease from *Geobacillus stearothermophilus* purchased from Sigma Aldrich Inc., Fluka, Chemical Co. (St. Louis, MO, USA).

### Isolation and growth culture conditions of protease-producing white-rot-fungi and enzyme production

Fruiting bodies of a wild *Pleurotus* and *Shiitake* were collected from different biotope sites, from the symptomatic wood of the camphor trees *Quercus faginea* (L.) at the El Feïdja National Park (Aïn Draham, Jendouba, Tunisia) to isolate protease-producing fungi. Mycelium pure cultures of the *Pleurotus* P10 (designated as CTM10057) were obtained in Petri dishes, containing potato dextrose agar (PDA) medium at 28 ± 2 °C and pH 6.0. The plates were then incubated for 7 d at 28 ± 2 °C. To avoid bacterial growth, the plates were supplemented with chloramphenicol 25 μg/mL. The protease-producing fungal strain was tested onto skimmed milk agar plates containing (g/L): peptone, 5; yeast extract, 3; skimmed milk 250 mL; and bacteriological agar, 20 at pH 7.4. The protease-producing strain was propagated on PDA plates at 28 ± 2 °C, and the inoculates were prepared from 6-day-old colonies by flooding with 10 mL of sterile distilled water and scraping off the agar plates. 2 mL of homogenized mycelium were used for inoculation of 1000-mL Erlenmeyer flask with baffles, containing 100 mL of culture medium. This basal medium incorporates a complete potato dextrose broth (PDB) liquid culture medium supplemented with (g/L): 15 lentil flour; 3 yeast extract; 10 glucose; 1 $KH_2PO_4$; 1 $K_2HPO_4$ at pH 5.6. Cultures were incubated for 3 d at 28 °C on a rotary shaker with a shaking speed of 180 rpm. Afterwards, the culture broths were centrifuged at 11000×*g* for 30 min to remove mycelia and medium debris. The cell-free supernatant was designated as crude enzyme.

### PCR amplification and ITS of 18S rDNA gene sequencing and molecular phylogenetic investigation

The genomic DNA of the isolate fungal strain CTM10057 was extracted from freeze-dried mycelia according to CTAB method as investigated previously by Li [27]. The gene encoding ITS region of rDNA was amplified by the fungi identification PCR kit by means of universal primers ITS1 and ITS4 as described elsewhere [13]. The PCR product was placed into the PCR2.1 vector by means of a pCR-Blunt ended cloning kit (GE Healthcare/Amersham Biosciences Corp., Piscataway, NJ, USA). The nucleotide sequences of both strands pertaining to the cloned ITS rDNA gene sequence were sequenced in both directions via universal primers (U-19 and T7) using the automated DNA sequence ABI PRISM® 3100-Avant Genetic Analyzer (Applied Biosystems, Foster City, CA, USA). The nucleotide sequence alignment was conducted several times, using the BioEdit (v 7.0.2) software. The analyses of phylogenetic and molecular evolutionary were carried out using Molecular Evolutionary Genetics Analysis (v 4.1) program.

### Determination of protease activity

The SPPS activity was measured spectrophotometrically at 660 nm using casein as substrate [28] at 70 °C and 100 mM glycine-NaOH buffer added with 2 mM $CaCl_2$ at pH 9.5 (i.e., Buffer A), under the standard assay conditions described previously [26].

The SPPS activity, in detergent solution, was measured at 450 nm using *N,N*-dimethylated casein as substrate at 40 °C and pH 9 as reported elsewhere [29].

### SPPS purification procedure

To remove microbial cells, five hundred mL of a 72-h culture of *Pleurotus sajor-caju* strain CTM10057 were centrifuged. The supernatant containing extracellular protease was used as the crude enzyme preparation and submitted to the following purification steps. First, the filtrate was incubated for 20 min at 80 °C and insoluble material was removed by centrifugation at 11000×*g* for 20 min. The ammonium sulfate was added to clear supernatant to reach 35% saturation, then centrifuged at

11000×*g* for 30 min. The saturation of the obtained supernatant was done up to 55% with solid $(NH4)_2SO_4$, re-centrifuged, re-suspended in a minimal volume of 50 mM PIPES at pH 6.1 (i.e., Buffer B), and dialyzed overnight against repeated changes of the same buffer. The sample, thus, obtained was deposited on a UNO Q-6 column (12 mm × 53 mm), using FPLC system, previously equilibrated with buffer B. The proteins were eluted with the same buffer, containing an increasing concentration of NaCl of 0 to 500 mM at a rate of 0.5 mL/min. Fractions of each peak were collected manually and analyzed by measuring absorbance at 280 nm and the proteolytic activity on casein. Pooled fractions, containing protease activity, were further resolved by HPLC system using ZORBAX PSM 300 HPSEC (6.2 mm × 250 mm), Agilent Laboratories, pre-equilibrated with 25 mM HEPES buffer at pH 7.5, were supplemented with 2 mM $CaCl_2$ (i.e., Buffer C). Protein was separated by molecular weight at a flow rate of 0.5 mL/min with buffer C and the protease activity was detected using a UV spectrophotometric detector at 280 nm.

### Proteins measurement, electrophoresis, and analytical methods

Protein concentration was measured as reported by Bradford [30], using BSA as reference. Previously reported by Laemmli [31], the subunit molecular weight of the purified SPPS enzyme was evaluated by SDS-PAGE under reducing conditions, using 12% separating gel (pH 8.8) and 5% stacking gel (pH 6.8). The Native-PAGE or non-reducing conditions of the purified SPPS was performed using 12% resolving gel (stacking gel was omitted) in Tris-glycine buffer (pH 8.5) at 4 °C. The molecular weight of the purified SPPS protein was determined in comparison with standard protein markers and also confirmed by size exclusion HPLC, using ZORBAX PSM 300 HPSEC (Agilent Technologies, Lawrence, Kansas, MO, USA), pre-equilibrated with the buffer C. The protein bands were visualized with Coomassie Brilliant Blue R-250 (Bio-Rad Laboratories, Inc., Hercules, CA, USA) staining. Substrate-PAGE, or casein zymography staining, was performed by incorporating azo-casein into the separating gel before polymerization as detailed by Jaouadi et al. [32]. The gels were rinsed twice for 30 min in Triton X-100 (2.5%), after electrophoresis, in order to remove the SDS. The occurrence of clearance zone around the protein band on the blue background elucidated the SPPS protease activity.

### Edman degradation for amino-acid sequencing

To sequence $NH_2$-terminal, the bands of purified SPPS enzyme were cut from SDS gels and transferred to a ProBlott membrane (Applied Biosystems, Foster City, CA, USA). Accordingly, in order to perform such analysis, we used automated Edman's degradation at the Faculty of Science of Sfax (FSS), University of Sfax, by means of a protein sequencer (Applied Biosystems Protein sequencer ABI Procise 492/610A).

### Biochemical characterization of the purified SPPS enzyme

#### *Effects of inhibitors and metallic ions on enzyme stability*

The effects of some inhibitors and various divalent metallic ions on protease stability were examined by means of pre-incubating the purified SPPS enzyme for 1 h at 40 °C, in the presence of each inhibitor or metallic ions, in order to characterize the SPPS-owned class. Performing enzyme assays required standard assay conditions.

#### *Influence of pH on the activity and stability of SPPS*

The influence of pH was carried out with casein at 10 g/L as a substrate. Protease activities were studied over a pH range [2.0–13.0] at 70 °C. For the measurement of pH stability, the purified SPPS enzyme was pre-incubated in buffers at different pH in the range of 2.0–13.0 for 24 h at 40 °C. Aliquots were withdrawn periodically and the residual proteolytic activities were measured as aforementioned.

#### *Effects of temperature on the enzyme activity and stability*

With respect to proteolysis activity, its optimum temperature was determined via performing SPPS at pH 9.5 at different temperatures from 40 to 100 °C, with and without 2 mM $CaCl_2$. At a pH of 9.5, the thermostability was determined by incubating SPPS at 80, 90, and 100 °C for 12 h of incubation with and without 2 mM $CaCl_2$. Aliquots were withdrawn at desired time intervals to test the remaining activity under standard conditions. Considered as control (100%), the non-heated SPPS was left at room temperature. Half-life times of the purified SPPS protease were determined after incubation at 80, 90, and 100 °C for 12 h.

#### *Effect of various polyol(s) and /or calcium on SPPS thermostability*

The effect of variety of polyols, with concentration of 100 mg/mL, on SPPS thermostability was carried out by its incubation at 90 °C for 1 h with and without different polyols. The combination of sorbitol (100 mg/mL) and calcium (2 mM) on the thermostability of SPPS enzyme was monitored at 90 °C for 12 h. Protease assays were carried out under the standard experimental assay conditions.

### Kinetic study of the SPPS, SPTC, Flavourzyme® 500 L, and Thermolysin type X enzymes

#### *Substrate specificities profile*

The substrate profile specificity of SPPS was determined using natural and modified protein substrates as well as esters substrates. Enzymatic activities were determined on each substrate based on standard conditions, previously described elsewhere [33, 34].

***Kinetic measurements***

The kinetic constants for protease activity were determined by performing casein as substrate at different concentrations, ranging from 0.10 to 50 mg/mL for 20 min. The pH and temperature values used in the kinetic study were adjusted to the optimum conditions for each enzyme (SPPS, pH 9.5 and 70 °C; SPTC, pH 9.0 and 60 °C; Flavourzyme® 500 L, pH 11.0 and 60 °C; and Thermolysin type X, pH 8.0 and 70 °C). The purified enzymes used were SPPS; SPTC; Flavourzyme® 500 L, and Thermolysin, at a final concentration of proteins 1.5 mg/mL using natural (casein) as specific substrates for all proteases. Each assay was carried out in triplicate, and Lineweaver–Burk plots were used in order to rate kinetic parameters. The Hyper32 v.1.0. software was used to calculate Kinetic constants, Michaelis–Menten constant ($K_m$), and maximal reaction velocity ($V_{max}$) values.

**Performance evaluation of the purified SPPS, SPTC, Flavourzyme® 500 L, and Thermolysin type X enzymes**

***Determination of hydrolysis degree with casein***

Casein hydrolysis was carried out at 70 °C and pH 9.5 for SPPS; at 60 °C and pH 9.0 for SPTC; at 70 °C and pH 8.0 for Thermolysin type X; and at 60 °C and pH 11.0 for Flavourzyme® 500 L. So, an amount of 5 g of casein was dissolved in 100 mL of 100 mM buffer per used enzyme and then treated with 500 U of the purified enzymes (SPPS, SPTC, Flavourzyme® 500 L, and Thermolysin type X). For each case, the degree of hydrolysis (DH) was calculated based on the amount of base (NaOH) added so that the pH remains constant during hydrolysis [35] as previously reported by Zaraî Jaouadi [34]. The DH is defined as the percentage of peptide bonds cleaved according to the total number of peptide bonds, i.e. it is the ratio of free amino-acids to those in peptide linkage and the relative size of the peptide fragments [36]. By totally decomposing the protein into its constituent elements, this degree of hydrolysis is estimated to be 100%. Since enzymes are specific biocatalysts, no enzyme alone is capable of hydrolyzing each peptide bond of a protein up to 100% DH. Indeed, the DH of an enzyme provides us with an idea of its functionality. Hydrolysis helped to keep the pH titrimetrically constant by using the pH-Stat (Metrohm 718 Stat Titrino, Herisau, Switzerland) set at the desired pH value of each enzyme via constantly adding 5 N of NaOH (caustic soda) [35, 37].

***Effects of organic-solvents on protease stability***

Various organic solvents, with different log $P_{ow}$ values at 50% (v/v), were tested by shaking at 150 strokes per min and 37 °C for 72 h to evaluate their effects on SPPS and Thermolysin type X protease stabilities. The polarity or hydrophobicity of an organic solvent was quantified by a parameter termed "log $P_{ow}$" value. Generally, organic solvents, using low log $P_{ow}$ values, show more biotoxicity and result in the inhibition of the biocatalyst compared with high log $P_{ow}$ solvents [38]. The relative and residual caseinolytic activities were assayed under each assay condition at 70 °C and pH 9.5 (for SPPS) and at 60 °C and pH 8.0 (for Thermolysin type X). Devoid of any organic solvent, the activity of SPPS was taken as 100%.

***Compatibility of SPPS enzyme with some commercial detergents***

The stability of the alkaline proteases (SPPS, Flavourzyme® 500 L, and SPTC) in the presence of miscellaneous laundry detergents was examined by incubating enzymes (500 U/mL) for 1 h at 40 °C with various common detergent preparations (7 mg/mL), and the residual activities were determined. Prior to the addition of each purified enzyme, the endogenous enzymes contained in used detergents were deactivated by incubating the diluted detergents at 65 °C for 1 h. Under similar incubation conditions, the enzyme activity of a control (devoid of any detergent) was taken as 100%.

***Washing performance evaluation of SPPS enzyme***

The application of the proteases SPPS and Flavourzyme® 500 L at 500 U/mL as a detergent bio-additive in Dixan detergent (7 mg/mL) were studied on white cotton cloth pieces stained with blood and evaluated as already detailed by the authors [26].

***Culture collection depository number and nucleotide sequence accession number***

The fungal culture of P10 isolate was deposited in the « Collection Tunisienne de Microorganismes (CTM) » at the Service of Culture Collection Maintenance of the Centre of Biotechnology of Sfax (Sfax, Tunisia) under the authentic culture collection number: CTM < TUN>: 10057 (designated as CTM10057). The data reported in this work for the nucleotide sequences of the 18S rDNA (750 bp) gene were deposited in the GenBank/ENA/EMBL databases under accession number: MH806376.

## Results

### Screening of alkaline protease-producing strains

The new basidiomycete fungus, *Pleurotus* P10, was newly isolated from the symptomatic wood of the camphor trees *Quercus faginea* (L.) at the El Feïdja National Park, Aïn Draham (Jendouba, Tunisia). It was registered in the CTM-CBS under the authentic culture number: "CTM10057" and was retained, for further and advanced study, as the highest protease producer fungal strain on skimmed milk agar after 24 h (with diameter of clear zone/ diameter of colony growth > 3 mm) and on PDB liquid medium after 72 h (Fig. 1) (with protease activity = 10,500 U/mL) at 28 ± 2 °C.

### Identification and molecular phylogeny of the microorganism

In order to identify CTM10057 strain, the internal transcribed spacer region of 18S rDNA (750 bp) was amplified, cloned in PCR2.1 vector and sequenced. The ITS nucleotide sequence was analyzed with the GenBank database using BLAST program and showed 99 and 98% homology with ITS of *Pleurotus sajor-caju* MUCL:40167 (accession no.: JN645075) and *Trametes marianna* BJFC12714 (accession no.: KC848334) strains, respectively. The phylogenetic tree was established by the neighbor-joining method (data not shown). The multiple alignment and phylogenetic tree revealed that CTM10057 strain is *Pleurotus sajor-caju* (accession no.: MH806376).

### Purification of SPPS enzyme

The SPPS enzyme was purified from the culture supernatant according to the procedure described in the Methods section. The supernatant was obtained by the centrifugation of a 72-h old culture of the *Pleurotus sajor-caju* strain CTM10057 (Fig. 1) using broth (500 mL) as a crude enzyme solution. The protein elution profile obtained at the final purification step indicated that the protease was eluted at 110–160 mM NaCl (Fig. 2a). Fractions corresponding protease activity were loaded in HPLC column.

The results of the purification procedure are summarized in Table 1. SPPS purified enzyme preparation contained about 15% of the total activity of the crude and had a specific activity of 79,000 U/mg (Table 1).

### Molecular weight determination of SPPS enzyme

To analyze the homogeneity and molecular weight of the purified SPPS enzyme, Native-PAGE and SDS-PAGE were performed. As shown in Fig. 2b, the molecular mass of the SPPS was almost 65 kDa, as assessed by Native-PAGE. Moreover, the SPPS preparation was a homogeneous pure enzyme with high purity confirmed by a symmetrical peak with retention time ($R_t$) of 11.6 min, matching with protein of ~ 65 kDa (Fig. 2c) reached after native condition in HPLC chromatography (Fig. 2c). Under reducing conditions (SDS-PAGE), a unique monomeric protein band estimated to 65 kDa was obtained for the purified SPPS enzyme (Fig. 2d) corresponding to that determined by gel filtration. Purity of the SPPS protease was also evaluated by zymography. The zymogram activity staining confirms the presence of a clear zone of protease activity for the purified sample and total cell extract co-migrating with proteins, having molecular masses of 65 kDa (Fig. 2e).

These observations strongly suggested that SPPS was a monomeric protein comparable to those previously reported for other fungi proteases [25, 26, 33, 39].

### $NH_2$-terminal amino-acid sequence determination of SPPS enzyme

The sequence of the first 26 $NH_2$-terminal amino-acid residues of SPPS from the *Pleurotus sajor-caju* strain CTM10057 was determined as: GPEDPALPPDSESTH-VITGVEKLHAQ. It showed uniformity, thus indicating that it was isolated in a pure form. This sequence was submitted to comparisons with existing protein sequences in the GenBank non-redundant protein

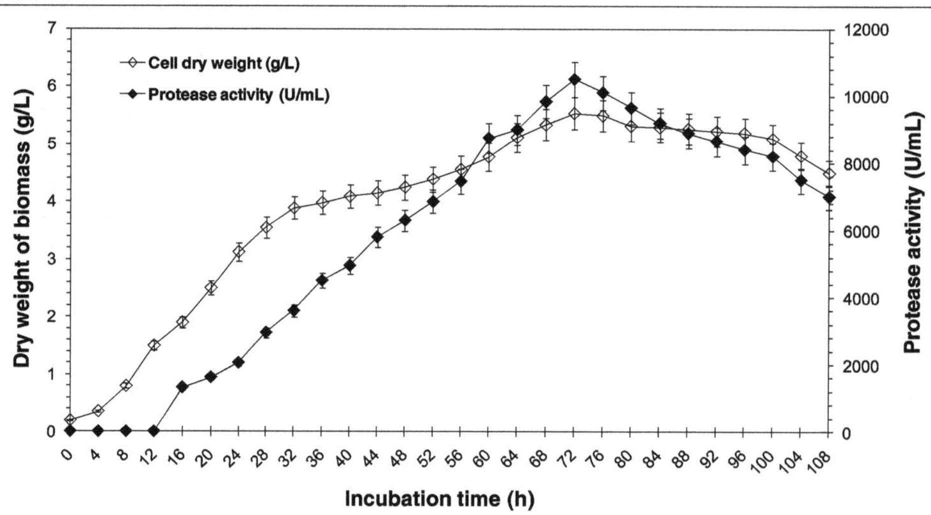


**Fig. 1** Time course of *Pleurotus sajor-caju* strain CTM10057 biomass (dry weight of mycelia) (◊) and protease activity (production) (♦) on optimized culture medium. Cell growth was performed by measuring the dry matter. Protease activity was determined in culture filtrates obtained after removal of cells by centrifugation, as described in Methods section. Each point represents the mean ($n = 3$) ± standard deviation

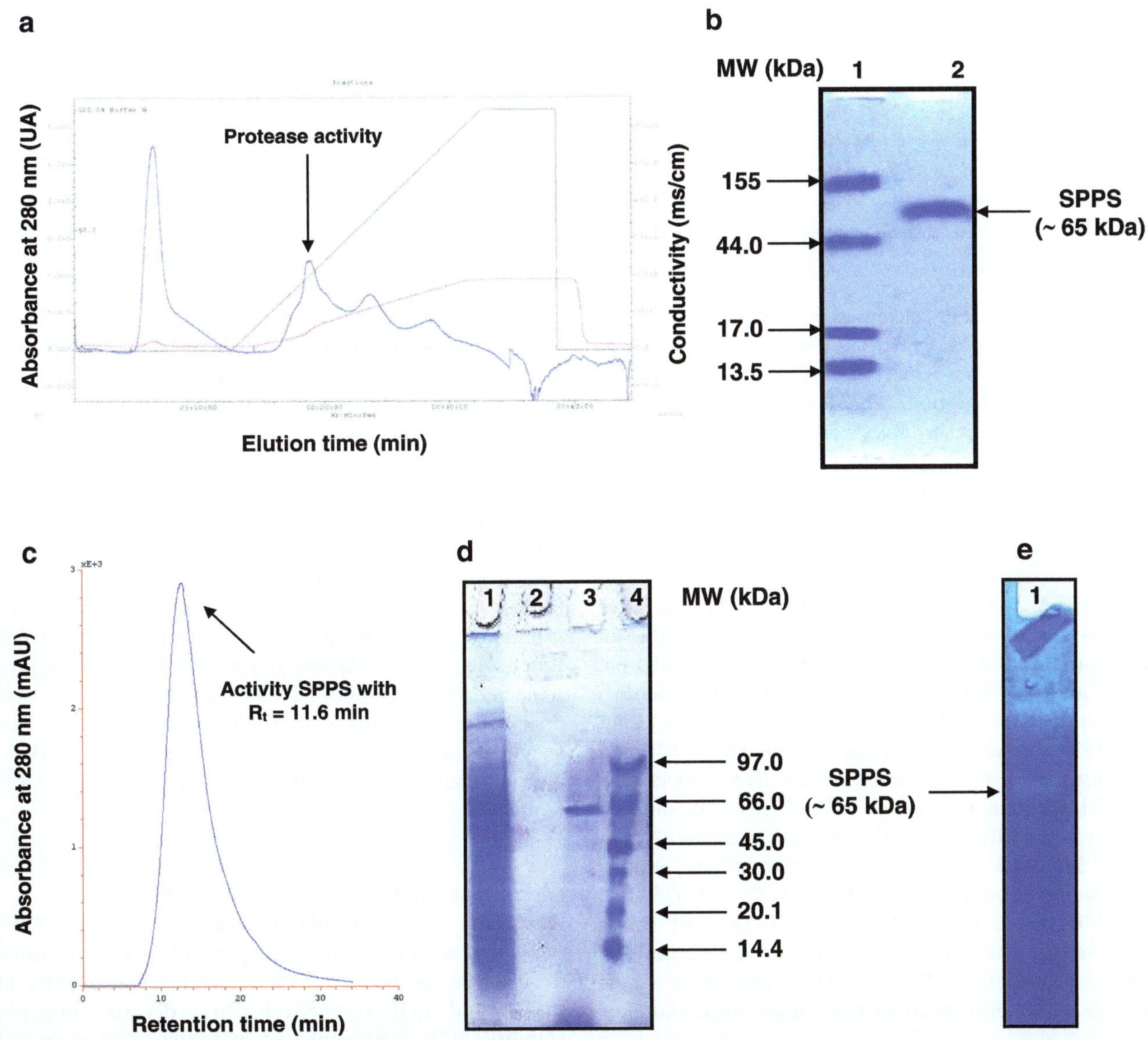


**Fig. 2** Purification and identification of the protease SPPS from *Pleurotus sajor-caju* strain CTM 10057. **a** Chromatography profile of the partial protease purification on FPLC system using UNO Q-6. The column (12 mm × 53 mm) (Bio-Rad Laboratories, USA) was equilibrated with buffer C. The adsorbed material was eluted with a linear NaCl gradient (0 to 500 mM in buffer C at a flow rate of 30 mL/h, as described in Methods section. **b** The assessment of homogeneity and molecular weight analysis of purified SPPS protein on Native-PAGE, Lane 1, protein markers (molecular masses in kDa). Lane 2, purified SPPS enzyme from *Pleurotus sajor-caju* strain CTM 10057 (30 μg). **c** Chromatography profile of the purified protease SPPS on HPLC system using ZORBAX PSM 300 HPSEC. The column (6.2 mm × 250 mm) (Agilent Technologies, Lawrence, Kansas, MO, USA) was equilibrated with buffer C and native protein markers of 670, 158, 44, 17, and 13.5 kDa, show a single peak of 65 kDa, approximately. Proteins were separated by isocratic elution at a flow rate of 30 mL/h with buffer C and detected using a UV-VIS Spectrophotometric detector (Knauer, Berlin, Germany) at 280 nm. The pure SPPS enzyme, with retention time ($R_t$) of 11.6 min, contains protease activity. **d** SDS-PAGE 12% of the purified protease. Lane 1, Total cell extract. Lane 2, Empty. Lane 3, Purified SPPS (30 μg) obtained after HPLC-ZORBAX PSM 300 HPSEC chromatography at $R_t$ = 11.6 min. Lane 4: Amersham LMW protein marker (GE Healthcare Europe GmbH, Freiburg, Germany). **e** Zymogram caseinolytic activity staining of the purified protease SPPS (30 μg)

database and the Swiss-Prot database, using the BLASTP and tBlastn search programs (Table 2). The alignment of the $NH_2$-terminal amino-acid shows that this sequence has homology with the subtilisin-like proteases; or this group of subtilisin-like proteases is found in bacteria, mammals, insects, and in some yeasts. This protease class exhibits significant homology with the protease of basidomycetes fungi, which consist of 300 to 400 amino-acid residues and do not have disulfide bridges as well as being characterized by the absence of cysteine and cystine residues. They are involved in various functions depending on the species, having in most cases nutritive

**Table 1** Flow sheet purification of SPPS from *Pleurotus sajor-caju* strain CTM10057

| Purification step [a] | Total activity (units) [b] × $10^4$ | Total protein (mg) [b,c] | Specific activity (U/mg of protein) [b] | Activity recovery rate (%) | Purification factor (fold) |
|---|---|---|---|---|---|
| Crude extract | 525 ± 13 | 590 ± 14 | 8823 | 100 | 1 |
| Heat treatment (20 min at 80 °C) | 430 ± 10 | 240 ± 6 | 17,917 | 82 | 2 |
| $(NH_4)_2SO_4$ fractionation (35–55%)-dialysis | 375 ± 9 | 160 ± 4 | 23,437 | 71.4 | 3 |
| FPLC (UNO Q-6) | 152 ± 3 | 40 ± 1 | 38,000 | 28.5 | 4.3 |
| HPLC (ZORBAX PSM 300 HPSEC) | 79 ± 2 | 10 ± 0.5 | 79,000 | 15 | 9 |

[a] Experiments were conducted three times and ± standard errors are reported
[b] One unit (U) of protease activity was defined as the amount of enzyme required to release 1 μg tyrosine per minute under the experimental conditions used
[c] Amounts of protein were estimated by the method of Bradford [30]

functions. The effect of calcium as a stabilizer was demonstrated in some cases. It appears that the $Ca^{2+}$ ion binds and gives the enzyme a more or less rigid structure, thus decreasing auto-digestion. The $NH_2$-terminal sequence displayed a medium to significant similarity (64–90%) compared to that of fungi serine proteases. Table 2 represents the $NH_2$-terminal sequence of SPPS, which was compared to other closely related proteolytic sequences, noted to share greatest homology with fungal proteolytic enzymes, especially those from *Pleurotus ostreatus* strain ATCC® MYA-2306™ (90% similarity with SPPS). The latter contains one modified amino-acid. The Glu3 residue in SPPS was substituted by Asp3. Besides, SPPS showed 85, 79, 75, and 64% similarities with proteases from *Schizopora paradoxa* strain KUC8140, *Phlebia centrifuga* strain FBCC195, *Phanerochaete chrysosporium* strain S1, and *Metarhizium anisopliae* strain ARSEF 549, respectively. The $NH_2$-terminal amino-acid of SPPS sequence differed from protease PR1C from *Metarhizium anisopliae* strain ARSEF 549 (64% similarity with SPPS) by fifteen amino-acid residues at positions G1L, P2S, E3S, D4R, A7D, L8D, P8G, P9A, S11I, E12F, T14P, I17M, E21K, and K22L. These results and the comparison with the closest enzymes indicated that SPPS proteolytic enzyme is a new member of the serine protease from fungal strains family.

### Biochemical proprieties of SPPS enzyme

#### *Influence of enzyme inhibitor and metal ions*

In order to determine the nature of SPPS enzyme, the effect of several enzyme inhibitors, (chelating agent and a specific group reagent) [7] on enzyme stability was investigated (Table 3). Trypsin-like and chymotrypsin competitive reagents did not have any influence on the stability of the purified SPPS. However, SPPS was thoroughly inhibited by the serine protease inhibitors, indicating that this enzyme belongs to that subclass. Metallo-enzyme inhibitors EDTA and EGTA sparsely affect the SPPS stability. EDTA at 1 mM and EGTA at 2 mM inhibited SPPS only by 91 and 80% of its activity, respectively. The relative stability of this enzyme in the presence of EDTA and EGTA is beneficial and auxiliary for use as detergent bio-additive, since detergents formulation contain high portion of chelating agents, which function as water softeners and further facilitate the proteinacious stain removal. Other metallic ions were also assayed for their effects on SPPS activity (Table 3). This fungi protease enzyme is totally activated by $Ca^{2+}$, $Mg^{2+}$, and $Fe^{2+}$ but strongly inhibited by $Cd^{2+}$, $Ni^{2+}$, and $Hg^{2+}$ (2 mM).

#### *Effects of pH on protease activity and stability*

According to the results presented in Fig. 3a, SPPS possesses an optimum activity at pH 9.5. This fungi protease keeps 60 and 71% of its original activity at pH 5.0 and 11.0, respectively. In order to study pH stability, SPPS was incubated in buffers at various pH values. The pH stability profile revealed that SPPS was highly stable in the pH range of 6.0–11.0. The half-life times of SPPS at pH 6.0, 7.0, 8.0, 9.0, 10.0, and 11.0 were 22, 20, 18, 16, 10, and 6 h, respectively (Fig. 3b). SPPS is an alkaline protease stable in

**Table 2** Alignment of the $NH_2$-terminal amino-acid sequence of the purified SPPS from *Pleurotus sajor-caju* strain CTM10057 with the sequences of other fungal proteases

| Enzyme | Origin | $NH_2$-terminal amino-acid [a,b] | Identity (%) |
|---|---|---|---|
| SPPS (this work) | *Pleurotus sajor-caju* strain CTM10057 | GPEDPALPPDSESTHVITGVEKLHAQ | – |
| PoSI | *Pleurotus ostreatus* strain ATCC® MYA-2306™ | GPDDPALPPD | 90 |
| PA_PoS1 | *Schizopora paradoxa* strain KUC8140 | GPTDPAIPPDSESTHVLTGV | 85 |
| VPR | *Phlebia centrifuga* strain FBCC195 | DDPAIPPDTESTHVLTGVDKLHAQ | 79 |
| PcSI | *Phanerochaete chrysosporium* strain S1 | GKSDPAVPADTESTH | 75 |
| PR1C | *Metarhizium anisopliae* strain ARSEF 549 | LSSRADDGADIFSPHVMTQVKL | 64 |

[a] Amino-acid sequences for comparison were obtained using the program BLASTP (NCBI, NIH, USA) database with default parameters
[b] Residues not identical with SPPS protease from *Pleurotus sajor-caju* strain CTM10057 are indicated in black box

**Table 3** Effects of various inhibitors, reducing agents, and metallic ions on SPPS enzyme stability. Protease activity measured in the absence of any inhibitor or reducing agent was taken as control (100%). The non-treated and dialyzed enzyme was considered as 100% for metallic ion assay. Residual activity was measured at pH 9.5 at 70 °C

| Reagent | | Specificity/origin | Concentration | Relative protease stability (%) [a] |
|---|---|---|---|---|
| *Inhibitor/reducing agents* | Control | | – | 100 ± 2.5 |
| | PMSF or DFP | Serine protease specific inhibitors | 5 mM | 0 ± 0.0 |
| | Benzamidine | Competitive inhibitor of serine proteases | 10 mM | 103 ± 2.6 |
| | Aprotinin | | 5 mM | 101 ± 2.5 |
| | SBTI | Soybean trypsin inhibitor | 3 mg/mL | 101 ± 2.5 |
| | TPCK | Chymotrypsin specific inhibitor | 1 mM | 94 ± 2.2 |
| | TLCK | Trypsin-like specific inhibitor | 1 mM | 102 ± 2.5 |
| | DTNB | Cysteine or thiol-containing protease inhibitors | 5 mM | 41 ± 0.8 |
| | NEM | | 5 mM | 74 ± 1.8 |
| | MIA | | 10 mM | 97 ± 2.4 |
| | EPNP | | 2 mM | 87 ± 2.1 |
| | Iodoacetamide | | 50 µg/mL | 65 ± 1.4 |
| | β-ME | Thiol reducing agent for cleaving protein disulfide bonds (cystin) | 1 mM | 66 ± 1.6 |
| | LD-DTT | | 50 µM | 61 ± 1.5 |
| | EDTA | Specific metallo-protease inhibitors | 1 mM | 91 ± 2.2 |
| | EGTA | | 2 mM | 80 ± 2.1 |
| | 1,10-Phenanthroline monohydrate | | 10 mM | 102 ± 2.5 |
| | Leupeptine | Acid protease specific inhibitors | 10 µg/mL | 101 ± 2.5 |
| | Pepstatine A | | 10 mM | 104 ± 2.7 |
| *Metallic ions* | Control | | – | 100 ± 2.5 |
| | $Ca^{2+}$ | $CaCl_2$ | 2 mM | 270 ± 6.0 |
| | $Mg^{2+}$ | $MgCl_2$ | 2 mM | 180 ± 3.8 |
| | $Fe^{2+}$ | $FeCl_2$ | 2 mM | 145 ± 3.4 |
| | $Cu^{2+}$ | $CuCl_2$ | 2 mM | 99 ± 2.5 |
| | $Mn^{2+}$ | $MnCl_2$ | 2 mM | 109 ± 2.6 |
| | $Co^{2+}$ | $CoCl_2$ | 2 mM | 61 ± 1.2 |
| | $Ba^{2+}$ | $BaCl_2$ | 2 mM | 75 ± 1.8 |
| | $Zn^{2+}$ | $ZnCl_2$ | 2 mM | 40 ± 0.8 |
| | $Hg^{2+}$ or $Cd^{2+}$ or $Ni^{2+}$ | $HgCl_2$, $CdCl_2$, or $NiCl_2$ | 2 mM | 0 ± 0.0 |

[a] Values represent means of three independent replicates, and ± standard errors are reported

*SBTI* soybean trypsin inhibitor, *TPCK* Nα-*p*-tosyl-L-phenylalanine chloromethyl ketone, TLCK Nα-*p*-tosyl-L-lysine chloromethyl ketone, *DTNB* 5,5′-dithio-bis-2-nitro benzoic acid, *NEM* N-ethylmalemide, *MIA* monoiodoacetic acid, *EPNP* 1.2-epoxy-3-(*p*-nitrophenoloxy) propane, *β-ME* β-mercaptoethanol, *LD-DTT* LD-dithiothreitol

alkaline solution. This propriety is solicited in protease and used as bio-additive in detergent formulations.

### *Effects of temperature on protease activity and stability*

The optimum temperature for SPPS at pH 9.5 was 60 °C devoid of $CaCl_2$ and 70 °C using 2 mM $Ca^{2+}$ as shown in Fig. 3c. The half-life times of SPPS in the absence of any additive were 8, 3, and 1 h at 80, 90, and 100 °C, respectively. The addition of different concentrations of $CaCl_2$ (1 to 10 mM) improved the thermostability of SPPS. The maximal thermostability was reached with 2 mM $Ca^{2+}$. As shown in Fig. 3d, the half-life times of SPPS at 80, 90, and 100 °C increased to 10, 5, and 3 h in the presence 2 mM $CaCl_2$. Compared to other proteases isolated from this genus announced in the review of Inácio et al. [1], SPPS has the highest optimum temperature and highest pH. This fungus protease is an alkaline protease acid tolerant and it is highly stable and active at high temperature and pH. SPPS is an acid resistant protease. This enzyme can have a biotechnology perspective in food applications. The high activity and stability

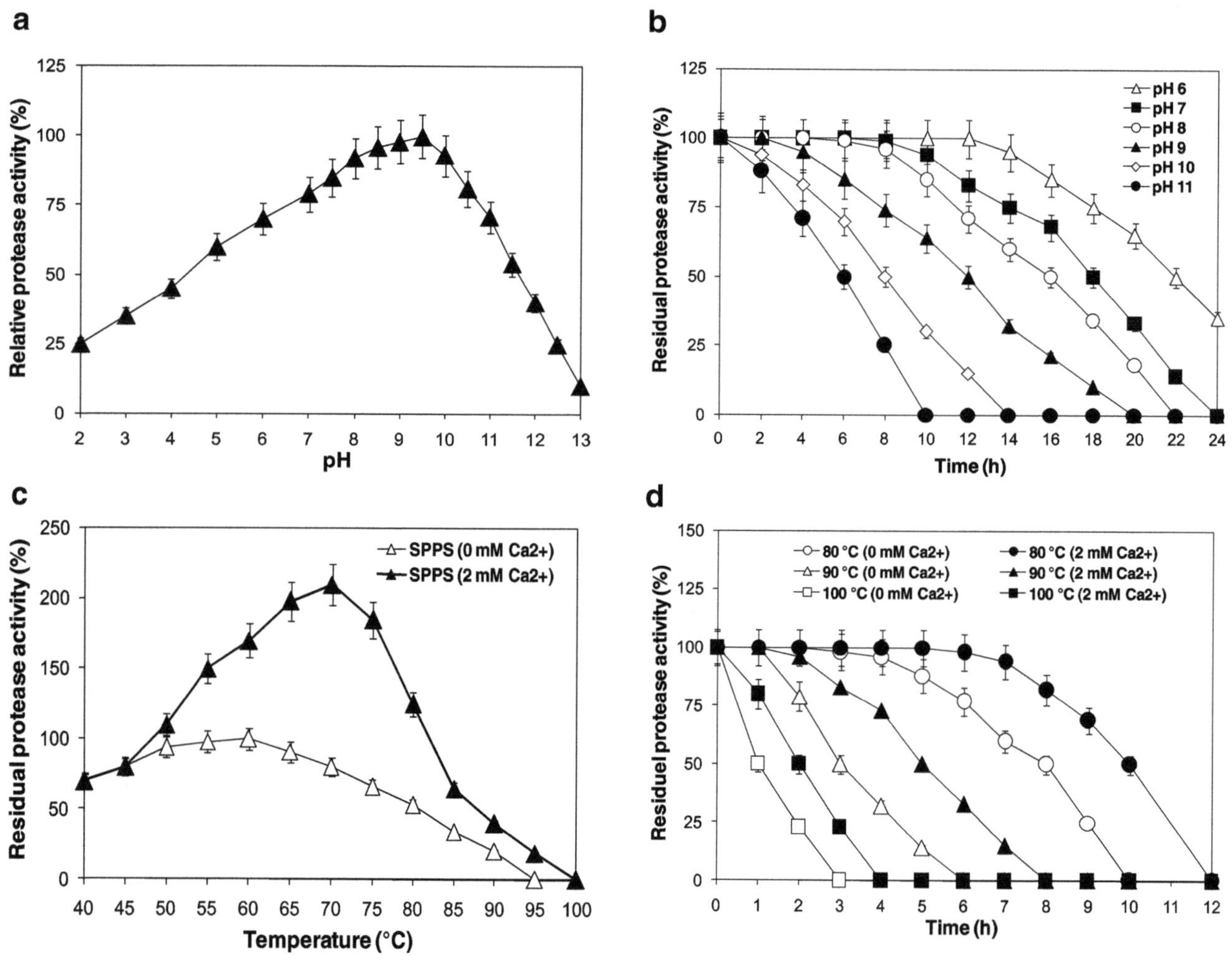


**Fig. 3** Physico-chemical proprieties of the purified protease SPPS from *Pleurotus sajor-caju* strain CTM 10057. Effects of pH on the activity (**a**) and stability (**b**) of SPPS. The activity of the enzyme at pH 9.5 was taken as 100%. Effects of the thermoactivity (**c**) and thermostability (**d**) of SPPS. The enzyme was pre-incubated for 12 h in the presence and absence of $CaCl_2$ (2 mM) at temperatures ranging from 80, 90, and 100 °C. The activity of the non-heated enzyme was taken as 100%. Each point represents the mean ($n = 3$) ± standard deviation

exhibited by SPPS at high pH solutions is, in fact, a very important attribute that provides further support for its potential strong candidacy for future application in detergent formulations since laundry detergents generally operate at a pH ranging from 7.0 to11.0 [6].

### *Effect of some polyols on the thermostability of SPPS*

As detailed earlier, the modification of the microenvironment of enzymes via the addition of polyols in aqueous solutions generally improves its thermostability [40]. In the respect to determine the stabilizing effect, several polyols on SPPS enzyme thermal stability was investigated at 90 °C for 1 h (Fig. 3a). Data demonstrated that sorbitol was the best additive used. The combination effect of sorbitol and $CaCl_2$ was also assayed at 90 °C for 12 h. The results displayed in Fig. 4b show that the half time is 9 h (in the presence of sorbitol and $CaCl_2$ compared to 3 h in the absence of the two additives) (Fig. 4b). Thus, the combination effect of sorbitol and $CaCl_2$ enhance the SPPS thermal stability. This improvement of SPPS thermostability in the presence of calcium and polyol is presumably attributed to the strengthening of interactions inside protein molecules and the binding of $Ca^{2+}$ and sorbitol to the autolysis site.

## Kinetic study of the SPPS, SPTC, Flavourzyme® 500 L, and Thermolysin type X enzymes

### *Substrate specificities*

The substrate specificity of proteases is often attributed to the amino-acid residues preceding the peptide bond they hydrolyze. The relative hydrolysis rates of various substrates were investigated to elucidate the amino-acid preference/substrate specificity of SPPS (Table 4). The specificity of cleavage to ester is well detailed in

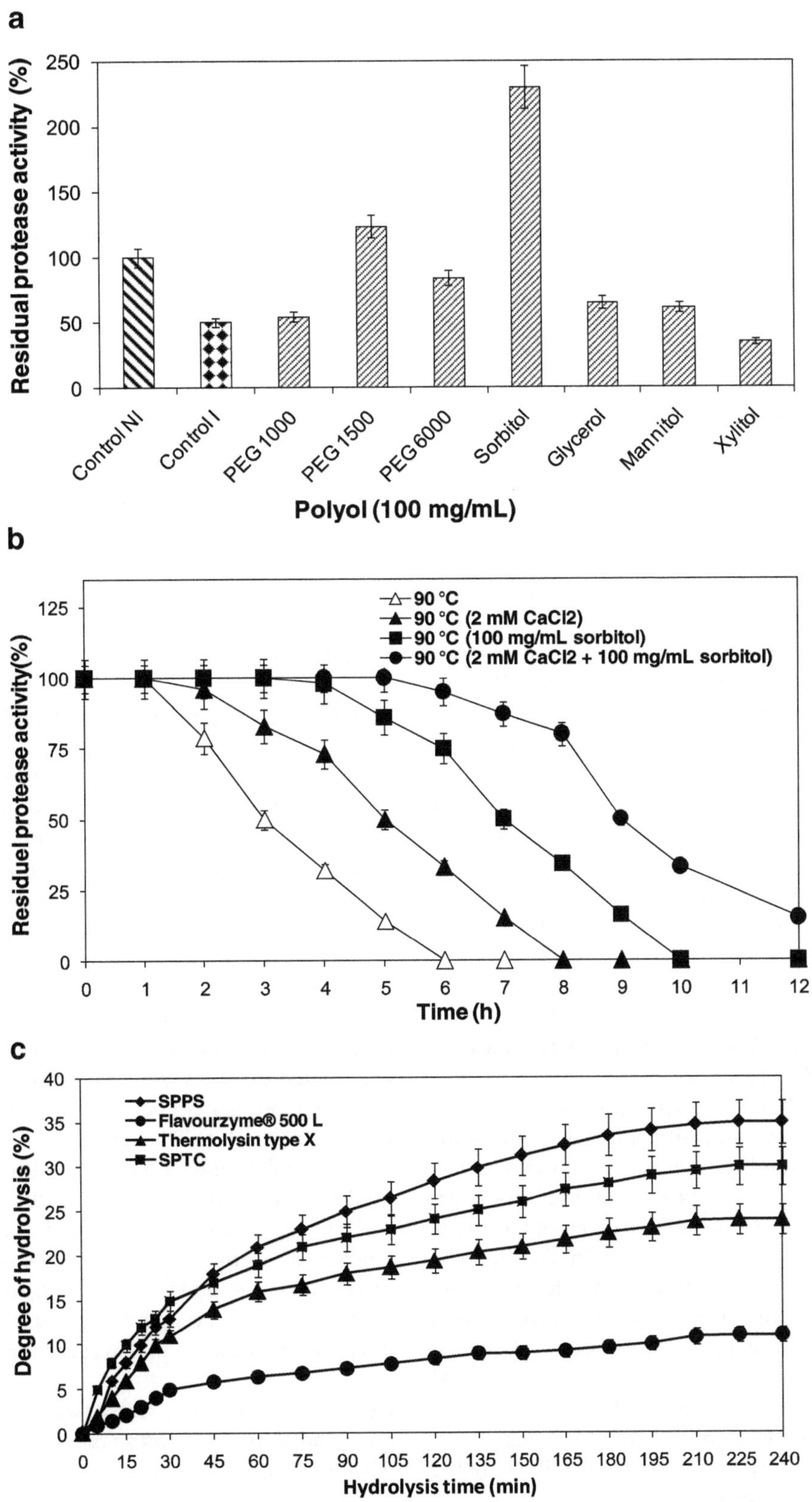


**Fig. 4** **a** Effect of polyols on protease SPPS stability. **b** The combined effect of $Ca^{2+}$ and sorbitol 10% (v/v) on protease SPPS stability at 90 °C. **c** Hydrolysis curves of casein protein treated with various purified enzymes. The purified proteases used were: SPPS (♦), SPTC (■), Flavourzyme® 500 L (●), and Thermolysin type X (▲)

literature. The purified SPPS was noted to exhibit esterase and amidase activities on BTEE and ATEE, but not on TAME, BAEE, and BCEE. It also displayed a preference for aromatic and hydrophobic amino-acid residues, such as Ala, Phe, Met, Leu, and Tyr and the carboxyl side of the splitting point in the P1 position (Table 4). This indicates that the SPPS enzyme acts on hydrophobic amino–acid residues attached to the carboxyl side of the cleavage sites of the chromogenic substrate. These characteristics, were in line with the other previously depicted subtilisins from bacterial [28, 32] and fungal [26, 33, 41, 42] origins. This indicated that the SPPS protease was slightly similar to subtilisins, not only in terms of specificity for position P1, but also with regard to the effects of amino-acids residues related to the cleavage site of the enzyme.

***Determination of kinetic parameters***

The kinetic parameters determined from Lineweaver–Burk plots are shown in Table 5 for SPPS, SPTC, Flavourzyme® 500 L, and Thermolysin type X purified enzymes using casein as substrate. The four used enzymes exhibited the classical kinetics of Michaelis–Menten. The returns indicate that the $K_m$ and $V_{\max}$ parameters for the SPPS enzyme, measured with casein, were 0.295 mg/mL and 79,000 μmol/mg.min$^{-1}$. The deduced catalytic efficiency ($k_{cat}/K_m$) of the SPPS enzyme is 1.45, 3.33, and 8.48 times compared to SPTC; Flavourzyme® 500 L, and Thermolysin type X, respectively.

The protease from *Pleurotus sajor-caju* strain CTM10057 displayed a $K_m$ value (0.275 mg/mL).

**Performance evaluation of the purified SPPS, SPTC, Flavourzyme® 500 L, and Thermolysin type X enzymes**

***Determination of hydrolysis degree***

Enzymatically hydrolyzed proteins possess functional properties, such as low viscosity, increased whipping ability, and high solubility, which make them favorable for food industry uses [43]. The hydrolysis degree of SPPS is achieved in curves of casein protein, after 240 min of incubation are shown in Fig. 4c. The purified enzymes were used at the same levels of activity for the production of protein hydrolysates from casein and for the subsequent comparisons of hydrolytic efficiencies. As shown in Fig. 4c, the purified SPPS was the most efficient, with 35% protease used during hydrolysis, followed by SPTC (30%) and Thermolysin type X (24%) with Flavourzyme® 500 L being the least efficient (11%) one.

These findings indicate that SPPS enzyme can be usefully employed for the preparation of protein hydrolysates.

***Effects of organic-solvents on protease activity and stability of SPPS and Thermolysin type X***

The effects of various organic solvents (at 50% v/v, with different log $P_{ow}$) on the stability of the purified SPPS and Thermolysin type X enzymes were determined by measuring residual activity under the optimum condition of each used enzyme after 72 h of incubation at 37 °C (Fig. 5a). When it is compared to Thermolysin type X, SPPS is stable in both hydrophobe and hydrophile organic solvents. It was also noted to be more stable and tolerant in the presence of the acetonitrile. The SPPS protease has then an elevated tolerance to organic solvents, whatever their hydrophobicity is. It can be used yet in homogeneous and heterogeneous catalysis. It was well informative that enzyme deactivation can be assigned to the change of medium hydrophobicity [44]. Logically, polar solvent can penetrate into protein and changing structural for the interaction between the active site and substrate than non polar solvents. Thus, SPPS can be active in aqueous and non aqueous solutions. Accordingly, SPPS can be a potential candidacy for future application as a biocatalyst for the synthesis of peptide reactions in low water activity systems.

***Stability and compatibility of SPPS and Flavourzyme® 500 L with laundry detergents***

Recent researches have aimed at finding a substitute for the costly toxic chemicals by bio-products. Correspondingly, the use of proteases in the detergent field provides environmental and energetic advantages. In terms of energy, proteases use consists in washing at lower temperatures (10–20 °C). Moreover, the environmental benefit of such proteases incorporates less chemical surfactants use in order to preserve the environment. However, the desired detergent protease must be very active and stable in alkaline solution with respect to detergents [2]. This experience explores the application of SPPS protease in detergent industry. The data presented in Fig. 5b show that, compared to Flavourzyme® 500 L, SPPS was extremely stable and compatible with the commercial liquid detergents used, retaining 100% of its initial activity with Dixan, Nadhif, Detch, and Class detergents at a concentration of 7 mg/mL vs 100% of the initial activity of Flavourzyme® 500 L with OMO, Ariel, and Det.

These results can approve the usefulness of SPPS as a cleaning bio-additive in detergent formulations used.

***Cleaning potential of detergent additive SPPS to remove the blood strain from cotton fabrics***

Alkaline serine proteases are usually sought in detergence industry [45]. They highly increase the cleaning efficacy of laundry detergent. Accordingly, a comparative test using Flavourzyme® 500 L in Dixan was carried out (Fig. 5c). In fact, the blood strain was completely

**Table 4** Substrate specificity profile of SPPS enzyme from *Pleurotus sajor-caju* strain CTM10057

| Substrate | | Concentration | Absorbance (nm) [a] | Relative protease activity (%) [b] |
|---|---|---|---|---|
| *Natural protein* | Casein | 30 g/L | 660 | 100 ± 2.5 |
| | Gelatin | 30 g/L | 660 | 80 ± 2.0 |
| | Keratin | 30 g/L | 660 | 37 ± 0.9 |
| | Ovalbumin | 30 g/L | 660 | 20 ± 0.5 |
| | BSA | 30 g/L | 660 | 10 ± 0.2 |
| *Modified protein* | Azo-casein | 25 g/L | 440 | 100 ± 2.5 |
| | Albumin azure | 25 g/L | 440 | 50 ± 1.8 |
| | Keratin azure | 25 g/L | 595 | 30 ± 1.2 |
| | Collagen type I | 20 g/L | 440 | 0 ± 0.0 |
| | Collagen type II | 20 g/L | 440 | 0 ± 0.0 |
| *Ester* | ATEE | 10 mM | 253 | 100 ± 2.5 |
| | BTEE | 10 mM | 253 | 96 ± 2.4 |
| | BAEE | 10 mM | 253 | 0 ± 0.0 |
| | BCEE | 10 mM | 253 | 0 ± 0.0 |
| | TAME ↓ | 10 mM | 253 | 0 ± 0.0 |
| *Synthetic peptide (pNA)* | Suc-F-*p*NA | 5 mM | 410 | 49 ± 1.5 |
| | Benz-Y-*p*NA | 5 mM | 410 | 0 ± 0.0 |
| | M-*p*NA | 5 mM | 410 | 0 ± 0.0 |
| | Ac-L-*p*NA | 5 mM | 410 | 0 ± 0.0 |
| | P-*p*NA | 5 mM | 410 | 0 ± 0.0 |
| | V-*p*NA | 5 mM | 410 | 0 ± 0.0 |
| | Ac-A-*p*NA | 5 mM | 410 | 62 ± 1.6 |
| | Benz-R-*p*NA | 5 mM | 410 | 0 ± 0.0 |
| | Suc-YLV-*p*NA | 5 mM | 410 | 0 ± 0.0 |
| | Suc-AAA-*p*NA | 5 mM | 410 | 50 ± 1.3 |
| | Suc-AAV-*p*NA | 5 mM | 410 | 64 ± 1.7 |
| | Suc-AAF-*p*NA | 5 mM | 410 | 62 ± 1.6 |
| | Benz-FVR-*p*NA | 5 mM | 410 | 0 ± 0.0 |
| | Suc-FAAF-*p*NA | 5 mM | 410 | 100 ± 2.5 |
| | Suc-AAPF-*p*NA | 5 mM | 410 | 97 ± 2.5 |
| | Suc-AAVA-*p*NA | 5 mM | 410 | 92 ± 2.4 |
| | Suc-AAPM-*p*NA | 5 mM | 410 | 77 ± 1.9 |
| | Suc-AAPL-*p*NA | 5 mM | 410 | 65 ± 1.7 |
| | Suc-LLVY-*p*NA | 5 mM | 410 | 57 ± 1.6 |
| | Ac-YVAD-*p*NA | 5 mM | 410 | 0 ± 0.0 |

[a] Values represent means of three independent replicates, and ± standard errors are reported

[b] The activity of each substrate was determined by measuring absorbance at specified wave lengths according to the relative method reported elsewhere [33, 34]

BTEE, *N*-benzol-L-tyrosine ethyl ester; ATEE, *N*-acetyl-L-tyrosine ethyl ester monohydrate; BAEE, *N*-benzol-L-arginine ethyl ester; BCEE, *S*-benzyl-L-cysteine ethyl ester hydrochloride; TAME, $N_\alpha$-*p*-tosyl- L-arginine methyl ester hydrochloride; *N*-succinyl-F-*p*-nitroanilide; *N*-benzoyl-Y-*p*-nitroanilide; M-*p*-nitroanilide; *N*-acetyl-L-*p*-nitroanilide; P-*p*-nitroanilide trifluoroacetate salt; V-*p*-nitroanilide hydrochloride; *N*-acetyl-A-*p*-nitroanilide; *N*-benzoyl-R-*p*-nitroanilide; *N*-succinyl-YLV-*p*-nitroanilide; *N*-succinyl-AAA-*p*-nitroanilide; *N*-succinyl-AAF-*p*-nitroanilide; *N*-succinyl-AAV-*p*-nitroanilide; *N*-benzoyl-FVR-*p*-nitroanilide; *N*-succinyl-FAAF-*p*-nitroanilide; *N*-succinyl-AAPF-*p*-nitroanilide; *N*-succinyl-AAVA-*p*-nitroanilide; *N*-succinyl-AAPM-*p*-nitroanilide; *N*-succinyl-AAPL-*p*-nitroanilide; *N*-succinyl-LLVY-*p*-nitroanilide; and *N*-acetyl-YVAD-*p*-nitroanilide

removed. It is noted that SPPS isolated from *Pleurotus sajor-caju* strain CTM10057 is more efficient than Flavourzyme® 500 L in Dixan. Thus, SPPS can be considered as a remarkable bio-additive in Dixan.

## Discussion

This study is concerned with the production, purification and characterization of a new serine fungi protease. This protease is produced by *Pleurotus sajor-caju* strain

**Table 5** Kinetic parameters of the purified proteases: SPPS, SPTC, Flavourzyme® 500 L, and Thermolysin type X for hydrolysis of natural protein (casein)

| Enzyme | Origin | $K_m$ (mg/mL) [a] | $V_{max}$ (μmol/mg.min$^{-1}$) [a] | $k_{cat}$ (min$^{-1}$) | $k_{cat}/K_m$ (mL.mg$^{-1}$.min$^{-1}$) |
|---|---|---|---|---|---|
| SPPS | *Pleurotus sajor-caju* CTM 10057 | 0.275 ± 0.01 | 79,000 ± 695 | 52,667 | 191,516 |
| SPTC | *Trametes cingulata* CTM10101 | 0.475 ± 0.02 | 94,075 ± 856 | 62,717 | 132,036 |
| Flavourzyme® 500 L | *Aspergillus oryzea* | 0.546 ± 0.04 | 47,125 ± 566 | 31,417 | 57,540 |
| Thermolysin type X | *Geobacillus stearothermophilus* | 0.613 ± 0.05 | 20,764 ± 323 | 13,843 | 22,582 |

[a] Values represent means of three independent replicates, and ± standard errors are reported

CTM10057. The purification procedure strated by heat treatment (80 °C for 20 min) followed by ammonium sulfate precipitation (35–55%)-dialysis, then UNO Q-6 FPLC ion-exchange chromatography and finally HPLC-ZORBAX PSM 300 HPSEC gel filtration chromatography. SPPS contained about 15% of the total activity of the crude and had a specific activity of 79,000 U/mg (Table 1) using casein as substrate. This specific activity is higher than *Aspergillus oryzae* strain CH 93 (15,86 U/mg) [14] and *Aspergillus clavatus* strain ES1 (37,600 U/mg) [25]. According to native and SDS page study, SPPS was a homogeneous enzyme with high purity as it exhibits a unique elution symmetrical peak at retention time ($R_t$) = 11.6 min corresponding to a single protein of nearly 65 kDa by HPLC gel filtration chromatography (Fig. 2b). This molecular weight was higher compared to those previously mentioned with respect to other fungal proteases like *Pleurotus sajor-caju* (48 kDa) [39]; *Pleurotus citrinopileatus* (28 kDa) [46]; *Aspergillus clavatus* ES1 (30 kDa) [25], and *Aspergillus oryzae* CH93 (47.5 kDa) [14]. Moreover, the electrophoretic analyses and size exclusion chromatography showed that SPPS is a monomer comparable to the other previously reported fungi proteases [25, 26, 33, 39]. These results strongly suggested that SPPS enzyme from strain CTM10057 was a novel protease. The highest homology was noted with the protease isolated from *Pleurotus* genus (90%). SPPS belongs to the serine protease family according to the inhibitory test in bibliography, 22% of basidiomycete's protease belongs to serine protease [19]. Besides, this protease's class is widespread in eukaryotes microorganisms [13]. This enzyme is totally activated by $Ca^{2+}$, $Mg^{2+}$, and $Fe^{2+}$, however, strongly inhibited by $Cd^{2+}$, $Ni^{2+}$, and $Hg^{2+}$ (2 mM). This is similar to other serine proteases such as ES1 [25], SAPB [32], serine protease isolated from *Pleurotus sajor-caju* increased by $Ca^{2+}$ and $Mg^{2+}$ [39], protease isolated from *Aspergillus oryzae* strain CH93 and protease isolated from *Aspergillus tamari* strain URM 4634 [47]. Serine proteases are used to incorporate two calcium binding sites [34, 48] and their addition increase thermostability. It plays an important role in maintaining their 3D structure. In fact, $Ca^{2+}$ was previously reported to improve serine protease activity and stability [32]. These ions can preserve the active confirmation at higher temperature [49]. This kind of protease (serine protease) is well known in the field of detergency. Insensitivity to chelators would be a valuable property for potential application in detergent formulations since these agents are added to soften water and remove stain [6]. The SPPS activity was totally inhibited by mercury ions, which could be ascribed to the exert toxicity of these ions, especially towards thiol serine protease [50, 51].

SPPS is an alkaline protease stable in alkaline solution. This propriety is also searched in protease and used as bio-additive in the detergent preparation. Proteolysis enzymes with appropriate specificity and stability to temperature and pH are necessary for the physiology of ligninolytic fungi [52]. The thermo-stability of SPPS in very high temperature is a substantial feature preferred in proteases used as bio-additive in laundry detergent [9]. Extremely high temperature can destabilize the non-covalent interactions of protein [8, 53]. It has further been suggested that the protective role of polyols is ascribed to their capability to form hydrogen bonds that support and stabilize the native conformation of the enzyme to make it more resistant to heat treatment. The impact of polyols on protease activity is well studied. Other researchers have also shown the improvement of thermostability of proteases in the presence of polyols [28, 54]. The protective impact of polyols is dispersed based on these latter's hydroxyl groups belonging. The addition of some polyols is pivotal in order to maintain hydrophobic interactions within protein molecules. It has further been suggested that the protective role of polyols is assigned to their capability to form hydrogen bonds that support and stabilize the native conformation of the enzyme to make it more resistant to heat treatment.

The stability of microbial proteases by polyols has been broadly depicted by several works, such as SPACG from *Caldicoprobacter guelmensis* strain D2C22$^T$ [54], and SAPHM from *Bacillus licheniformis* strain K7A [33]. The highest activity on protein was noted with casein, followed by gelatin, and albumin azure. Other serine proteases such as BAP from *Beauveria* sp. strain MTCC 5184 possess the same substrate specificity. It was more active against casein compared to hemoglobin and

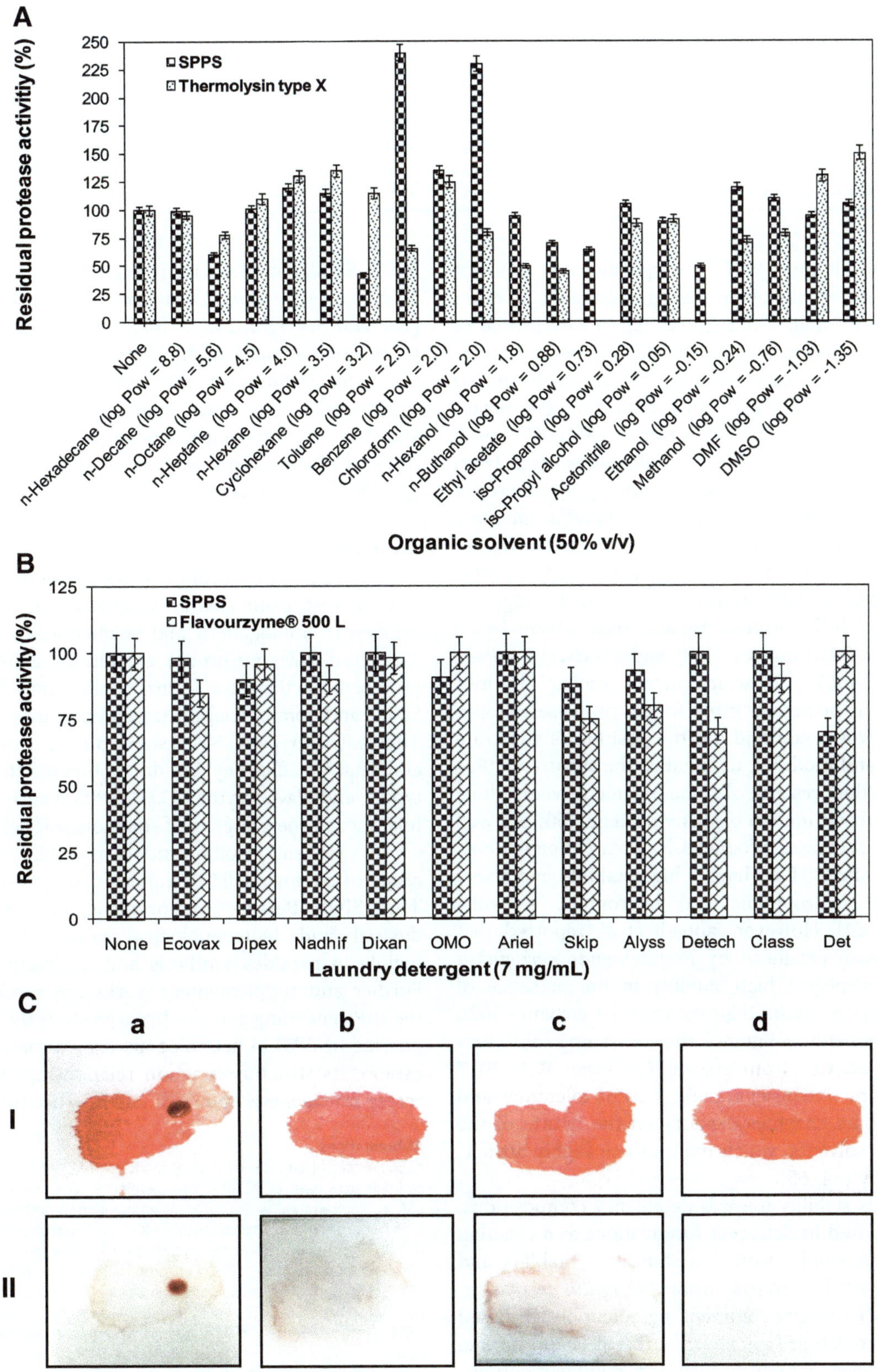


**Fig. 5** (See legend on next page.)

(See figure on previous page.)
**Fig. 5 A** Effect of different organic solvent on SPPS and Thermolysin type X enzyme stabilities; **B** Stability of SPPS and Flavourzyme® 500 L purified proteases in the presence of commercial liquid [Nadhif from Henkel-Alki (Tunisia), Ariel from Procter & Gamble (Switzerland), Skip from Unilever (France), Class from EJM (Tunisia), and Dipex from Klin Productions (Tunisia)] and solid [Dixan from Henkel (France), Ecovax from Klin Productions (Tunisia), OMO from Unilever (France), Alyss from EJM (Tunisia), Detech from SOTUP (Tunisia), and Det from Sodet (Tunisia)] laundry detergents at 7 mg/mL. Enzyme activity of the control sample, which contained no additive and incubated under similar conditions, was taken as 100%. Vertical bars indicate standard error of the mean ($n = 3$). **C** Washing performance analysis test of SPPS enzyme in the presence of the commercial detergent Dixan. Cloth stained with blood washed with: (**a**) tap water, (**b**) Dixan detergent (7 mg/mL), (**c**) Dixan added with Flavourzyme® 500 L (commercial enzyme, 500 U/mL), and (**d**) Dixan added with SPPS (500 U/mL). I: untreated cloths (control) and II: treated cloths

bovine serum albumin [51]. This specificity of cleavage to natural protein has already been elaborated in other proteases from fungi. The substrate specificity profile of SPPS indicates that it is subtilisin-like. However, the enzyme activity of protease produced by *Aspergillus flavus* strain AP2 was the highest with gelatin [55]. The protease from *Pleurotus sajor-caju* strain CTM10057 has a less $K_m$ value (0.275 mg/mL) as proteases isolated from *Pleurotus citrinopileatus* (3.44 mg/mL) [46], *Amanita virgineoides* (3.74 mg/mL) [56], and *Helvella lacunose* (3.81 mg/mL) [57], i.e. SPPS protease evinced a higher affinity towards casein than the above-mentioned proteases. Nevertheless, SPPS has a $K_m$ value (0.275 mg/mL) comparable with the protease isolated from *Mucor bacilliformis* ($K_m = 0.185$ mg/mL) [58] and *Cordyceps sobolifera* (0.41 mg/mL) [59], which means similar affinities towards casein possessed three fungal proteases. Hopefully, future works can find out the position of substrate binding site and modify it to enhance the substrate affinity. Thereby, the presence of some organic solvents altering the structure-function of many proteins [60]. As well detailed, the proteases produced by *Pseudomonas aeruginosa* PseA and PST-01 have a high stability and tolerance toward hydrophobic and hydrophilic solvents, respectively [61]. However, Jaouadi et al. reported that AMPP protease produced by *Pseudomonas aeruginosa* CTM50182 displayed high stability in the presence of various organic solvents than commercial enzymes [62]. On the other hand, *n*-butanol showed strong inhibition in protease activity from *Aspergillus niger* WA 2017 [63]. Thus, the intrinsic organic solvent-tolerance and stability of some enzymes depends on the nature of the used organic solvents, which makes them highly attractive to industry [64, 65].

Owing to its stability towards detergency compositions, SPPS can be used in detergent formulations as a cleaning bio-additive. A similar work has shown the stability and compatibility of the protease from *Aspergillus terreus* gr. in the presence of some comment ingredient bleach-based detergent formulation [16], showing that it is appropriate candidate for cleaning power of laundry detergent compositions [66] . The alkaline protease isolated from *Pleurotus sajor-caju* CTM10057 was completely stable in the presence of various detergent additives than Flavourzyme® 500 L. Similar studies have revealed that alkaline proteases produced by *Bacillus licheniformis* [33], *Aspergillus terreus* [16], and *Aspergillus niger* [67] demonstrated good stability in commercial laundry detergents. In alkaline condition, the stability of the protease SPPS indicates a significant enzyme compatibility with laundry detergent and, therefore, making it a potential candidate for use in detergent formulations as cleaning additive that facilitates proteinaceous stains release.

## Conclusions

Accordingly, a novel fungal protease with a great industrial interest, from *Pleurotus sajor-caju* CTM 10057 was purified to homogeneity and biochemically characterized. Therefore, this mushroom constitutes a new source of proteolytic activity for future applications. Compared to SPTC and commercial proteases Flavourzyme® 500 L and Thermolysin type X, SPPS showed high hydrolysis degree and catalytic efficiency. Besides, compared to Thermolysin type X and Flavourzyme® 500 L, SPPS displayed significant tolerance in the presence of some organic solvents, as well as an excellent stability and compatibility with a wide range of commercialized laundry detergents, respectively. Thus, SPPS offers new and promising opportunities for industrial and biotechnological perspective bioprocess, mainly in peptides synthesis and detergent formulation. Further and supplementary works are needed to isolate the *spPS* encoding gene, to hyperproduce the recombinant enzyme (rSPPS) in heterologous-expression system and to explore its structure-function relationships using site-directed mutagenesis approach and 3D structure modeling.

### Abbreviations

ATEE : *N*-acetyl-L-tyrosine ethyl ester monohydrate; BAEE: *N*-benzyol-L-arginine ethyl ester; BCEE: *S*-benzyl-L-cysteine ethyl ester hydrochloride; BSA: Bovine serum albumin; BTEE: *N*-benzol-L-tyrosine ethyl ester; DFP: Diiodopropyl fluorophosphate; DMC: *N,N*-dimethylated casein; DTNB: 5,5′-dithio-bis-2-nitro benzoic acid; EPNP: 1.2-epoxy-3-(*p*-nitrophenoloxy) propane; FPLC: Fast protein liquid chromatography; HPLC: High performance liquid chromatography; iodoacetamide: β-ME, β-mercaptoethanol; LD-DTT: LD-dithiothreitol; MIA: Monoiodoacetic acid; NEM: *N*-ethylmalemide; PAGE: Polyacrylamide gel electrophoresis; PCR: Polymerase chain reaction; PDB: Potato dextrose broth; PEG: Propylene glycol; PMSF: Phenylmethyl sulfonyfluoride; rDNA: ribosomal DNA; SBTI: Soybean trypsin inhibitor; SDS: Soduim dodecyl sulphate; SPPS: Serine alkaline protease from *Pleurotus sajor-caju* strain CTM10057; TAME: $N_\alpha$-*p*-tosyl-L-arginine methyl ester hydrochloride; TLCK: *Nα*-*p*-tosyl-L-lysine chloromethyl ketone; TPCK: *Nα*-*p*-tosyl-L-phenylalanine chloromethyl ketone

**Acknowledgments**
The authors would like to express their gratitude to Mr. K. Walha, Mrs. N. Kchaou, and Mrs. N. Masmoudi (Analysis Unit-CBS) for their technical assistance with the HPLC and FPLC and Pr. H. Mejdoub and Mr. C. Bouzid (USCR/SP, FSS, University of Sfax) for the $NH_2$-terminal amino-acid sequencing of the SPPS protein. We extend our thanks to Dr. W. Saibi and Pr. H. Belghith (CBS) for their constructive discussions and suggestions and Mrs. S. Kchaou (CTM Unit-CBS) for the gift of the filamentous fungi strain CTM10057. Special thanks are also due to Pr. W. Hariz from the English Department at the FSS, University of Sfax (Sfax, Tunisia) for constructively proofreading and language polishing services. This work is part of a Doctoral Thesis in Chemical-Process Engineering by Miss Maroua OMRANE BENMRAD at the National Engineering School of Gabes (ENIG) at the University of Gabes (Gabes, Tunisia).

**Authors' contributions**
BJ: Study design as well as experiments collection, analysis, and interpretation; MOB, SM, NZJ, HR, and MBE: Performing experiments; MOB, SM, NZJ, MBE, and HR: Data analysis and interpretation; MOB and SM: Manuscript writing; BJ, NK, SS, and SB: Manuscript revision and editing. All authors read and approved the final manuscript.

**Funding**
This study was supported by the Tunisian Ministry of Higher Education and Scientific Research under the Contract Program (Grant n°.: LMBEE_CBS/code: LR15CBS06_2015–2019). The funding entity had no contribution in the study design and data collection, analysis, and interpretation as well as the manuscript writing.

**Ethics approval and consent to participate**
Not applicable.

**Consent for publication**
Not applicable.

**Competing interests**
The authors declare that they have no competing interests.

**Author details**
[1]Laboratory of Microbial Biotechnology and Engineering Enzymes (LMBEE), Centre of Biotechnology of Sfax (CBS), University of Sfax, Road of Sidi Mansour Km 6, P.O. Box 1177, 3018 Sfax, Tunisia. [2]Biotech ECOZYM Start-up, Business Incubator, Centre of Biotechnology of Sfax (CBS), University of Sfax, Road of Sidi Mansour Km 6, P.O. Box 1177, 3018 Sfax, Tunisia. [3]Laboratory of Environmental Bioprocesses (LEBP), LMI COSYS-Med, Centre of Biotechnology of Sfax (CBS), University of Sfax, Road of Sidi Mansour Km 6, P.O. Box 1177, 3018 Sfax, Tunisia. [4]Research Group of Agro-Food Processing Engineering (GP2A), Laboratory of Applied Fluid Mechanics, Process Engineering and Environment, National School of Engineers of Sfax (ENIS), University of Sfax, Road of Soukra Km 4, P.O. Box 1173, 3038 Sfax, Tunisia.

**References**

1. Inácio FD, Ferreira RO, Araujo CAVd, Brugnari T, Castoldi R, Peralta RM, Souza CGMd. Proteases of wood rot fungi with emphasis on the genus *Pleurotus*. BioMed Res Int. 2015;290161:1-10. http//dx.doi.org/10.1155/2015/290161.
2. Hirata A, Hori Y, Koga Y, Okada J, Sakudo A, Ikuta K, Kanaya S, Takano K. Enzymatic activity of a subtilisin homolog, Tk-SP, from *Thermococcus kodakarensis* in detergents and its ability to degrade the abnormal prion protein. BMC Biotechnol. 2013;13(1):19.
3. Singh R, Kumar M, Mittal A, Mehta PK. Microbial enzymes: industrial progress in 21st century. 3 Biotech. 2016;6(2):174.
4. Kudryavtseva O, Dunaevsky YE, Kamzolkina O, Belozersky M. Fungal proteolytic enzymes: features of the extracellular proteases of xylotrophic basidiomycetes. Microbiology. 2008;77(6):643–53.
5. Yin C, Zheng L, Chen L, Tan Q, Shang X, Ma A. Cloning, expression, and characterization of a milk-clotting aspartic protease gene (*Po-Asp*) from *Pleurotus ostreatus*. Appl Biochem Biotechnol. 2014;172(4):2119–31.
6. Gupta R, Beg Q, Lorenz P. Bacterial alkaline proteases: molecular approaches and industrial applications. Appl Microbiol Biotechnol. 2002;59(1):15–32.
7. Kulkarni N, Shendye A, Rao M. Molecular and biotechnological aspects of xylanases. FEMS Microbiol Rev. 1999;23(4):411–56.
8. Li Q, Yi L, Marek P, Iverson BL. Commercial proteases: present and future. FEBS Lett. 2013;587(8):1155–63.
9. PMd S, MLdA B, Caprara CC, Md F, RPCd A, Silveira D, Fonseca YM, Ferreira Filho EX, Pessoa Junior A, Magalhães PO. A biotechnology perspective of fungal proteases. Braz J Microbiol. 2015;46(2):337–46.
10. Baweja M, Singh PK, Sadaf A, Tiwari R, Nain L, Khare SK, Shukla P. Cost effective characterization process and molecular dynamic simulation of detergent compatible alkaline protease from *Bacillus pumilus* strain MP27. Process Biochem. 2017;58:199–203.
11. Olsen HS, Falholt P. The role of enzymes in modern detergency. J Surfactant Deterg. 1998;1(4):555–67.
12. Zhao HY, Feng H. Engineering *Bacillus pumilus* alkaline serine protease to increase its low-temperature proteolytic activity by directed evolution. BMC Biotechnol. 2018;18(1):34.
13. Mienda B, Yahya A, Galadima I, Shamsir M. An overview of microbial proteases for industrial applications. Res J Pharm Biol Chem Sci. 2014;5(1):388–96.
14. Salihi A, Asoodeh A, Aliabadian M. Production and biochemical characterization of an alkaline protease from *Aspergillus oryzae* CH93. Int J Biol Macromol. 2017;94:827–35.
15. Sharma KM, Kumar R, Panwar S, Kumar A. Microbial alkaline proteases: optimization of production parameters and their properties. J Genet Eng Biotechnol. 2017;15(1):115–26.
16. Niyonzima FN, More SS. Purification and characterization of detergent-compatible protease from *Aspergillus terreus* gr. 3. Biotech. 2015;5(1):61–70.
17. Haddar A, Bougatef A, Agrebi R, Sellami-Kamoun A, Nasri M. A novel surfactant-stable alkaline serine-protease from a newly isolated *Bacillus mojavensis* A21. Purification and characterization. Process Biochem. 2009;44(1):29–35.
18. Mechri S, Kriaa M, Ben Elhoul Berrouina M, Omrane Benmrad M, Zaraî Jaouadi N, Rekik H, Bouacem K, Bouanane-Darenfed A, Chebbi A, Sayadi S. Optimized production and characterization of a detergent-stable protease from *Lysinibacillus fusiformis* C250R. Int J Biol Macromol. 2017;101:383–97.
19. Sabotič J, Trček T, Popovič T, Brzin J. Basidiomycetes harbour a hidden treasure of proteolytic diversity. J Biotechnol. 2007;128(2):297–307.
20. Jalomo-Khayrova E, Mares RE, Munoz PLA, Melendez-Lopez SG, Rivero IA, Ramos MA. Soluble expression of an amebic cysteine protease in the cytoplasm of *Escherichia coli* SHuffle express cells and purification of active enzyme. BMC Biotechnol. 2018;18(1):20.
21. Li Y, Pan Y, She Q, Chen L. A novel carboxyl-terminal protease derived from *Paenibacillus lautus* CHN26 exhibiting high activities at multiple sites of substrates. BMC Biotechnol. 2013;13:89.
22. Breene WM. Nutritional and medicinal value of specialty mushrooms. J Food Protect. 1990;53:883–94.
23. Choi JH, Kim DW, Kim S, Kim SJ. Purification and partial characterization of a fibrinolytic enzyme from the fruiting body of the medicinal and edible mushroom *Pleurotus ferulae*. Prep Biochem Biotechnol. 2017;47(6):539–46.
24. Palmieri G, Bianco C, Cennamo G, Giardina P, Marino G, Monti M, Sannia G. Purification, characterization, and functional role of a novel extracellular protease from *Pleurotus ostreatus*. Appl Environ Microbiol. 2001;67(6):2754–9.
25. Hajji M, Kanoun S, Nasri M, Gharsallah N. Purification and characterization of an alkaline serine-protease produced by a new isolated *Aspergillus clavatus* ES1. Process Biochem. 2007;42(5):791–7.
26. Omrane Benmrad M, Moujehed E, Ben Elhoul M, Zaraî Jaouadi N, Mechri S, Rekik H, Kourdali S, El Hattab M, Badis A, Sayadi S, Bejar S, Jaouadi B. A novel organic solvent- and detergent-stable serine alkaline protease from *Trametes cingulata* strain CTM10101. Int J Biol Macromol. 2016;91:961–72.
27. Li F. Wen ha, Liu X, Zhou F, Chen G. gene cloning and recombinant expression of a novel fungal immunomodulatory protein from *Trametes versicolor*. Protein Expres Purif. 2012;82(2):339–44.
28. Mechri S, Ben Elhoul Berrouina M, Omrane Benmrad M, Zaraî Jaouadi N, Rekik H, Moujehed E, Chebbi A, Sayadi S, Chamkha M, Bejar S, Jaouadi B.

Characterization of a novel protease from *Aeribacillus pallidus* strain VP3 with potential biotechnological interest. Int J Biol Macromol. 2017;94(Part A): 221–32.

29. Boulkour Touioui S, Zaraî Jaouadi N, Boudjella H, Ferradji FZ, Belhoul M, Rekik H, Badis A, Bejar S, Jaouadi B. Purification and biochemical characterization of two detergent-stable serine alkaline proteases from *Streptomyces* sp. strain AH4. World J Microbiol Biotechnol. 2015;31(7):1079–92.
30. Bradford MM. A rapid and sensitive method for the quantitation of microgram quantities of protein utilizing the principle of protein-dye binding. Anal Biochem. 1976;72(1–2):248–54.
31. Laemmli UK. Cleavage of structural proteins during the assembly of the head of bacteriophage T4. Nature. 1970;227(5259):680–5.
32. Jaouadi B, Ellouz-Chaabouni S, Rhimi M, Bejar S. Biochemical and molecular characterization of a detergent-stable serine alkaline protease from *Bacillus pumilus* CBS with high catalytic efficiency. Biochimie. 2008;90(9):1291–305.
33. Hadjidj R, Badis A, Mechri S, Eddouaouda K, Khelouia L, Annane R, El Hattab M, Jaouadi B. Purification, biochemical, and molecular characterization of novel protease from *Bacillus licheniformis* strain K7A. Int J Biol Macromol. 2018;114:1033–48.
34. Zaraî Jaouadi N, Rekik H, Badis A, Trabelsi S, Belhoul M, Yahiaoui AB, Ben Aicha H, Toumi A, Bejar S, Jaouadi B. Biochemical and molecular characterization of a serine keratinase from *Brevibacillus brevis* US575 with promising keratin-biodegradation and hide-dehairing activities. PLoS One. 2013;8(10):e76722.
35. Adler-Nissen J. Enzymic hydrolysis of food proteins. London: Elsevier Applied Science Publishers Ltd.; 1986.
36. Frost DV, Heinsen J. Determination of the degree of hydrolysis of partial acid hydrolysates of casein and fibrin. J Biol Chem. 1945;161(2):517–21.
37. Zaraî Jaouadi N, Jaouadi B, Aghajari N, Bejar S. The overexpression of the SAPB of *Bacillus pumilus* CBS and mutated *sapB*-L31I/T33S/N99Y alkaline proteases in *Bacillus subtilis* DB430: new attractive properties for the mutant enzyme. Bioresour Technol. 2012;105:142–51.
38. Colja L, Sjef B, Kees V, Cees V. Rules for optimization of biocatalysis in organic solvents. Biotechnol Bioeng. 1987;30(1):81–7.
39. Ravikumar G, Gomathi D, Kalaiselvi M, Uma C. A protease from the medicinal mushroom *Pleurotus sajor-caju*: production, purification and partial characterization. Asian Pac J Trop Biomed. 2012;2(1):S411–7.
40. Steuer C, Heinonen KH, Kattner L, Klein CD. Optimization of assay conditions for dengue virus protease: effect of various polyols and nonionic detergents. J Biomol Screen. 2009;14(9):1102–8.
41. Omrane Benmrad M, Moujehed E, Ben Elhoul M, Mechri S, Bejar S, Zouari R, Baffoun A, Jaouadi B. Production, purification, and biochemical characterization of serine alkaline protease from *Penicillium chrysogenium* strain X5 used as excellent bio-additive for textile processing. Int J Biol Macromol. 2018;119:1002–16.
42. Deng Y, Liu X, Katrolia P, Kopparapu NK, Zheng X. A dual-function chymotrypsin-like serine protease with plasminogen activation and fibrinolytic activities from the GRAS fungus, *Neurospora sitophila*. Int J Biol Macromol. 2018;109:1338–43.
43. Adler-Nissen J. Determination of the degree of hydrolysis of food protein hydrolysates by trinitrobenzenesulfonic acid. J Agric Food Chem. 1979;27(6): 1256–62.
44. Annamalai N, Rajeswari MV, Sahu SK, Balasubramanian T. Purification and characterization of solvent stable, alkaline protease from *Bacillus firmus* CAS 7 by microbial conversion of marine wastes and molecular mechanism underlying solvent stability. Process Biochem. 2014;49(6):1012–9.
45. Coral G, Arikan B, Unaldi M, Guvenmez H. Thermostable alkaline protease produced by an *Aspergillus niger* strain. Ann Microbiol. 2003;53(4):491–8.
46. Cui L, Liu Q, Wang H, Ng T. An alkaline protease from fresh fruiting bodies of the edible mushroom *Pleurotus citrinopileatus*. Appl Microbiol Biotechnol. 2007;75(1):81–5.
47. da Silva OS, de Almeida EM, de Melo AHF, Porto TS. Purification and characterization of a novel extracellular serine-protease with collagenolytic activity from *Aspergillus tamarii* URM4634. Int J Biol Macromol. 2018;117: 1081–8.
48. Ben Elhoul M, Zaraî Jaouadi N, Rekik H, Omrane Benmrad M, Mechri S, Moujehed E, Kourdali S, El Hattab M, Badis A, Bejar S. Biochemical and molecular characterization of new keratinoytic protease from *Actinomadura viridilutea* DZ50. Int J Biol Macromol. 2016;92:299–315.
49. Kumar CG, Takagi H. Microbial alkaline proteases: from a bioindustrial viewpoint. Biotechno Adv. 1999;17(7):561–94.
50. Ben Elhoul M, Zaraî Jaouadi N, Rekik H, Bejar W, Boulkour Touioui S, Hmidi M, Badis A, Bejar S, Jaouadi B. A novel detergent-stable solvent-tolerant serine thiol alkaline protease from *Streptomyces koyangensis* TN650. Int J Biol Macromol. 2015;79:871–82.
51. Shankar S, Rao M, Laxman RS. Purification and characterization of an alkaline protease by a new strain of *Beauveria* sp. Process Biochem. 2011;46(2):579–85.
52. Datta A. Purification and characterization of a novel protease from solid substrate cultures of *Phanerochaete chrysosporium*. J Biol Chem. 1992;267(2): 728–32.
53. Dill KA. Dominant forces in protein folding. Biochemistry. 1990;29(31):7133–55.
54. Bouacem K, Bouanane-Darenfed A, Laribi-Habchi H, Elhoul MB, Hmida-Sayari A, Hacene H, Ollivier B, Fardeau ML, Jaouadi B, Bejar S. Biochemical characterization of a detergent-stable serine alkaline protease from *Caldicoprobacter guelmensis*. Int J Biol Macromol. 2015;81:299–307.
55. Hossain M, Das F, Marzan L, Rahman MS, Anwar M. Some properties of protease of the fungal strain *Aspergillus flavus*. Int J Agric Biol. 2006;8:162–4.
56. Yuan X, Zhu M, Tian G, Zhao Y, Zhao L, Ng TB, Wang H. Biochemical characteristics of a novel protease from the basidiomycete *Amanita virgineoides*. Biotechnol Appl Biochem. 2017;64(4):532–40.
57. Zhang G, Wang H, Zhang X, Ng T. Helvellisin, a novel alkaline protease from the wild ascomycete mushroom *Helvella lacunosa*. J Biosci Bioeng. 2010; 109(1):20–4.
58. Venera GD, Machalinski C, Zum'arraga H, Biscoglio MJ. de Bonino je. Further characterization and kinetic parameter determination of a milk-clotting protease from *Mucor bacilliformis*. Appl Biochem Biotechnol. 1997;68(3):207–16.
59. Wang SX, Liu Y, Zhang GQ, Zhao S, Xu F, Geng XL, Wang HX. Cordysobin, a novel alkaline serine protease with HIV-1 reverse transcriptase inhibitory activity from the medicinal mushroom *Cordyceps sobolifera*. J Biosci Bioeng. 2012;113(1):42–7.
60. Barberis S, Quiroga E, Morcelle S, Priolo N, Luco JM. Study of phytoproteases stability in aqueous-organic biphasic systems using linear free energy relationships. J Mol Catal B-Enzy. 2006;38(2):95–103.
61. Ogino H, Watanabe F, Yamada M, Nakagawa S, Hirose T, Noguchi A, Yasuda M, Ishikawa H. Purification and characterization of organic solvent-stable protease from organic solvent-tolerant *Pseudomonas aeruginosa* PST-01. J Biosci Bioeng. 1999;87(1):61–8.
62. Jaouadi B, Zaraî Jaouadi N, Rekik H, Naili B, Beji A, Dhouib A, Bejar S. Biochemical and molecular characterization of *Pseudomonas aeruginosa* CTM50182 organic solvent-stable elastase. Int J Biol Macromol. 2013;60:165–77.
63. Abdel Wahab WA, Ahmed SA. Response surface methodology for production, characterization and application of solvent, salt and alkali-tolerant alkaline protease from isolated fungal strain *Aspergillus niger* WA 2017. Inter J Biol Macromol. 2018;115:447–58.
64. Barberis S, Quiroga E. Arribére MaC, Priolo N. peptide synthesis in aqueous–organic biphasic systems catalyzed by a protease isolated from *Morrenia brachystephana* (Asclepiadaceae). J Mol Catal B- Enzy. 2002;17(1):39–47.
65. Ogino H, Uchiho T, Doukyu N, Yasuda M, Ishimi K, Ishikawa H. Effect of exchange of amino acid residues of the surface region of the PST-01 protease on its organic solvent-stability. Biochem Biophys Res Commun. 2007;358(4):1028–33.
66. Beena A, Geevarghese P, Jayavardanan K. Detergent potential of a spoilage protease enzyme liberated by a psychrotrophic spore former isolated from sterilized skim milk. Am J Food Technol. 2012;7(2):89–95.
67. Devi MK, Banu AR, Gnanaprabha G, Pradeep B, Palaniswamy M. Purification, characterization of alkaline protease enzyme from native isolate *Aspergillus niger* and its compatibility with commercial detergents. Indian J Sci Technol. 2008;1(7):1–6.

16

# Production of high titer of citric acid from inulin

Magdalena Rakicka*, Jakub Wolniak, Zbigniew Lazar and Waldemar Rymowicz

**Abstract**

**Background:** Citric acid is considered as the most economically feasible product of microbiological production, therefore studies on cheap and renewable raw materials for its production are highly desirable. In this study citric acid was synthesized by genetically engineered strains of *Yarrowia lipolytica* from widely available, renewable polysaccharide – inulin. Hydrolysis of inulin by the *Y. lipolytica* strains was established by expressing the inulinase gene (*INU1* gene; GenBank: X57202.1) with its native secretion signal sequence was amplified from genomic DNA from *Kluyveromyces marxianus* CBS6432. To ensure the maximum citric acid titer, the optimal cultivation strategy–repeated-batch culture was applied.

**Results:** The strain *Y. lipolytica* AWG7 INU 8 secreted more than 200 g $dm^{-3}$ of citric acid during repeated-batch culture on inulin, with a productivity of 0.51 g $dm^{-3}$ $h^{-1}$ and a yield of 0.85 g $g^{-1}$.

**Conclusions:** The citric acid titer obtained in the proposed process is the highest value reported in the literature for *Yarrowia* yeast. The obtained results suggest that citric acid production from inulin by engineered *Y. lipolytica* may be a very promising technology for industrial citric acid production.

**Keywords:** Inulin, Citric acid, Repeated-batch culture, *Yarrowia lipolytica*

## Background

Production of citric acid (CA), the intermediate of Krebs cycle, is one of the oldest technology of organic acid production applied at industrial scale [1]. Initially CA was extracted from Italian lemons which what was later replaced by its biosynthesis using *Aspergillus niger* [2]. Afterwards, the discovery of CA secretion by the yeast *Yarrowia lipolytica* caused a rapid progress in research on that process [3]. The global CA production in 2015 reached 2 Mln tones and 3.7% of annual growth is expected till 2020 (https://ihsmarkit.com/products/citric-acid-chemical-economics-handbook.html). Therefore, CA is the most abundantly produced chemical on an industrial scale and the most widely used organic acid, it has also "generally regarded as safe" (GRAS) status [4]. The main function of this acid is its application as food acidulant. It is also used for prevention of oxidative deterioration in flavor or color [5]. CA and its derivatives are also ingredients in detergents, personal care products as well as they are widely used in pharmaceutical and biomedical industry. Due to many applications of this valuable compound, there is a great demand on developing very efficient technology for its production. Nowadays, from an industrial point of view, the most important challenges when developing novel technologies are to design the production process with high titer, productivity and yield, simultaneously fulfilling the main principles of cleaner production, energy saving and sustainable development [6].

One of the criteria when developing a new biotechnology process is the cultivation system. The most common systems used industrially are batch or fed-batch cultures. To achieve better results, the repeated-batch culture (RBC) could be used. This cultivation system allows for better dynamics and higher efficiency of the biosynthesis process by extending the effective production phase in comparison to traditional batch culture [7]. The RBC was already successfully used for lipid, ethanol and erythritol production [7–9].

Other important criterion of efficient technology using microorganisms is the substrate applied in the process. The raw materials used for biotechnological processes must be cheap, renewable and not competing with food

* Correspondence: magdalena.rakicka@upwr.edu.pl
Department of Biotechnology and Food Microbiology, Wroclaw University of Environmental and Life Sciences, 37 Chełmońskiego St, 51-630 Wroclaw, Poland

production. Ideally, such a feedstock should be waste or by-products from other industries.

The nonconventional *Y. lipolytica* yeast is rapidly emerging as a valuable host for the production of a variety of lipids, organic acids, polyols and other metabolites [10]. Nowadays, both, lab-scale as well as industrial production of chemicals using *Y. lipolytica*, are being improved through genetic engineering. Recently, Sabra et al. [11] focused on transcriptome and fluxome characterization of citrate producing strain ACA DC 50109 and enhanced citrate secretion in a glucose based medium to 55 g dm$^{-3}$. Furthermore, improvement of succinic acid production (110.7 g dm$^{-3}$) was achieved through deletion of succinate dehydrogenase and CoA-transferase genes followed by overexpression of TCA cycle genes [12]. More recent work characterized biological pathway of erythritol biosynthesis and improved its production through overexpression of transketolase or erythrose reductase genes in *Y. lipolytica* [13, 14]. Furthermore, β-carotene production in *Y. lipolytica* was also investigated [15]. Overexpression of the optimum promoter-gene pairs for each transcriptional unit in lipid overproducing strain followed by optimization of cultivation method significantly improved β-carotene production in comparison to strain without lipid overproduction.

In the processes of CA production using *Y. lipolytica* many research groups focused on glycerol, by-product from biodiesel or soap production [16–18]. However, some other waste substrates were also applied: olive mill waste-water [19], molasses [20] or pretreated cellulose [21]. The plant biomass constitutes very promising raw material due to high sugar content and its availability. Besides cellulose, some plants, such as chicory, Jerusalem artichoke or dahlia, are known to store energy in the form on inulin (IN), a fructose polymer, which is accumulated in large quantity in roots and rhizoids [22]. Inulin is classified as dietary fibre and promotes the growth of intestinal bacteria [23]. IN has significant number of pharmaceutical and food applications. It is frequently used as sugar or fat substitute in different types of food and as an excipient and stabilizer in many pharmaceuticals [24, 25]. Furthermore, IN was found to have also anti-cancer [26] and immuno-modulatory properties [27]. The global IN market was valued at USD 1674.3 Mln in 2017, and is expected to reach USD 5099.2 Mln by 2025. Interestingly, Europe is the largest IN market which accounts for approximately 42% of the global IN market (http://www.a-cutemarketreports.com/report/artichoke-inulin-market--professional-survey-report). Instantly growing IN market results in increasing amounts of inulin-rich wastes after its production that must be managed. One of the possibilities to utilize these raw materials is microbiological production of bioethanol, single-cell protein, lipids, CA, butanediol or lactic acid [28–32].

Due to the always high demand for CA and increasing market of IN, the main goal of this study was to improve biosynthesis of CA by engineered *Y. lipolytica* strains, using IN as carbon source. To enhance the CA titer, yield and productivity from IN, repeated-batch culture was applied.

## Results

### Overexpression of *INU1* gene in *Y. lipolytica* strains

It is known phenomenon, that CA secretion by *Y. lipolytica* occurs when nitrogen and to some level phosphorous are deficient in the medium. The abilities to secrete large amount of this acid are also strain dependent [33–35]. In the current study, new inulinase expressing derivatives of *Y. lipolytica* AWG7, able to grow on IN, were investigated for CA production from this raw material. The parental AWG7 strain was selected as efficient CA producer, with high purity of CA over ICA. During continuous culture with glycerol as carbon source, it secreted over 97 g dm$^{-3}$ of CA from 150 g dm$^{-3}$ of glycerol [36]. In the current study, growth of the wild strain and its inulinase positive derivatives was analyzed in YNB medium supplemented with fructose or IN (Fig. 1). Furthermore, control experiment with glucose was also performed to analyze if the random *INU1* gene integration into *Y. lipolytica* genome, did not affect growth of the resulted transformants (Fig. 1). No difference was observed for all analyzed strains in both, YNB medium with glucose and fructose (Fig. 1a, b). Although the integration cassette for *INU1* gene expression was inserted into the genome using retrotransposable zeta sequences [37], what causes randome integration and may result in different inulinase expression, no differences was observed among the analyzed transformants growing in medium with IN (Fig. 1c). Slow and not very significant growth of the wild strain AWG7 may be the result of trace amounts of free fructose present in the medium (Fig. 1c). The good growth of all analyzed transformants in medium with IN is an indirect proof for efficient inulinase expression and hydrolysis of IN. Similar growth pattern for all strains was the basis for their further analysis for CA biosynthesis.

### Selection of the best CA producer using batch cultures

In the next step of the study, all the engineered strains were analyzed for their abilities to secrete CA while growing in medium with IN. In control experiment, the AWG7 strain grew in synthetic medium with fructose and glucose. The obtained results are presented in Table 1. Inulin concentration of 100 g dm$^{-3}$ was used at the beginning of the process. The culture was carried out until the available substrate (fructose released from inulin) was exhausted from the medium. Fructose concentration was monitored by HPLC method. Six out of seven analyzed transformants were able to secrete more

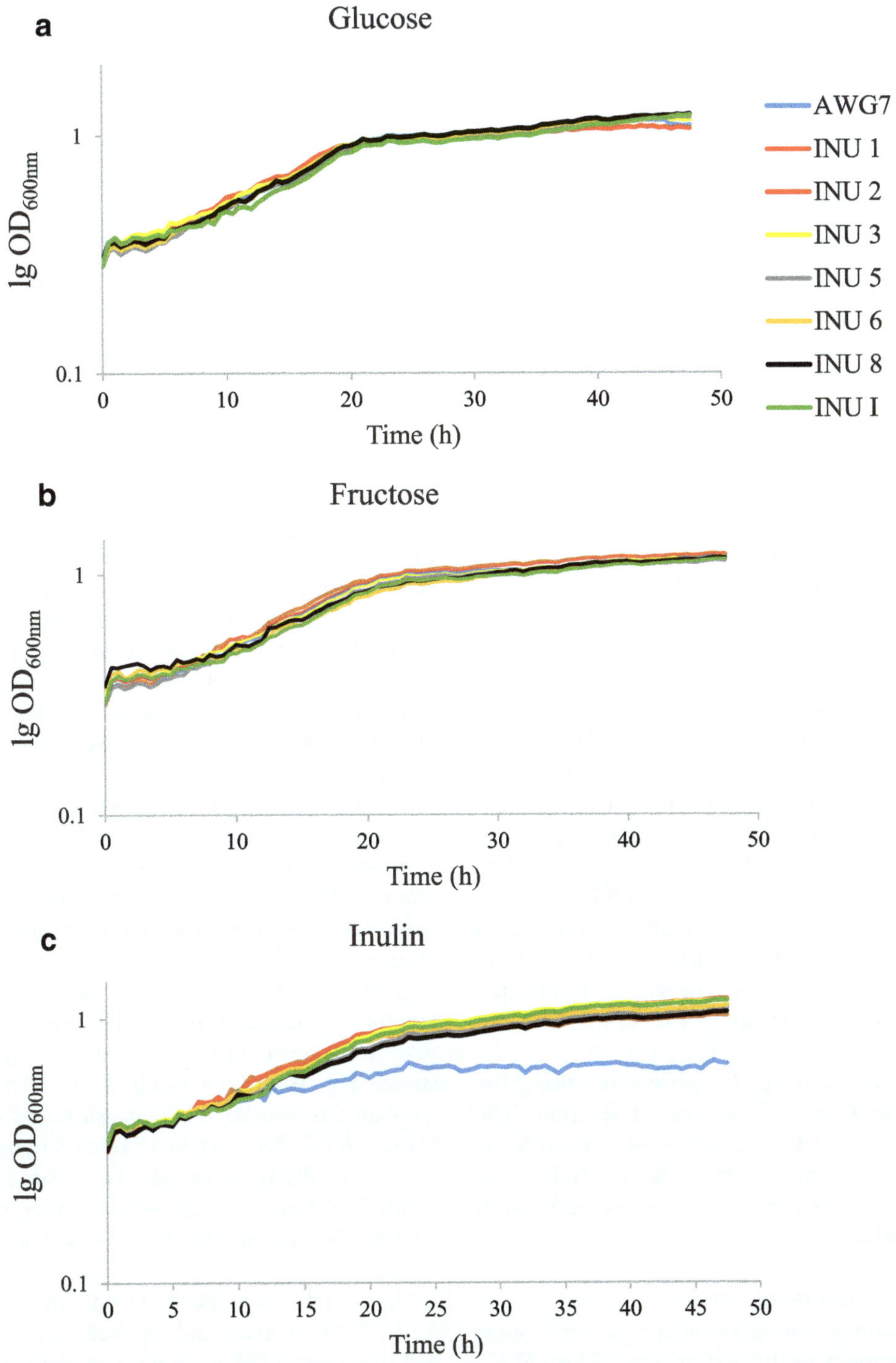


**Fig. 1** Growth curves of various *Y. lipolytica* strains growing on glycose (**a**), fructose (**b**) and inulin (**c**): AWG7 (blue line), AWG7 INU 1 (dark blue line), AWG7 INU 2 (red line), AWG7 INU 3 (yellow line), AWG7 INU 5 (grey line), AWG7 INU 6 (orange line), AWG7 INU 8 (black line), AWG7 INU I (green line). The strains were grown on an YNB/glucose, fructose or inulin medium. Quintuple experiments were performed at 28 °C under constant agitation using Bioscreen C (Oy Growth Curves Ab Ltd.)

than 60 g $dm^{-3}$ of CA from 100 g $dm^{-3}$ of IN. The best CA producing strain - AWG7 INU8 - secreted 75.5 g $dm^{-3}$. It is very important to notice, that parental AWG7 strain secreted only 48.7 g $dm^{-3}$ of CA in fructose based medium, which was significantly lower not only compared to glucose based medium but also to all its derivatives growing in medium with IN (fructose polymer). What is also important to mention is the fact

**Table 1** Parameters of citric acid synthesis by various *Y. lipolytica* strains in batch culture

| Strain | | *Yarrowia lipolytica* | | | | | |
|---|---|---|---|---|---|---|---|
| | Substrat | Biomass g dm$^{-3}$ | Citric acid | Isocitric acid | Mannitol | $Q_{CA}$ g dm$^{-3}$ h$^{-1}$ | $Y_{CA}$ g g$^{-1}$ |
| AWG7 Control 1 | Fructose | 12.0 ± 0.8 | 48.7 ± 2.0 | 2.0 ± 0.4 | 1.1 ± 0.2 | 0.42 ± 0.7 | 0.49 ± 0.01 |
| AWG7 Control 2 | Glucose | 13.8 ± 1.0 | 72.9 ± 0.8 | 2.5 ± 0.9 | 0.0 | 0.83 ± 0.5 | 0.73 ± 0.01 |
| AWG7 INU 1 | Inulin | 9.0 ± 2.0 | 64.6 ± 1.6 | 1.4 ± 0.2 | 6.6 ± 1.1 | 0.72 ± 0.9 | 0.64 ± 0.02 |
| AWG7 INU 2 | Inulin | 10.5 ± 0.8 | 71.6 ± 0.9 | 1.4 ± 0.5 | 5.3 ± 1.0 | 0.79 ± 0.4 | 0.72 ± 0.02 |
| AWGZ INU 3 | Inulin | 12.0 ± 0.6 | 69.2 ± 1.4 | 1.4 ± 0.7 | 4.1 ± 0.4 | 0.72 ± 0.5 | 0.70 ± 0.01 |
| AWG7 INU I | Inulin | 11.0 ± 1.1 | 58.0 ± 4.1 | 1.4 ± 0.5 | 0.0 | 0.50 ± 0.7 | 0.60 ± 0.02 |
| AWG7 INU 5 | Inulin | 13.0 ± 0.9 | 65.6 ± 1.8 | 1.4 ± 0.5 | 3.7 ± 0.4 | 0.93 ± 0.3 | 0.66 ± 0.01 |
| AWG7 INU 6 | Inulin | 16.5 ± 1.2 | 64.6 ± 2.8 | 1.4 ± 0.4 | 6.1 ± 1.3 | 0.88 ± 0.5 | 0.65 ± 0.02 |
| AWG 7 INU 8 | Inulin | 15.8 ± 0.9 | 75.5 ± 1.5 | 1.5 ± 0.4 | 0.0 | 0.80 ± 0.7 | 0.76 ± 0.02 |

$Q_{CA}$ productivity of citric acid, $Y_{CA}$ yield of citric acid

that the concentration of the undesirable ICA remained for all transformants at very low level (less than 2 g dm$^{-3}$). The processes using different inulinase expressing strains differed significantly in terms of CA productivity and yield. While the best CA producer secreted this acid with the productivity of 0.8 g dm$^{-3}$ h$^{-1}$, the highest value of this parameter was reached by AWG7 INU5 strain (0.93 g dm$^{-3}$ h$^{-1}$). Additionally, although the CA yield was very different depending on the strain, all the obtained values were satisfactory, reaching 0.76 g g$^{-1}$ for AGW7 INU 8 strain. This value was 35% higher than the CA yield obtained by the parental AWG7 strain on fructose. In the process with IN as carbon source some of the analyzed strains produced also small amounts of mannitol into the culture broth, however, concentration of this by product remained at low level and did not exceed 7 g dm$^{-3}$.

Significant differences were also observed among the analyzed strains in terms of biomass production. The concentration of cells in the culture ranged from 9.0 g dm$^{-3}$ to 16.5 g dm$^{-3}$. The parental AWG7 strain grew only slightly better in medium with glucose compared to fructose based medium.

### RBC as a toll of process intensification

Two best CA producers identified in the previous stage of the study was chosen for further analysis. The AWG7 INU8 transformant was used due to its highest CA yield and AWG7 INU5 strain was chosen due to its highest CA productivity. Both strains were cultured using RBC system to increase the titer, yield and productivity obtained during traditional batch cultures. The kinetics of the RBC processes (changes of biomass, CA and fructose over time) for both transformants was presented on Fig. 2. There was no technical difficulties observed during the quintuple RBC experiment as well as no microbiological contamination occurred. CA biosynthesis using this system is stable and feasible new process. The whole process remained stable during 450 to 600 h. Medium exchange in the bioreactor (40% of working volume each time) was carried out four times (Figs. 2 and 3). The rate of medium exchange was chosen based on the previous work on erythritol production in RBC system [6].

The RBC processes for both strains differed slightly in terms of biomass biosynthesis (Fig. 1). Higher concentration of cells was observed for AWG7 INU5 strain (Fig. 1b), however, the biomass concentration dropped after the first medium exchange for both strains (Fig. 1). IN hydrolysis was occurred rapidly in both processes which can be seen as high increase of fructose concentration in the medium. Higher fructose utilization rate was reached by AWG7 INU5 strain (Fig. 1b) which resulted in 150 h shorter process compared to AWG7 INU8 transformant (Fig. 1a). In contrary, AWG7 INU8 strain secreted 30 g dm$^{-3}$ more of CA into the medium (203 g dm$^{-3}$) compared to AWG7 INU5 strain (170 g dm$^{-3}$). Furthermore, after two medium exchange, the concentration of CA was increasing for both strains.

The overall CA yield of the RBC process was higher for AWG7 INU 8 strain and reached 0.51 g g$^{-1}$ (Table 2). Albeit, the overall CA yield did not obtain very high values, the maximum partial yield reached 0.85 g g$^{-1}$ for AWG7 INU 8 strain in 2nd exchange of the medium (Table 2). The overall CA productivity was almost the same for both strains and reached 0.3 g dm$^{-3}$ h$^{-1}$ (Table 2). The highest CA productivity (0.51 g dm$^{-3}$ h$^{-1}$) during the RBC process was noted for AWG7 INU8 strain during the 2nd medium exchange (Table 2). The concentration of the undesired ICA during the whole RBC process for both transformants did not exceed 2 g dm$^{-3}$ (Table 2).

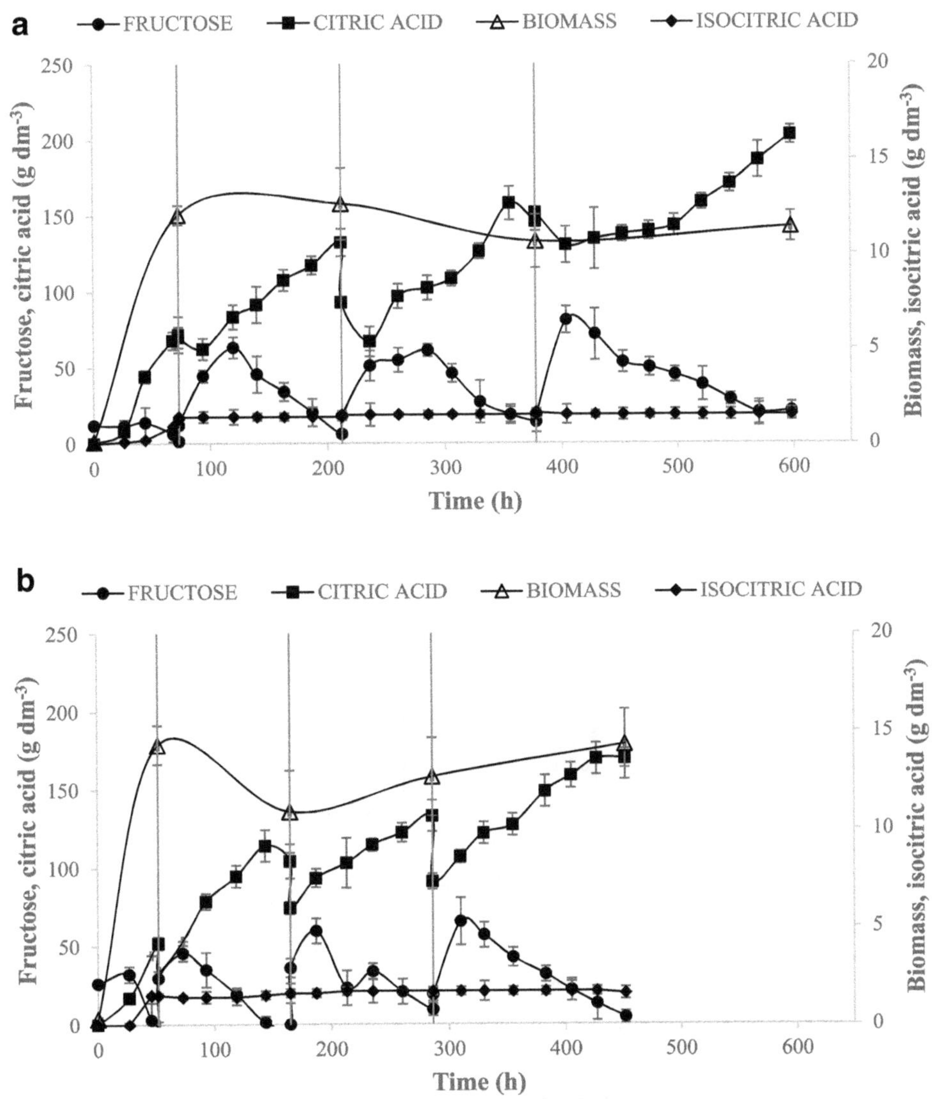


**Fig. 2** Time courses of biomass production (Δ), citric acid (■) and isociric (♦) acids titer and fructose consumption (•) during repeated batch culture of AWG7 INU 8 (**a**) and AWG7 INU 5 (**b**) strains of *Y. lipolytica*. 40% of medium was replacement for four times. Each variant of RBC was repeated two times

### Activity of inulinase displayed on the yeast cells

During RBC on IN, the activity of inulinase was measured. Both strains showed the activity of expressed enzyme (Fig. 4). The initial activity of inulinase was increased until the first replacement of the medium was done for both strains (Fig. 4) and then decreased when concentration of released fructose was decreased. The highest inulinase activity was noted for AWG7 INU 8 strain which reached 120 U $g^{-1}$ of cell dry mass (Fig. 4a). For AWG7 INU5 strain maximum activity of inulinase was reached after 320 h of the culture (Fig. 4b).

## Discussion

The CA is mostly produced organic acid using from molasses and fungi *A. niger.* Although the process is well known and used for decades, it shows some disadvantages which must be eliminated – it is multi-stages and dependent on the purity of the substrate. Therefore, the whole process generates some environmentally undesired wastes [17]. As shown previously, CA biosynthesis using *Y. lipolytica* may become an efficient alternative method if its production [36, 38]. The process with yeast is more favorable due to the higher resistance of the cells to high substrate concentration and their high resistance to metal ions. Furthermore, yeast are easy to culture and to establish stable continuous processes of desired metabolites production. Additional advantage of using yeast, especially *Y. lipolytica*, is the availability of genetic tools to engineer their metabolism and boost the production processes [39].

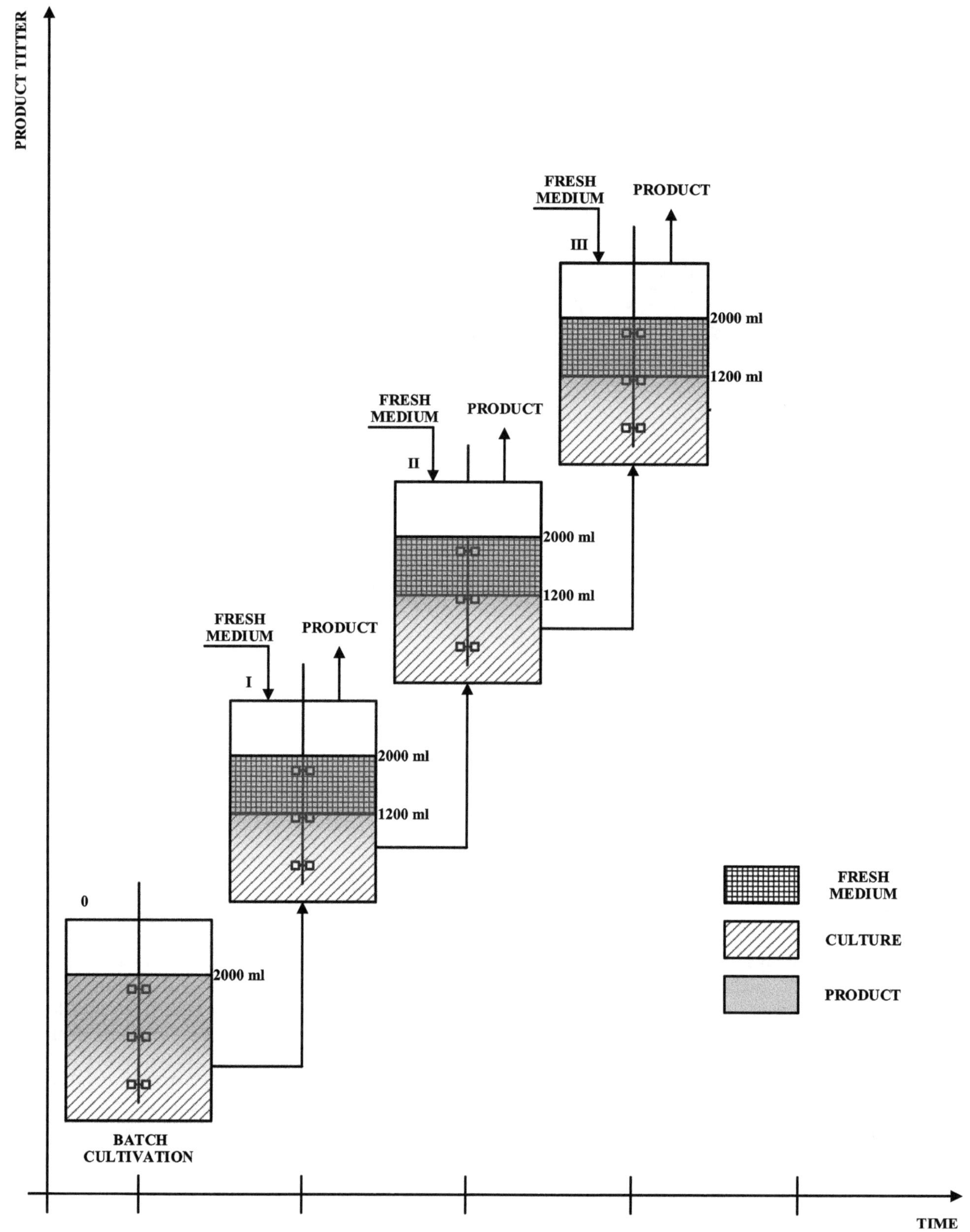


**Fig. 3** The scheme of increasing product titer during the repeated-batch culture used in this study

**Table 2** Citric acid productivity and yield from inulin for *Y. lipolytica* strains expressing inulinase gene in repeated batch culture

| Strain | $Q_{ov}$ ($g\,dm^{-3}\,h^{-1}$) | $Q_H$ | $Y_{ov}$ ($g\,g^{-1}$) | $Y_H$ |
|---|---|---|---|---|
| AWG7 INU8 | 0.34 ± 0.02 | 0.51 ± 0.04 | 0.51 ± 0.02 | 0.85 ± 0.03 |
| AWG7 INU5 | 0.37 ± 0.08 | 0.49 ± 0.07 | 0.42 ± 0.01 | 0.61 ± 0.01 |

$Q_{ov}$ overall productivity of citric acid production, $Y_{ov}$ overall yield of citric acid production during RBC, $Q_H$ the highest productivity noticed during the process, $Y_H$ the highest yield noticed during the process

The CA production by *Y. lipolytica* were based mainly on glucose and crude glycerol [33, 36, 40, 41]. However, the global increase in IN and prebiotic food manufacturing caused growing interest in utilization of IN for biotechnological processes which may help to reduce the amount of IN-rich wastes generated during the production process [30, 31, 42]. Application of IN for the processes with *Y. lipolytica* in concert with the use of RBC system could help to develop more efficient CA production technology. The use of the RBC system for CA biosynthesis is a very desirable solution. It offers application of high concentrated media without inhibition of cells growth due to the high concentration of biomass in the bioreactor. Furthermore, the secreted acid is partially removed from the tank which eliminated its toxicity to the cells.

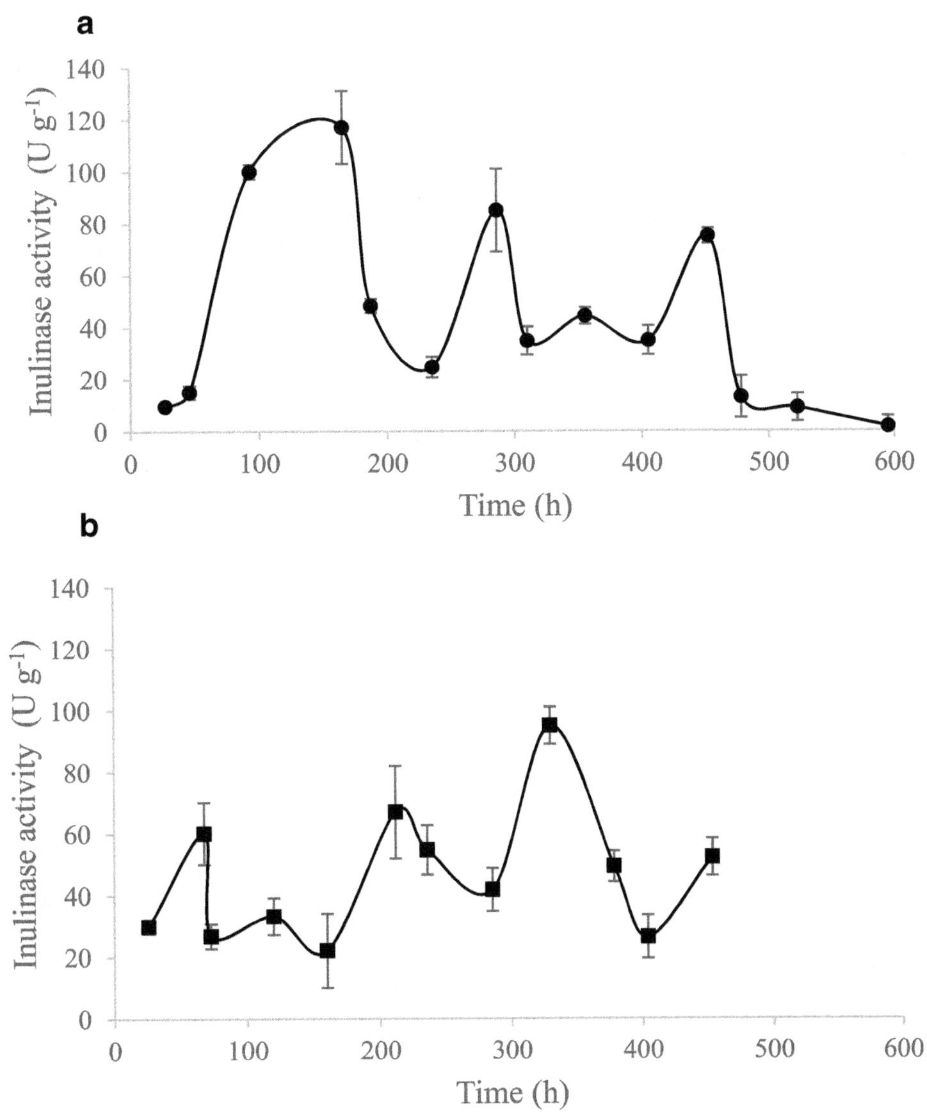


**Fig. 4** Time course of inulinase production by *Y. lipolytica* AWG7 INU 8 (**a**) and AWG7 INU5 (**b**) transformants during repated-batch culture on inulin

In the literature, many examples of CA production by *Y. lipolytica* using different cultivation systems can be found. Summary of the available results is presented in Table 3. The highest CA titer was obtained in the presented work by *Y. lipolytica* AWG7 INU 8 transformant during RBC process with inulin. The CA titer obtained by AWG7 INU 5 strain is also higher to these obtained for other strains growing on sunflower oil using batch cultures or on glycerol using fed-batch systems [16, 43, 44]. Furthermore, studies with the parental AWG7 strain using RBC system with glycerol [44] also did not reach the concentration of CA obtained by its inulinase positive transformants studied in the current work (Table 3). Nevertheless, the parental strain used in the previous work presented higher productivity during the RBC process [44]. In the study conducted by Liu et al. [42], genetic modified strain of *Y. lipolytica* growing on inulin in batch culture, produced CA almost with the same yield (0.84 g $g^{-1}$) as the strain AWG7 INU8 used in this study in RBC system.

Additional advantage of the process with inulinase positive transformants of AWG7 strain growing on inulin is the low level of the undesired ICA (below 2 g $dm^{-3}$, Tables 2 and 3). The formation of ICA is strain and substrate dependent and differs among different cultivation systems [43]. In general, the concentration of ICA secreted by different wild strains of *Y. lipolytica* growing on carbohydrates or glycerol ranges from 8 to 16% [45, 46]. Other *Y. lipolytica* strain - Wratislavia K1, also derivative of Polish A-101 strain, during fed-batch culture on IN secreted 6.6–12.9 g $dm^{-3}$ [31]. The high CA/ICA ratio obtained in the RBC process with IN during the presented study will facilitate the process of its purification from the post-culture medium. The transformant AWG7 INU8 can produce 120 U $g^{-1}$ of inulinase activity. Liu et al. [30] determined the activity of immobilized inulinase with 6× His-tag to be 22.6 U $mg^{-1}$ of cell dry mass after 96 h of the cell growth for *Y. lipolytica* expressing the inulinase gene.

## Conclusion

In conclusion, the growing IN market and increasing amount of IN-rich wastes prompt to find new ways of their utilization. The presented study showed great potential of genetically engineered strains of *Y. lipolytica* AWG7 expressing inulinase to efficiently convert IN to highly desired CA. Furthermore, application of the RBC system allowed to increase the CA titer to 200 g $dm^{-3}$, which is, to our knowledge, the highest concentration of CA reported for *Y. lipolytica*.

## Methods

### Strains

The parental strain *Y. lipolytica* AWG7 used in this study was isolated from *Y. lipolytica* A-101-1.31 strain after its exposure to UV irradiation [47]. The strain belongs to the Yeast Culture Collection of the Department of Biotechnology and Food Microbiology, Wroclaw University of Environmental and Life Sciences, Poland. The strain has been deposited in a culture collection CIRM - Levures (INAG 36017) under number CLIB 81. This strain is a very efficient CA producer and is not able to form mycelium (data not shown). The new derivatives of AWG7 strain (AWG7 INU 1, AWG7 INU 2, AWG7 INU 3, AWG7 INU I, AWG7 INU 5, AWG7 INU 6, AWG7 INU 8) expressing the inulinase gene

**Table 3** Citric acid production by various strains of *Yarrowia lipolytica*

| Strain | System | Substrate | Citric acid (g $dm^{-3}$) | Q (g $dm^{-3}$ $h^{-1}$) | Y (g $g^{-1}$) | Reference |
|---|---|---|---|---|---|---|
| W29 | batch | glucose | 18 | 0.13 | 0.62 | [18] |
| N15 | batch | sunflower oil | 150 | 1.32 | 1.56 | [43] |
| NG40/UV7 | batch | glycerol | 115 | 0.6 | 0.64 | [33] |
| 187/1 | batch | rapeseed oil | 135 | 0.94 | 1.55 | [53] |
| H222-S4(p67ICL1)T5 | fed-batch | sucrose | 140 | 0.73 | 0.82 | [46] |
| A 101–1.31 | fed-batch | glycerol | 155 | 1.05 | 0.6 | [35] |
| AWG7 | repeated-batch | glycerol | 154 | 0.94 | 0.6 | [35] |
| AWG7 | repeated-batch | glycerol | 154 | 1.05 | 0.78 | [46] |
| A101–1.22 | repeated-batch | glycerol | 124 | 0.85 | 0.6 | [41] |
| SWJ-1b | batch | inulin | 68.9 | 0.22 | 0,68 | [30] |
| SWJ-1b 30 | batch | inulin | 84 | 0.39 | 0.84 | [42] |
| AWG7 | continuous | glycerol | 97.8 | 0.49 | 0.98 | [36] |
| AWG7 INU 5 | repeated-batch | inulin | 170 | 0.49[a] | 0.61[b] | This study |
| AWG7 INU 8 | repeated-batch | inulin | 203 | 0.51[a] | 0.85[b] | This study |

[a] the highest productivity noticed during the process' [b] the highest yield noticed during the process
Q – productivity of citric acid, Y – yield of citric acid production

from *Kluyveromyces marxianus* CBS6432 (*INU1* gene; GenBank: X57202.1) were constructed according to the procedure described previously [31]. *Y. lipolytica* transformants were selected on the YPD medium with 400 μg $cm^{-3}$ of hygromycin B. Verification of the transformants was performed both, by PCR as well as growth detection on solid minimal medium with 1% of IN. Seven randomly selected transformants were chosen for further analysis of citric acid production in bioreactors. All yeast strains used in this study were stored at − 80 °C and refreshed before use by 24 h culture in YPD medium at 28 °C, 2.33 Hz.

#### Media and culture conditions

IN (Mlyn Oliwski, Gdańsk, Poland), glucose and fructose (Sigma-Aldrich) were used as sources of carbon and energy. Inoculum medium consisted of glucose – 50 g, yeast extract – 3 g (Sigma-Aldrich), malt extract – 3 g (Sigma-Aldrich) and peptone – 5 g (Sigma-Aldrich) in distilled water – 1 $dm^3$. The yeast were grown in 0.3-$dm^3$ flasks containing 0.05 $dm^3$ of the medium on a rotary shaker (CERTOMAT IS, Sartorius Stedim Biotech), at 30 °C, 140 rpm for 72 h, as described previously [48, 49]. In order to inoculate media for bioreactor experiments, 0.2 $dm^3$ of inoculum (10% of the working volume) was used.

#### Determination of IN utilization using microcultures

Inoculation medium (YPD) contained: glucose – 20 g, yeast extract – 10 g and bacteriological peptone – 20 g in 1 $dm^3$ of distilled water. The inoculation cultures were grown for 24 h at 28 °C, 2.33 Hz on a rotary shaker (CERTOMAT IS, Sartorius Stedim Biotech).

The growth of *Y. lipolytica* AWG7 and its derivatives was analyzed in Bioscreen C system (Oy Growth Curves Ab Ltd., Finland), in Yeast Nitrogen Base medium (Sigma-Aldrich) supplemented with 2% (w $v^{-1}$) of glucose, fructose or IN. The overnight inoculation cultures were centrifuged and washed twice with sterile water. The yeast strains grew in 150 μl of the appropriate medium using 100-well microplates. The optical density ($OD_{600}$) of the cells was standardized to 0.15. The experiments were performed in quintuples at 28 °C under constant, intensive agitation. The growth of cells was monitored by measuring the OD at 420–600 nm every 30 min for 48 h. Rakicka et al. [49] have published this methodology previously.

#### CA biosynthesis in bioreactors

The differences in CA production among the 8 investigated strains were tested using batch cultures in medium containing: inulin – 100 g, $NH_4Cl$ – 2 g, $KH_2PO_4$–0.25 g, $MgSO_4 \times 7H_2O$ – 1.0 g, yeast extract – 1 g in 1 $dm^3$ of tap water [35]. The parental strain (AWG7) was cultivated on glucose and fructose as a control experiment.

The intensification of CA production was performed by applying the repeated batch strategy (RBC) for only two best CA producers – AWG7 INU 5 and AWG7 INU 8 transformants. The medium used for CA biosynthesis was as described above. After the whole substrate was exhausted from the medium, 800 $cm^3$ of the culture was withdrawn, and replaced by the same volume of fresh medium. This procedure was repeated three times (Fig. 3). During each RBC culture the working volume at the beginning of the process was maintained at 2 $dm^3$. The end of each RBC cycle was determined when the concentration of fructose dropped below 20 g $dm^{-3}$.

All bioreactor cultures were performed in 5-$dm^3$ stirred-tank reactor (BIOSTAT B-PLUS, Sartorius, Germany) with the working volume of 2.0 $dm^3$ at 28 °C. Aeration and stirring rates were set at 0.6 $m^{-1}$ and 13.3 Hz, respectively. The pH 5.5 was maintained by additions of 5 M NaOH solution, as described previously [48, 50]. All cultures were cultivated until complete exhaustion of the carbon source. Bioreactor with the appropriate medium was autoclaved at 121 °C for 20 min. All cultures were conducted in two biological replicates.

#### Analytical methods

Ten milliliters of culture broth was centrifuged (5 min, 2700 RCF). The biomass was washed twice with distilled water and filtered on 0.45 mm pore-size membranes. The biomass concentration was determined gravimetrically after drying at 105 °C and expressed in grams of cell dry weight per liter (g $dm^{-3}$). The concentration of substrate (glucose, fructose) and CA was measured in the supernatants by high-performance liquid chromatography (Dionex- Thermo Fisher Scientific, UK) using a Carbohydrate H+ Column (Thermo Scientific, Waltham, MA) coupled to a UV detector (k = 210 nm) and refractive index detector (Shodex, Ogimachi, Japan). The column was eluted with 25 mM trifluoroacetic acid at 65 °C and a flow rate of 0.6 $cm^3$ $min^{-1}$. The concentration of isocitric acid (ICA) in the supernatant was determined using an Isocitrate Assay Kit (Sigma-Aldrich). This methodology has been described previously by Rakicka et al. [48, 50].

Inulinase activity of the strains was determined according to Gong et al. [51]. The reaction mixture containing 0.1 $cm^3$ of the supernatant, 0.9 $cm^3$ of phosphate buffer (0.1 M, pH 6.0) and 1% (mass per volume) of IN (Sigma) was incubated at 50 °C for 15 min. The reaction was inactivated immediately by keeping the reaction mixture at 100 °C for 10 min. The amount of reducing sugar in the reaction mixture was assayed by the method introduced by Miller [52]. One unit of inulinase activity (U) was defined as the amount of enzyme that releases 1 μM of reducing sugar per minute nder the assay conditions applied in this study. The specific inulinase activity was calculated as units per gram ($U\,g^{-1}$) of cell dry mass.

### Calculation of fermentation parameters and list of abbreviations

The yield of CA production from IN (Y), expressed in $g\,g^{-1}$, was calculated using the formula:

$$Y = \frac{CA}{IN}$$

The productivity of CA in batch and repeated-batch culture (Q), expressed in $g\,dm^{-3}\,h^{-1}$, was calculated as:

$$Q = \frac{CA}{\Delta t}$$

In all formulas, CA stands for citric acid concentration, IN stands for the consumed inulin concentration in the culture ($g\,dm^{-3}$), t is the duration of the fermentation process (h).

#### Abbreviations

CA: Citric acid; ICA: Isocitric acid; IN: Inulin; RBC: Repeated batch culture

#### Acknowledgements

Not applicable.

#### Funding

This study was supported by Wroclaw Centre of Biotechnology, programme The Leading National Research Centre (KNOW) for years 2014–2018.

#### Authors' contributions

Conceived, designed and performed the experiments: MR, JW, ZL, WR Analyzed the data: MR, ZL, and WR. Wrote the paper: MR, ZL. All authors reviewed the manuscript. All authors read and approved the final manuscript.

#### Ethics approval and consent to participate

Not applicable.

#### Consent for publication

Not applicable.

#### Competing interests

The authors declare that they have no competing interests.

#### References

1. Sauer M, Porro D, Mattanovich D, Branduardi P. Microbial production of organic acids: expanding the markets. Trends Biotechnol. 2008;26(2):100–8.
2. Papagianni M. Advances in citric acid fermentation by *Aspergillus niger*: biochemical aspects, membrane transport and modelin. Biotechnol Adv. 2007;25(3):244–63.
3. Cavallo E, Charreau H, Cerrutti P, Foresti ML. *Yarrowia lipolytica*: a model yeast for citric acid production. FEMS Yeast Res. 2017;17:84.
4. Shahidi F, Zhong Y. Antioxidants: regulatory status. Bailey's Industrial Oil Fat Products. 2005;1:491–512.
5. Soccol CR, Vandenberghe LPS, Rodrigues C, Pandey A. New perspectives for citric acid production and application. Food Technol Biotech. 2006;44(2): 141–9.
6. Nemet A, Varbanov PS, Klemes JJ. Cleaner production, process integration and intensification. Clean Technol Envir. 2016;18(7):2019–35.
7. Mirończuk AM, Dobrowolski A, Rakicka M, Rywińska A, Rymowicz W. Newly isolated mutant of *Yarrowia lipolytica* MK1 as a proper hostfor efficient erythritol biosynthesis from glycerol. Process Biochem. 2015;50:61–8.
8. Araujo K, Romero A, Chirinos P, Paéz G, Mármol Z, Rincón M. Ethanol production from whey using fed-batch culture with repeated cycles. Interciencia. 2015;40(5):305–10.
9. Dashti MG, Abdeshahian P. Batch culture and repeated-batch culture of *Cunninghamella bainieri* 2A1 for lipid production as a comparative study. Saudi J Biol Sci. 2016;23(2):172–80.
10. Abdel-Mawgoud AM, Markham KA, Palmer CM, Liu N, Stephanopoulos G, Alper HS. Metabolic engineering in the host *Yarrowia lipolytica*. Metab Eng. 2018;50:192–208.
11. Sabra W, Bommareddy RR, Maheshwari G, Papanikolaou S, Zeng AP. Substrates and oxygen dependent citric acid production by *Yarrowia lipolytica*: insights through transcriptome and fluxome analyses. Microb Cell Factories. 2017;16:1–14.
12. Cui Z, Gao C, Li J, Hou J, Lin CSK, Qi Q. Engineering of unconventional yeast *Yarrowia lipolytica* for efficient succinic acid production from glycerol at low pH. Metab Eng. 2017;42:126–33.
13. Janek T, Dobrowolski A, Biegalska A, Mirończuk AM. Characterization of erythrose reductase from *Yarrowia lipolytica* and its influence on erythritol synthesis. Microb Cell Factories. 2017;16:1–13.
14. Mirończuk AM, Biegalska A, Dobrowolski A. Functional overexpression of genes involved in erythritol synthesis in the yeast *Yarrowia lipolytica*. Biotechnol Biofuels. 2017;10:1–12.
15. Larroude M, Celińska E, Back A, Thomas S, Nicaud JM, Ledesma-Amaro R. A synthetic biology approach to transform *Yarrowia lipolytica* into a competitive biotechnological producer of β-carotene. Biotechnol Bioeng. 2018;115:464–72.
16. André A, Chatzifragkou A, Diamantopoulou P, Sarris D, Philippoussis A, Galiotou-Panayotou M, Komaitis M, Papanikolaou S. Biotechnological conversions of bio-diesel-derived crude glycerol by *Yarrowia lipolytica* strains. Eng Life Sci. 2009;9(6):468–78.
17. Rywińska A, Rymowicz W. High-yield production of citric acid by *Yarrowia lipolytica* on glycerol in repeated-batch bioreactors. J Ind Microbiol Biotechnol. 2010;37:431–5.
18. Kamzolova S, Fatykhova A, Dedyukhina E, Anastassiadis SG, Golovchenko NP, Morgunov IG. Citric acid production by yeast grown on glycerol-containing waste from biodiesel industry. Food Technol Biotech. 2011;49:65–6.
19. Papanikolaou S, Chatzifragkou A, Fakas S, Galiotou-Panayotou M, Komaitis M, Nicaud J-M, Aggelis G. Biosynthesis of lipids and organic acids by *Yarrowia lipolytica* strains cultivated on glucose. Eur J Lipid Sci Tech. 2009;111(12):1221–32.
20. Żarowska B, Wojtatowicz M, Rymowicz W, Robak M. Production of citric acid on sugar beet molasses by single and mixed cultures of *Yarrowia lipolytica*. Electron J Pol Agric Univ Biotechnology. 2001;4(2):1–8.
21. Liu X, Lv J, Zhang T, Deng Y. Citric acid production from hydrolysate of pretreated straw cellulose by *Yarrowia lipolytica* SWJ-1b using batch and fed-batch cultivation. Prep Biochem Biotech. 2015;45:825–35.
22. Muthuselvi S, Sathishkumar T, Kumaresan K, Rajeshkumar M. Improved inulinase activity by *Penicillium purpurogenum* grown in microwave pretreated coffee spent by L16 orthogonal design of experiment. Innov Rom Food Biotechnol. 2012;11:44–50.
23. Slavin J. Fiber and prebiotics: mechanisms and health benefits. Nutrients. 2013;5(4):1417–35.
24. Barclay T, Ginic-Markovic M, Cooper P, Petrovsky N. Inulin - a versatile polysaccharide with multiple pharmaceutical and food chemical uses. J Excipients Food Chem. 2010;1(3):27–50.
25. Ronkart SN, Paquot M, Blecker CS, Fougnies C, Lynn D, Lambrechts JC, Norberg B, Deroanne C. Impact of the crystallinity on the physical properties of inulin during water sorption. Food Biophys. 2009;4:49–58.

26. Korbelik M, Cooper PD. Potentiation of photodynamic therapy of cancer by complement: the effect of gamma-inulin. Br J Cancer. 2007;96:67–72.
27. Cooper PD, Steele EJ. The adjuvantiity of gamma inulin. Immunol Cell Biol. 1988;66:345–52.
28. Chi ZM, Zhang T, Cao T-S, Liu X-Y, Cui W, Zhao C-H. Biotechnological potential of inulin for bioprocesses. Bioresour Technol. 2011;102(6):4295–303.
29. Gao J, Yuan W, Li Y, Xiang R, Hou S, Zhong S, Bai F. Transcriptional analysis of *Kluyveromyces marxianus* for ethanol production from inulin using consolidated bioprocessing technology. Biotechnol Biofuels. 2015;8:115.
30. Liu XY, Chi Z, Liu GL, Wang F, Madzak C, Chi ZM. Inulin hydrolysis and citric acid production from inulin using the surface-engineered *Yarrowia lipolytica* displaying inulinase. Metab Eng. 2010;12(5):469–76.
31. Rakicka M, Lazar Z, Rywińska A, Rymowicz W. Efficient utilization of inulin and glycerol as fermentation substrates in erythritol and citric acid production using *Yarrowia lipolytica* expressing inulinase. Chem Pap. 2016; 70(11):1452–9.
32. Hapeta P, Rakicka M, Dulermo R, Gamboa-Meléndez H, Cruz-Le Coq A-M, Nicaud J-M, Lazar Z. Transforming sugars into fat – lipid biosynthesis using different sugars in *Yarrowia lipolytica*. Yeast. 2017;34:293–304.
33. Morgunov IG, Kamzolova SV, Lunina JN. The citric acid production from raw glycerol by *Yarrowia lipolytica* yeast and its regulation. Appl Microbiol Biotechnol. 2013;97:73–87.
34. Lazar Z, Walczak E, Robak M. Simultaneous production of citric acid and invertase by *Yarrowia lipolutica* SUC+ transformants. Bioresour Technol. 2011;102:6982–9.
35. Rywińska A, Rymowicz W, Żarowska B, Skrzypiński A. Comparison of citric acid production from glycerol and glucose by different strains of *Yarrowia lipolytica*. World J Microbiol Biotechnol. 2010;26:1217–24.
36. Rywińska A, Juszczyk P, Wojtatowicz M, Rymowicz W. Chemostat study of citric acid production from glycerol by *Yarrowia lipolytica*. J Biotechnol. 2011;152:54–7.
37. Schmid-Berger N, Schmid B, Barth G. Ylt1, a highly repetitive retrotransposon in the genome of the dimorphic fungus *Yarrowia lipolytica*. J Bacteriol. 1994;176:2477–82.
38. Papanikolaou S, Galiotou-Panayotou M, Fakas S, Komaitis M, Aggelis G. Citric acid production by *Yarrowia lipolytica* cultivated on olive-mill wastewater-based media. Bioresour Technol. 2008;99(7):2419–28.
39. Fu G, Lu Y, Chi Z, Liu G-L, Zhao S-F, Jiang H, Chi Z-M. Cloning and characterization of a pyruvate carboxylase gene from *Penicillium rubens* and overexpression of the gene in the yeast *Yarrowia lipolytica* for enhanced citric acid production. Mar Biotechnol. 2016;18:1–14.
40. Papanikolaou S, Muniglia L, Chevalot I, Aggelis G, Marc I. *Yarrowia lipolytica* as a potential producer of citric acid from raw glycerol. J App Microbiol. 2002;92(4):737–44.
41. Rymowicz W, Fatykhova AR, Kamzolova SV, Rywińska A, Morgunov IG. Citric acid production from glycerol-containing waste of biodiesel industry by *Yarrowia lipolytica* in batch, repeated batch, and cell recycle regimes. Appl Microbiol Biot. 2010;87(3):971–9.
42. Liu XY, Chi Z, Liu GL, Madzak C, Chi ZM. Both decrease in ACL1 gene expression and increase in ICL1 gene expression in marine-derived yeast *Yarrowia lipolytica* expressing INU1 gene enhance citric acid production from inulin. Mar Biotechnol. 2013;15(1):26–36.
43. Kamzolova S, Finogenova T, Morgunov IG. Microbiological production of citric and isocitric acids from sunflower oil. Food Technol Biotech. 2008;46: 51–9.
44. Rywińska A, Rymowicz W. Citric acid production from raw glycerol by *Yarrowia lipolytica* Wratislavia 1.31. In: Aggelis G, editor. Microbial conversions of raw glycerol. New York: Nova Science Publishers Inc.; 2009. p. 19–30.
45. Finogenova TV, Morgunov IG, Kamzolova SV, Chernyavskaya OG. Organic acid production by the yeast *Yarrowia lipolytica*: a review of prospects. Appl Biochem Microbio. 2005;1:418–25.
46. Förster A, Jacobs K, Juretzek T, Mauersberger S, Barth G. Overexpression of the ICL1 gene changes the product ratio of citric acid production by *Yarrowia lipolytica*. Appl Microbiol Biot. 2007;77:861–9.
47. Rywińska A, Wojtatowicz M, Wielebińska A. Obtaining of fil$^-$ mutants in *Yarrowia lipolytica* yeast for citric acid production. Acta Sci Pol Biotechnologia. 2003;2(1–2):11–20.
48. Rakicka M, Rukowicz B, Rywińska A, Lazar Z. RymowiczW. Technology of efficient continuous erythritol production from glycerol. J Clean Prod. 2016; 139:905–13.
49. Rakicka M, Biegalska A, Rymowicz W, Dobrowolski A, Mirończuk AM. Polyol production from waste materials by genetically modified *Yarrowia lipolytica*. Bioresour Technol. 2017;243:393–9.
50. Rakicka M, Rywińska A, Lazar Z, Rymowicz W. Two-stage continuous culture – technology boosting erythritol production. J Clean Prod. 2016;168:420–7.
51. Gong F, Sheng J, Chi Z, Li J. Inulinase production by a marine yeast Pichia guilliermondii and inulin hydrolysis by the crude inulinase. J Ind Microbiol Biotechnol. 2007;34:179–85.
52. Miller GL. Use of dinitrosalicylic acid reagent for determination of reducing sugar. Anal Chem. 1959;31:426–8.
53. Kamzolova S, Morgunov IG, Perevoznikova OA, Shishkanova NV, Stottmeister U, Finogenova T. Lipase secretion and citric acid production in *Yarrowia lipolytica* yeast grown on animal and vegetable fat. Food Technol Biotech. 2005;43(2):113–22.

17

# Identification and characterization of the *GmRD26* soybean promoter in response to abiotic stresses: potential tool for biotechnological application

Elinea O. Freitas[1,2†], Bruno P. Melo[1,3†], Isabela T. Lourenço-Tessutti[1], Fabrício B. M. Arraes[1,4], Regina M. Amorim[1], Maria E. Lisei-de-Sá[1,5], Julia A. Costa[1,6], Ana G. B. Leite[1,2], Muhammad Faheem[1,7], Márcio A. Ferreira[8], Carolina V. Morgante[1,9], Elizabeth P. B. Fontes[3] and Maria F. Grossi-de-Sa[1,6*]

**Abstract**

**Background:** Drought is one of the most harmful abiotic stresses for plants, leading to reduced productivity of several economically important crops and, consequently, considerable losses in the agricultural sector. When plants are exposed to stressful conditions, such as drought and high salinity, they modulate the expression of genes that lead to developmental, biochemical, and physiological changes, which help to overcome the deleterious effects of adverse circumstances. Thus, the search for new specific gene promoter sequences has proved to be a powerful biotechnological strategy to control the expression of key genes involved in water deprivation or multiple stress responses.

**Results:** This study aimed to identify and characterize the *GmRD26* promoter (p*GmRD26*), which is involved in the regulation of plant responses to drought stress. The expression profile of the *GmRD26* gene was investigated by qRT-PCR under normal and stress conditions in Williams 82, BR16 and Embrapa48 soybean-cultivars. Our data confirm that *GmRD26* is induced under water deficit with different induction folds between analyzed cultivars, which display different genetic background and physiological behaviour under drought. The characterization of the *GmRD26* promoter was performed under simulated stress conditions with abscisic acid (ABA), polyethylene glycol (PEG) and drought (air dry) on *A. thaliana* plants containing the complete construct of p*GmRD26::GUS* (2.054 bp) and two promoter modules, p*GmRD26A::GUS* (909 pb) and p*GmRD26B::GUS* (435 bp), controlling the expression of the β-glucuronidase (*uidA*) gene. Analysis of GUS activity has demonstrated that p*GmRD26* and p*GmRD26A* induce strong reporter gene expression, as the p*AtRD29* positive control promoter under ABA and PEG treatment.

**Conclusions:** The full-length promoter p*GmRD26* and the p*GmRD26A* module provides an improved *uidA* transcription capacity when compared with the other promoter module, especially in response to polyethylene glycol and drought treatments. These data indicate that p*GmRD26A* may become a promising biotechnological asset with potential use in the development of modified drought-tolerant plants or other plants designed for stress responses.

**Keywords:** Stress-responsive promoter, Drought tolerance, Abscisic acid, Promoter modules analysis, Gene-promoter characterization

* Correspondence: mariafatimagrossidesa@gmail.com; fatima.grossi@embrapa.br
[†]Elinea O. Freitas and Bruno P. Melo contributed equally to this work.
[1]Embrapa Genetic Resources and Biotechnology, Brasília, DF, Brazil
[6]Catholic University of Brasilia - Post-Graduation Program in Genomic Sciences and Biotechnology, Brasília, DF, Brazil

## Background

Drought is one of the most limiting and severe abiotic stresses for field crops because it causes significant losses in plants production on a global scale [1, 2]. Under drought conditions, plants trigger many physiological, biochemical, and molecular responses. The sign of abiotic stress is perceived by cellular receptors and secondary messengers, culminating in the gene expression reprogramming to improve plant tolerance, adaptation, and survival. In signal transduction cascade, transcription factors (TFs) emerge as one of the most important messengers in plant adaptation, because they are capable of modifying specific gene expression, encompassing different physiological changes [3, 4]. For example, abiotic factors such as drought, salinity, and heat (high evaporation) alter the osmotic balance in plants' cell, inducing the biosynthesis of abscisic acid (ABA), a vital phytohormone involved in the expression of drought-related genes [5, 6].

Several TFs are involved in water stress tolerance, including ABA-responsive element (ABRE), nitrogen assimilation control (NAC), dehydration-responsive element binding (DREB), basic leucine zipper (bZIP), myeloblastosis (MYB) and myelocytomatosis (MYC) proteins. All of these TFs are mediators of the classic ABA-dependent or ABA-independent signaling pathways [7–11]. These transcription factors bind preferentially to the dehydration-responsive element (DRE) core sequence (A/GCCGAC) of gene-responsive promoters and regulate several stress-induced genes [12]. The DRE sequence is present in the *A. thaliana RD29A* (*AtRD29*) promoter region and is used extensively to drive expression in a stress-inducible manner in different plants, such as tobacco [13], potato [14], and soybean [15].

The expression of *DREB1A* under the control of the *AtRD29* promoter in *A. thaliana* increased the survival rate of plants stressed with freezing, drought, high salinity, and high temperature [16]. Similar results were also observed in tobacco plants [13]. In both cases, the use of inducible promoter *AtRD29* displays a higher gene expression than the Cauliflower mosaic virus 35S (CaMV35S) constitutive promoter, as it reduces the pleiotropic effects on growth due to the overexpression of *DREB1A* [13, 16]. The expression of genes of interest under the control of the *AtRD29* promoter has been widely used to regulate drought tolerance-associated genes in different plant species [16–19]. However, compared with *AtRD29*, some specific inducible promoters can achieve higher levels of expression. An example is the promoter of the *Coffea arabica CaHB12* gene [20]. For this reason, the identification, isolation, and characterization of new specific promoters inducible by abiotic stress have been crucial to ensure the successful application of gene modification and, consequently, the development of new cultivars resistant to water deprivation stress.

In this study, we have investigated the gene expression profile of *GmRD26* in soybean (*Glycine max*), homologous to the *A. thaliana AtRD26* gene (ANAC072). *GmRD26* is highly induced by ABA, PEG, and drought. According to our gene expression analysis, it displays a similar expression profile in comparison with *AtRD26* and *GmNAC085*, a soybean *GmRD26* gene-paralogue already characterized [11, 21]. These genes belong to SNAC-A subfamily (ATAF), as well as *GmRD29*, the *AtRD29* orthologue, extensively used as a model of drought-inducible genes. Previous studies of the GENOSOJA project have demonstrated that *GmRD29* was not differentially expressed during severe water deprivation as *GmRD26* [22], selected as the focus for this study.

Many *SNAC-A* genes are involved with different abiotic stress responses and senescence progression. In soybean, 44% of NAC genes are differentially expressed (DE) during age triggered senescence, being 90% of genes from SNAC-A subfamily [21]. In *A. thaliana*, all SNAC-A gene members - *ANAC055* (AT3G15500), *ANAC019* (AT1G52890), *ANAC072/RD26* (AT4G27410), *ANAC002/ATAF1* (AT1G 01720), *ANAC081/ATAF2* (AT5G08790), *ANAC102* (AT5 G63790), and *ANAC032* (AT1G77450) - are induced by age triggered leaf senescence [21, 23]. *At*RD26 acts as a transcriptional activator in ABA-mediated dehydration response, positively regulating NYE1, which triggers chlorophyll degradation [24]. The *GmRD26* paralogue in soybean (*GmNAC085*) is also a positive regulator of leaf senescence, displaying high expression during age triggered senescence and classic senescence symptoms when transiently expressed in *Nicotiana benthamiana* [21].

We subsequently isolated and characterized the *GmRD26* promoter (p*GmRD26*). The transcriptional activity of p*GmRD26* and its modules were evaluated in transgenic *A. thaliana* plants under the stress conditions with abscisic acid (ABA), polyethylene glycol (PEG) and drought (air dry) to evaluate the activities of the different regions of the p*GmRD26*.

## Results

### Soybean *RD26* gene expression profile in distinct soybean lines under different stress conditions

To identify and characterize the orthologue gene of *A. thaliana AtRD26* (ANAC072) in soybean, an in-silico approach was applied. The *AtRD26* (AT4G27410) sequence was accessed and compared against the Williams 82 soybean reference genome. A putative *RD26* orthologue (*GmNAC043* – Glyma.06G248900) was identified by neighbour-joining analysis, which revealed that at least four genes (Glyma.13G279900, Glyma.12G221500, Glyma.06G248900, and Glyma.12G149100), closely related to *AtRD26*, are present in the soybean genome (Fig. 1). A comparative amino acid deduced sequence analysis of candidate genes was performed, and *GmNAC043*, called *GmRD26*, (Glyma.06G248900) displayed a relatively high amino acid similarity with *AtRD26*.

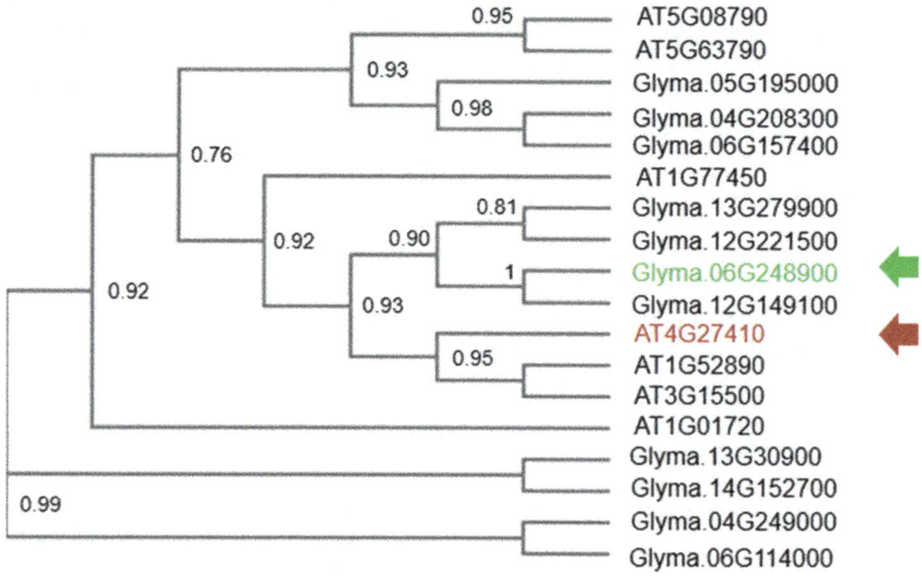


**Fig. 1** Phylogenetic reconstruction of ATAF soybean genes, members of the NAC transcription factor subfamily. The deduced amino acid sequences of soybean and Arabidopsis were used to perform a multiple alignment using BLASTP and ClustalW2. The phylogenetic tree was constructed using MEGA4.0 software via the neighbour-joining method with a consensus of 10.000 bootstraps. The red arrow indicates the orthologue *A. thaliana* reference gene (AT4G27410), and the green arrow indicates *GmRD26* (Glyma.06G248900)

To evaluate whether the *GmRD26* soybean transcription factor is induced during water stress, its expression pattern was analyzed in cDNA subtractive libraries related to dissection experiments available in the GENOSOJA LGE (Genomics and Expression Laboratory: GENOSOJA Project) database and from these analyses, the presence of *GmRD26* was confirmed. To evaluate the *GmRD26* expression profile and its relation to multiple stress responses in soybean, the transcript levels were analyzed in the leaves and roots of Williams 82 soybean seedlings by qRT-PCR. The expression pattern was also evaluated for *GmNAC085*, a paralogue of *GmRD26* gene whose stress induction profile is reported previously [21]. As expected, *GmRD26* is highly expressed under the use of PEG (10% m/v) in leaves but is also induced by ABA (150 mM) and drought in leaves and roots (Fig. 2). This gene expression profile is similar to the *GmNAC085* expression (Additional file 1: Figure S1). In addition, both related genes are repressed by tunicamycin (Tun) in leaves and roots, showing an expressive induction by salicylic acid (SA) (5 mM) treatment in roots (Fig. 2; Additional file 1: Figure S1).

The expression profile of *GmRD26* was also determined in two contrasting soybean genotypes in response to drought tolerance under simulated drought stress (Fig. 3) and ABA exogenous stimuli (Fig. 3a). It is expected that positive regulators of drought perception, signal transduction, and drought avoidance-associated genes are expressed higher in tolerant lineages than in susceptible lineages, as shown in the gene expression analysis results. In addition, the gene expression-folding is extensively high in BR16 and Embrapa48 cultivars when compared with the expression in Williams 82.

*GmRD26* was differentially expressed in both leaves and roots of contrasting cultivars, and the tissues display a similar induction pattern as observed in Williams 82 under PEG (10% m/v) stress (Figs. 2 and 3). In the roots, the gene expression was considerably lower than in the leaves. The difference between the cultivars is the gene expression levels: the susceptible cultivar BR16 had a significantly lower *GmRD26* transcript accumulation in comparison with the tolerant cultivar Embrapa48 at all times of stress progression (Fig. 3). The gene expression significantly increased beginning at 125 min, showing that the *GmRD26* gene is strongly induced under severe stress conditions. The ABA response was also analyzed. As observed in *Arabidopsis*, the results revealed that *GmRD26* is also up-regulated by ABA in both soybean cultivars and the mRNA levels are significantly higher in tolerant cultivar Embrapa48, as observed in drought treatment (Fig. 3a and b).

### Analysis of water deficit-responsive *cis*-elements frequency

To investigate the transcriptional activity of the *GmRD26* soybean promoter under different stress conditions, the full-length promoter sequence (2.054 bp) was analyzed using PLACE and Genomatix for *cis*-regulatory element mapping. The promoter sequence analysis revealed some conserved TATA- and CAAT-box regions that are essential for transcription initiation complex

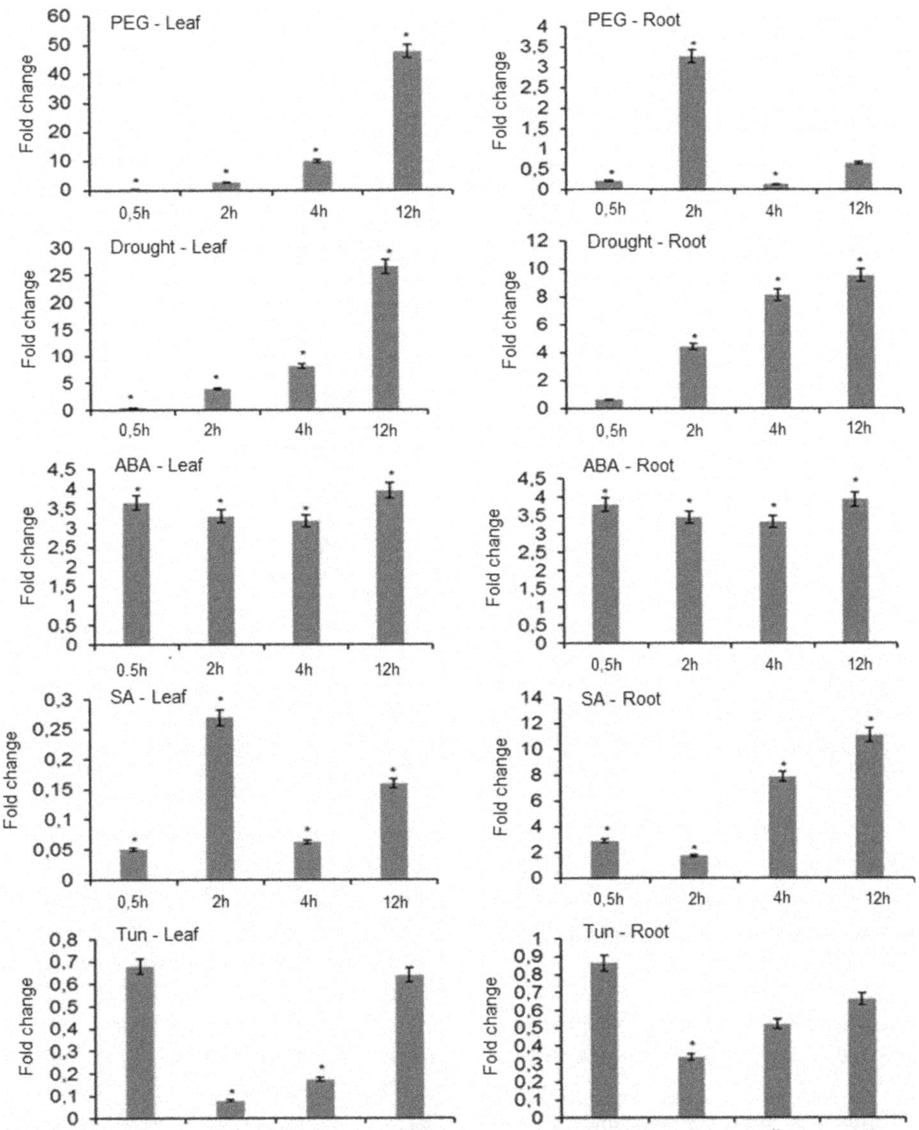


**Fig. 2** *GmRD26* expression profile in soybean (Williams 82) under multiple stresses. To determine the gene expression profile of the *GmRD26* gene, the soybean seedlings were submitted to different stress conditions (ABA, PEG, SA, Tun and drought), and the gene expression in leaves and roots was analyzed by qRT-PCR. The fold change values were calculated in relation to untreated plants (0 h), considering the relative expression in these plants as 1. *CYP2* and *ELF1A* were used as endogenous controls for normalization. The relative gene expression was calculated by the $2^{-\Delta\Delta Ct}$ method in biological triplicates ($n = 3$). The bars represent standard errors and the asterisks (*) indicate statistical significance determined by the Student's *t*-test ($p \leq 0.05$)

assembly and gene transcription in eukaryotes. Potential *cis*-regulatory element families such as the ABRE, DREB, G-box, MYC and MYB families, which can respond to many environmental signals, abiotic stresses and phytohormones were also found in the p*GmRD26* sequence (Fig. 4 and Table 1). The families' distribution in each promoter module used for *A. thaliana* genetic transformation is represented in Fig. 4b. Our analysis also revealed some specific drought-responsive *cis*-elements, MYB2AT and ACGTATERD1, as well as ABA-responsive ones, ABRERATCAL, ABREATCONSENSUS, DPBFCOREDCDC3, and EBOXBNNAPA Moreover, in the p*GmRD26* sequences, some doubly responsive elements, MYB2CONSENSUSAT, ABREZMRAB28, MYBCORE, and G-box, have been identified that respond to both drought and ABA. (Fig. 5 and Table 1). The most frequent *cis*-elements identified in the modular p*GmRD26A* (909 bp) and p*GmRD26B* (435 bp) were DPBFCOREDCDC3, ABRERATCAL, and ABREATCONSENSUS, required in ABA-signaling and MYCCONSENSUSAT, ACGTATERD1 and MYBCORE, involved in dehydration-responses (Table 1). High-salinity responsive *cis*-elements are also present.

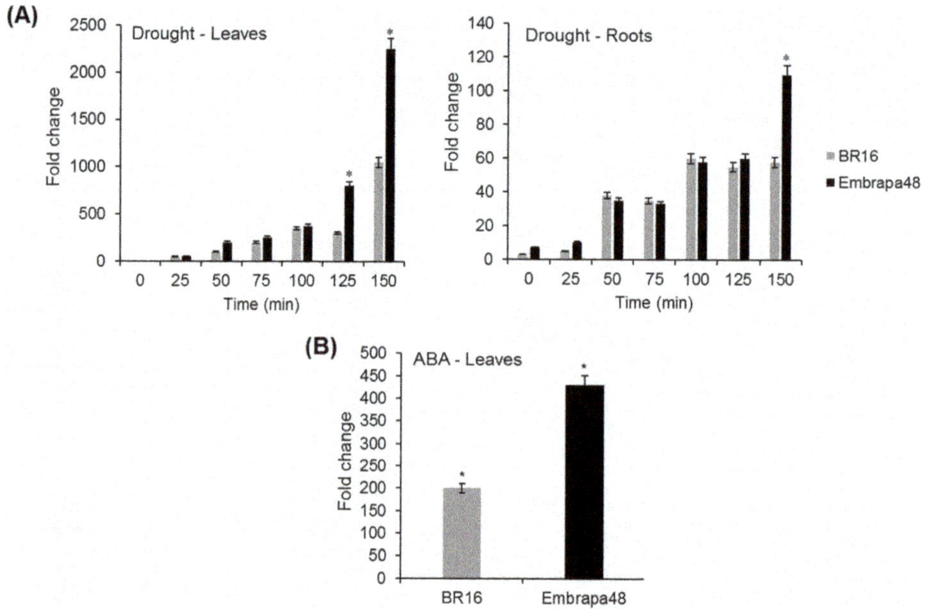


**Fig. 3** *GmRD26* expression profile in two soybean cultivars, the drought-susceptible BR16, and the drought-tolerant EMBRAPA48. **a** Expression profile of the *GmRD26* gene under drought conditions, the gene expression pattern was determined at 25, 50, 75, 100, 125, and 150 min after water deprivation. **b** Expression profile of the *GmRD26* soybean gene in the leaves of contrasting genotypes BR16 and EMBRAPA48 after 6 h of exogenous ABA stimuli. *CYP2* and *ELF1A* were used as endogenous controls for normalization. The relative gene expression was calculated by the $2^{-\Delta\Delta Ct}$ method in biological triplicates (n = 3). The bars indicate the standard errors and the asterisks (*) indicate statistical significance determined by Student's *t*-test ($p \leq 0.05$)

These stress-associated *cis*-elements were also found in *AtRD29* promoter, and it was observed that *GmRD26* promoter has nine of thirteen dehydration and ABA-responsive *cis*-elements, as found in p*AtRD29* promoter (Additional file 2: Table S1).

### GUS activity and expression in transgenic *A. thaliana* lineages under p*GmRD26* control during different stress treatments

Homozygous $T_2$ *A. thaliana* lineages carrying the full-length p*GmRD26::GUS* and the promoter modules p*GmRD26A::GUS* and p*GmRD26B::GUS* were used to analyse promoter induction under drought stress through GUS activity. The GUS histochemical assay was performed after 12 h of treatment with ABA and PEG in transgenic lineages and the controls p*AtRD29::GUS* (positive control) and non-treated plants (negative control). Plants carrying the p*GmRD26::GUS* and p*GmRD26A::GUS* displayed intense GUS activity in their foliar vascular tissue after ABA treatment, as well the positive control p*AtRD29* (Fig. 6A- a, e, m). In contrast, the p*GmRD26B::GUS* is not strongly inducible by ABA, according to its GUS activity (Fig. 6A - i), although ABRE elements are abundantly distributed in this promoter module. Under

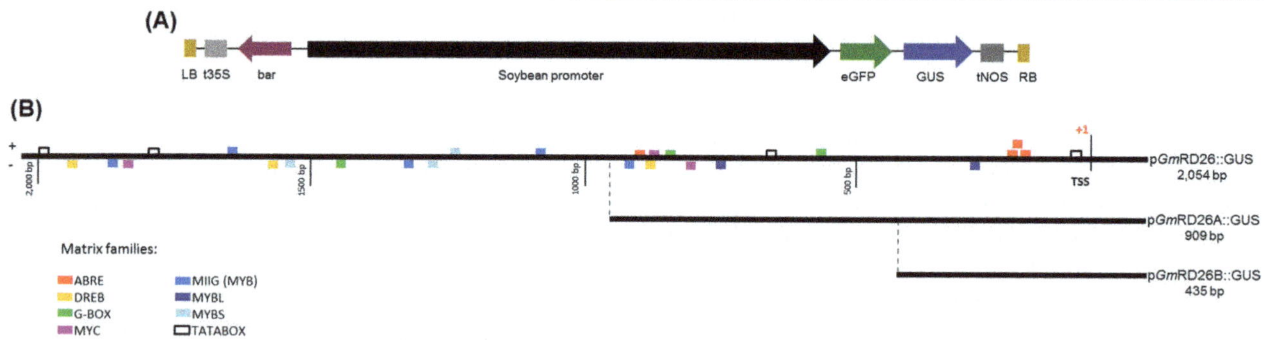


**Fig. 4** Schematic representation of the *GmRD26* promoter regions controlling the expression of the *GUS* reporter gene. **a** Schematic drawing of the soybean promoter expression cassette in the *pC1149::GUS* expression vector. **b** Diagram of the main *cis*-acting elements in the full-length p*GmRD26* (2.054 bp) promoter and the modular promoters p*GmRD26A* (909 bp) and p*GmRD26B* (435 bp). The families of *cis*-elements were identified using the Genomatix databases (*p*-value ≤0.05) and are represented by coloured boxes

**Table 1** *Cis*-regulatory elements related to drought in the p*GmRD26* soybean promoter

| *Cis*-regulatory element | Core sequence | Description | References |
|---|---|---|---|
| ACGTATERD1 | ACGT | Dehydration | [25] |
| MYCCONSENSUSAT | CANNTG | Dehydration, ABA and Cold | [26, 27] |
| ACGTABREMOTIFA2OSEM | ACGTGKC | Dehydration and ABA | [28] |
| DRE2COREZMRAB17 | ACCGAC | Dehydration and ABA | [29] |
| MYB2CONSENSUSAT | YAACKG | Dehydration and ABA | [30] |
| ABREZMRAB28 | CCACGTGG | ABA-responsive | [31] |
| ABREATCONSENSUS | YACGTGGC | ABA-responsive | [32, 33] |
| MYBCORE | CNGTTR | Dehydration and ABA | [34, 35] |
| MYB1AT | WAACCA | Dehydration and ABA | [30] |
| MYB2AT | TAACTG | Dehydration | [34] |
| G-box | CACGTG | Dehydration, high salinity, ABA | [36] |
| EBOXBNNAPA | CANNTG | ABA-responsive | [26, 37] |
| DPBFCOREDCDC3 | ACACNNG | ABA-responsive | [38, 39] |
| ABRERATCAL | MACGYGB | ABA-responsive | [40] |

K = G/T; R = G/A; W = A/T; N = A/C/G/T; Y = *T C*

PEG treatment, the GUS activity pattern was similar (Fig. 6A - b, f, j, n), with a discrete decline in activity in the p*GmRD26*::*GUS* plants when compared with ABA and drought (air dry) treatment (Fig. 6A - b). In the p*GmRD26A*::*GUS* construct, a strong GUS-derived staining was observed in almost all the leaves surfaces in PEG treatment (Fig. 6A - f). Under the drought treatment, GUS activity was strongly detected in all analyzed leaves, mainly in the modular constructs p*GmRD26A*::*GUS*, p*GmRD26B*::*GUS* (Fig. 6A - g and k). The basal expression in the control plants (without stress conditions) was low but detectable (Fig. 6A - d, h, l, and p). In our study, *pGmRD26A* displays activity in all treatments, but this activity is higher under desiccation conditions (Fig. 6A - g), reinforcing the role of RD26 in desiccation-triggered protective mechanisms in plants.

To confirm the induction profile of p*GmRD26* revealed by histochemical assays, GUS activity was also monitored in transgenic lineage plants. Under ABA treatment, full-length p*GmRD26* encompassed the same results when compared to the p*AtRD29* positive control, and the module p*GmRD26A* displays the higher GUS activity (Fig. 6B). These results contrast with the PEG treatment, in which the full-length promoter and the p*GmRD26A* module exhibit higher activity when compared with the positive control, p*AtRD29*, and the p*GmRD26B* (Fig. 6B). When GUS activity was analyzed under drought treatment, p*GmRD26A* shows the same activity of the full-length promoter, higher than the smaller module p*GmRD26B*, but lower than the positive control. The activity of the p*GmRD26A* module was higher than the other fragments, and the p*AtRD29* control under ABA and PEG treatments (Fig. 6B). In addition, p*GmRD26A* transgenic lines display high levels of *uidA* mRNA after PEG treatment, while p*GmRD26* lines display high levels of *uidA* transcripts under ABA treatment. In the drought treatment, p*AtRD29* control lines presented higher expression level than the p*GmRD26* promoter and its modules. However, when we analyzed the differences between the three fragments after drought treatment, p*GmRD26A* showed higher expression levels compared to p*GmRD26* and p*GmRD26B* (Fig. 7).

The results of qPCR demonstrate that transcriptional GUS activity, driven by the promoters p*GmRD26* and p*GmRD26A* was similar but not the same during ABA treatment. According to our data, p*GmRD26* display a similar expression when compared with p*AtRD29* (positive control) and higher mRNA accumulation when compared with p*GmRD26A* module (Fig. 7a). This data is compatible with the histochemical assay. Under PEG treatment, the module p*GmRD26A* displayed a higher transcriptional activity, followed by the full-length p*GmRD26* and the positive control p*AtRD29* (Fig. 7b). The module p*GmRD26B* continued to displaying lower *GUS* transcriptional activity. As expected, under drought condition, p*AtRD29* displayed higher *GUS* expression, followed by considerable *GUS* expression driven by modules p*GmRD26A* and p*GmRD26B* and the full-length promoter p*GmRD26* (Fig. 7c).

## Discussion

In this study, we confirmed that the *GmRD26* gene is induced under different simulated drought conditions. In view of the potential of this gene as a target for the development of strategies for the genetic engineering of resistant plants, we decided to isolate and characterize the *GmRD26* promoter region. Our results show that p*GmRD26* and its modules activated the reporter gene *uidA* under different water deprivation stress conditions. These results are consistent with the characteristics of

**Fig. 5** Physical map of the *GmRD26* promoter. The transcription start site is highlighted in red and is designated as + 1. The TATA-box is highlighted in bold. The numbers on the left side indicate the distance from the transcription start site. The sequences were analyzed by Genomatix databases (p-value ≤0.05). The putative *cis*-elements provided in p*GmRD26* are indicated by a bar and their names. Sense acting motifs (5′– 3′) are indicated by a superior bar, while antisense acting motifs (3′– 5′) are indicated by an inferior bar. All the stress-responsive motives are represented by different colours

the *cis*-regulatory elements identified by in silico analyses of p*GmRD26* sequence. In Arabidopsis, *At*RD26 is an important member of ABA-dependent drought tolerance, and its overexpression is associated with a drought-tolerant phenotype [41].

The *GmRD26* gene, as well as its *A. thaliana* orthologue (*AtRD26*/ANAC072), belongs to the subfamily SNAC-A, whose members have some correlation of functional conservation with the *ATAF1* gene (AT1G01720), which has been shown to be a regulator of ABA biosynthesis and

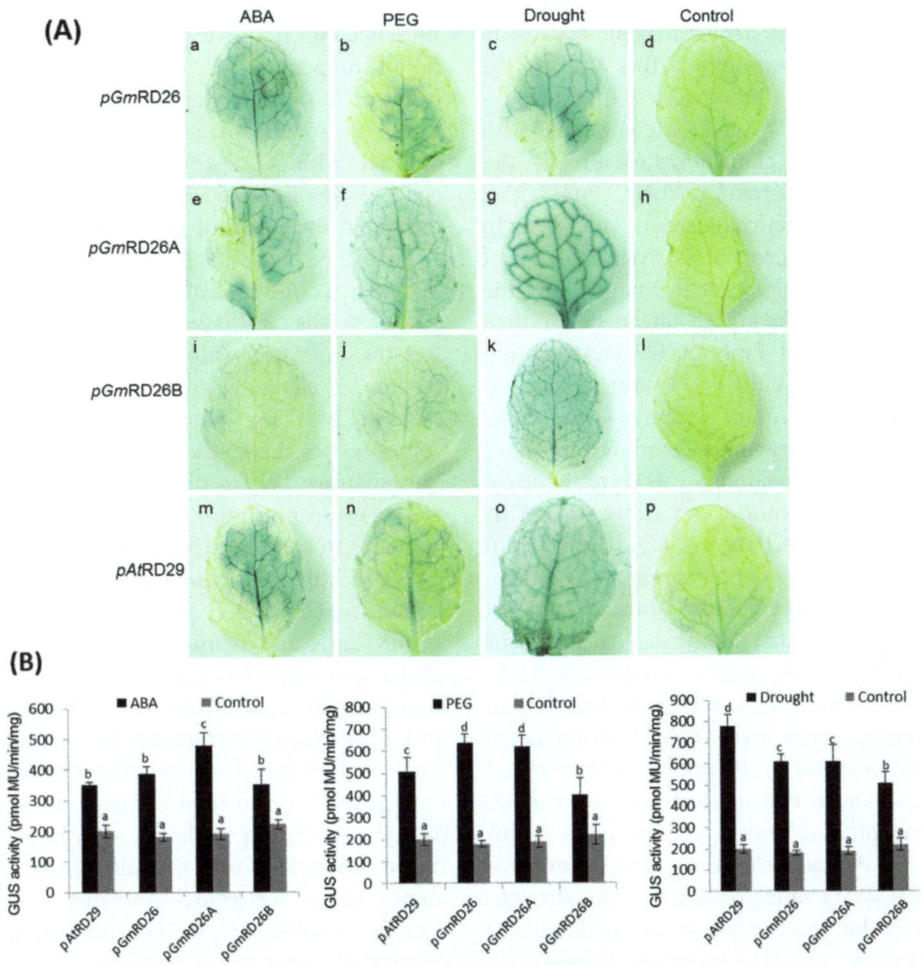


**Fig. 6** Histochemical and quantitative fluorimetric analysis of different *GmRD26* promoter deletion constructs in transgenic *A. thaliana* plants. The stress treatments for GUS activity analysis was performed on 4-week-old *A. thaliana* plants under 12 h treatments with ABA, PEG, or drought. **(A)** Histochemical localization of GUS activity in transgenic *A. thaliana* plants harboring promoter-*GUS* constructs. **(B)** The quantitative fluorimetric assay for GUS activity was carried out in three replicates. The soybean promoter and its modules were compared with the p*AtRD29* and wild type plants. Control samples consisted of untreated plants. Bars indicate standard error and different lowercase letters indicate significant differences at $P < 0.05$ on Tukey's Test. The data shown are representative of three independent lines (n = 3)

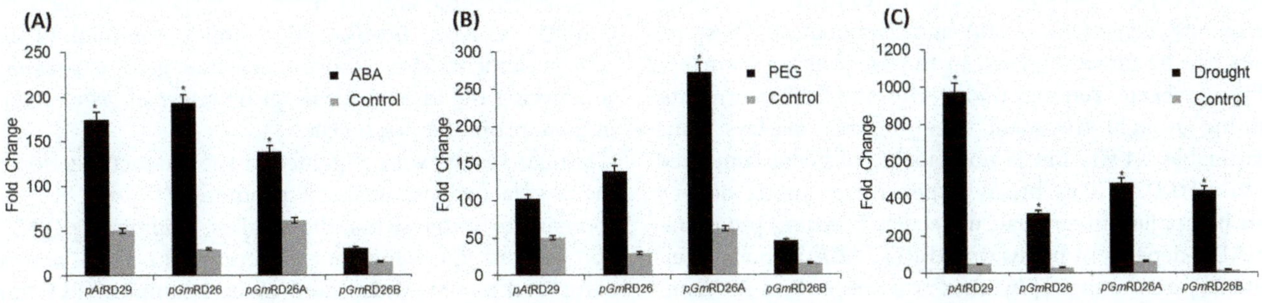


**Fig. 7** Transcriptional *GUS* activity in transgenic *A. thaliana* under 12 h of ABA (**a**), PEG (**b**) or drought (**c**) treatments. Levels of *uidA* mRNA of non-treated (control) and treated transgenic plants were measured using real-time PCR. The expression levels were normalized using *ACT2* and *GAPDH* as endogenous controls. The relative gene expression was calculated using the $2^{-\Delta\Delta Ct}$ method. The bars represent standard error and the asterisks (*) indicate statistical significance determined by Student's *t*-test ($p \leq 0.05$). The data shown are representative of three independent lines (n = 3)

responsive to water stress [42–46]. Analysis of *GmRD26* expression in two contrasting drought-tolerant soybean cultivars, BR16 and Embrapa48, and in the Williams 82 (soybean reference genome) demonstrated that the expression profile of the *GmRD26* gene is compatible with phylogenetic and molecular characteristics already described for orthologue genes, reinforced by *GmNAC085* expression profile, a phylogenetically close-related gene in soybean [42, 47]. Our gene expression analysis reveals that the induction fold of *GmRD26* is not the same along the three analyzed cultivars, displaying an unexpectedly high expression level in BR16 and Embrapa48. The different genetic background of these cultivars should proportion this difference, once the BR16 and Embrapa48 are commercial cultivars, obtained by genetic breeding programs. In our study, *GmRD26* was responsive to osmotic stress during PEG treatment, desiccation and exogenous ABA stimulation in both leaves and roots, while the tolerant soybean variety displays higher gene expression level than susceptible variety. The leaves exhibited a more significant folding variation, suggesting that the physiological mechanism triggered by *GmRD26*, mainly related to ABA-dependent responses, is more effective in the leaves than in the roots. In general, genes involved in ABA-mediated stress responses are involved in leaf morphophysiological changes, including stomatal closure, leaf area adjustment, photosynthesis, transpiration index, and osmolyte accumulation [48, 49]. In Arabidopsis, it was already demonstrated that *GmNAC085* overexpression confers drought tolerance, improving the plant physiological performance during water deprivation stress. The transgenic lineages display a more robust antioxidative response under stress and many readouts genes, involved in ABA-dependent signalling, are up-regulated [11]. These results, associated with the determined *GmRD26* and *GmNAC085* gene expression profile, may justify the drought inducibility of SNAC-A genes during abiotic stress and confirm their potential to drive expression of genes involved with plant adaptability.

Compared with p*AtRD29*, a previously characterized drought-associated promoter, the promoter p*GmRD26* was also enriched in stress-related *cis*-elements, responsive to salinity, dehydration, ABA and temperature. These results can be directly related to the excellent performance of the soybean promoter under ABA and PEG treatments. During drought treatment, the promoter displays some reasonable activity but is not capable of being compared with p*AtRD29*. This broad responsive promoter activity can be applied in soybean molecular breeding programs. In ABA-dependent pathways, ABREs (ABA-responsive elements) are the main phytohormone-responsive *cis*-element [43]. The occurrence of three ABRE motifs from five total ABA-responsive elements indicates a strong promoter induction under drought conditions, which can trigger increased drought-responsive gene expression by p*GmRD26* during stress. This effect is reinforced by the presence and frequency of the ACGT motif, a characteristic and important *cis*-element in drought-responsive promoters [50]. *Cis*-acting elements of the G-box family, found in several plant genes' promoters are known to interact with bZIP transcription factors, mediating responses to different stimuli. Studies comparing the patterns and evolution of the G-box family core (ACGT) in *O. sativa*, *S. bicolor*, *A. thaliana* and *G. max* suggest that this is the family with the most conserved elements between species and leads to responses to exogenous stresses, especially water and salt stress [51]. Other stress-responsive elements are also present in p*GmRD26*, such as MYCs/MYBs, which exhibit rapid induction in response to ABA treatment and water stress. These elements are targets of a large TF family in soya. MYC and MYB transcription factors are necessary for the early response to osmotic stress [43, 52].

In this study, we also the activity of the soybean promoter p*GmRD26* and two-promoter modules, p*GmRD26*A and p*GmRD26*B, in transgenic *A. thaliana* plants that were submitted to simulated (ABA and PEG) and real drought stress. The p*GmRD26*, p*GmRD26A*, and p*GmRD26B* promoters were induced by all stress treatment assays, showing greater or similar GUS activity than p*AtRD29* (positive control) under ABA, PEG, and drought treatment. Differences in induction intensity between p*GmRD26* and modules under different types of abiotic stress are probably related to the distribution of specific *cis*-elements in their sequences involved in the control of water stress response [53, 54]. The transcriptional activation of some genes depends not only on the promoters' *cis*-acting elements and their sequences but also on their position and the presence of enhancers, regulatory sequences and other synergistic *cis*-elements [55, 56]. It is important to highlight that some differences between transcriptional and translational activity are common on promoter's genes analysis. Our data demonstrate that the induction profile of p*GmRD26* is similar in the tested conditions, showing that the full-length promoter and its modules respond to the same conditions. Also, promoters' modules enriched in *cis*-acting elements drive more consistent gene expression, reinforcing the idea of a synergistic effect of *cis*-elements in gene promoter sequences.

Similar results were obtained in the characterization of the α-galactosidase soybean promoter (*GlymaGAL*) responsive to water stress; the smallest fragment, p*GAL*-1 kb, showed no significant difference in GUS activity compared to the control and treated samples (PEG and dry). The full-length fragment promoter, p*GAL*-2 kb, however, led to a significant increase in *GUS* expression. This increase in *GUS* expression of p*GAL*-2 kb was associated with a high number of ABRE, MYCATERD1, G-box, and DRE *cis*-elements [57]. Other studies have also reported the importance of distal promoter regions in responses associated with water stress in other species [58, 59].

## Conclusions

In this study, we analyzed the expression profile of *GmRD26* gene, which is expressed under simulated osmotic and drought conditions in soybean. The stressed soybean seedlings display a high *GmRD26* expression under ABA exogenous-stimuli in leaves and roots, as well under PEG and air dry treatment. This gene expression pattern raised the hypothesis of drought-inducible *cis*-elements enriched promoter of *GmRD26*. Our analysis showed that the *GmRD26* promoter region is enriched with essential *cis*-elements associated with drought stress, such as ABRE, DREB, MYB, MYC, and G-BOX. Molecular characterization of p*GmRD26* in *A. thaliana* has demonstrated that the full promoter (p*GmRD26*) and two different promoter-modules (p*GmRD26A* and p*GmRD26B*) are inducible under simulated osmotic and drought stress conditions, confirming the soybean gene expression profile. In addition, our data also revealed that the full-length promoter and the p*GmRD26A* module, with higher *cis*-acting elements incidence compared to the other module, displayed a slightly higher level of expression than p*GmRD26B* and the p*AtRD29*, an *A. thaliana* promoter used as a model to drought inducible gene studies, during ABA and PEG treatment. The complete characterization of p*GmRD26* and its modules suggests that the promoter or the fragment p*GmRD26A* may become a potential biotechnological tool capable of inducing expression of genes of interest under specific conditions, such as drought or other abiotic stresses related with an osmotic imbalance to improve the tolerance associated to physiological performance in genetically modified plants.

## Methods

### Identification of the drought marker gene *GmRD26* in soybean

The *A. thaliana* RD26 (AT4G27410) deduced amino acid sequence (available in TAIR database - https://www.arabidopsis.org/) was used to identify the closely related orthologue gene (*GmRD26*/*Gm*NAC043 - Glyma.06G248900) in the soybean genome (Williams.82 v2.2-available in Phytozome: https://phytozome.jgi.doe.gov) [60]. For sequence comparison, BLASTP (https://blast.ncbi.nlm.nih.gov) was used, and the alignment was confirmed using the online tool ClustalW2 (https://www.ebi.ac.uk/Tools/msa/clustalw2). To determine the phylogenetic relationship between the Arabidopsis and soybean genes, the neighbour-joining clustering method derived from a distance matrix from a Poisson model was used, and the tree was reconstructed using MEGA software [61].

To evaluate whether the selected putative soybean gene *GmRD26* is induced during drought stress, its expression pattern was analyzed from cDNA subtractive libraries related to dissection experiments available in the GENOSOJA database LGE (Genomics and Expression Laboratory: GENOSOJA Project) [22].

### Soybean plant growth conditions and stress treatments

For the *GmRD26* gene expression profile analysis, soybean (Williams 82) seeds were germinated in the soil and grown under greenhouse conditions (12 h of light, 25–35 °C, 70% relative humidity) until the V2-V3 development stage. To simulate multiple stress conditions, the seedlings were first transferred to Hoagland hydroponic solution for 24 h. After acclimation, the soybean roots were immersed in the same solution supplemented with 10% (w/v) PEG 8000 to induce osmotic stress, 5 μg/mL tunicamycin (Tun) to induce endoplasmic reticulum stress, 150 mM ABA and 5 mM salicylic acid (SA) to simulate drought and biotic stress conditions, respectively. For the drought treatment, the plants were removed from the hydroponic solution and placed on plates with cotton. Leaf discs and roots of treated and control (0 h - untreated plants were collected one time) seedlings were collected after 0.5 h, 2 h, 4 h and 12 h of stress treatment and immediately frozen in liquid nitrogen. All treatments were performed at a three-plants pool, and samples were collected in triplicate.

Embrapa48 and BR16 soybean cultivars were used to determine the expression profile during the ABA and drought treatments in drought responses contrasting soybean lineages. The BR16 variety is considered as a model of drought sensitivity, while Embrapa48 is considered as drought-tolerant cultivar. The seeds were germinated in watered germination test paper and then transferred to a hydroponic box system filled with Hoagland solution. The seedlings in stages V3-V4 were grown under the same conditions as the Williams 82 seedlings. The drought stress was generated by removing the plants from the hydroponic solution and placing them in empty boxes for different water deprivation periods: 0 min (T0 - control), 25 min (T25), 50 min (T50), 75 min (T75), 100 min (T100), 125 min (T125) and 150 min (T150). Roots and leaf disc samples from three plants of each cultivar were collected during the exposure to water and were immediately frozen in liquid nitrogen for RNA extraction and gene expression analysis.

The contrasting soybean cultivars were also submitted to exogenous ABA treatment. Plants germinated and grown under the same conditions were sprayed with water (control) or ABA solution (300 ppm). Three biological replicates were used, consisting of three plants per treatment. After 6 h, leaf discs were collected and immediately frozen in liquid nitrogen for RNA extraction.

### RNA extraction, cDNA synthesis and *GmRD26* gene expression analysis

The total RNA of soybean leaf and roots was extracted according to the TRIzol® manual (Invitrogen, USA). RNA quantification was performed using a NanoDrop™ Spectrophotometer ND-1000 (Thermo Scientific, USA) and the RNA integrity was assessed by 1% agarose gel electrophoresis. A total of 2 μg of RNA was used for cDNA synthesis with MMLV reverse transcriptase protocol (Invitrogen, USA).

The gene expression profile was determined by qPCR. The analysis was performed using an ABI 7500 Fast instrument, SYBR Green (Invitrogen, USA) reagent, specific primers (Additional file 3: Table S2) and three independent cDNA pools. All the analyses were performed using three biological and two technical replicates, originated from a five soybean plants pool. The reaction was performed as follow: 2 min at 50 °C, 10 min at 95 °C, and 40 cycles of 94 °C for 15 s and 60 °C for 1 min. The *CYP* and *ELF* soybean genes [62] were used as endogenous controls for expression normalization and relative gene expression calculated by the $2^{-\Delta\Delta Ct}$ method. The endogenous gene stability was determined by G-norm algorithm (https://genorm.cmgg.be/), from Q-base package, and the M-value is 0.89 and 0.91 for *CYP* and *ELF*, respectively. The *GmRD26* orthologue gene, *GmNAC085*, was used as a comparative control in Williams 82 for gene profile determination.

### Analysis of p*GmRD26* soybean *cis*-acting elements

The p*GmRD26* promoter sequence (2.054 bp) was obtained from the soybean genome available in the Phytozome database (https://phytozome.jgi.doe.gov) [60]. The *cis*-acting elements responsive to drought-, salinity-, osmotic- and ABA-induced stress were identified, analyzed, and mapped using the Genomatix (https://www.genomatix.de/online_help/matinspector/matinspector). For this study, we considered only the *cis*-elements statistically significant, with a $p$-value ≤ 0.05 [63, 64].

### Construction of p*GmRD26* plasmids

The full-length *GmRD26* soybean promoter region was considered as the 2.054 bp gene-promoter, and A and B promoter-modules contain 909 bp and 435 bp, respectively, considering the distribution of the drought-responsive *cis*-acting elements. The sequences were transcriptionally fused in frame to the *GUS* gene in a binary expression pC1407 vector backbone, synthesized by Epoch Biolabs (Sugar Land, TX, USA). The generated recombinant plasmids were called p*GmRD26::GUS* (2.054 bp), p*GmRD26A::GUS* (909 bp) and p*GmRD26B::GUS* (435 bp). The plasmids carry out the translational *GUS-GFP* fusion and BaR plant selection marker gene. The *AtRD29A* (*AtRD29*) promoter gene sequence [65] was cloned into the same plasmid as a positive control of drought-inducible promoters.

### Transgenic *A. thaliana* plants

The recombinant plasmids were introduced into *Agrobacterium tumefaciens* GV3101 strain, which was then transferred to *A. thaliana* ecotype Columbia (Col-0) by floral dip method [66]. Transgenic plants with a T-DNA insertion were identified by glufosinate-ammonium selection and confirmed by PCR. Three homozygous independent lines were obtained for each construction and $T_2$ plants expressing *GUS*-GFP used in abiotic stress treatments and promoter characterization.

### Drought, PEG and ABA treatment of *A. thaliana* transgenic lineages

The *A. thaliana* seeds were germinated in the soil and grown under growth chamber-controlled conditions (12 h photoperiod, 21 °C temperature and 70% relative humidity). After 10 days, the seedlings were sprayed three times at intervals of 5 days with glufosinate-ammonium (100 mg/L) for positive transgenic plant selection. Four weeks old transgenic plants were carefully removed of soil moisture, and their roots were immersed in Hoagland hydroponic solution supplemented with 5% (w/v), PEG (MW 8.000) and 50 μM ABA solution to simulate drought conditions. For the drought treatment, the plants were removed from the hydroponic solution and placed on open plates. The non-stressed controls consisted of plants that were kept in Hoagland hydroponic solution. Two leaves of three plants for each full-length or modular promoter were collected after 12 h of treatment and immediately frozen in liquid nitrogen and stored at − 80 °C for further extraction of RNA.

### Histochemical GUS assays

To detect GUS activity in transgenic *A. thaliana* lineages, fresh leaves were incubated for 12-16 h at 37 °C in 5-bromo-4-chloro-3-indolyl-β-D-glucuronic acid (X-Gluc) solution [67]. After X-Gluc incubation, the leaves were washed with water, and the chlorophyll was removed with ethanol (70% v/v) for approximately 10 h. The leaves were washed and then observed under Leica Wild Heerbrugg M3Z Stereozoom Microscope (Leica, Wetzlar, Germany). For each construct, leaves were collected from at least three different transgenic plants lineages.

### Fluorimetric GUS assay

The *A. thaliana* transgenic plants were grouped into three-plant pools and subjected to stress treatments

(PEG, ABA, and drought) as previously described. For plants pool protein extraction, extraction buffer with 100 mM $NaH_2PO_4$, 0.01% SDS, 10 mM EDTA, 0.1% sodium lauryl sarcosine, 0.1% Triton X-100 and 1 mM DTT was used. Protein extraction was performed with frozen tissue powder (~ 100 mg), and samples were manipulated on ice. The total soluble proteins were quantified by the Bradford method [68] and used for a fluorimetric assay. The fluorimetric GUS assay was performed in a 500 μL reaction consisting of 400 μL of protein extract and 100 μL of 10 mM 4-methylumbelliferyl β-D-glucuronide (MUG; Sigma, USA). The reaction was incubated for 1 h at 37 °C. At the start point, a 50 μL reaction aliquot was removed and added to 450 μL of 0.2 M $Na_2CO_3$ stop buffer. The fluorescence of 4-methylumbelliferone (4-MU) was monitored using a Versa Fluor Fluorometer (BioRad) with excitation at 365 nm and emission at 455 nm. Each sample was analyzed in triplicate, and values were calculated according to a reference range of MU. GUS activity was expressed in nanomoles of MU produced per minute per microgram of soluble protein.

All GUS fluorimetric assays were repeated at least three times. The results were expressed as the mean of independent experiments with the respective standard error. Different lowercase letters above the bars indicate significant differences at $p < 0.05$.

### *GUS* gene expression analysis

The *GUS* gene expression was analyzed in transgenic *Arabidopsis* plants expressing p*GmRD26*::*GUS*, p*GmRD26A*::*GUS*, p*GmRD26B*::*GUS* and p*AtRD29*::*GUS*. The gene expression level was monitored by qRT-PCR using three biological and two technical replicates, as previously described for soybean genes; the expression levels were normalized using *ACT2* (AT3G18780 [69]) and *GAPDH* (AT1G13340 [70]) as endogenous controls. The endogenous gene stability was determined by G-norm algorithm, from Q-base package, and the M-value is 0.86 and 0.79 for *ACT2* and *GAPDH*, respectively. The primers used are described in Additional file 3: Table S2.

## Supplementary information

**Additional file 1: Figure S1**. *GmNAC085* expression profile in soybean (Williams 82) under multiple stresses. To determine the gene expression profile of the *GmNAC085* gene, the soybean seedlings were submitted to different stress conditions (ABA, PEG, AS, Tun, and drought), and the gene expression were analyzed in leaves and roots by qRT-PCR. The fold change values were calculated in comparison of plants treated with untreated plants (0 h). *CYP2* and *ELF1A* were used as endogenous controls for normalization. The relative gene expression was calculated by the $2^{-\Delta\Delta Ct}$ method in biological triplicates ($n = 3$). The bars represent standard errors and the asterisks (*) indicate statistical significance determined by the Student's t-test ($P \leq 0.05$).

**Additional file 2: Table S1.** *Cis*-regulatory elements related to drought revealed in the p*GmRD26* soybean promoter and the *A. thaliana* promoter *RD29* (*p*-value of 0.05).

**Additional file 3: Table S2.** Primer sequences used in the qRT-PCR analysis.

#### Abbreviations

ABA: Abscisic acid; ABRE: ABA-responsive element; *AtRD29*: *Arabidopsis thaliana* responsive to desiccation 29; DREB: Dehydration-responsive element binding; MYB: Myeloblastosis; MYC: Myelocytomatosis; NAC: Nitrogen assimilation control; PEG: Polyethylene glycol; p*GmRD26*: *Glycine max* responsive to desiccation 26 promoter; TFs: Transcription factors

#### Acknowledgments

Not Applicable.

#### Authors' contributions

EOF and BPM executed the experiments and wrote the first draft of the manuscript; ITLT and FBMA executed qPCR and fluorimetric assay; RMA and MELS performed abiotic stress experiments on soybean; JAC and AGBL transformed *A. thaliana* with p*GmRD26* full-length promoter and modular promoters; MF and CVM performed statistical analysis, MAF, EPBF and MFGS designed the study, revised and finalized the manuscript. All contributing authors have read and approved the final manuscript.

#### Funding

This work was supported by EMBRAPA, CNPq (INCT 465480/2014-4), CAPES and FAPDF (INCT 0193-001.265/2017). CAPES provides fellowship grants to EOF, BPM, FBMA and JAC. CNPq provides a fellowship grant to AGBL. The funding body did not participate of the study design, data collection and analysis and manuscript writing.

#### Ethics approval and consent to participate

Not applicable.

#### Consent for publication

Not applicable.

#### Competing interests

The authors declare that they have no competing interests.

#### Author details

[1]Embrapa Genetic Resources and Biotechnology, Brasília, DF, Brazil. [2]Federal University of Brasília, Brasília, DF, Brazil. [3]Federal University of Viçosa, Viçosa, MG, Brazil. [4]Federal University of Rio Grande do Sul, Porto Alegre, RS, Brazil. [5]Agricultural Research Company of Minas Gerais State, Uberaba, MG, Brazil. [6]Catholic University of Brasilia - Post-Graduation Program in Genomic Sciences and Biotechnology, Brasília, DF, Brazil. [7]National University of Medical Sciences, Rawalpindi, Punjab, Pakistan. [8]Federal University of Rio de Janeiro, Rio Janeiro, RJ, Brazil. [9]Embrapa Semi-Arid, Petrolina, PE, Brazil.

#### References

1. Sheffield J, Wood EF, Roderick ML. Little change in global drought over the past 60 years. Nature. 2012;491:435-8.
2. Leng G, Hall J. Crop yield sensitivity of global major agricultural countries to droughts and the projected changes in the future. Sci Total Environ. 2019; 654:811-21.
3. Tuteja N. Cold, salinity and drought stress. Plant Stress Biol. 2009;149:137-59.

4. Fang Y, Xiong L. General mechanisms of drought response and their application in drought resistance improvement in plants. Cell Mol Life Sci. 2015;72:673–89.
5. Nakashima K, Yamaguchi-Shinozaki K. ABA signaling in stress-response and seed development. Plant Cell. 2013;32:959–70.
6. Takahashi F, Takehiro S, Yuriko O, Shigeyuki B, Yuki K, Naoshi D, et al. A small peptide modulates stomatal control via abscisic acid in long-distance signalling. Nature. 2018;556:235–8.
7. Abe H, Yamaguchi-Shinozaki K, Urao T, Iwasaki T, Hosokawa D, Shinozaki K. Role of MYC and MYB homologs in drought- and abscisic acid-regulated gene expression. Plant Cell. 1997;9:1859–68.
8. Zhao L, Hu Y, Chong K, Wang T. ARAG1, an ABA-responsive DREB gene, plays a role in seed germination and drought tolerance of rice. Ann Bot. 2010;105:401–9.
9. Yoshida T, Fujita Y, Sayama H, Kidokoro S, Maruyama K, Mizoi JK, Shinozaki Y. AREB1, AREB2, and ABF3 are master transcription factors that cooperatively regulate ABRE-dependent ABA signaling involved in drought stress tolerance and require ABA for full activation. Plant J. 2010;61:672–85.
10. Li J, Xu Y, Niu Q, He L, Teng Y, Bai S. Abscisic acid (ABA) promotes the induction and maintenance of pear (*Pyrus pyrifolia* white pear group) flower bud endodormancy. Int J Mol Sci. 2018;19:1–15.
11. Nguyen KH, Mostofa MG, Watanabe Y, Tran CD, Rahman MM, Tran LSP. Overexpression of *GmNAC085* enhances drought tolerance in Arabidospis by regulating glutathione biosynthesis, redox balance and glutathione dependent detoxification of reactive oxygen species and methylglyoxal. Environ Exp Botany. 2018;161:242–54.
12. Mizoi J, Shinozaki K, Yamaguchi-Shinozaki K. AP2/ERF family transcription factors in plant abiotic stress responses. Biochim Biophys Acta. 1819;2012:86–96.
13. Kasuga M, Miura S, Shinozaki K, Yamaguchi-Shinozaki K. A combination of the Arabidopsis DREB1A gene and stress-inducible rd29A promoter improved drought- and low-temperature stress tolerance in tobacco by gene transfer. Plant Cell Physiol. 2004;45:346–50.
14. Behnam B, Kikuchi A, Celebi-Toprak F, Yamanaka S, Kasuga M, Yamaguchi-Shinozaki K, Watanabe KN. The Arabidopsis DREB1A gene driven by the stress-inducible rd29A promoter increases salt-stress tolerance in proportion to its copy number in tetrasomic tetraploid potato (*Solanum tuberosum*). Plant Biotechnol. 2006;23:169–77.
15. Engels C, Fuganti-Pagliarini R, Marin SRR, Marcelino-Guimarães FC, Oliveira MCN, Kanamori N, et al. Introduction of the rd29A: *At*DREB2A CA gene into soybean (*Glycine max* L. Merril) and its molecular characterization in leaves and roots during dehydration. Genet Mol Biol. 2013;36:556–65.
16. Kasuga M, Liu Q, Miura S, Yamaguchi-Shinozaki K, Shinozaki K. Improving plant drought, salt and freezing tolerance by gene transfer of a single stress-inducible transcription factor. Nat Biotechnol. 1999;17:287–91.
17. Bihmidine S, Lin J, Stone JM, Awada T, Specht JE, Clemente TE. Activity of the Arabidopsis RD29A and RD29B promoter elements in soybean under water stress. Planta. 2013;237:55–64.
18. Kidokoro S, Watanabe K, Ohori T, Moriwaki T, Maruyama K, Mizoi J. Soybean DREB 1/CBF type transcription factors function in heat and drought as well as cold stress responsive gene expression. Plant J. 2015;81:505–18.
19. Sarker S, Biswas S, Shahed A, Razzaque S, Seraj ZI. Cloning, characterization and analysis of the Arabidopsis RD29A promoter for its inducible expression in rice under salinity and drought stress. Biores Commun. 2016;2(1):139–45.
20. Nobres P, Patreze CM, Waltenberg FP, Marcelo FC, Tavano EC, Mendes BMJ, et al. Characterization of the promoter of the homeobox gene *Ca*HB12 in *Coffea arabica*. Trop Plant Biol. 2016;9(1):50–62.
21. Melo BP, Fraga OT, Silva JCF, Ferreira DO, Brustolini OJB, Carpinetti PA, Machado JPB, Reis PAB, Fontes EPB. Revisitinh the soybean GmNAC superfamily. Front Plant Sci. 2018;9:1864.
22. Rodrigues FA, Marcolino J, Carvalho JDFC, Nascimento LCD, Neumaier N, Farias JRB, et al. Subtractive libraries to prospect differentially expressed genes in soybean plants submitted to water deficit. Genet Mol Biol. 2012;35: 304–14.
23. Takasaki H, Maruyama K, Takahashi F, Fujita M, Yoshida T, Nakashima K, et al. SNAC-as, stress-responsive NAC transcription factors, mediate ABA-inducible leaf senescence. Plant J. 2015;84:1114–23.
24. Li S, Gao J, Yao L, Ren L, Zhu X, Gao S, et al. The role of ANAC072 in the regulation of chlorophyll degradation during age- and dark-induced leaf senescence. Plant Cell Rep. 2016;35:1729.
25. Simpson SD, Nakashima K, Narusaka Y, Seki M, Shinozaki K, Yamaguchi-Shinozaki K. Two different novel cis-acting elements of e1, a clpA homologous Arabidopis gene function in induction by dehydration stress and dark-induced senescence. Plant J. 2003;33:259–70.
26. Hartmann U, Sagasser M, Mehrtens F, Stracke R, Weisshaar B. Differential combinatorial interactions of cis-acting elements recognized by R2R3-MYB, BZIP, and BHLH factors control light-responsive and tissue-specific activation of phenylpropanoid biosynthesis genes. Plant Mol Biol. 2005;57:155–71.
27. Agarwal M, Hao Y, Kapoor A, Dong CH, Fujii H, Zheng X, et al. A R2R3 type MYB transcription factor is involved in the cold regulation of CBF genes and in acquired freezing tolerance. J Biol Chem. 2006;281:37636–45.
28. Hattori T, Totsuka M, Hobo T, Kagaya Y, Yamamoto TA. Experimentally determined sequence requirement of ACGT-containing abscisic acid response element. Plant Cell Physiol. 2002;43:136–40.
29. Kizis D, Pages M. Maize DRE-binding proteins DBF1 and DBF2 are involved in rab17 regulation through the drought-responsive element in an ABA-dependent pathway. Plant J. 2002;30:679–89.
30. Abe H, Urao T, Ito T, Seki M, Shinozaki K, Yamaguchi-Shinozaki K. Arabidopsis AtMYC2 (bHLH) and AtMYB2 (MYB) function as transcriptional activators in abscisic acid signaling. Plant Cell. 2003;15:63–78.
31. Suzuki M, Ketterling MG, McCarty DR. Quantitative statistical analysis of cis regulatory sequences in ABA/VP1 and CBF/DREB1 regulated genes of Arabidopsis. Plant Physiol. 2005;139:437–47.
32. Kang JY, Choi HI, Im MY, Kim SY. Arabidopsis basic leucine zipper proteins that mediate stress-responsive abscisic acid signaling. Plant Cell. 2002;14: 343–57.
33. Choi HI, Park HJ, Park JH, Kim S, Im MY, Seo HH, et al. Arabidopsis calcium-dependent protein kinase AtCPK32 interacts with ABF4, a transcriptional regulator of abscisic acid-responsive gene expression and modulates its activity. Plant Physiol. 2005;139:1750–61.
34. Urao T, Yamaguchi-Shinozaki K, Urao S, Shinozaki K. An Arabidopsis myb homolog is induced by dehydration stress and its gene product binds to the conserved MYB recognition sequence. Plant Cell. 1993;5:1529–39.
35. Solano R, Nieto C, Avila J, Canas L, Diaz I, Paz-Ares J. Dual DNA binding specificity of a petal epidermis-specific MYB transcription factor (MYB.Ph3) from *Petunia hybrida*. EMBO J. 1995;14:1773–84.
36. Yang T, Poovaiah BW. Arabidopsis calmodulin-binding/CGCG box DNA-binding protein family involved in multiple signaling pathways in plants. J Biol Chem. 2002;277:45049–58.
37. Stalberg K, Ellerstom M, Ezcurra I, Ablov S, Rask L. Disruption of an overlapping E-box/ABRE motif abolished high transcription of the napA storage-protein promoter in transgenic *Brassica napus* seeds. Planta. 1996; 199:515–9.
38. Kim SY, Chung HJ, Thomas TL. Isolation of a novel class of bZIP transcription factors that interact with ABA-responsive and embryo-specification elements in the Dc3 promoter using a modified yeast one-hybrid system. Plant J. 1997;11:1237–51.
39. Finkelstein RR, Lynch TJ. The Arabidopsis abscisic acid response gene ABI5 encodes a basic leucine zipper transcription factor. Plant Cell. 2000;12:599–609.
40. Kaplan B, Davydov O, Knight H, Galon Y, Knight MR, Fluhr R, et al. Rapid transcriptome changes induced by cytosolic $Ca^{2+}$ transients reveal ABRE-related sequences as $Ca^{2+}$-responsive cis elements in Arabidopsis. Plant Cell. 2006;18:2733–48.
41. Fujita M, Fujita Y, Maruyama K, Seki M, Hiratsu K, Ohme-Takagi M, et al. A dehydration-induced NAC protein, RD26, is involved in a novel ABA-dependent stress-signaling pathway. Plant J. 2004;39:863–76.
42. Thu NBA, Hoang XLT, Doan H, Nguyen TH, Bui D, Thao NP, Tran LS. Differential expression analysis of a subset of *Gm*NAC genes in shoots of two contrasting drought-responsive soybean cultivars DT51 and MTD720 under normal and drought conditions. Mol Biol Rep. 2014;41:5563.
43. Shinozaki K, Yamaguchi-Shinozaki K. Gene networks involved in drought stress response and tolerance. J Exp Bot. 2007;58:221–7.
44. Tran LS, Quach TN, Guttikonda SK, Aldrich DL, Kumar R, Neelakandan A, et al. Molecular characterization of stress-inducible *Gm*NAC genes in soybean. Mol Gen Genomics. 2010;281:647–64.
45. Lu PL, Chen NZ, An R, Su Z, Qi BS, Ren F, et al. A novel drought-inducible gene, ATAF1, encodes a NAC family protein that negatively regulates the expression of stress-responsive genes in Arabidopsis. Plant Mol Biol. 2007;63: 289–305.
46. Jensen MK, Lindemose S, Mais F, Reimer JJ, Nielsen M, Perera V, et al. ATAF1 transcription factor directly regulates abscisic acid biosynthetic gene NCED3 in *A. thaliana*. FEBS Open Bio. 2013;29:321–7.

47. Ye Y, Ding Y, Jiang Q, Wang F, Sun J, Zhu C. The role of receptor-like protein kinases (RLKs) in abiotic stress response in plants. Plant Cell Rep. 2017;36:235–42.
48. Zhu J-K. Salt and drought stress signal transduction in plants. Rev Plant Biol. 2002;53:247–73.
49. Nakashima K, Takasaki H, Mizoi J, Shinozaki K, Yamaguchi-Shinozaki K. NAC transcription factors in plant abiotic stress responses. Biochim Biophys Acta. 1819;2012:97–103.
50. Guiltinan M, Marcotte W, Quatrano R. A plant leucine zipper protein that recognizes an abscisic acid response element. Science. 1990;250:267–71.
51. Mehrotra R, Sethi S, Zutshi I, Bhalothia P, Mehrotra S. Patterns and evolution of ACGT repeat cis-element landscape across four plant genomes. BMC Genomics. 2013;14:2–11.
52. Nakashima K, Tran LS, Van Nguyen D, Fujita M, Maruyama K, Todaka D. Functional analysis of a NAC- type transcription factor OsNAC6 involved in abiotic and biotic stress-responsive gene expression in rice. Plant J. 2006;51: 617–30.
53. Yamaguchi-Shinozaki K, Shinozaki K. Organization of cis-acting regulatory elements in osmotic- and cold-stress-responsive promoters. Trends Plant Sci. 2005;10:88–94.
54. Bhargava S, Sawant K. Drought stress adaptation: metabolic adjustment and regulation of gene expression. Plant Breed. 2013;132:21–32.
55. Hernandez-Garcia CM, John JJ. Identification and validation of promoters and cis-acting regulatory elements. Plant Sci. 2014;217:109–19.
56. Martinez MC, Goossens A. Combinatorial transcriptional control of plant specialized metabolism. Trends Plant Sci. 2018;23:324–36.
57. Conforte AJ, Guimarães-Dias F, Neves BAC, Bencke-Malato M, Whipps DF, Alves-Ferreira M. Isolation and characterization of a promoter responsive to salt, osmotic and dehydration stresses in soybean. Gen Mol Biol. 2017;40: 226–37.
58. Behnam B, Iuchi S, Fujita M, Fujita Y, Takasaki H, Osakabe Y, et al. Characterization of the promoter region of an Arabidopsis gene for 9-cis-epoxycarotenoid dioxygenase involved in dehydration-inducible transcription. DNA Res. 2013;20:315–24.
59. Imtiaz M, Yang Y, Liu R, Xu Y, Khan MA, Wei Q, Gao J, Hong B. Identification and functional characterization of the BBX24 promoter and gene from Chrysanthemum in a. Plant Mol Biol. 2015;89:1–19.
60. Goodstein DM, Shu S, Howson R, Neupane R, Hayes RD, Fazo J, et al. Phytozome: a comparative platform for green plant genomics. Nucleic Acids Res. 2012;40:1178–86.
61. Tamura K, Dudley J, Nei MS. MEGA4: molecular evolutionary genetics analysis (MEGA) software ver. 4.0. Mol Biol Evol. 2007;24:1596–9.
62. Jian B, Liu B, Bi Y, Hou W, Wu C, Han T. Validation of internal control for gene expression study in soybean by quantitative real-time PCR. BMC Mol Biol. 2008;9:1–14.
63. Cartharius K, Frech K, Grote K, Klocke B, Haltmeier M, Klingenhoff A, et al. MatInspector and beyond: promoter analysis based on transcription factor binding sites. Bioinformatics. 2005;21:2933–42.
64. Ho Sui SJ, Mortimer JR, Arenillas DJ, Brumm J, Walsh CJ, Kennedy BP, Wasserman WW. OPOSSUM: identification of over-represented transcription factor binding sites in co-expressed genes. Nucleic Acids Res. 2005;33:3154–64.
65. Yamaguchi-Shinozaki K, Shinozaki K. Characterization of the expression of a desiccation-responsive rd29 gene of *A. thaliana* and analysis of its promoter in transgenic plants. Mol Gen Genet. 1993;236:331–40.
66. Clough SJ, Bent AF. Floral dip: a simplified method for agrobacterium-mediated transformation of *A. thaliana*. Plant J. 1998;16:735–43.
67. Jefferson RA. The *GUS* reporter gene system. Nature. 1989;342:837–8.
68. Bradford MM. A rapid and sensitive method for the quantitation of microgram quantities of protein utilizing the principle of protein-dye binding. Anal Biochem. 1976;72:248.
69. An YQ, McDowell JM, Huang S, McKinney EC, Chambliss S, Meagher RB. Strong constitutive expression of the Arabidopsis ACT2/ACT8 actin subclass in vegetative tissues. Plant J. 1996;10:107–21.
70. Bustin SA. Absolute quantification of mRNA using real-time reverse transcription polymerase chain reaction assays. J Mol Endocrinol. 2000;25:169–93.

# 18

# LC-MSMS based screening of emerging pollutant degradation by different peroxidases

Khadega A. Almaqdi[1], Rana Morsi[1], Bahia Alhayuti[1], Farah Alharthi[1] and S. Salman Ashraf[2*]

**Abstract**

**Background:** The presence of a wide range of bioactive organic pollutants in wastewater and municipal water sources is raising concerns about their potential effects on humans. Not surprisingly, various approaches are being explored that can efficiently degrade these persistent organic pollutants. Use of peroxidases has recently been recognized as a novel remediation approach that may have potential advantages over conventional degradation techniques. However, testing the abilities of different peroxidases to degrade diverse emerging pollutants is tedious and cumbersome.

**Results:** In the present study, we present a rapid and robust approach to easily test the degradability of 21 different emerging pollutants by five different peroxidases (soybean peroxidase, chloroperoxidase, lactoperoxidase, manganese peroxidase, and horseradish peroxidase) using an LC-MSMS approach. Furthermore, this approach was also used to examine the role of a redox mediator in these enzymatic degradation assays. Our results show that some of the organic pollutants can be easily degraded by all five of the peroxidases tested, whereas others are only degraded by a specific peroxidase (or when a redox mediator was present) and there are some that are completely resistant to degradation by any of the peroxidases tested (even in the presence of a redox mediator). The degradation of furosemide and trimethoprim by soybean peroxidase and chloroperoxidase, respectively, was investigated in detail by examining the transformation products generated during their degradation. Some of the products generated during enzymatic breakdown of these pollutants have been previously reported by others, however, we report many new transformation products.

**Conclusions:** LC-MSMS approaches, like the one described here, can be used to rapidly evaluate the potential of different peroxidases (and redox requirements) to be used as bioremediation agents. Our preliminary result shows peroxidases hold tremendous potential for being used in a final wastewater treatment step.

**Keywords:** Emerging pollutants, Soybean peroxidase, Chloroperoxidase, Lactoperoxidase, Manganese peroxidase, Horseradish peroxidase, Redox mediator, Furosemide and trimethoprim

## Background

It is now well-established that "contaminants of emerging concerns" or "emerging pollutants" are increasingly being detected in our water supply. These emerging pollutants comprise an extensive array of diverse compounds and their transformation products, such as nonsteroidal anti-inflammatory drugs (NSAIDs), analgesics, antibiotics, textile dyes, hormones, personal care products and pesticides [1]. A recent study of pesticide contamination due to agriculture activities found significant concentrations of Fluometuron (317.6 μg/L), Chlorpyrifos (0.42 μg/L), and Prometryn (0.48 μg/L) in surface waters of Lake Vistonis Basin, in Greece [2]. Similarly, significant levels of pharmaceuticals (e.g. Lincomycin, Sulfamethoxazole, and Tetracycline) have been detected in U.S. streams as early as 1999 [3]. Not surprisingly, these compounds are suspected to cause a wide array of adverse ecological or human health effects and have become the focus of various

* Correspondence: syed.ashraf@ku.ac.ae
[2]Department of Chemistry, College of Arts and Sciences, Khalifa University, P O Box 127788, Abu Dhabi, UAE

government as well as academic research groups [4]. For example, the presence of perfluorinated compounds in the serum has been correlated with breast cancer risk in Greenlandic Inuit women [5]. Additionally, it has been reported that pollutants such as perfluorooctanoate and perfluorooctane sulfonate may be linked to decreased human reproductive abilities [6]. Scientific literature is full of reports of various physical and chemical approaches can that be employed for the removal of these emerging pollutants [7–10]. However, more research is still needed to develop more efficient, economical, and 'environmental-friendly' and 'greener' remediation approaches.

During the past few years, the role of oxidoreductive enzymes in 'green processes' has become more established and not surprisingly various enzyme systems have been employed for the efficient degradation of diverse organic pollutants [11, 12]. Amongst the various advantages offered by enzymatic degradation approach, the most important ones are the mild and less toxic reagents and conditions that are normally employed in their use as well as their ability to degrade a wide range of substrates. The main potential disadvantage with the use of enzymes is their relatively high cost, however, this can be ameliorated using recombinant DNA technology to mass-produce cheaper enzymes. Literature survey shows that various types of pollutants have been degraded by two different classes of enzymes such as laccases and peroxidases such as Soybean Peroxidase (SBP), Manganese Peroxidase (MnP), Lignin Peroxidase (LiP) and Horseradish Peroxidase (HRP) [13–15]. Additionally, peroxidases from other plant sources such as cauliflower, white radish, and turnip, have been used for the degradation of various organic compounds [16–19]. Besides this, peroxidases from bamboo shoots and lemon peel have also been used for degrading dyes [20, 21]. The addition of redox mediator (RM) to the system has shown to enhance the degradation process to produce less toxic substances [22–25]. Despite the relatively large number of reports showing the application of peroxidases for remediation purposes, only a very few studies have carried out detailed and systemic studies comparing the efficiencies of different oxidoreductases (e.g. peroxidases and laccases) towards degrading a wide range of emerging pollutants. This shortage of systematic studies further highlights the cumbersome and tedious nature of these 'peroxidase-degradability screening' studies.

The current work describes a sensitive and robust approach using LC-MSMS that was developed to simultaneously quantify a large number of emerging pollutants and to easily examine their degradability by different peroxidases. This approach was also used to examine the effect of redox mediators for efficient peroxidase-mediated degradation of emerging pollutants. Additionally, we report on the transformation products generated during the enzymatic breakdown of furosemide (with SBP) and trimethoprim (with CPO) in the presence of redox mediator HOBT. Interestingly, many of the intermediates observed have not been previously reported for the degradation of these emerging pollutants by other remediation methods.

## Results

### Development of a sensitive LC-MSMS based method for the quantification of 21 emerging pollutants

HPLC and LC-MS-based methods are widely reported for the detection and quantification of various individual organic compounds, including emerging pollutants. However, since we wanted to simultaneously study the degradation of a large number of different emerging pollutants (24 of them), we first developed a sensitive, robust, and easy LC-MSMS method, using the Multiple Reaction Monitoring (MRM) approach. The MRM method uses tandem mass spectrometers to specifically monitor the "precursor to product transition" generated when a specific emerging pollutant (precursor ion) is fragmented into a specific product ion. Since the detection is based on a specific "precursor → product transition" which is unique to a specific compound, it allows for simultaneous detection of a large number of compounds without having them completely resolved in the liquid chromatography part of the LC-MSMS method [26]. Figure 1 schematically shows the steps that are taken in developing these MRM-based assays (for sulfamethoxazole, for example) – starting with confirming the parental mass of sulfamethoxazole (253 Da) in the LC-MS (when run in "Total Ion Chromatogram" (TIC) mode. This precursor (parent) ion $(M+H)^+$ species (254 m/z) is then fragmented in the LC-MS by increasingly higher collision energy values (0 V, 10 V, 20 V, and 30 V were used for sulfamethoxazole). When a significantly high and strong signal for a specific product ion is observed (e.g. 156 m/z), that specific collision energy (e.g. 20 V) and the precursor → product transition (254→ 156) are then used for the MRM method. There are numerous examples of the use of such MRM-based analyses for organic compounds in various matrices [26, 27].

Table 1 shows the categories and structures of these different emerging pollutants, as well as their MRM and mass-spectrometry parameters. Also shown in

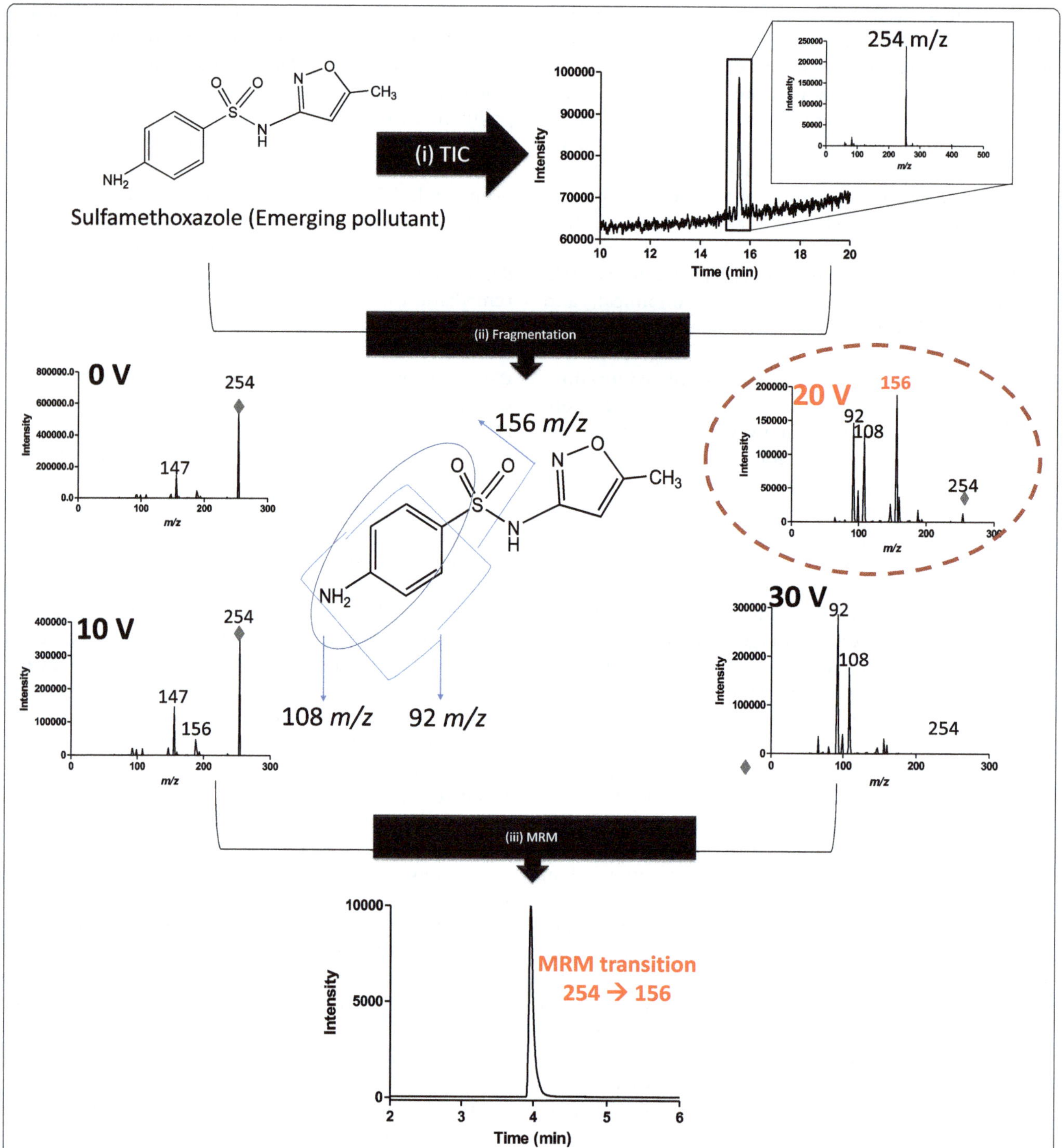


**Fig. 1** Schematic diagram of the MRM (LC-MSMS) method development for Sulfamethoxazole

Table 1 are the retention times on a $C_{18}$ column for these compounds (explained in more detailed in the Methods section). Figure 2 shows the typical chromatogram obtained when a mixture of these 21 emerging pollutants were analyzed using the developed method. The extracted individual chromatograms of these emerging pollutants (concentration 2 ppm) are shown in Fig. 3, which shows the specificity and sensitivity of the analytical assay.

### Degradation of emerging pollutants by five different peroxidases

It has been previously reported that different peroxidases may have different remediation efficiencies for

**Table 1** Names and chemical structures of the 21 emerging pollutants used in this study and their LC-MSMS parameters

| | Category | Emerging pollutant | Structure | Retention Time (min) | MRM Transition (m/z) | Dwell time (ms) | Frag-mentor Voltage (V) | Collision Energy (V) | Polarity |
|---|---|---|---|---|---|---|---|---|---|
| 1 | Antibiotic | Roxithromycin | 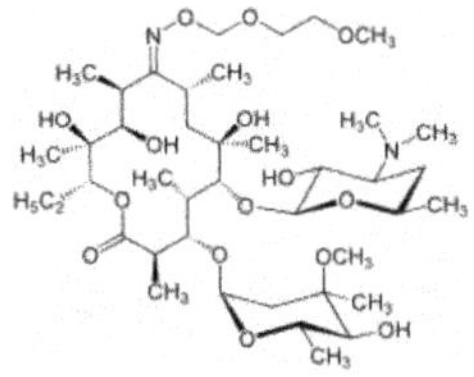 | 11.6 | 837 → 680 | 200 | 135 | 20 | Positive |
| 2 | | Lincomycin-HCl | 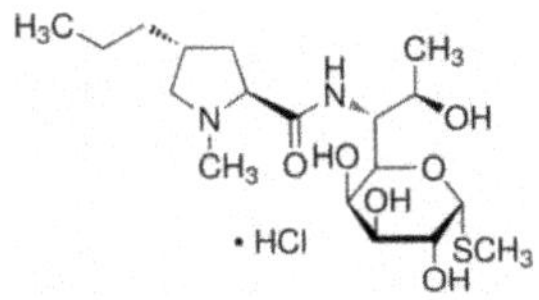 | 7.6 | 407 → 359 | 200 | 135 | 20 | Positive |
| 3 | | Norfloxacin | 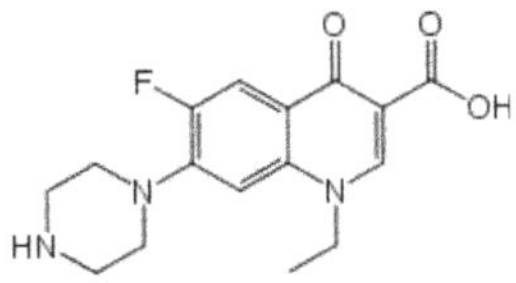 | 8.2 | 320 → 302 | 200 | 135 | 20 | Positive |
| 4 | | Trimethoprim | 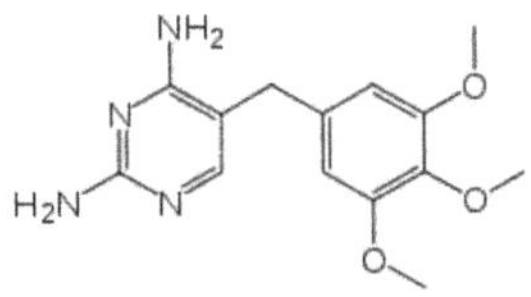 | 7.9 | 291 → 230 | 200 | 135 | 20 | Positive |
| 5 | | Sulfamethoxazole | 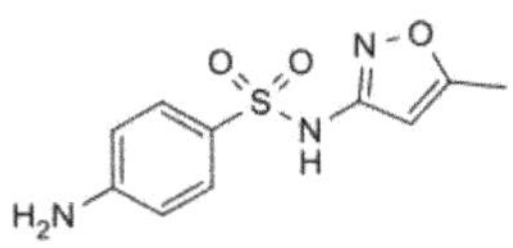 | 9.3 | 254 → 156 | 200 | 135 | 20 | Positive |
| 6 | Antidepressant | Venlafaxine-HCl | 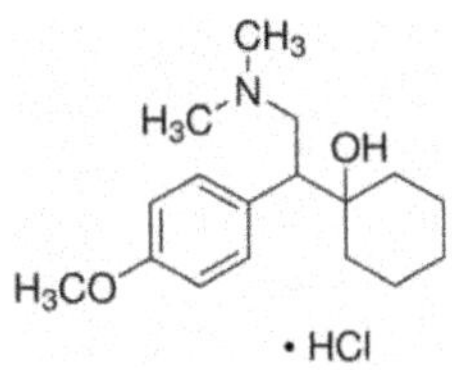 | 9.4 | 278 → 260 | 200 | 135 | 10 | Positive |
| 7 | Antioxidant | Caffeic acid | 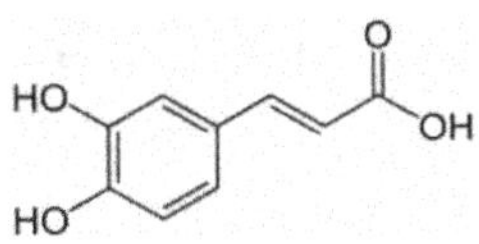 | 7.8 | 181 → 163 | 200 | 135 | 20 | Positive |
| 8 | Anti-seizure drug | Phenytoin | 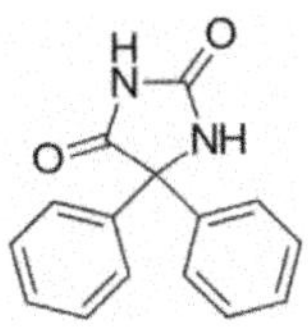 | 11.1 | 253 → 182 | 200 | 135 | 10 | Positive |
| 9 | Diuretic drug | Hydrochloro-thiazide | 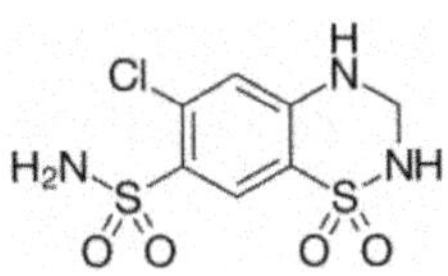 | 6.4 | 296 → 269 | 70 | 140 | 20 | Negative |
| 10 | | Furosemide | 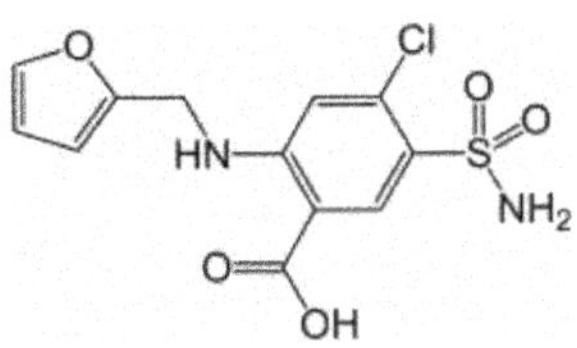 | 11 | 329 → 285 | 70 | 140 | 15 | Negative |

**Table 1** Names and chemical structures of the 21 emerging pollutants used in this study and their LC-MSMS parameters *(Continued)*

| | Category | Emerging pollutant | Structure | Retention Time (min) | MRM Transition (m/z) | Dwell time (ms) | Frag-mentor Voltage (V) | Collision Energy (V) | Polarity |
|---|---|---|---|---|---|---|---|---|---|
| 11 | Beta-blocker (high blood pressure drug) | Atenolol | 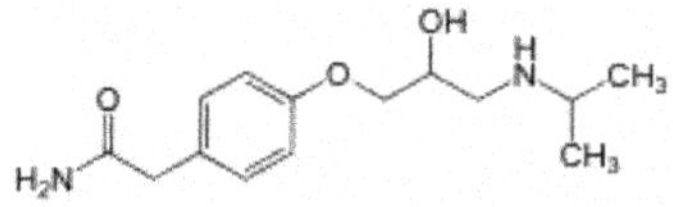 | 7.1 | 267 ➔ 190 | 200 | 135 | 20 | Positive |
| 12 | Fungicide | Thiabendazole | 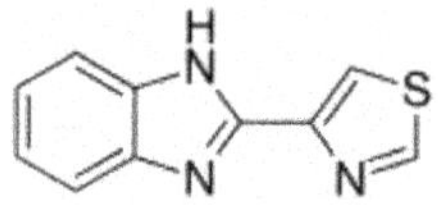 | 7.6 | 202 ➔175 | 200 | 135 | 30 | Positive |
| 13 | Herbicide | Prometryn | 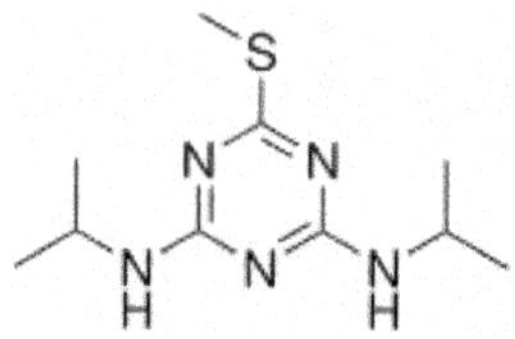 | 11.6 | 242 ➔ 158 | 200 | 135 | 30 | Positive |
| 14 | | MCPA | 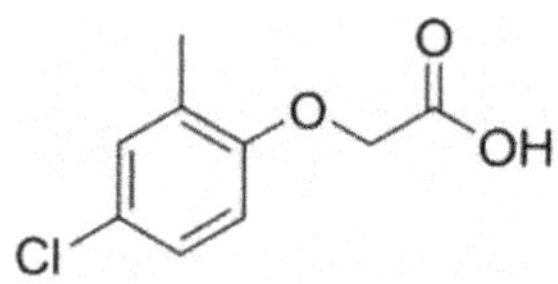 | 12 | 201 ➔ 125 | 200 | 47 | 13 | Positive |
| 15 | | Fluometuron | 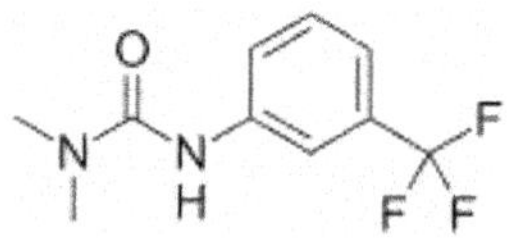 | 11.7 | 233 ➔ 72 | 200 | 135 | 30 | Positive |
| 16 | Histamine $H_2$ receptor antagonist | Cimetidine | 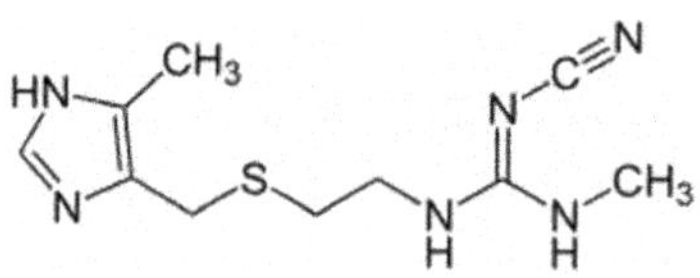 | 6.9 | 253 ➔ 159 | 200 | 135 | 10 | Positive |
| 17 | Insect repellent | DEET | 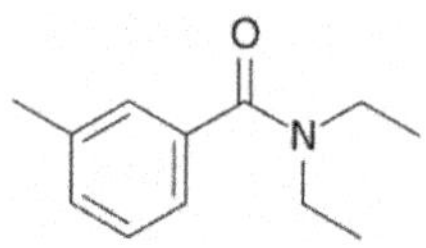 | 11.9 | 192 ➔ 119 | 200 | 135 | 30 | Positive |
| 18 | Nonsteroidal anti-inflammatory drug (NSAID) | Meloxicam | 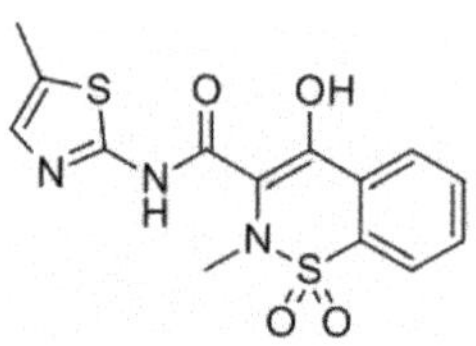 | 12.8 | 352 ➔ 115 | 200 | 135 | 6 | Positive |
| 19 | | Ibuprofen | 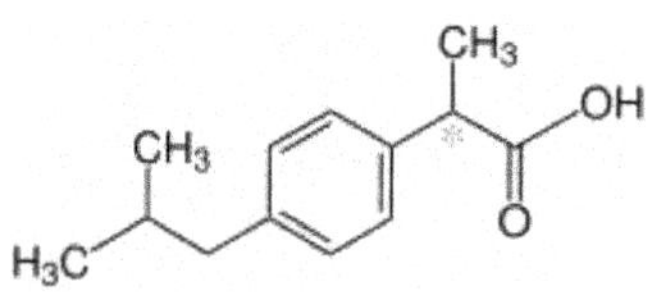 | 14.4 | 207 ➔ 161 | 200 | 135 | 20 | Positive |
| 20 | Stimulant | Caffeine | 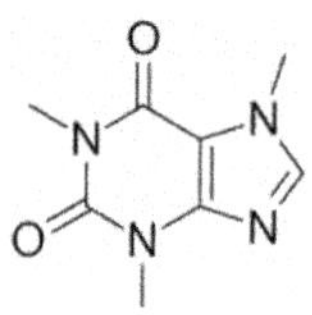 | 7.8 | 195 ➔ 138 | 200 | 135 | 30 | Positive |
| 21 | Vulcanization agent (for rubber) | 2-Mercapto Benzothiazole (MBT) | | 10.6 | 168 ➔ 135 | 200 | 135 | 30 | Positive |

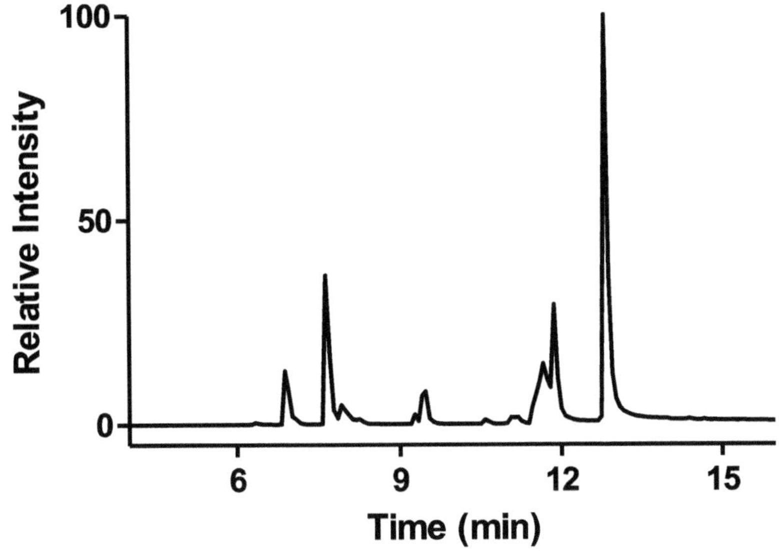


**Fig. 2** Total MRM scan of a mixture of 21 emerging pollutants (EPs). [EPs] = 2 ppm, [$H_2O_2$] = 0.1 mM, [HOBT] = 0.1 mM, pH = 4

different organic pollutants [28, 29]. Therefore, mixtures of 21 emerging pollutants were separately treated with Soybean Peroxidase (SBP), Chloroperoxidase (CPO), Lactoperoxidase (LPO), Manganese Peroxidase (MnP), or Horseradish Peroxidase (HRP), as described in more detail in the Methods section. Figure 4a and b show the residual amounts of Meloxicam, a nonsteroidal anti-inflammatory drug, that has been found in many water bodies [30], upon treatment by the above five named peroxidases. As can be seen from Fig. 4b, all of the peroxidases tested were able to degrade Meloxicam nicely, with SBP being slightly more efficient than the others. This is consistent with our previous studies showing that SBP could efficiently degrade related thiazole compounds [28]. However, it seems that not all organic pollutants could be equally degraded by the peroxidases tested. For example, Fig. 5a and b show the percentage of Roxithromycin remaining after treatment with the peroxidases. As can be seen, SBP was not able to degrade this compound and CPO and MnP were also not efficient in degrading it, showing only around 25% degradation. However, HRP could easily degrade it and showed almost 95% degradation (in 30 min), as could LPO as well, showing about 80% degradation. These and other differences in the efficacies of different peroxidases to degrade specific emerging pollutants are summarized in Table 2. As can be seen from the table, 8 of the 21 emerging pollutants tested could be degraded very efficiently (> 75% in 30 min) by at least one of the peroxidases, with two more (Thiabendazole and Meloxicam) showing fairly good enzymatic degradation (> 50%).

### Role of redox mediators for efficient degradation of emerging pollutants by peroxidases

In the present study, we also evaluated the role of a redox mediator for the efficient degradation of our chosen 21 emerging pollutants by the five peroxidases. As can be seen in Fig. 6a and b, Hydrochlorothiazide showed only marginal degradation by HRP + $H_2O_2$, resulting in about 10% degradation in 30 min. However, the presence of 0.1 mM HOBT increased the degradation rate dramatically (to about 75% degradation in 30 min). Similar results were observed with SBP-mediated degradation of Sulfamethoxazole, which also required HOBT for efficient degradation [31]. However, such a dramatic enhancement of organic pollutant degradation by the redox mediator, HOBT, was not universal. For example, SBP + $H_2O_2$ could degrade about 55% of Thiabendazole in 30 min, however, the addition of HOBT had no beneficial effect on the degradation of this fungicide (Fig. 6c and Additional file 1: Figure 6CS). Interestingly, the inclusion of HOBT could also hinder the peroxidase-mediated degradation of these compounds. This is dramatically seen for Manganese Peroxidase mediated degradation of Thiabendazole, where the presence of HOBT completely inhibited its degradation (Fig. 6d and Additional file 1: Figure 6DS). In fact, these inhibitory effects of redox mediators are not unexpected as these redox mediators can bind to and react with peroxidases with high affinity and thereby compete with organic pollutants for binding. In fact, our lab has been previously reported that CPO-mediated chlorination of ThT showed a significant decrease in the presence of HOBT [28]. Nevertheless, 8 of the compounds tested appeared to be completely

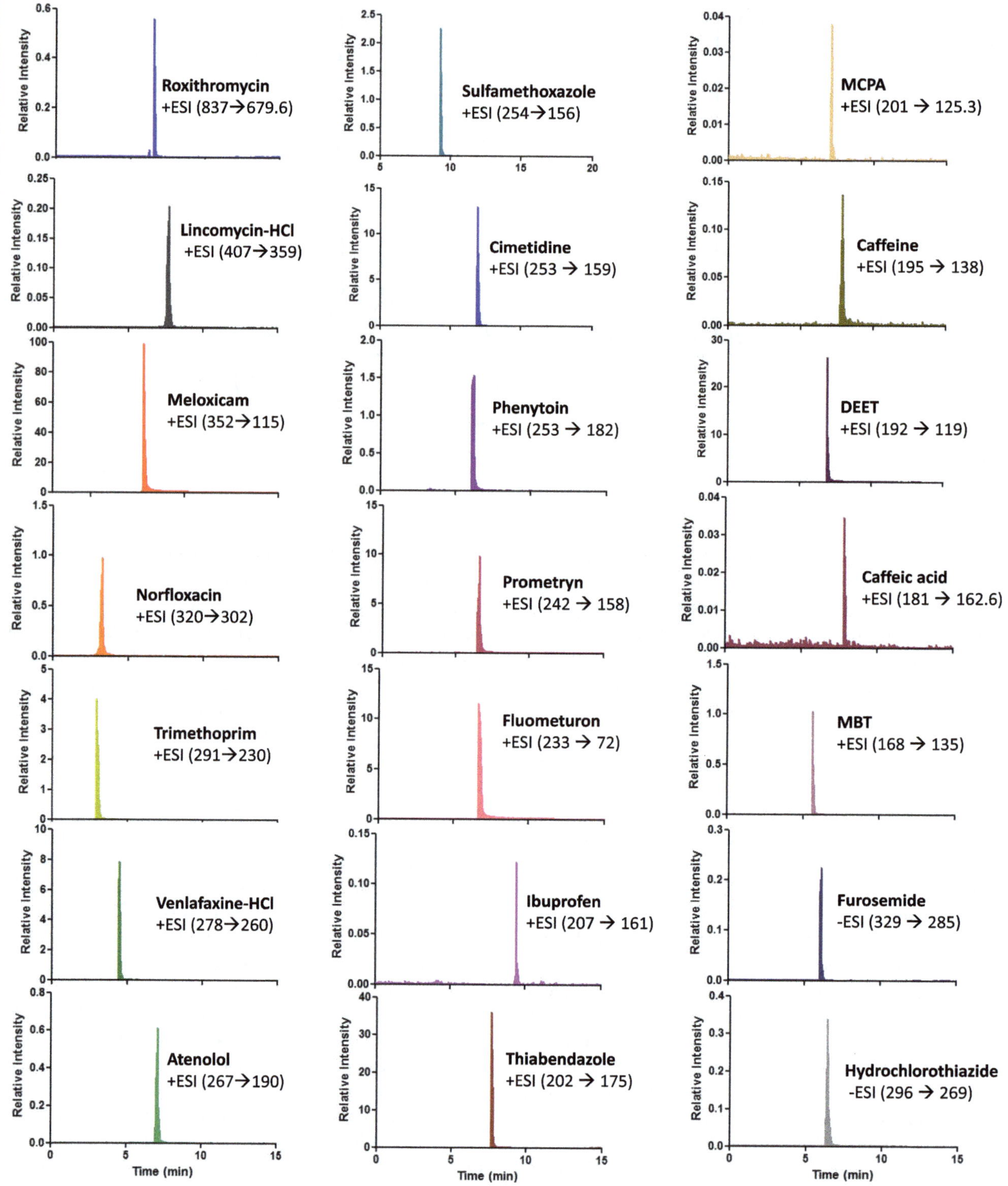


**Fig. 3** Individual extracted MRM scans for the 21 emerging pollutants (EPs). [EPs] = 2 ppm, $[H_2O_2] = 0.1$ mM, [HOBT] = 0.1 mM, pH = 4

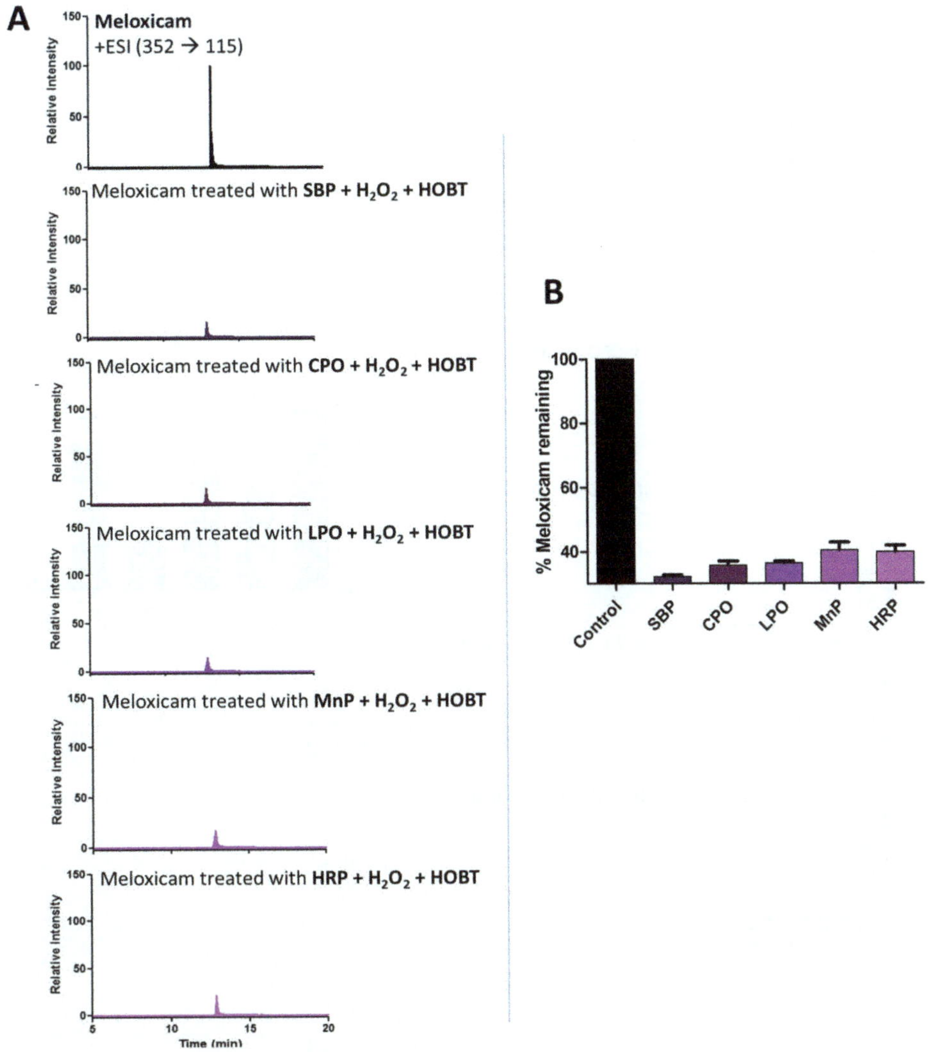


**Fig. 4 a** MRM scans of Meloxicam treated with SBP, CPO, LPO, MnP and HRP enzymes. **b** % of Meloxicam remaining after treatment with SBP, CPO, LPO, MnP and HRP enzymes. [Meloxicam] = 2 ppm, [Enzyme] = 0.36 μM, [$H_2O_2$] = 0.1 mM added 3 times of 10 min interval, [HOBT] = 0.1 mM, pH = 2 for CPO, pH = 4 for SBP, pH = 5 for MnP and pH = 6 for LPO and HRP

recalcitrant to degradation by any of the five peroxidases tested, even in the presence of the redox mediator, HOBT (Table 2).

### SBP-mediated degradation of emerging pollutants in a real wastewater sample

We also wanted to use our newly developed approach to carry out a preliminary and exploratory study to screen a real wastewater sample for the presence of emerging pollutants and to test their degradation by one of the peroxidases. LC-MSMS analysis of a sample of pretreatment wastewater feed from the local municipal wastewater treatment plant showed the presence of two different emerging pollutants, namely, meloxicam and DEET. The presence of DEET in wastewater feed has been previously reported in China by Sui et al. [32], whereas meloxicam and many other emerging pollutant have been detected in surface water as well as in wastewater treatment plant in Serbia [33].

Treatment of the local municipal wastewater sample treatment produced results very similar to those obtained with neat solutions (Table 2). As can be seen in Fig. 7, SBP was able to degrade about 50% of the meloxicam in the wastewater sample in 30 min, whereas no significant degradation was observed for the DEET present in the sample (Additional file 1: Figure 7AS and figure 7BS). Although no optimization steps were carried out in this exploratory study with real wastewater, these preliminary experiments confirmed the results obtained with neat pollutants. Furthermore, the results also support the potential applicability of peroxidase enzymes to

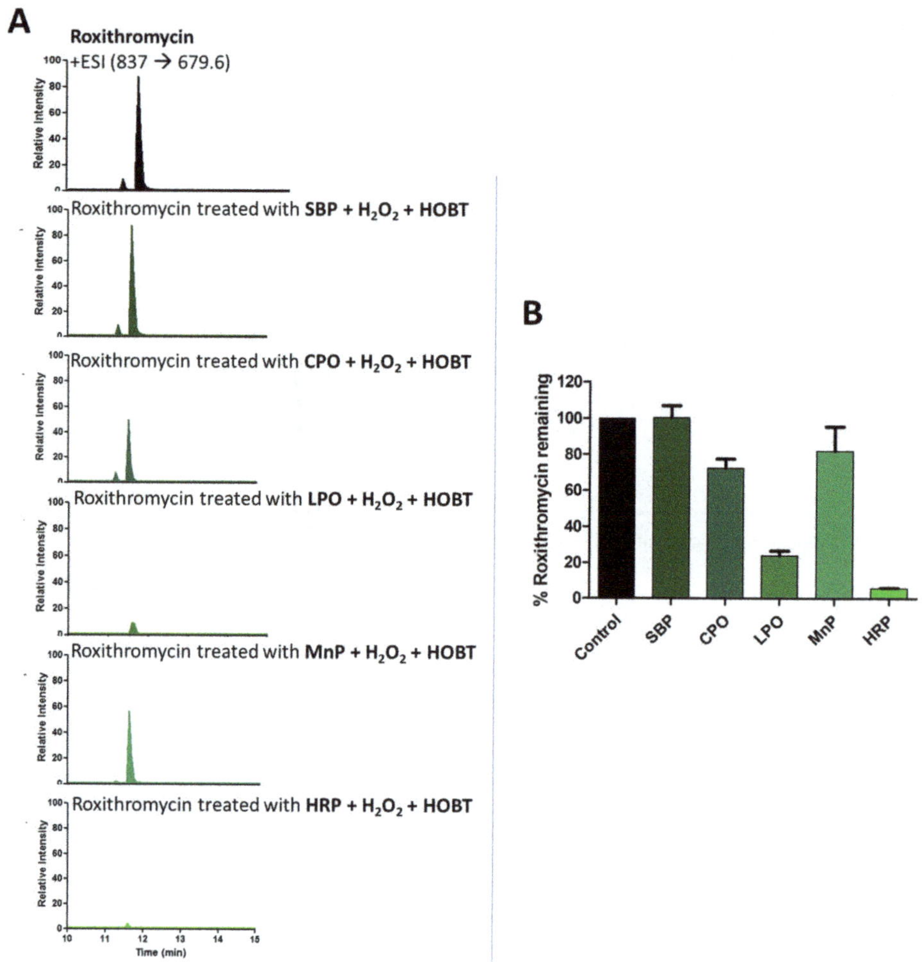


**Fig. 5 a** MRM scans of Roxithromycin treated with SBP, CPO, LPO, MnP and HRP enzymes. **b** % of Roxithromycin remaining after treatment with SBP, CPO, LPO, MnP and HRP enzymes. [Roxithromycin] = 2 ppm, [Enzyme] = 0.36 μM, [$H_2O_2$] = 0.1 mM added 3 times of 10 min interval, [HOBT] = 0.1 mM, pH = 2 for CPO, pH = 4 for SBP, pH = 5 for MnP and pH = 6 for LPO and HRP

degrade some organic pollutants in complex matrixes, such as wastewater samples.

### Peroxidase-mediated degradation of furosemide and trimethoprim and transformation product identification

Although the use of mixture of emerging pollutants allowed for rapid and simultaneous screening of peroxidase-mediated degradation of various compounds, no specific details could be obtained about the nature of the degradation products generated. Therefore, we decided to carry out detailed studies with two of the emerging pollutants (furosemide and trimethoprim) to examine the transformation products generated upon their degradation. Addition of SBP + $H_2O_2$ to neat furosemide could degrade about 40% of the diuretic drug, interestingly this could be drastically improved to 100% degradation when the redox mediator HOBT was added to the reaction mixture. This is more clearly seen in the LC-MS total ion chromatogram (Fig. 8), where the furosemide peak decreases upon the addition of SBP + $H_2O_2$ and is completely gone when HOBT was added. These results are consistent with what was observed for furosemide in a mixture with 20 other compounds (Table 2). Furthermore, it can be seen in the chromatogram for the "furosemide + SBP + $H_2O_2$ + HOBT" sample, that decrease of the furosemide peak was accompanied by the appearance of several minor peaks, suggesting the generation of degradation intermediates. The insets in Fig. 8 show the new transformation products to have the m/z ratios of 249, 205 and 118. Fig. 9a shows a summary of the degradation scheme of furosemide by

**Table 2** A summary of the % of EPs remaining after treatment with five different enzymes SPO, CPO, LPO, MnP and HRP with and without HOBT. Key: %EP remaining 0–25% (+++), 25–50% (++), 50–75% (+), More than 75% (–)

| NO. | Category | Emerging pollutant | SBP Only | SBP + HOBT | CPO Only | CPO + HOBT | LPO Only | LPO + HOBT | MnP Only | MnP + HOBT | HRP Only | HRP + HOBT |
|---|---|---|---|---|---|---|---|---|---|---|---|---|
| 1 | Antibiotic | Roxithromycin | - | - | - | + | - | +++ | + | - | + | +++ |
| 2 | Antibiotic | Lincomycin-hydrochloride | - | - | - | - | - | ++ | - | - | - | +++ |
| 3 | Antibiotic | Norfloxacin | - | - | - | - | - | + | - | - | - | + |
| 4 | Antibiotic | Trimethoprim | - | - | - | ++ | - | - | - | - | - | - |
| 5 | Antibiotic | Sulfamethoxazole | + | +++ | - | - | + | +++ | - | - | - | +++ |
| 6 | Antidepressant | Venlafaxine-hydrochloride | - | - | - | - | - | - | - | - | - | + |
| 7 | Antioxidant | Caffeic acid | +++ | +++ | + | + | +++ | +++ | +++ | +++ | +++ | +++ |
| 8 | Anti-seizure drug | Phenytoin | - | - | - | - | - | - | - | - | - | - |
| 9 | Diuretic drug | Hydrochlorothiazide | - | - | - | - | - | + | - | - | - | +++ |
| 10 | Drug for treating fluid build-up due to heart failure, liver scarring, or kidney disease | Furosemide | ++ | +++ | ++ | +++ | - | +++ | - | - | - | +++ |
| 11 | Drug for treating high blood pressure | Atenolol | - | - | - | - | - | - | - | - | - | - |
| 12 | Fungicide | Thiabendazole | ++ | ++ | ++ | - | ++ | ++ | ++ | - | ++ | ++ |
| 13 | Herbicide | Prometryn | - | - | - | - | - | - | - | - | - | - |
| 14 | Herbicide | MCPA | - | - | - | - | - | - | - | - | - | - |
| 15 | Herbicide | Fluometuron | - | - | - | - | - | - | - | - | - | - |
| 16 | Histamine $H_2$ receptor antagonist | Cimetidine | ++ | ++ | ++ | - | ++ | - | ++ | - | ++ | +++ |
| 17 | Insect repellents | DEET | - | - | - | - | - | - | - | - | - | - |
| 18 | Nonsteroidal anti-inflammatory drug (NSAID) | Meloxicam | ++ | ++ | + | ++ | ++ | ++ | + | ++ | ++ | ++ |
| 19 | Nonsteroidal anti-inflammatory drug (NSAID) | Ibuprofen | - | - | - | - | - | - | - | - | - | - |
| 20 | Stimulant | Caffeine | - | - | - | - | - | - | - | - | - | - |
| 21 | Sulfur vulcanization of rubber | Mercaptobenzothiazole (MBT) | +++ | +++ | ++ | ++ | +++ | +++ | +++ | +++ | +++ | +++ |

SBP, with the proposed structure of the 249 m/z intermediate. A number of research groups have shown that furosemide can be degraded into smaller breakdown products by various other remediation approaches, including photodegradation [34, 35], electro-Fenton and bioremediation [36]. Table 3 summarizes of these degradation studies, showing the various furosemide breakdown products that have been previously reported. It is interesting to note the 249 m/z species that we observed in our degradation experiments has also been reported earlier by [34, 35].

A similar study was carried out for the degradation of the antibiotic trimethoprim by Chloroperoxidase - the only peroxidase (of the five tested) that could degrade this compound. As shown in Table 2, degradation of trimethoprim by CPO needed the presence of the redox mediator HOBT (data not shown). LC-MSMS analysis of the degradation reaction showed that 6 different transformation products could be detected (Additional file 1: Figure 9BS). Table 3 and Fig. 9b show a summary of the results of the enzymatic degradation of trimethoprim by CPO. Comparison of the transformation products generated during CPO-mediated degradation of trimethoprim with previously published degradation studies of this pollutant [37–39] showed that 3 of the previously published trimethoprim breakdown products were also detected in our study with Chloroperoxidase.

## Discussion

As has been previously reported, Soybean Peroxidase (SBP) and Chloroperoxidase (CPO) show different efficiency and specificity in degrading a given thiazole pollutant [28, 29]. We wanted to extend that observation to three additional peroxidases and examine a total of 21 different emerging pollutants. Our results clearly showed remarkable differences in

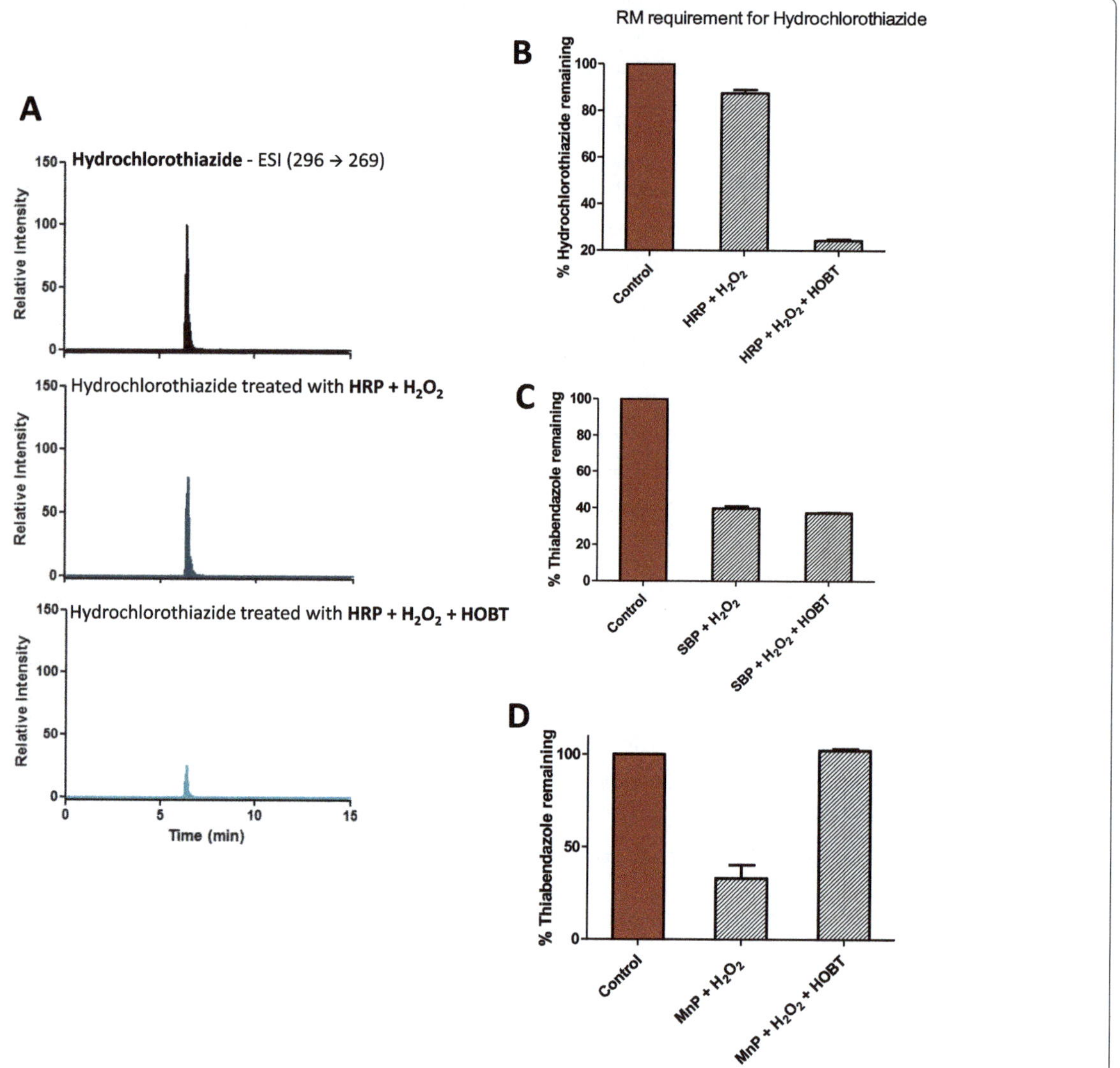


**Fig. 6** Effect of redox mediator on peroxidase-mediated pollutant degradation. **a** MRM scans of Hydrochlorothiazide treated with HRP enzyme with and without HOBT. **b** % of Hydrochlorothiazide remaining after treatment with HRP enzyme with and without HOBT. [Hydrochlorothiazide] = 2 ppm, [HRP] = 0.36 μM, [$H_2O_2$] = 0.1 mM added 3 times of 10 min interval, [HOBT] = 0.1 mM, pH = 6. **c** % of Thiabendazole remaining after treatment with SBP enzyme with and without HOBT. **d** % of Thiabendazole remaining after treatment with MnP enzyme with and without HOBT. [Thiabendazole] = 2 ppm, [$H_2O_2$] = 0.1 mM added 3 times of 10 min interval, [HOBT] = 0.1 mM, pH = 4 with SBP enzyme and pH = 5 with MnP enzyme, [enzyme] = 0.36 μM

the specificity and degradability of a range of organic pollutants by different peroxidases. It is tempting to speculate that such differences are primarily related to binding efficiencies of these compounds in the active sites of the peroxidases. Perhaps, future detailed structural studies of the pollutant-peroxidase complexes will shed some light on these interesting observations.

It is also well established that small, diffusible redox mediators can significantly enhance the oxidative degradation of organic pollutants by peroxidases. For example, it has been previously shown that Thioflavin T, a model thiazole pollutant, could only be degraded by SBP in the presence of the redox mediator, 1-hydroxybenzotriazole (HOBT) [28]. Interestingly, as reported earlier [28], addition of HOBT could also inhibit the enzymatic remediation of a few compounds, thus indicating that for some pollutants, HOBT can act as a competitor inhibitor.

The surprising observation that some of the emerging pollutants (8 of the 21 tested) could not be easily degraded by enzyme-mediated oxidative reactions (Table 2), suggest that these enzymatically recalcitrant compounds may require other remediation approaches, such as advanced oxidative processes (AOPs). In fact, we have recently shown that both UV-$H_2O_2$ photolytic AOP and peroxidase-

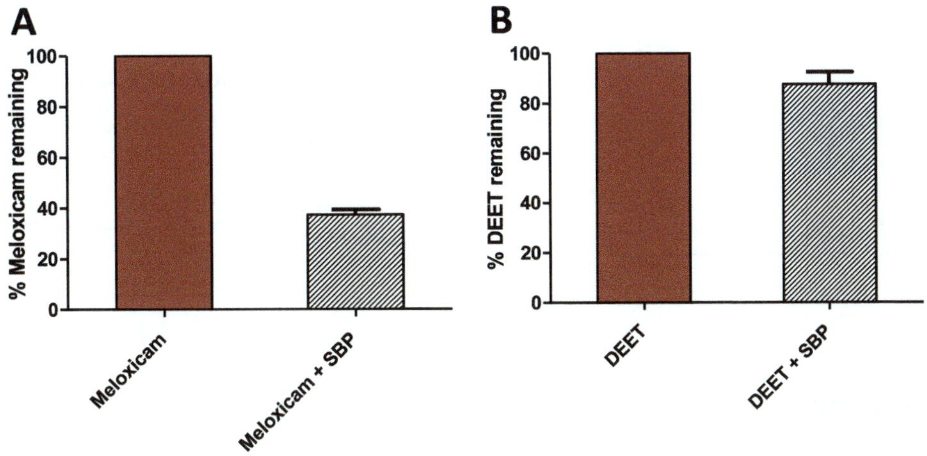


**Fig. 7** Degradation of pollutants spiked in real wastewater. **a** % of meloxicam remaining after treatment with SBP enzyme. **b** % of DEET remaining after treatment with SBP enzyme. [SBP] = 1.56 μM, [$H_2O_2$] = 0.1 mM, [HOBT] = 50 μM, pH = 4

mediated approaches could both be used for the efficient degradation of Sulfamethoxazole, however, they appear to be based on different mechanistic degradation pathways [31]. Perhaps, the enzyme-based degradation approaches could be used in tandem with other AOP-based treatments in wastewater treatment plants to ensure complete and efficient degradation of diverse kinds of emerging pollutants. Our additional experiments with two selected pollutants showed that enzymatically degraded pollutants generated some of the same intermediates that have been previously reported by other remediation methods. However, in the present study, we reported two new transformation products from furosemide that have not been reported previously, namely the m/z 205 and m/z 118. Similarly, we also present three new trimethoprim transformation products (obtained during CPO-mediated degradation) that had not been reported earlier (m/z 120, 225, and 269).

## Conclusions

In summary, we present here a rapid, robust and easy approach to test several oxidative enzymes (peroxidases, laccases, etc.) for their abilities to be used as remediating agents in degrading a large number of emerging pollutants. Our experiments with five different peroxidases show that different enzymes show varying abilities to degrade specific organic compounds. The described LC-MSMS approach was also be used to examine the requirement for redox mediators (HOBT) for peroxidase-mediated degradation of organic pollutants. Additionally, we report that surprisingly, 8 of the 21 emerging pollutants appear to be completely recalcitrant to oxidative degradation by the five peroxidases tested. Finally, we show that degradation studies of two different pollutants (neat and individually) by the peroxidases showed similar results as in a mixture. We also used LC-MSMS to identify some of the furosemide and trimethoprim products generated during their degradation studies using SBP and CPO, respectively.

## Methods

### Reagents and enzymes

All emerging pollutants were obtained from Sigma-Aldrich. Solvents used in LC-MS like LC-MS grade water, acetonitrile, and formic acid as well as Hydrogen peroxide was purchased from Sigma-Aldrich. Universal buffers were used in all experiments (0.2 M potassium phosphate ($K_2HPO4$) and 0.1 M citrate acid). The specific enzymes activity for SBP, CPO, LPO MnP and HRP were 2700 IU/mg (1 mg/mL, 26 μM), 1296 IU/mg (17 mg/mL, 405 μM), (10 mg/mL, 26 μM), 200 IU/g (1 mg/mL, 26 μM) and 279 IU/mg (1 mg/mL, 26 μM) respectively. The enzymes (SBP, CPO, and LPO) were purchased from Bio-Research Products (North Liberty, USA). The enzymes (MnP and HRP) they were purchased from Sigma-Aldrich.

### LC-MSMS method development

After treating the 21 emerging pollutants with the five different enzymes SBP, CPO, LPO, MnP, and HRP the samples were analyzed using LCMS. The samples were filtered before injecting them in the LCMS using a 0.45 μm cellulose acetate syringe filter. The column used for analysis was $C_{18}$ column (ZORBAX Eclipse Plus). The column had the following characteristics 1.8 μm particle size, 2.1 mm inner diameter and its length was 50 mm. For the C18 column, its temperature was maintained at 35 °C. The Mass Spectrometry used was 6420 Triple Quad detector (Agilent Technologies). The flow rate for the mobile phase in the column was 0.4 mL/min. The method developed used two mobile phases: 1.

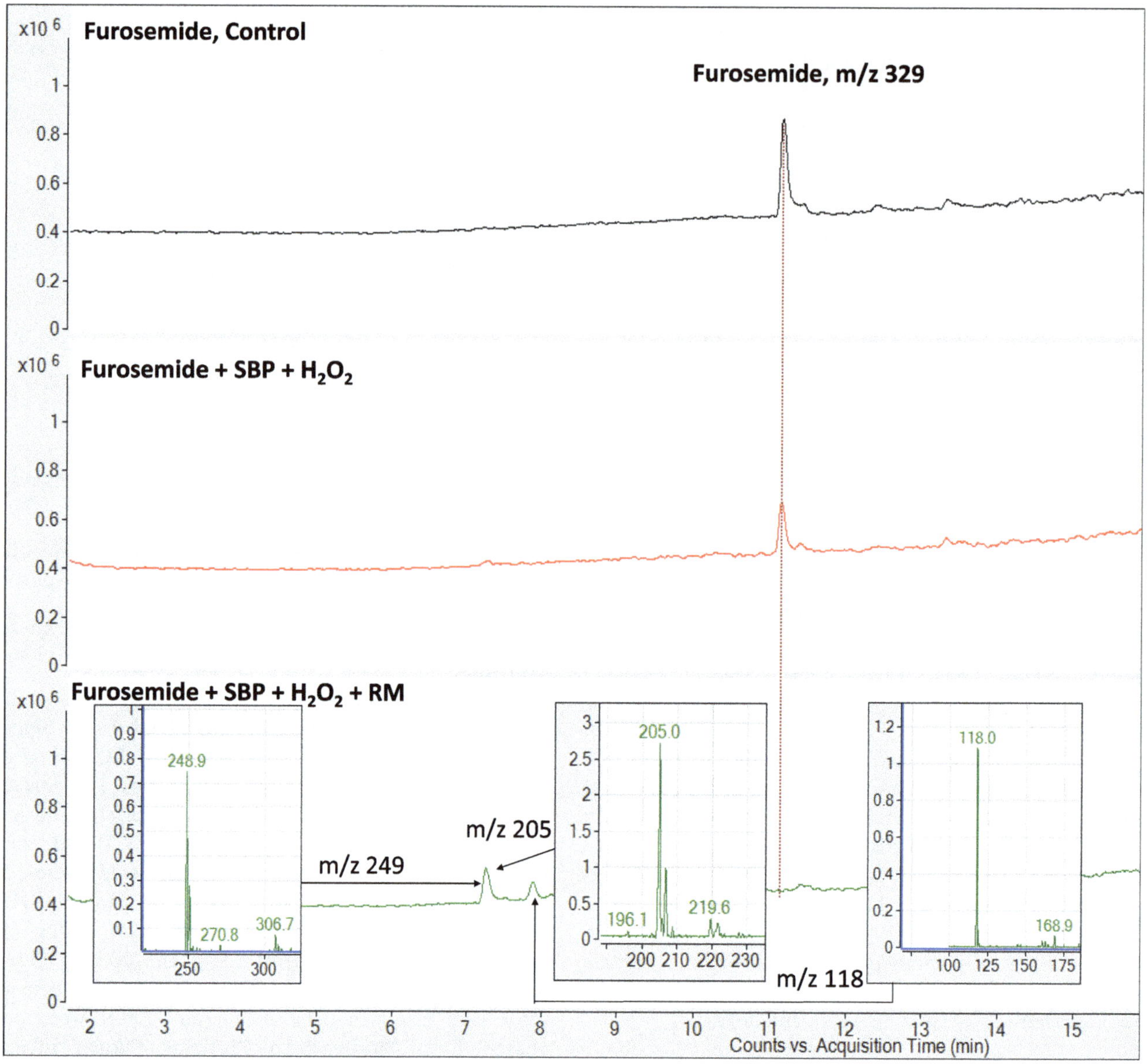


**Fig. 8** LC-MS total ion chromatograms of furosemide treated with SBP, SBP + $H_2O_2$, and SBP + $H_2O_2$ + HOBT

(mobile phase A) which was LCMS grade water with 0.1% LCMS grade formic acid 2. (mobile phase B) which was 100% LCMS grade acetonitrile. The method in the multiple reaction monitoring (MRM) analysis was set as follows: 2.5 min of 100% A and 0% B, followed by a 0–80% gradient of B from 2.5–15 min, then at 15.1 min A was 10% and B was 90% for 3 min and finally 2 min of 95% A and 5% B. Positive and negative polarity mode was used for the LC-MSMS experiments depending on the EPs. For the MSMS system, nitrogen gas was used in fragmentation, the capillary voltage, the gas flow, the gas temperature, and the nebulizer pressure were kept at 4000 V, 8 L/min, 3000C and 45 psi, respectively.

**Emerging pollutant degradation and analysis**

Twenty-one different emerging pollutants were treated with five different enzymes in the presence and absence of redox mediator. The degradation experiments were done as follows: SBP, CPO, LPO, MnP and HRP enzymes (0.36 μM) were added to 21 EPs (2 ppm) + $H_2O_2$ (0.1 mM). The experiments were carried out in universal buffer, pH = 2 for CPO, pH = 4 for SBP, pH = 5 for MnP and pH = 6 for LPO and HRP. With the redox mediator experiment, 1-hydroxybenzotriazole (0.1 mM) was added to the reaction mixture.

For degradation of emerging pollutants in wastewater, 3 mL of wastewater sample was adjusted to pH 4 using a

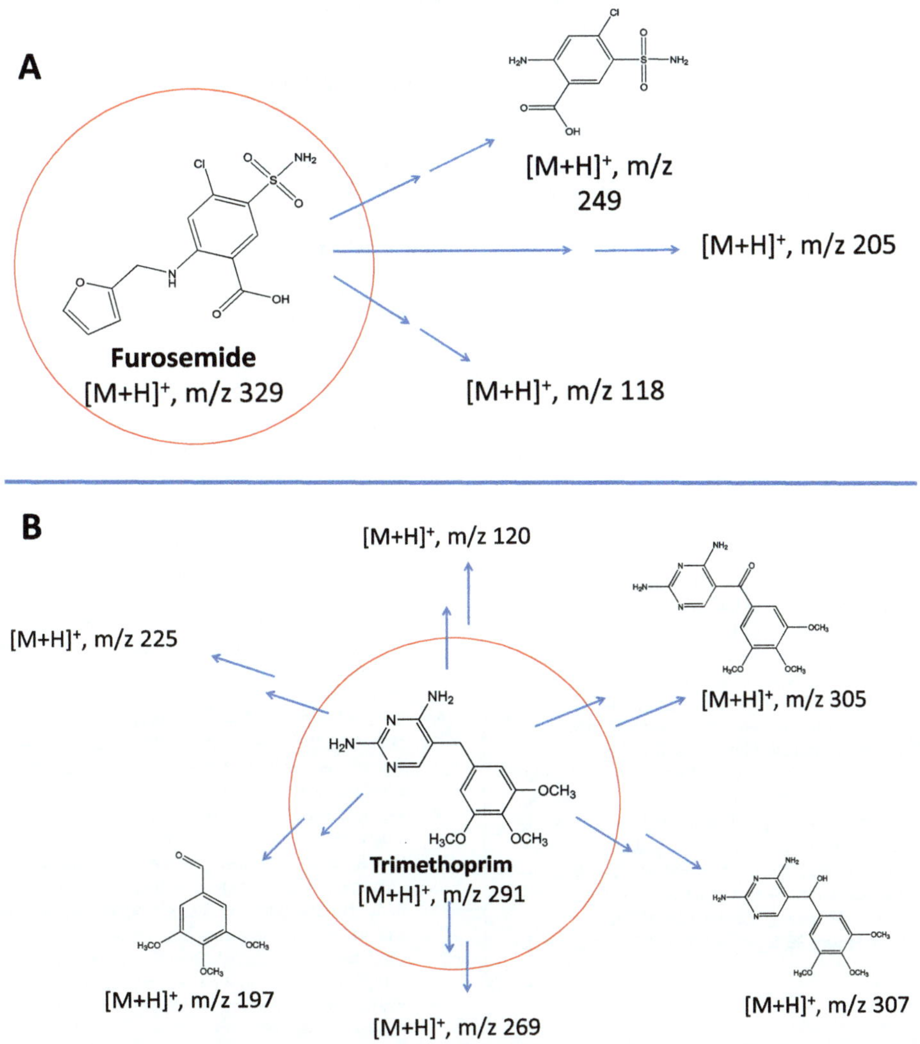


**Fig. 9 a** Proposed degradation pathways of furosemide by SBP + $H_2O_2$ + HOBT. **b** Proposed degradation pathways of trimethoprim by CPO + $H_2O_2$ + HOBT

**Table 3** Summary of transformation products generated upon the degradation of furosemide and trimethoprim, using AOPs (previous studies) or enzymatic treatment (present study)

| | This study | Previous studies | Reference |
|---|---|---|---|
| Furosemide transformation products | | | |
| Enzymatic (Soybean Peroxidase) | m/z 118, 205, **249** | | Present study |
| Photodegradation | | m/z **249**, 311, 352 and 555 | [34] |
| Photodegradation | | m/z 231, 251, **249**, 295, 311 and 329 | [35] |
| Electro-Fenton + Bioconversion | | m/z 251,329 and 345 | [36] |
| Trimethoprim transformation products | | | |
| Enzymatic (Horseradish Peroxidase) | m/z 120, **197**, 225, 269, **305**, **307** | | Present study |
| Solar-photodegradation | | m/z 139, 141, 155, **197**, 213, **305**, **307** and 337 | [37] |
| UVA/LED/TiO2 photocatalysis | | m/z 139, 141, **305**, **307**, 323, 325 and 341 | [38] |
| Fenton | | m/z 143, 195, 279, 324 and 325 | [39] |

buffer and then treated with SBP, $H_2O_2$ and HOBT for 30 min ([SBP] = 1.56 μM, [$H_2O_2$] = 0.112 mM, [HOBT] = 0.05 mM, pH = 4). The sample was then filtered and analyzed on LC-MSMS as described above.

Emerging pollutant degradation was represented as "% remaining" and calculated using the "Area Under the Curve" (AUC) of the peaks in the LC-MS-MS spectra (MRM mode), as follows:

$$\%\text{compound remaining} = (\text{AUC}_i/\text{AUC}_f)\text{x}100$$

Where $AUC_i$ = AUC of the compound peak in the presence of peroxidase, HOBT and buffer and.

$AUC_f$ = AUC of the compound peak in the presence of peroxidase, buffer, HOBT, and $H_2O_2$.

## Additional file

**Additional file 1. Figure 6C S**: Effect of redox mediator on peroxidase mediated pollutant degradation. Thiabendazole remaining after treatment with SBP enzyme with and without HOBT. [Thiabendazole ] = 2 ppm, [H 2 O 2 ] = 0.1 mM added 3 times of 10 min interval, [HOBT] = 0.1 mM , pH = 4 with SBP enzyme and pH = 5 with MnP enzyme, [enzyme] = 0.36μM. **Figure 6D S**: Effect of redox mediator on peroxidase-mediated pollutant degradation. Thiabendazole remaining after treatment with MnP enzyme with and without HOBT. [Thiabendazole] = 2 ppm, [H2O2] = 0.1 mM added 3 times of 10 min interval, [HOBT] = 0.1 mM, pH = 4 with SBP enzyme and pH = 5 with MnP enzyme, [enzyme] = 0.36μM. **Figure 7A S**: Degradation of pollutants spiked in real wastewater. Meloxicam remaining after treatment with SBP enzyme. [SBP] = 1.56μM, [H2O2] = 0.1 mM, [HOBT] = 50 μM, pH = 4. **Figure 7B S**: Degradation of pollutants spiked in real wastewater. DEET remaining after treatment with SBP enzyme. [SBP] = 1.56μM, [H2O2] = 0.1 mM, [HOBT] = 50 μM, pH = 4. **Figure 9B S**: Degradation of Trimethoprim by CPO + H2O2 + HOBT.

### Abbreviations

AOP: Advanced Oxidation Process; CPO: Chloroperoxidase; $H_2O_2$: Hydrogen Peroxide; HOBT: 1-hydroxybenzotriazole; HRP: Horseradish peroxidase; LC-MSMS: Liquid Chromatography-tandem Mass Spectrometry; LiP: Lignin peroxidase; LPO: Lactoperoxidase; MnP: Manganese peroxidase; MRM: Multiple Reaction Monitoring; NSAIDs: Nonsteroidal anti-inflammatory drugs; RM: Redox mediator; SBP: Soybean peroxidase

### Acknowledgments

The authors thank Dr. Iltaf Shah for the use of his lab resources and his kind help with troubleshooting LC-MSMS issues.

### Authors' contributions

KAA developed the LC-MSMS method as well as carried out the initial degradation studies using the mixtures of pollutants. RM, BA, and FA carried out the degradation studies of the two pure pollutants as well as transformation product identification. SSA conceived and supervised the experiments and wrote the manuscript. All authors have read and approved the final version of the manuscript.

### Funding

This work was partially funded by UAE University (National Water Center grant #31R078) to SSA as well as UAE University Graduate College funding for KAA and RM. The funding bodies had no input in the design, experimentation, or content and writing up of the study.

### Ethics approval and consent to participate

Not applicable.

### Consent for publication

Not applicable.

### Competing interests

The authors declare that they have no competing interests.

### Author details

[1]Department of Chemistry, College of Science, UAE University, Al Ain, UAE. [2]Department of Chemistry, College of Arts and Sciences, Khalifa University, P O Box 127788, Abu Dhabi, UAE.

### References

1. Gavrilescu M, Demnerová K, Aamand J, Agathos S, Fava F. Emerging pollutants in the environment: present and future challenges in biomonitoring, ecological risks and bioremediation. New Biotechnol. 2015;32:147–56.
2. Papadakis E-N, Tsaboula A, Kotopoulou A, Kintzikoglou K, Vryzas Z, Papadopoulou-Mourkidou E. Pesticides in the surface waters of Lake Vistonis Basin, Greece: occurrence and environmental risk assessment. Sci Total Environ. 2015;536:793–802.
3. Kolpin DW, Furlong ET, Meyer MT, Thurman EM, Zaugg SD, Barber LB, et al. Pharmaceuticals, hormones, and other organic wastewater contaminants in U.S. streams, 1999-2000: a national reconnaissance. Environ Sci Technol. 2002;36:1202–11.
4. la Farré M, Pérez S, Kantiani L, Barceló D. Fate and toxicity of emerging pollutants, their metabolites and transformation products in the aquatic environment. TrAC Trends Anal Chem. 2008;27:991–1007.
5. Bonefeld-Jorgensen EC, Long M, Bossi R, Ayotte P, Asmund G, Krüger T, et al. Perfluorinated compounds are related to breast cancer risk in Greenlandic Inuit: a case control study. Environ Health Glob Access Sci Source. 2011;10:88.
6. Loos R, Wollgast J, Huber T, Hanke G. Polar herbicides, pharmaceutical products, perfluorooctanesulfonate (PFOS), perfluorooctanoate (PFOA), and nonylphenol and its carboxylates and ethoxylates in surface and tap waters around Lake Maggiore in northern Italy. Anal Bioanal Chem. 2007;387:1469–78.
7. Liu Z, Kanjo Y, Mizutani S. Removal mechanisms for endocrine disrupting compounds (EDCs) in wastewater treatment — physical means, biodegradation, and chemical advanced oxidation: a review. Sci Total Environ. 2009;407:731–48.
8. Robinson T, McMullan G, Marchant R, Nigam P. Remediation of dyes in textile effluent: a critical review on current treatment technologies with a proposed alternative. Bioresour Technol. 2001;77:247–55.
9. Bokare AD, Choi W. Review of iron-free Fenton-like systems for activating H2O2 in advanced oxidation processes. J Hazard Mater. 2014;275:121–35.
10. Poyatos JM, Muñio MM, Almecija MC, Torres JC, Hontoria E, Osorio F. Advanced oxidation processes for wastewater treatment: state of the art. Water Air Soil Pollut. 2009;205:187–204.
11. Boopathy R. Factors limiting bioremediation technologies. Bioresour Technol. 2000;74:63–7.
12. Rauf MA, Salman AS. Survey of recent trends in biochemically assisted degradation of dyes. Chem Eng J. 2012;209:520–30.
13. Bautista LF, Morales G, Sanz R. Biodegradation of polycyclic aromatic hydrocarbons (PAHs) by laccase from Trametes versicolor covalently immobilized on amino-functionalized SBA-15. Chemosphere. 2015;136:273–80.
14. Husain Q. Peroxidase mediated decolorization and remediation of wastewater containing industrial dyes: a review. Rev Environ Sci Biotechnol. 2009;9:117–40.
15. Coconi-Linares N, Ortiz-Vázquez E, Fernández F, Loske AM, Gómez-Lim MA. Recombinant expression of four oxidoreductases in Phanerochaete chrysosporium improves degradation of phenolic and non-phenolic substrates. J Biotechnol. 2015;209:76–84.
16. Satar R, Husain Q. Catalyzed degradation of disperse dyes by calcium alginate-pectin entrapped bitter gourd (Momordica charantia) peroxidase. J Environ Sci. 2011;23:1135–42.
17. Silva MC, Corrêa AD, Amorim MTSP, Parpot P, Torres JA, Chagas PMB. Decolorization of the phthalocyanine dye reactive blue 21 by turnip peroxidase and assessment of its oxidation products. J Mol Catal B Enzym. 2012;77:9–14.

18. Satar R, Husain Q. Applications of Celite-adsorbed white radish (Raphanus sativus) peroxidase in batch process and continuous reactor for the degradation of reactive dyes. Biochem Eng J. 2009;46:96–104.
19. Jamal F, Qidwai T, Pandey PK, Singh D. Catalytic potential of cauliflower (Brassica oleracea) bud peroxidase in decolorization of synthetic recalcitrant dyes using redox mediator. Catal Commun. 2011;15:93–8.
20. Nouren S, Bhatti HN. Mechanistic study of degradation of basic violet 3 by Citrus Limon peroxidase and phytotoxicity assessment of its degradation products. Biochem Eng J. 2015;95:9–19.
21. Hsu S-K, Chung Y-C, Chang C-T, Sung H-Y. Purification and characterisation of two acidic peroxidase isoforms from the sheaths of bamboo shoots. Int J Food Sci Technol. 2012;47:1872–81.
22. Husain M, Husain Q. Applications of redox mediators in the treatment of organic pollutants by using Oxidoreductive enzymes: a review. Crit Rev Environ Sci Technol. 2007;38:1–42.
23. Shiraishi T, Sannami Y, Kamitakahara H, Takano T. Comparison of a series of laccase mediators in the electro-oxidation reactions of non-phenolic lignin model compounds. Electrochim Acta. 2013;106:440–6.
24. Bibi I, Bhatti HN, Asgher M. Comparative study of natural and synthetic phenolic compounds as efficient laccase mediators for the transformation of cationic dye. Biochem Eng J. 2011;56:225–31.
25. Telke AA, Kagalkar AN, Jagtap UB, Desai NS, Bapat VA, Govindwar SP. Biochemical characterization of laccase from hairy root culture of Brassica juncea L. and role of redox mediators to enhance its potential for the decolorization of textile dyes. Planta. 2011;234:1137–49.
26. Gros M, Petrović M, Barceló D. Development of a multi-residue analytical methodology based on liquid chromatography–tandem mass spectrometry (LC–MS/MS) for screening and trace level determination of pharmaceuticals in surface and wastewaters. Talanta. 2006;70:678–90.
27. Alneyadi A, Shah I, Ashraf SS. An innovative bioanalytical research project course to train undergraduate students on liquid chromatography–mass spectrometry. Biochem Mol Biol Educ. 2019;47:228–38.
28. Alneyadi AH, Ashraf SS. Differential enzymatic degradation of thiazole pollutants by two different peroxidases – a comparative study. Chem Eng J. 2016;303:529–38.
29. Al-Maqdi KA, Hisaindee SM, Rauf MA, Ashraf SS. Comparative degradation of a Thiazole pollutant by an advanced oxidation process and an enzymatic approach. Biomolecules. 2017;7:64.
30. Rodriguez-Mozaz S, Chamorro S, Marti E, Huerta B, Gros M, Sànchez-Melsió A, et al. Occurrence of antibiotics and antibiotic resistance genes in hospital and urban wastewaters and their impact on the receiving river. Water Res. 2015;69:234–42.
31. Al-Maqdi KA, Hisaindee S, Rauf MA, Ashraf SS. Detoxification and degradation of sulfamethoxazole by soybean peroxidase and UV + H2O2 remediation approaches. Chem Eng J. 2018;352:450–8.
32. Sui Q, Huang J, Deng S, Yu G, Fan Q. Occurrence and removal of pharmaceuticals, caffeine and DEET in wastewater treatment plants of Beijing, China. Water Res. 2010;44:417–26.
33. Petrovic M, Skrbic B, Živančev J, Ferrando-Climent L, Barcelo D. Determination of 81 Pharmaceutical Drugs by High Performance Liquid Chromatography Coupled to Mass Spectrometry with Hybrid Triple Quadrupole-linear Ion Trap in Different Types of Water in Serbia. Sci Total Environ. 2013;468-469C:415–28.
34. Katsura S, Yamada N, Nakashima A, Shiraishi S, Furuishi T, Ueda H. Identification of furosemide Photodegradation products in water-acetonitrile mixture. Chem Pharm Bull (Tokyo). 2015;63:617–27.
35. Jakimska A, Śliwka-Kaszyńska M, Reszczyńska J, Namieśnik J, Kot-Wasik A. Elucidation of transformation pathway of ketoprofen, ibuprofen, and furosemide in surface water and their occurrence in the aqueous environment using UHPLC-QTOF-MS. Anal Bioanal Chem. 2014;406:3667–80.
36. Laurencé C, Rivard M, Martens T, Morin C, Buisson D, Bourcier S, et al. Anticipating the fate and impact of organic environmental contaminants: a new approach applied to the pharmaceutical furosemide. Chemosphere. 2014;113:193–9.
37. Sirtori C, Agüera A, Gernjak W, Malato S. Effect of water-matrix composition on trimethoprim solar photodegradation kinetics and pathways. Water Res. 2010;44:2735–44.
38. Cai Q, Hu J. Decomposition of sulfamethoxazole and trimethoprim by continuous UVA/LED/TiO2 photocatalysis: Decomposition pathways, residual antibacterial activity and toxicity. J Hazard Mater. 2017;323(Pt A):527–36.
39. Wang S, Wang J. Trimethoprim degradation by Fenton and Fe (II)-activated persulfate processes. Chemosphere. 2018;191:97–105.

# Isolating promoters from *Corynebacterium ammoniagenes* ATCC 6871 and application in CoA synthesis

Yingshuo Hou[1,2], Siyu Chen[1], Jianjun Wang[1], Guizhen Liu[3], Sheng Wu[1*] and Yong Tao[1*]

**Abstract**

**Background:** *Corynebacterium ammoniagenes* is an important industrial organism that is widely used to produce nucleotides and the potential for industrial production of coenzyme A by *C. ammoniagenes* ATCC 6871 has been shown. However, the yield of coenzyme A needs to be improved, and the available constitutive promoters are rather limited in this strain.

**Results:** In this study, 20 putative DNA promoters derived from genes with high transcription levels and 6 promoters from molecular chaperone genes were identified. To evaluate the activity of each promoter, red fluorescence protein (RFP) was used as a reporter. We successfully isolated a range of promoters with different activity levels, and among these a fragment derived from the upstream sequence of the 50S ribosomal protein L21 ($P_{rpl21}$) exhibited the strongest activity among the 26 identified promoters. Furthermore, type III pantothenate kinase from *Pseudomonas putida* (PpcoaA) was overexpressed in *C. ammoniagenes* under the control of $P_{rpl21}$, CoA yield increased approximately 4.4 times.

**Conclusions:** This study provides a paradigm for rational isolation of promoters with different activities and their application in metabolic engineering. These promoters will enrich the available promoter toolkit for *C. ammoniagenes* and should be valuable in current platforms for metabolic engineering and synthetic biology for the optimization of pathways to extend the product spectrum or improve the productivity in *C. ammoniagenes* ATCC 6871 for industrial applications.

**Keywords:** *Corynebacterium ammoniagenes*, Transcriptome, Promoter, Coenzyme A

## Background

*Corynebacterium ammoniagenes* (formerly known as *Brevibacterium ammoniagenes*) is a Gram-positive, non-pathogenic soil bacterium with high guanine–cytosine (GC) DNA content. Due to the divergent evolution of *C. ammoniagenes*, different strains are used to produce different metabolites. For example, *C. ammoniagenes* ATCC 6872 is used in the production of nucleotides and nucleosides [1], while *C. ammoniagenes* ATCC 6871 shows a relatively high synthesis capacity of coenzyme A (CoA), which is a ubiquitous and essential cofactor found in all three domains of life and is involved in numerous metabolic pathways [2–4].

Sequencing of the *C. ammoniagenes* genome has helped to identify key enzymes for diverting carbon flow from metabolic pathways to other products [5, 6]. However, a major breakthrough in genetic manipulation to precisely regulate the expression of individual biosynthetic genes is still in process. Progress in the manipulation of *C. ammoniagenes* genes has been achieved through the development of effective transformation protocols and cloning vectors [7]. Most vectors are based on the related *C. glutamicum* endogenous cryptic plasmids that have incorporated *E. coli* elements for transfer between the species [8–10]. Almost all these vectors have adopted *E. coli* promoters or *C. ammoniagenes* native promoters to express foreign proteins in *C. ammoniagenes* [11–13]. Although some of these

* Correspondence: shengwu@im.ac.cn; taoyong@im.ac.cn
[1]CAS Key Laboratory of Microbial Physiological and Metabolic Engineering, Institute of Microbiology, Chinese Academy of Sciences, Beijing 100101, People's Republic of China
Full list of author information is available at the end of the article

promoters are active in *C. ammoniagenes*, the activities are quite low [14].

To efficiently engineer *C. ammoniagenes*, it is necessary to screen endogenous promoters. Paik and coworkers isolated 22 representative promoters from the genome of *C. ammoniagenes* ATCC 6872 and the activity of overexpressed chloramphenicol transacetylase from the strongest promoter IJ73 was 2.85 U [7]. Similar studies have been performed by Park and coworkers, where they overexpressed the attenuator binding protein in *C. ammoniagenes* ATCC 6872 using the endogenous promoter CJ1 [15]. However, the above promoters have not been universally applied in *C. ammoniagenes* ATCC 6871 (see text), probably due to the divergent evolution of *C. ammoniagenes* genomes. For instance, the sequence identities of promoters IJ59 and IJ73 from *C. ammoniagenes* ATCC 6872 with the homologous promoters from *C. ammoniagenes* ATCC 6871 are 93 and 89%, respectively. Therefore, due to the importance of CoA, it is necessary to find promoters that are effective in *C. ammoniagenes* ATCC 6871.

Endogenous promoters can be obtained by genomic dissection or promoter trapping et.al [16]. Nonetheless, with the advent of genomics and transcriptomics, we can isolate promoters more rationally. Studies have shown that strong promoters are usually obtained from the promoters of essential genes whose transcription levels are presumed to be high and constant [17]. In addition, we found that some commonly-used strong promoters such as $P_{dnak}$ in *Gluconobacter oxydans* and $P_{gro}$ in *Corynebacterium glutamicum* are promoters of molecular chaperones [18, 19]. In this study, the promoters for 20 genes with high transcription levels and 6 molecular chaperone-encoding genes are identified by bioinformatic methods and their activities were further examined experimentally. Among them, $P_{rpl21}$ was found to be the strongest promoter, which was used in the biosynthesis of CoA, increasing the yield by 4.4 times.

## Results

### Screening promoters and construction of probe-vector

*C. ammoniagenes* ATCC 6871 cells grown in LB medium and fermentation medium to the exponential phase ($OD_{600nm} \approx 2.0$) were collected and sent to Mega genomics (Beijing, China) for RNA-seq. A total of 2411 genes were identified from the transcriptomic data. Considering that there are many cases in which genes are in the same operon, genetic loci analysis was performed and the results showed that 1508 genes in 547 operons were expressed. In general, genes in the same operons share a common promoter, so the first gene in each operon was selected. Thus, the total number of remaining genes was 1450 (2411−1508 + 547). Transcription abundance analysis was used to select 71 genes with transcription levels in the top 100 in both RNA-seq profiles. Sequences upstream of the top 20 genes were predicted and scored using the online promoter prediction tool as described below. The annotations and RPKM values for these genes are shown in Table 1. A transcriptomic data analysis flow chart and the transcription levels of genes are shown in Fig. 1.

Previous studies have shown that most promoters of molecular chaperones exhibit high transcription levels in prokaryotes, so all 6 annotated molecular chaperones in the *C. ammoniagenes* ATCC 6871 genome (Accession Number: NZ_CP009244) (two GroEL, two DnaJ, one DnaK, and one Co-chaperone GroES) were identified and their promoters were predicted as described in materials and methods. The information of these genes and transcriptomic RPKM values are also listed in Table 1. By predicting the promoter regions, 26 promoters were constructed according to the rules described below and their corresponding RFP expression plasmids were named as pXMJ190-$P_n$. The sequences of the 26 promoters are listed in Additional file 1: Table S3.

For a comprehensive comparison, two conserved homologous promoters CJ1 and IJ59 that are active in *C. ammoniagenes* ATCC 6872 were aligned against the genome of *C. ammoniagenes* ATCC 6871 by BLAST and were cloned into pXMJ190 to give pXMJ190-CJ1 and pXMJ190-IJ59 [7, 15], respectively. Moreover, previous studies have shown promoters that work well in *C. glutamicum* may also be active in *C. ammoniagenes* [20]. Thus, a strong endogenous *C. glutamicum* promoter named $P_{gro}$ (the promoter of *groES* gene) was cloned as pXMJ190-$P_{gro}$ [19]. In addition, the most widely used *tac* promoter and conserved SD sequence was also selected and cloned into pXMJ190 as pXMJ190-$P_{tac}$ [21]. The pXMJ190 was used as negative control. A flowchart for construction of the probe vectors to identify promoters is shown in Fig. 2. All of the above constructed plasmids were validated by sequencing and then transformed into *C. ammoniagenes* ATCC 6871.

### Analysis and comparison of promoters in C. ammoniagenes ATCC 6871

To analyze the activities of our cloned promoters, we first made a visual assessment of all *C. ammoniagenes* ATCC 6871 strains harboring pXMJ190, pXMJ190-CJ1, pXMJ190-IJ59, pXMJ190-$P_{gro}$, pXMJ19-$P_{tac}$ and pXMJ190-$P_n$ series plasmids on LB plates containing 20 μg/ml chloramphenicol. Red fluorescence was observed by microscopy, with different intensities in different colonies, indicating that there was variable expression of RFP from the 26 putative promoters (Fig. 3a). Among these promoters, $P_{rpl21}$, $P_{rpl10}$, $P_{groELB}$, and $P_{dnaK}$ exhibited strong activity in *C. ammoniagenes* ATCC 6871, with $P_{rpl21}$ being the strongest promoter. The remaining promoters had very low or non-existent red fluorescence.

**Table 1** Identified molecular chaperone genes or those with high level expression from RNA-seq data

| Name | Downstream Product | Average RPKM | Score |
|---|---|---|---|
| $P_{rrlA}$ | 23S ribosomal RNA | 988,873 | 0.99 |
| $P_{rrlB}$ | 23S ribosomal RNA | 913,696 | 0.97 |
| $P_{tmr}$ | Transfer-messenger RNA | 853,640 | 0.96 |
| $P_{fer}$ | Ferritin | 46,832 | 0.92 |
| $P_{raiA}$ | Ribosomal subunit interface protein | 44,446 | 0.91 |
| $P_{usp}$ | Universal stress protein | 28,915 | 0.83 |
| $P_{rpl29}$ | 50S ribosomal protein L29 | 24,596 | 0.89 |
| $P_{nat}$ | N-acetyltransferase | 18,830 | 0.96 |
| $P_{atpG}$ | F0F1 ATP synthase subunit gamma | 16,820 | 0.80 |
| $P_{tuf}$ | Elongation factor Tu | 14,228 | 0.98 |
| $P_{rpl21}$ | 50S ribosomal protein L21 | 13,153 | 0.81 |
| $P_{fnt}$ | NarK/NasA family nitrate transporter | 12,650 | 0.94 |
| $P_{gpdI}$ | Type I glyceraldehyde-3-phosphate dehydrogenase | 12,223 | 0.83 |
| $P_{hfp}$ | Hemoglobin flavoprotein | 12,078 | 0.86 |
| $P_{acn}$ | Aconitate hydratase | 10,439 | 0.84 |
| $P_{rpl10}$ | 50S ribosomal protein L10 | 10,398 | 1.00 |
| $P_{abm}$ | Antibiotic biosynthesis monooxygenase | 9935 | 0.73 |
| $P_{fbaII}$ | Class II fructose-bisphosphate aldolase | 9740 | 0.77 |
| $P_{rpl11}$ | 50S ribosomal protein L11 | 9473 | 0.82 |
| $P_{sbp}$ | ABC transpoter, substrate-binding protein | 9120 | 0.99 |
| $P_{groELA}$ | Molecular chaperone GroEL | 6594 | 0.89 |
| $P_{groELB}$ | Molecular chaperone GroEL | 6261 | 0.87 |
| $P_{dnaJA}$ | Molecular chaperone DnaJ | 239 | 0.85 |
| $P_{dnaJB}$ | Molecular chaperone DnaJ | 703 | 0.87 |
| $P_{groES}$ | Co-chaperone GroES | 2215 | 1.00 |
| $P_{dnaK}$ | Molecular chaperone DnaK | 4523 | 0.98 |

To quantify the activities of the cloned promoters, growth-normalized fluorescence intensities were measured using a microplate assay and the data are shown in Fig. 3b and Additional file 1: Table S2, the results showed that we successfully isolated 20 endogenous promoters with different activities. Among the 20 promoters ($P_{rrlA}$-$P_{sbp}$) from genes with high transcription levels, $P_{tmr}$, $P_{nat}$, $P_{atpG}$, $P_{gpdI}$, $P_{acn}$ are essentially inactive, while $P_{groELA}$ from the 6 molecular chaperone promoters was also inactive. Meanwhile, four strong promoters, including the $P_{rpl21}$ with 43,433 RFU/OD intensity and followed by $P_{dnaK}$ (19,125 RFU/OD), $P_{groELB}$ (8166 RFU/OD) and $P_{rpl10}$ (6490 RFU/OD) were identified whose fluorescence levels were consistent with those observed on plates. Thus, a range of promoters with different transcriptional activities, including four strong promoters, were identified and can be used directly in further applications. According to the obtained data, only two promoters ($P_{rpl21}$ and $P_{rpl10}$) from the 20 genes with high levels of transcription had strong activities, while two promoters ($P_{groELB}$ and $P_{dnaK}$) sourced from the six molecular chaperone genes exhibited relatively high activities, which may indicate that isolating promoters from molecular chaperones may be a more efficient strategy. Although most of the promoters identified from the RNA-seq profiles exhibited low level expression or were silent, the red fluorescence intensity of $P_{rpl21}$ was almost 2.3 times that of the highest promoter ($P_{dnaK}$) derived from upstream regions of the molecular chaperones. This suggests isolating promoters from genes with high transcription levels may yield strong promoters.

When it comes to the known promoters, the promoter CJ1 had some certain activity (4171 RFU/OD), the IJ59 (2866 RFU/OD) promoter had almost non activity, and the activity of $P_{gro}$ (5872 RFU/OD) was slightly higher than the $P_{tac}$ (5683 RFU/OD) promoter but still lower than promoter $P_{rpl10}$ (6490 RFU/OD). This demonstrates the promoters that work well in one species or homologous strains may not possess the same characteristics in another.

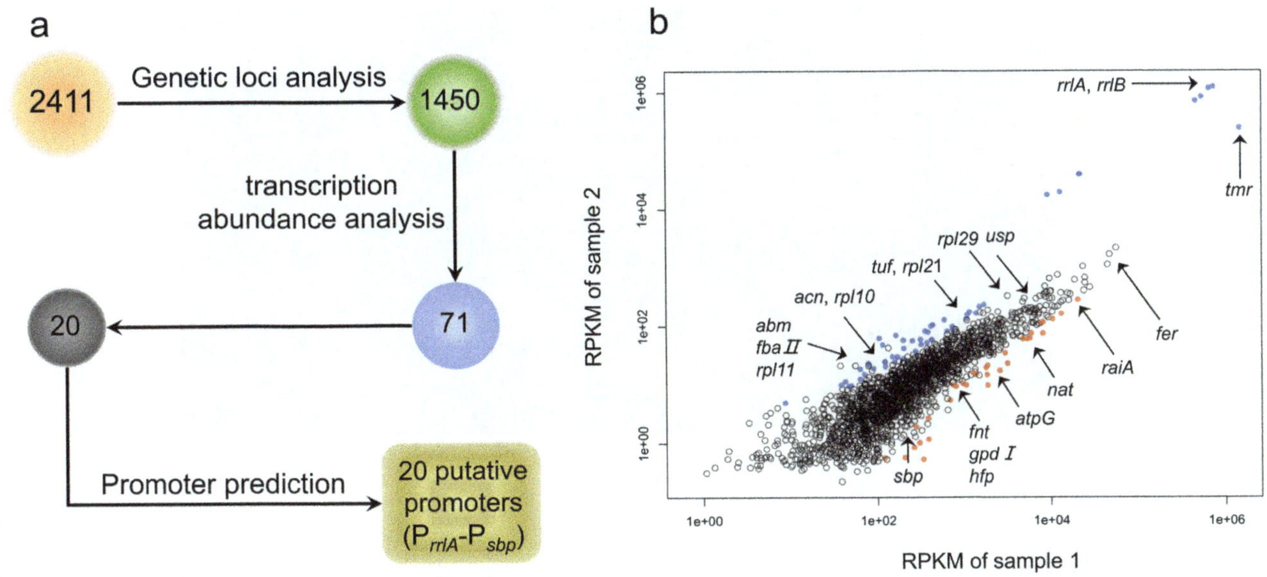


**Fig. 1** Rational selection of promoters form RNA-seq profile. **a**: The flowchart for selection of promoters from the transcriptomic data. **b**: The transcription levels of selected 20 genes in two different samples

### Application of $P_{rpl21}$ in C. ammoniagenes for improving the production of CoA

It has been reported that CoA, a ubiquitous and essential cofactor in biochemical reactions, can be produced by *C. ammoniagenes* ATCC 6871 [25]. In the CoA biosynthetic pathway, the committed step catalyzed by pantothenate kinase (coaA) is subject to feedback inhibition by CoA and acyl-CoAs [26]. Bacterial coaA proteins are categorized based on their amino acid sequences into three types, namely type I, II, and III. *C. ammoniagenes* carries a type I coaA which is highly regulated by CoA and its derivatives. In contrast, type II

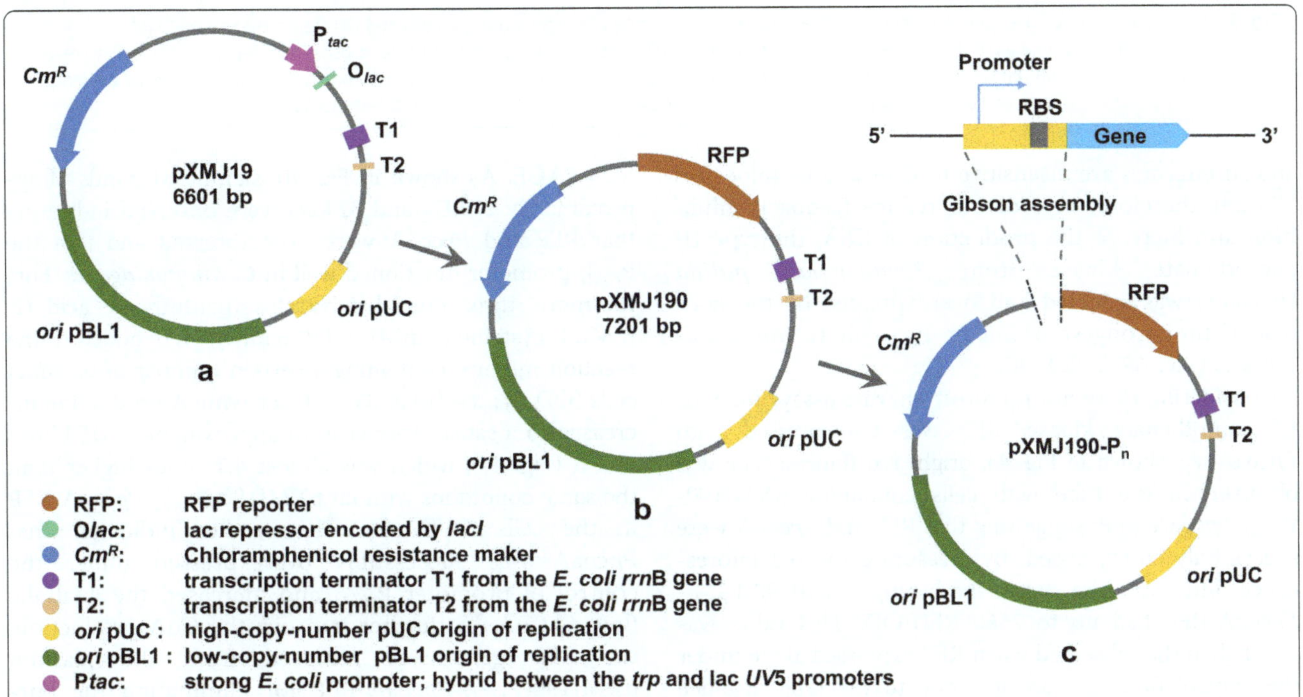


**Fig. 2** Flowchart for the construction of probe-vector that used for screening promoters. **a**: The shuttle vector pXMJ19 serves as the backbone. **b**: The probe-vector pXMJ190 with reporter of RFP and deletion of $P_{tac}$ and $O_{lac}$. **c**: pXMJ190-$P_n$ represent vectors with the 26 putative promoters inserted in the upstream of RFP

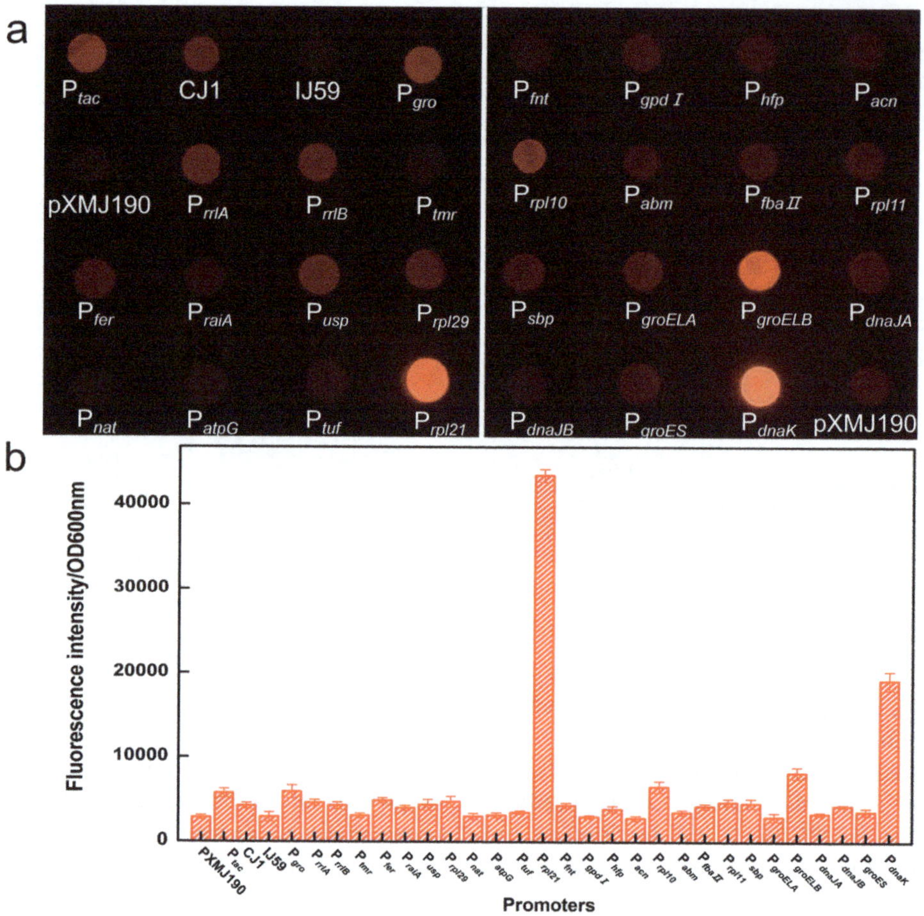


**Fig. 3** Different activities of selected promoters. **a**: Observed red fluorescence of bacteria using a LUYOR-3430 stereo microscope with a fluorescence excitation source (LUYOR, USA) set at 501 nm. **b**: Measured fluorescence intensities of bacteria using a Synergy H4 microplate reader. Cells were washed once with PBS (pH 7.4) and then resuspended in PBS (pH 7.4) at an $OD_{600\ nm}$ of approximately 1.0. The excitation wavelength for RFP was set at 554 nm and emission was set at 586 nm. Error bars show the standard deviation of three measurements

and III enzymes are insensitive to CoA and its thioesters [27, 28], therefore, in order to reduce feedback inhibition and increase the production of CoA, the type III pantothenate kinase from *Pseudomonas putida* (*Pp*coaA) was selected and overexpressed by the control of the strongest promoter $P_{rpl21}$ in *C. ammoniagenes* ATCC 6871 [29, 30].

Considering there are no direct enzyme assays for type III pantothenate kinase, RFP was co-expressed with *Pp*coaA. As shown in Fig. 4a, bright red fluorescence was observed in the tube with cells containing pXMJ190-$P_{rpl21}$-*Pp*coaA-RFP, suggesting that RFP and *Pp*coaA were successfully co-expressed. By measuring the red fluorescence intensity, the strain harboring pXMJ190-$P_{rpl21}$-*Pp*coaA-RFP had up to 7560 RFU/OD. This value was lower than that observed when RFP expressed alone under the control of $P_{rpl21}$, possibly due to the long distance from the transcription initiation site and the increased cellular burden caused by co-expression of two proteins. The expression of RFP and *Pp*coaA was also be observed by SDS-PAGE. As shown in Fig. 4b, significant bands of approximately 27 kDa and 32 kDa were observed indicating that RFP and *Pp*coaA were overexpressed and that the $P_{rpl21}$ promoter functioned well in *C. ammoniagenes*. Furthermore, three crucial substrates (pantothenic acid (2 mM), L-cysteine (2 mM), ATP (6 mM)) were added to the reaction mixture containing a certain number of bacterial cells ($OD_{600\ nm} \approx 40$) at 39 °C. Coenzyme A production increased to a satisfactory yield of approximately 315 U/mL in 6 h (Fig. 4c), which was almost 4.1 times higher than the same conditions without pXMJ190-$P_{rpl21}$-*Pp*coaA-RFP in the cells (76 U/mL). The results indicated that *Pp*coaA was successfully overexpressed under the control of promoter $P_{rpl21}$ and increased the anabolic flow of CoA. To further improve the CoA production, the RFP gene was removed from the plasmid pXMJ190-$P_{rpl21}$-*Pp*coaA-RFP for eliminating the burden of RFP expression in cells. With the *Pp*coaA overexpressed alone in *C. ammoniagenes*, the CoA production was reached $332 \pm 13$ U/mL, which was

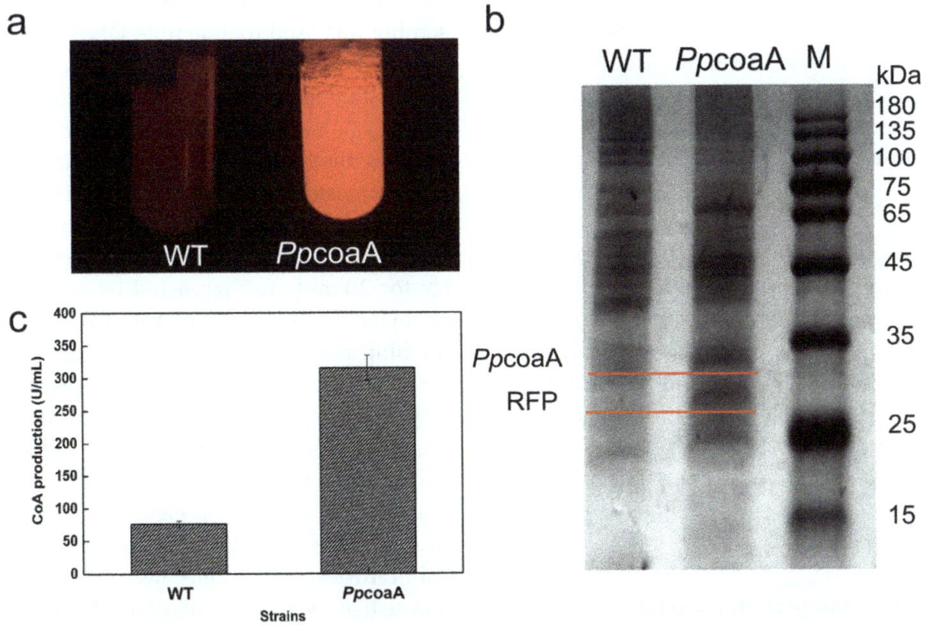


**Fig. 4** Application of $P_{rpl21}$ in *C. ammoniagenes* for improving the production of CoA. **a**: The comparison of fluorescence intensities between cells harboring pXMJ190-$P_{rpl21}$-*Ppc*oaA-RFP and wild type cells. **b**: SDS–PAGE analysis of RFP and *Ppc*oaA co-expression in *C. ammoniagenes*. Samples were prepared with an equal concentration of cells, and 40 μg of cell lysate were loaded per lane. Lane M: protein marker. **c**: The CoA production of cells harboring pXMJ190-$P_{rpl21}$-*Ppc*oaA-RFP and control. Error bars show the standard deviation of three measurements

slightly increased than the *Ppc*oaA and RFP co-expressional system.

## Discussion

*C. ammoniagenes* ATCC 6871 is a producer of CoA, however, its genetic operation has been rarely reported, and its industrial application potential still needs to be improved. In order to further increase the production of CoA, endogenous promoters from *C. ammoniagenes* ATCC 6871 were screened in this study. Firstly, 20 putative promoters based on genes with high transcription levels and 6 promoters upstream of molecular chaperones were identified and characterized for transcriptional activity by using RFP as the reporter gene. Among the 26 putative promoters, 20 with different transcriptional activities were isolated including 15 from the high transcription level genes and 5 from molecular chaperone genes. The output efficiency demonstrates that isolating promoters from molecular chaperones (5/6) might be a more efficient strategy than using RNA-seq (15/20). Therefore, isolating promoters from molecular chaperone genes is more effortless and cost-effective than RNA-seq. These results may ascribe to that the RPKM values of the genes only represent the abundance of RNA and do not equivalent to its promoter activities, which might be caused by multiple factors, such as the codon usage of genes, copy numbers of genes, the half-lives of mRNA and limitations inherent in transcriptomics [22–24]. It thus, isolating promoters based on RNA-Seq requires further verification by experiments.

Nevertheless, we should note that the red fluorescence intensity of the highest promoter ($P_{rpl21}$) identified from RNA-seq was almost 2.3 times than the highest promoter ($P_{dnaK}$) derived from molecular chaperone genes, so isolating promoters based on the genes transcriptional data may be more likely to yield promoters with highest activity. To our best knowledge, $P_{rpl21}$ from the upstream sequence of the 50S ribosomal protein L21 is the strongest *C. ammoniagenes* ATCC 6871 promoter reported so far. The promoters obtained in this work will enrich the available promoter toolkit for *C. ammoniagenes* and should be valuable in current platforms for metabolic engineering and synthetic biology for the optimization of pathways to extend the product spectrum or improve the productivity in *C. ammoniagenes* ATCC 6871.

To further verify the capacity of $P_{rpl21}$, the key gene *Ppc*oaA in the biosynthetic pathway of CoA was overexpressed in *C. ammoniagenes* ATCC 6871 with the aim of reducing the feedback inhibition and increasing the production of CoA. The CoA production of manipulated *C. ammoniagenes* was approximate 4.4 times to the control, which confirmed that *Ppc*oaA overexpressed and functionalized successfully in *C. ammoniagene*. These results prove that the selected promoter for the overexpression of foreign genes in *C. ammoniagenes* could be used as an efficient tool for improving the yield of major products in *C. ammoniagenes*. However, eliminating the burden of co-expressing RFP only slightly increase the CoA production, possibly due to the undiscovered limitations exist in the CoA synthetic pathway. Thus, more efforts should be focus on the resolving of rate-limiting steps of CoA synthesis in *C. ammoniagenes*.

## Conclusions

In summary, this study provides a rational strategy to isolate endogenous promoters from *C. ammoniagenes* ATCC 6871, which may be helpful in other similar scientific research. Through this strategy, we successfully isolated a range of promoters with different transcriptional activities and the strongest one was applied to improve the CoA production in *C. ammoniagenes* ATCC 6871, raising hope for further improving the industrial production level of CoA.

## Methods

### Bacterial strains, media, and growth conditions

The *C. ammoniagenes* ATCC 6871 was purchased from China General Microbiological Culture Collection Center (CGMCC, Beijing, China). *E. coli* Top10 (Tsingke, Beijing, China) was cultivated in Luria-Bertani (LB) broth or on LB plates with 2% (w/v) agar at 37 °C as a host for transformation. *C. ammoniagenes* ATCC 6871 grown in NCM medium at 30 °C was used as the transformation host for *E. coli*/Corynebacteria vectors used in this study. After transformation by electroporation (1.8 kV), transformants were plated on BHIS plates with 20 μg/mL chloramphenicol at 30 °C. Fermentation medium (glucose 100 g/L, peptone 12 g/L, yeast powder 8 g/L, $MgSO_4{\cdot}7H_2O$ 10 g/L, $KH_2PO_4$ 10 g/L, $K_2HPO_4$ 10 g/L, urea 7 g/L, pH 7.5–7.8) was used to cultivate strains producing CoA. Cultivation for expression analysis was performed in at least biological duplicates.

### Recombinant DNA techniques

*C. ammoniagenes* genomic DNA was isolated using a genomic DNA isolation kit (Tiangen, Beijing, China). Kits for plasmid isolation, extraction of DNA from agarose gels and PCR product purification were also purchased from Tiangen (Beijing, China). I-5 2 × High-Fidelity Master Mix and Trelief™ SoSoo Cloning Kit were purchased from Tsingke (Beijing, China) and used for routine molecular biology applications.

### Construction of probe-vector pXMJ190

The shuttle vector pXMJ19 (a kind gift from Professor Dong, Fig. 2a) was used as the backbone for vectors constructed in this study. To eliminate interference from the promoter already present in pXMJ19, the *tac* promoter and *lac* operator were removed. A gene encoding red fluorescent protein (RFP) was inserted into the MCS to form a probe-vector named pXMJ190 (Fig. 2b). Gibson assembly was used to assemble DNA fragments upstream of the reporter gene *rfp*, resulting in seamless ligation between fragments and the probe-vector [31, 32].

### Genetic manipulation of C. ammoniagenes

To increase transformation efficiency, recipients were grown in 100 mL of NCM medium at 30 °C until an $OD_{600\ nm}$ of approximately 1.0. Cells were incubated on ice for 20 min and harvested by centrifugation in a polypropylene tube at 4000 rpm for 10 min at 4 °C. After washing twice in cold distilled water and two washes in ice-cold 10% glycerol, cells were resuspended in 1 mL of 10% glycerol. For electroporation, 100 μL of the freshly prepared electro-competent cells were mixed with 3 μL plasmid (50 ng/μL in $ddH_2O$) in a cold sterile electroporation cuvette (1 mm electrode gap) and pulsed immediately with a MicroPulser electroporator (Bio-Rad Laboratories, Inc., Hercules, CA). The electroporator was usually set at a voltage of 1.8 kV. Cells were subsequently resuspended in 0.9 mL of BHIS, heated at 46 °C for 6 min and withdrawn immediately for recovery by incubating for 3 h at 30 °C and then plated for selection of transformants.

### Construction and analysis of selected promoters

The transcriptional levels of genes can be estimated by Reads Per Kilobase per Million mapped reads (RPKM), so the genes were ranked by RPKM value based on transcriptional data from *C. ammoniagenes* and the top 20 genes were selected. All of the genes were identified from the genome and their promoters were predicted as below. Six annotated molecular chaperones in the genome were sorted and their promoters were predicted as outlined below. According to previous studies, the promoter-5′-UTR junction influences mRNA and protein levels [23, 33]. Therefore, we integrated the corresponding 5′-UTR into each promoter.

Promoter sequences were predicted using the Neural Network Promoter Prediction online tool (http://www.fruitfly.org/seq_tools/promoter.html) [34]. The promoter prediction score threshold was set to 0.7. Moreover, due to the possibility of tandem promoters, all eligible promoter sequences within 300 bp upstream of the start codon were considered [35]. For correct transcription initiation, the complete "promoter" consisted of 5′-UTR, predicted promoter region and Shine-Dalgarno sequence, which typically correspond to the region 60 bp upstream of a predicted promoter and 1 bp upstream of the start codon. The 60 bp extension upstream of the promoter accounts for potential UP-elements [36, 37].

Promoters were amplified from *C. ammoniagenes* ATCC 6871 and *C. glutamicum* ATCC 13032 genomic DNA with the corresponding primers. Oligonucleotide primers used in this work are listed in Additional file 1: Table S1. PCR products were ligated into the vector pXMJ190 using Gibson assembly as shown in Fig. 2c. All plasmids were constructed in *E. coli* TOP10 and then transformed into *C. ammoniagenes* ATCC 6871 for subsequent analysis.

### Fluorescence intensity assay

To evaluate RFP expression under the control of the various promoters, *C. ammoniagenes* strains harboring various vectors were grown overnight on LB plates containing 20 μg/ml chloramphenicol. Fluorescence was observed using a LUYOR-3430 stereo microscope with a fluorescence excitation source (LUYOR, USA) set at 501 nm and matching lenses to detect RFP. Pictures were captured with a camera.

For more accurate comparisons, the fluorescence intensities of bacteria harboring different plasmids were measured using a Synergy H4 microplate reader (BioTek, USA). In order to exclude other interfering factors, harvested cells were washed once with PBS (pH 7.4) and then resuspended in PBS (pH 7.4) at an $OD_{600\ nm}$ of approximately 1.0. The excitation wavelength for RFP was set at 554 nm and emission was set at 586 nm. Fluorescence intensities were normalized by $OD_{600\ nm}$ and were used to indicate the activities of the promoters. Bacteria harboring pXMJ19-$P_{tac}$ were induced with 1 mM IPTG.

### Construction of the recombinant plasmid pXMJ190-$P_{rpl21}$-PpcoaA-RFP

To further examine the function of the isolated promoters, a type III pantothenate kinase from *P. putida* KT2440 (*Pp*coA) and RFP were co-expressed under the control of the strongest promoter $P_{rpl21}$. The *PpcoaA* gene was amplified by PCR from the genomic DNA of *P. putida* KT 2440, and the Shine-Dalgarno sequence for translation of RFP was calculated using the RBS calculator online tool (https://www.denovodna.com/software/). All primers used in this section are listed in Additional file 1: Table S1. The recombinant plasmid pXMJ190-$P_{rpl21}$-*Pp*coaA-RFP was constructed with Gibson assembly and positive colonies were confirmed by DNA sequencing (Tsingke, China). Recombinant plasmids were transformed into *C. ammoniagenes* ATCC 6871 for further experiments.

### SDS-PAGE analysis

*C. ammoniagenes* strains were precultured in 10 mL LB medium at 30 °C and shaken at 220 rpm for 24 h. Ten percent of the culture was inoculated in a 250 mL shake flask containing 100 mL fermentation medium. After 24 h cultivation, cell samples were harvested by centrifugation at 12,000 rpm for 10 min. Forty micrograms of cell lysate were loaded per lane. The *Pp*coaA and RFP expression was analyzed by 15% (v/w) polyacrylamide gel electrophoresis (PAGE) with cell-free extract under denaturing conditions. Mini-Protean III Electrophoresis System (Bio-Rad, USA) was utilized to perform the operation. Coomassie Brilliant Blue R-250 (0.2%, w/v) was utilized to stain protein on the gel.

### Analysis of CoA production

Coenzyme A content was determined according to the modified phosphotransacetylase method [38, 39]. All reagents were purchased from National Institutes for Food and Drug Control. Briefly, 3.0 mL of Tris buffer (pH 7.6), 0.1 mL of acetyl phosphate dilithium salt (15.2 g/L) and 0.1 mL of the test solution were added into a 1 cm cuvette and mixed. Absorbance at 233 nm was recorded as $E_0$; and then 0.01 mL of the phosphotransacetylase (30–40 U/mL) solution was added, mixed well and the highest absorbance within 3 to 5 min was taken as $E_1$. Finally, another 0.01 mL of phosphotransacetylase solution was added, mixed well and the absorbance was read as $E_2$. The number of CoA units per milliliter was calculated as $U = (2E_1\text{-}E_0\text{-}E_2) \times 5.55 \times 413$.

## Supplementary information

**Additional file 1: Table S1.** Primer sequences used in this study. **Table S2.** Comparison of different promoter activities in *C. ammoniagenes* ATCC 6871. **Table S3.** The sequences of identified promoters.

#### Abbreviations

CoA: coenzyme A; IPTG: Isopropyl β-D-1-thiogalactopyranoside

#### Acknowledgements

We are grateful to Professor Zhiyang Dong (Institute of Microbiology, Chinese Academy of Sciences, China) for generously providing the plasmid pXMJ19. This work was supported by Grant from National Natural Science Foundation of China (31570077 to S.W.).

#### Authors' contributions

YH, SW, YT designed all the experiments; YH and SC carried out the experiments, YH organized and interpreted the data, and drafted the manuscript. JW, GL, SW and YT contributed to discussion, revised and corrected the manuscript. All authors read and approved the final manuscript.

#### Funding

This study was funded by National Natural Science Foundation of China (31570077). The funding body did not play any role in the design, execution, analysis, and interpretations of data or in writing the manuscript.

#### Ethics approval and consent to participate

Not applicable.

#### Consent for publication

Not applicable.

**Competing interests**
The authors declare that they have no competing interests.

**Author details**
[1]CAS Key Laboratory of Microbial Physiological and Metabolic Engineering, Institute of Microbiology, Chinese Academy of Sciences, Beijing 100101, People's Republic of China. [2]University of Chinese Academy of Sciences, Beijing 100101, People's Republic of China. [3]Kaiping Genuine Biochemical Pharmaceutical Co. Ltd, Kaiping, People's Republic of China.

**References**

1. Kamada N, Yasuhara A, Takano Y, Nakano T, Ikeda M. Effect of transketolase modifications on carbon flow to the purine-nucleotide pathway in *Corynebacterium ammoniagenes*. Appl Microbiol Biotechnol. 2001;56(5–6):710–7.
2. Liu L, Yang H, Shin HD, Li J, Du G, Chen J. Recent advances in recombinant protein expression by *Corynebacterium*, *Brevibacterium*, and *Streptomyces*: from transcription and translation regulation to secretion pathway selection. Appl Microbiol Biotechnol. 2013;97(22):9597–608.
3. Shimosaka T, Tomita H, Atomi H. Regulation of coenzyme A biosynthesis in the Hyperthermophilic bacterium *Thermotoga maritima*. J Bacteriol. 2016; 198(14):1993–2000.
4. Sibon OC, Strauss E. Coenzyme A. to make it or uptake it? Nat Rev Mol Cell Biol. 2016;17(10):605–6.
5. Fujio T, Maruyama A. Enzymatic production of pyrimidine nucleotides using *Corynebacterium ammoniagenes* cells and recombinant *Escherichia coli* cells: enzymatic production of CDP-choline from orotic acid and choline chloride (part I). Biosci Biotechnol Biochem. 1997;61(6):956–9.
6. Ishige T, Honda K, Shimizu S. Whole organism biocatalysis. Curr Opin Chem Biol. 2005;9(2):174–80.
7. Paik JE, Lee BR. I isolation of transcription initiation signals from *Corynebacterium ammoniagenes* and comparison of their gene expression levels in *C. ammoniagenes* and *Escherichia coli*. Biotechnol Lett. 2003;25(16): 1311–6.
8. Ozaki A, Katsumata R, Oka T, Furuya A. Transfection of *Corynebacterium glutamicum* with temperate phage φCG1. Agric Biol Chem. 1984;48(10): 2597–601.
9. Santamaria R, Gil JA, Mesas JM, Martin JF. Characterization of an endogenous plasmid and development of cloning vectors and a transformation system in *Brevibacterium lactofermentum*. J Gen Microbiol. 1984;130(9):2237–46.
10. Sonnen H, Thierbach G, Kautz S, Kalinowski J, Schneider J, Puhler A, Kutzner HJ. Characterization of pGA1, a new plasmid from *Corynebacterium glutamicum* LP-6. Gene. 1991;107(1):69–74.
11. Billman-Jacobe H, Wang L, Kortt A, Stewart D, Radford A. Expression and secretion of heterologous proteases by *Corynebacterium glutamicum*. Appl Environ Microbiol. 1995;61(4):1610–3.
12. Salim K, Haedens V, Content J, Leblon G, Huygen K. Heterologous expression of the *Mycobacterium tuberculosis* gene encoding antigen 85A in *Corynebacterium glutamicum*. Appl Environ Microbiol. 1997;63(11):4392–400.
13. Tsuchiya M, Morinaga Y. Genetic control Systems of *Escherichia Coli* can Confer Inducible Expression of cloned genes in Coryneform Bacteria. Nat Biotechnol. 1988;6(4):428–30.
14. Lee SM, Lee JY, Park KJ, Park JS, Ha UH, Kim Y, Lee HS. The regulator RamA influences cmytA transcription and cell morphology of *Corynebacterium ammoniagenes*. Curr Microbiol. 2010;61(2):92–100.
15. Park JU, Jo JH, Kim YJ, Chung SS, Lee JH, Lee HH. Construction of heat-inducible expression vector of *Corynebacterium glutamicum* and *C. ammoniagenes*: fusion of lambda operator with promoters isolated from *C. ammoniagenes*. J Microbiol Biotechnol. 2008;18(4):639–47.
16. Deaner M, Alper HS. Promoter and terminator discovery and engineering. Adv Biochem Eng Biot. 2018;162:21–44.
17. Li S, Wang J, Li X, Yin S, Wang W, Yang K. Genome-wide identification and evaluation of constitutive promoters in streptomycetes. Microb Cell Fact. 2015;14:172.
18. Hu Y, Wan H, Li J, Zhou J. Enhanced production of L-sorbose in an industrial *Gluconobacter oxydans* strain by identification of a strong promoter based on proteomics analysis. J Ind Microbiol Biot. 2015;42(7): 1039–47.
19. Schafer A, Schwarzer A, Kalinowski J, Puhler A. Cloning and characterization of a DNA region encoding a stress-sensitive restriction system from *Corynebacterium glutamicum* ATCC 13032 and analysis of its role in intergeneric conjugation with *Escherichia coli*. J Bacteriol. 1994;176(23):7309–19.
20. Itaya H, Kikuchi Y. Secretion of *Streptomyces mobaraensis* pro-transglutaminase by coryneform bacteria. Appl Microbiol Biotechnol. 2008; 78(4):621–5.
21. Jakoby M, Kramer R, Burkovski A. Nitrogen regulation in *Corynebacterium glutamicum*: isolation of genes involved and biochemical characterization of corresponding proteins. FEMS Microbiol Lett. 1999;173(2):303–10.
22. Jacobs E, Mills JD, Janitz M. The role of RNA structure in posttranscriptional regulation of gene expression. J Genet Genomics. 2012;39(10):535–43.
23. Mutalik VK, Guimaraes JC, Cambray G, Mai QA, Christoffersen MJ, Martin L, Yu A, Lam C, Rodriguez C, Bennett G, Keasling JD, Endy D, Arkin AP. Quantitative estimation of activity and quality for collections of functional genetic elements. Nat Methods. 2013;10(4):347–53.
24. Rytter JV, Helmark S, Chen J, Lezyk MJ, Solem C, Jensen PR. Synthetic promoter libraries for *Corynebacterium glutamicum*. Appl Microbiol Biotechnol. 2014;98(6):2617–23.
25. Shimizu S, Esumi A, Komaki R, Yamada H. Production of coenzyme A by a mutant of *Brevibacterium ammoniagenes* resistant to Oxypantetheine. Appl Environ Microbiol. 1984;48(6):1118–22.
26. Vallari DS, Jackowski S, Rock CO. Regulation of pantothenate kinase by coenzyme A and its thioesters. J Biol Chem. 1987;262(6):2468–71.
27. Brand LA, Strauss E. Characterization of a new pantothenate kinase isoform from *Helicobacter pylori*. J Biol Chem. 2005;280(21):20185–8.
28. Leonardi R, Chohnan S, Zhang YM, Virga KG, Lee RE, Rock CO, Jackowski S. A pantothenate kinase from *Staphylococcus aureus* refractory to feedback regulation by coenzyme A. J Biol Chem. 2005;280(5):3314–22.
29. Ogata Y, Chohnan S. Prokaryotic type III pantothenate kinase enhances coenzyme A biosynthesis in *Escherichia coli*. J Gen Appl Microbiol. 2015; 61(6):266–9.
30. Ogata Y, Katoh H, Asayama M, Chohnan S. Role of prokaryotic type I and III pantothenate kinases in the coenzyme A biosynthetic pathway of *Bacillus subtilis*. Can J Microbiol. 2014;60(5):297–305.
31. Engler C, Kandzia R, Marillonnet S. A one pot, one step, precision cloning method with high throughput capability. PLoS One. 2008;3(11):e3647.
32. Gibson DG, Young L, Chuang RY, Venter JC, Hutchison CA III, Smith HO. Enzymatic assembly of DNA molecules up to several hundred kilobases. Nat Methods. 2009;6(5):343–5.
33. Kosuri S, Goodman DB, Cambray G, Mutalik VK, Gao Y, Arkin AP, Endy D, Church GM. Composability of regulatory sequences controlling transcription and translation in *Escherichia coli*. Proc Natl Acad Sci U S A. 2013;110(34): 14024–9.
34. Reese MG. Application of a time-delay neural network to promoter annotation in the *Drosophila melanogaster* genome. Comput Chem. 2001; 26(1):51–6.
35. Ming YM, Wei ZW, Lin CY, Sheng GY. Development of a *Bacillus subtilis* expression system using the improved Pglv promoter. Microb Cell Factories. 2010;9:55.
36. Estrem ST, Gaal T, Ross W, Gourse RL. Identification of an UP element consensus sequence for bacterial promoters. Proc Natl Acad Sci U S A. 1998; 95(17):9761–6.
37. Zhao Z, Liu X, Zhang W, Yang Y, Dai X, Bai Z. Construction of genetic parts from the *Corynebacterium glutamicum* genome with high expression activities. Biotechnol Lett. 2016;38(12):2119–26.
38. Stadtman ER, Novelli GD, Lipmann F. Coenzyme A function in and acetyl transfer by the phosphotransacetylase system. J Biol Chem. 1951;191(1):365–76.
39. Tsuchiya Y, Pham U, Gout I. Methods for measuring CoA and CoA derivatives in biological samples. Biochem Soc Trans. 2014;42(4):1107–11.

# 20

# Disruption of a putative mitochondrial oxaloacetate shuttle protein in *Aspergillus carbonarius* results in secretion of malic acid at the expense of citric acid production

Lei Yang[1†], Tore Linde[1,2†], Abeer H. Hossain[3], Mette Lübeck[1], Peter J. Punt[3] and Peter S. Lübeck[1*]

**Abstract**

**Background:** In filamentous fungi, transport of organic acids across the mitochondrial membrane is facilitated by active transport via shuttle proteins. These transporters may transfer different organic acids across the membrane while taking others the opposite direction. In *Aspergillus niger*, accumulation of malate in the cytosol can trigger production of citric acid via the exchange of malate and citrate across the mitochondrial membrane. Several mitochondrial organic acid transporters were recently studied in *A. niger* showing their effects on organic acid production.

**Results:** In this work, we studied another citric acid producing fungus, *Aspergillus carbonarius*, and identified by genome-mining a putative mitochondrial transporter MtpA, which was not previously studied, that might be involved in production of citric acid. This gene named *mtpA* encoding a putative oxaloacetate transport protein was expressed constitutively in *A. carbonarius* based on transcription analysis. To study its role in organic acid production, we disrupted the gene and analyzed its effects on production of citric acid and other organic acids, such as malic acid. In total, 6 transformants with gene *mtpA* disrupted were obtained and they showed secretion of malic acid at the expense of citric acid production.

**Conclusion:** A putative oxaloacetate transporter gene which is potentially involved in organic acid production by *A. carbonarius* was identified and further investigated on its effects on production of citric acid and malic acid. The *mtpA* knockout strains obtained produced less citric acid and more malic acid than the wild type, in agreement with our original hypothesis. More extensive studies should be conducted in order to further reveal the mechanism of organic acid transport as mediated by the MtpA transporter.

**Keywords:** *Aspergillus carbonarius*, Mitochondrial transport protein, Citric acid, Malic acid, Metabolic engineering

* Correspondence: psl@bio.aau.dk
[†]Lei Yang and Tore Linde contributed equally to this work.
[1]Department of Chemistry and Bioscience, Section for Sustainable Biotechnology, Aalborg University, A.C. Meyers Vaenge 15, DK-2450 Copenhagen, SV, Denmark
Full list of author information is available at the end of the article

## Background

Bio-based production of organic acids by microorganisms as e.g. filamentous fungi in a biorefinery has a high potential as a substitution of chemicals produced from crude oil [1]. The organic acids (e.g. malic acid and succinic acid) can easily be converted and used as building blocks for deriving different commodity and specialty chemicals and in the past decades, filamentous fungi as *Aspergillus niger* and *Aspergillus oryzae* have been used in industrial production of organic acids. Recently, *Aspergillus carbonarius* was reported to be an efficient organic acid producer and as such may have a potential for bio-based production of C4-dicarboxylic acids. When several genetic modifications were made to improve carbon flux towards dicarboxylic acid production (fumaric, succinic and malic acid), an increased production of citric acid was often observed [2–4]. This phenomenon might be a result of the transport of organic acids across the mitochondrial membrane where e.g. malic acid is transported to the mitochondria in exchange with citric acid. Recently, we identified a plasma membrane $C_4$-dicarboxylate transporter, which was highly involved in the dicarboxylic acid production [5]. Overexpression of the transporter led to a significant increase of $C_4$-dicarboxylic acid production and decreased citric acid production.

In the well-known organic acid producer, *A. niger*, which is phylogenetically related to *A. carbonarius*, the roles of mitochondrial transport of organic acid have been investigated in more details. *A. niger* is able to produce high amounts of organic acids (e.g. citric acid) from a broad range of substrates [6–8]. Considerable research efforts have been taken to reveal the mechanism of citric acid accumulation by the fungus and to redirect carbon flux from the production of citric acid towards other types of organic acids, e.g. itaconic acid production by genetically altered *A. niger* strains [9, 10]. In *A. niger*, citrate is produced in the mitochondria through the Tricarboxylic acid cycle (TCA cycle). Citrate is then transported out of the mitochondria into the cytosol by membrane bound citrate transporters [11]. A common belief is that these transporters are integral membrane proteins and function as antiporters where they exchange citrate from the mitochondria with other organic acids from the cytosol (e.g. cytosolic malate) [8, 12, 13]. It has been suggested that the onset of citric acid production is initiated in relation to the concentration of malate in the cytosol, which is then transported into the mitochondria in exchange for citrate. It was observed that malate was accumulated in the cytosol directly preceding start of production of citrate [12], and an *A. niger* strain transformed with malate dehydrogenase which, in theory, should produce elevated amounts of malate instead produced significantly higher amounts of citrate [8]. This supports the hypothesis that concentration of malate (or possible fumarate or succinate) in the cytosol will not result in secretion of these organic acids but instead produce an increased amount of citric acid due to a possible organic acid transport mechanism where $C_4$-dicarboxylic acids are transported to the mitochondria in exchange with citric acid.

In *Saccharomyces cerevisiae*, it has been observed that the transport of citrate across the mitochondrial membrane is influenced not only by the concentration of malate, but also by the concentration of iso-citrate, succinate, and phosphoenolpyruvate [14, 15]. It is strongly suspected that the mitochondrial citrate transporter (CTP) is the primary transporter of citrate across the mitochondrial membrane in exchange of malate [16, 17]. In another *Aspergillus* species from the section Nigri, *Aspergillus luchuensis* the two putative CTP homologues *ctpA* and *ctpB* have been studied in some detail [18, 19]. The *ctpB* gene was shown to be non-expressed under citric acid production conditions, whereas *ctpA* was constitutively expressed in both *A. luchuensis* and *A. niger* [18, 20]. Although deletion of the *ctpA* gene resulted in serious growth effects of the resulting mutant strains, the effects on citric acid production were marginally [18, 19]. In addition to the mitochondrial citrate/malate antiporter, the role of two putative citrate/oxoglutarate antiporters was also studied in *A. luchuensis* [19] and one in *A. niger* [21]. In *S. cerevisiae* this antiporter is encoded by the *yhm2* gene responsible for exporting citrate from mitochondria to cytosol in exchange of oxoglutarate [22]. Also in this case one of the homologues, *yhmB*, was shown to be hardly expressed, while the closest homologue to *yhm2 (cocA)*, *yhmA* was also expressed constitutively in *A. luchuensis* and *A.niger*, [18, 20]. Deletion of *yhmB* showed no effect on citrate production, whereas *yhmA* and *cocA* deletion had a clear effect on citrate production, which was further aggravated in a conditional *ctpA*/*yhmA* double mutant, where malate and oxo-glutarate production was increased [19].

The hypothesis for the present study is based on the idea that when metabolic carbon-flux increases towards organic acids (e.g. dicarboxylic acids) other than citric acid in the cytosol in *A. carbonarius*, the accumulation of organic acids of interests will not happen due to transport of these acids (or their biosynthetic precursors) into the mitochondria by mitochondrial organic acid transporters in exchange of citrate (Fig. 1). If a transporter that is involved in exchange of organic acids is disrupted, the organic acid, which should be transported to the mitochondria in exchange with citrate, will remain in the cytosol, and in turn may be secreted to the exterior. The ability to reroute organic acid producing biocatalysts into production of more high-value acids would have great economic benefits since several organic acids are considered to be interesting building block chemicals [23]. In the present study, a putative mitochondrial

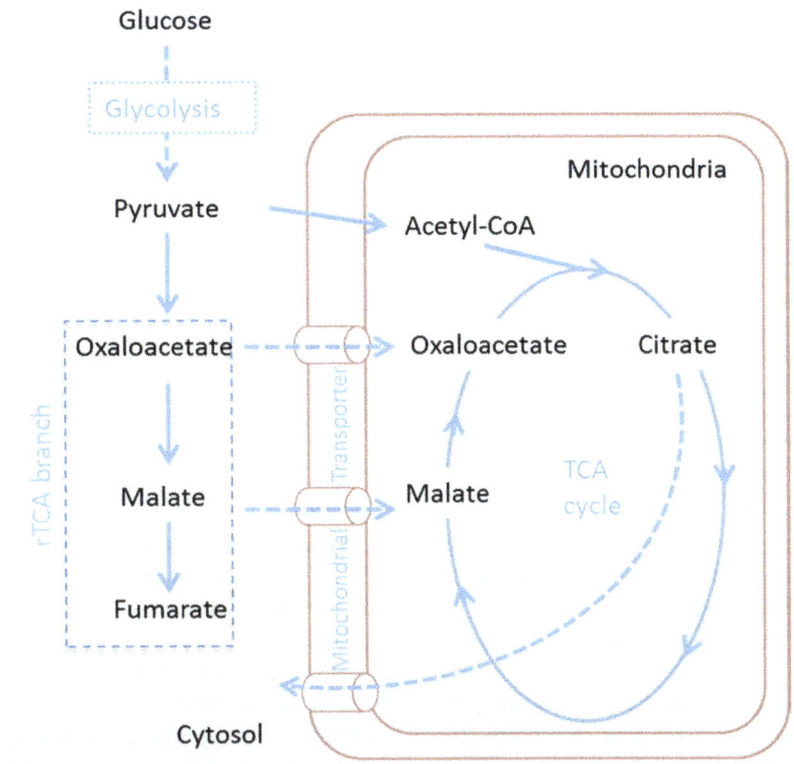


**Fig. 1** Metabolic pathway proposed for organic acid production by *A. carbonarius*

oxaloacetate transporter gene was identified in *A. carbonarius* ITEM5010 based on a bioinformatic approach using characterized fungal organic acid transporters. Oxaloacetate is the key intermediate in the reductive tricarboxylic acid branch (rTCA branch) and the TCA cycle, which are highly involved in production of a number of organic acids e.g. citric acid, malic acid and succinic acid. In our work, the role of the oxaloacetate transport in organic acid production was investigated by disrupting the transporter gene and examining the changes in organic acid production by the knockout strain.

## Results

### Identification and expression of *mtpA* gene

Mitochondrial transporters of carboxylic acids are not well-characterized proteins in *Aspergillus* sp. To date only a few mitochondrial transporters of carboxylic acids have been studied in filamentous fungi [19, 24]. The number of candidate transporter proteins in a single fungal species is high, in yeast for example at least 34 different mitochondrial transporter proteins are known [25], and 39 are annotated in the *A. niger* genome [26]. The putative mitochondrial transporter gene investigated in the present study was identified using genome mining based on characterized mitochondrial proteins of fungal origin. For this purpose, homologues of the three genes identified in *S. cerevisiae* as being involved in citrate or citrate-intermediate transport, CTP1, YHM2 and OAC1 were identified by BLAST analysis in the genome of the most well characterized black *Aspergillus* genome, *A. niger* NRRL3 [27]. This resulted in the identification of 5 putative homologues, two for CTP1, two for YHM2 and one for OAC1. In Table 1, an overview of the expression data in *A. niger* available for these five genes in various publically available RNAseq data sets is given. From this analysis it is clear that the CTP1 and YHM2 homologues, *ctpA* and *yhmA*, are expressed to much higher and constitutive levels compared to *ctpB* or *yhmB*, respectively, confirming what was also already described by Kirimura et al. (2016) and Kadooka et al. (2018) [18, 19] . The OAC1 homologue was expressed to a similar level in all the studies represented in Table 1 (*A. niger* An14g06860 [20]. The putative *A. carbonarius* OAC1 orthologue, which we refer as *mtpA* was selected for further analysis. In order to confirm the expression of the *mtpA* gene in *A. carbonarius*, transcriptional analysis was carried out using reverse transcription polymerase chain reaction (RT-PCR). As shown in Additional file 1: Figure S1, a shorter fragment (~ 200 bp) was obtained by polymerase chain reaction (PCR) amplification of cDNA compared with genomic DNA, since the primers spanned the terminal part of the gene, where the cDNA in contrast to the genomic DNA contained no intron, confirming *mtpA* expression under these conditions.

**Table 1** Transcription analysis of *A. niger* genes orthologous to the mitochondrial transporters identified in *A. carbonarius*

| *A. carbonarius* ITEM 5010 | Protein ID in *A. carbonarius* (JGI genome) | locus tag in *A. niger* | *A. niger* H915–1 | | | | | *A. niger* AB 1.13 | |
|---|---|---|---|---|---|---|---|---|---|
| | | | 6 h_FPKM | 12 h_FPKM | 24 h_FPKM | 36 h_FPKM | 48 h_FPKM | 75 h_RPKM | 75 h_RPKM |
| | | | 6 h | 12 h | 24 h | 36 h | 48 h | 75 h | 92 h |
| | | | citric acid titer at point of sampling (g/l) | | | | | | |
| | | | ≈ 3 | ≈ 18 | ≈ 50 | ≈ 86 | ≈ 118 | 2.5 | 7 |
| *ctp1 (ctpA)* | 139563 | An11g11230 | 387.65 | 248.98 | 245.78 | 253.95 | 268.37 | 139,82 | 141,61 |
| *ctp1 (ctpB)* | 515063 | An18g00070 | 3.70 | 0.83 | 3.24 | 0.54 | 1.18 | 0,54 | 0,56 |
| *yhmA* | 398055 | An09g06670 | 605.39 | 406.06 | 315.49 | 532.38 | 395.27 | 386,26 | 329,29 |
| *yhmB* | 399724 | An02g11090 | 36.37 | 55.64 | 9.04 | 41.48 | 76.31 | 0.07 | 0.06 |
| *mtpA* | 209833 | An14g06860 | 70.50 | 35.61 | 44.76 | 38.02 | 38.64 | 58,62 | 59,27 |

*A. niger* H915–1- a citric acid producing strain [28]; *A. niger* AB1.13 – a commonly laboratory strain [20]. RPKM (Reads Per Kilobase of target per Million mapped reads) and FPKM (Fragments Per Kilobase of target per Million mapped reads) values were calculated according to the method presented by Mortazavi et al. [29], in order to normalize data for gene length

### Generation of knockout strains and southern blotting

The putative organic acid transporter gene *mtpA* in *A. carbonarius* ITEM 5010 was knocked out using a bipartite approach to create knockout strains. Disruption of the target gene in the selected transformants was first verified by PCR (data not shown). This was done with a forward primer binding upstream to the gene in the genome paired with a reverse primer binding to the inserted hygromycin gene, so generation of the predicted PCR product indicated that the transporter gene was replaced by the hygromycin resistant gene (Fig. 2a). In total, 32 transformants were obtained, and of these, 6 transformants were verified as positive knockout giving a gene targeting efficiency of 20%, Southern blot hybridization was then performed on the selected transformant *mtpA-1*, and the hybridization of probe only occurred with the expected DNA fragment as shown in Fig. 2b, confirming successful gene disruption via homologous recombination without any ectopic integration in other sites of the genome.

### Organic acid production

All 6 putative *mtpA* disruption strains were grown in shake flask cultures and analyzed for organic acid production by High-performance liquid chromatography (HPLC) analysis. As shown in Fig. 3, citric acid and malic acid were analyzed during the cultivation. All the transformants produced less citric acid than the wild type after day 5, meanwhile, a low but significant amount of malic acid was also detected from the culture with transformants but not from the wild type. No other major organic acid peaks were detected in the HPLC profile. The transformant *mtpA-1* that was verified via southern blotting was also analyzed in a pH controlled fermentation. As shown in Table 2, it consumed 29 g/l glucose and produced 1.6 g/l citric acid and 0.35 g/l malic acid while the wild type consumed 31 g/l glucose and produced 2.6 g/l citric acid and no malic acid.

## Discussion

A putative mitochondrial oxaloacetate transporter gene *mtpA* was identified in *A. carbonarius* and knocked out. The *mtpA* mutant was investigated for its impact on production of citric acid and malic acid. Disruption of the *mtpA* gene resulted in a reduction in citric acid production and secretion of malic acid in the culture medium. The hypothesis of this research study was that the disruption of this organic acid transporter could reduce the production of citric acid and simultaneously increase accumulation of malic acid or other dicarboxylic acids. Oxaloacetate is used in both cytosolic rTCA branch and mitochondrial TCA cycle respectively to produce malic acid and citric acid (Fig. 1). The deletion of *mtpA* may stop the transport of oxaloacetate into TCA cycle in mitochondria and save more carbon flux towards downstream rTCA branch in cytosol, which could be the reason for the measured increase in malic acid. However, only mild effects on production of citric acid and malic acid have been obtained. Apart from the fact that these effects may be larger in high citric acid production strains like *A. niger* H915–1 [28], a reason for this could also be that the MtpA membrane protein is part of a more complex network of transmembrane proteins involved in antiport and transport of different organic acids over the mitochondrial membrane. Similarly modest effects on organic acid production were also reported in the case of knocking out mitochondrial citrate transporters in *A. niger* and *A. luchuensis*. Citric acid production was also only slightly affected in *ctpA* knockout strain suggesting that other transporters also play important roles in the export of citric acid during the production phase [18]. On the other hand, deletion of the citrate/oxoglutarate antiporter *yhmA* had a more serious effect on citric acid and malic acid production, but also showed serious pleiotropic effects, making conclusion about its role in citric acid production somewhat ambiguous [18, 19]. It should also be noted that the expression of

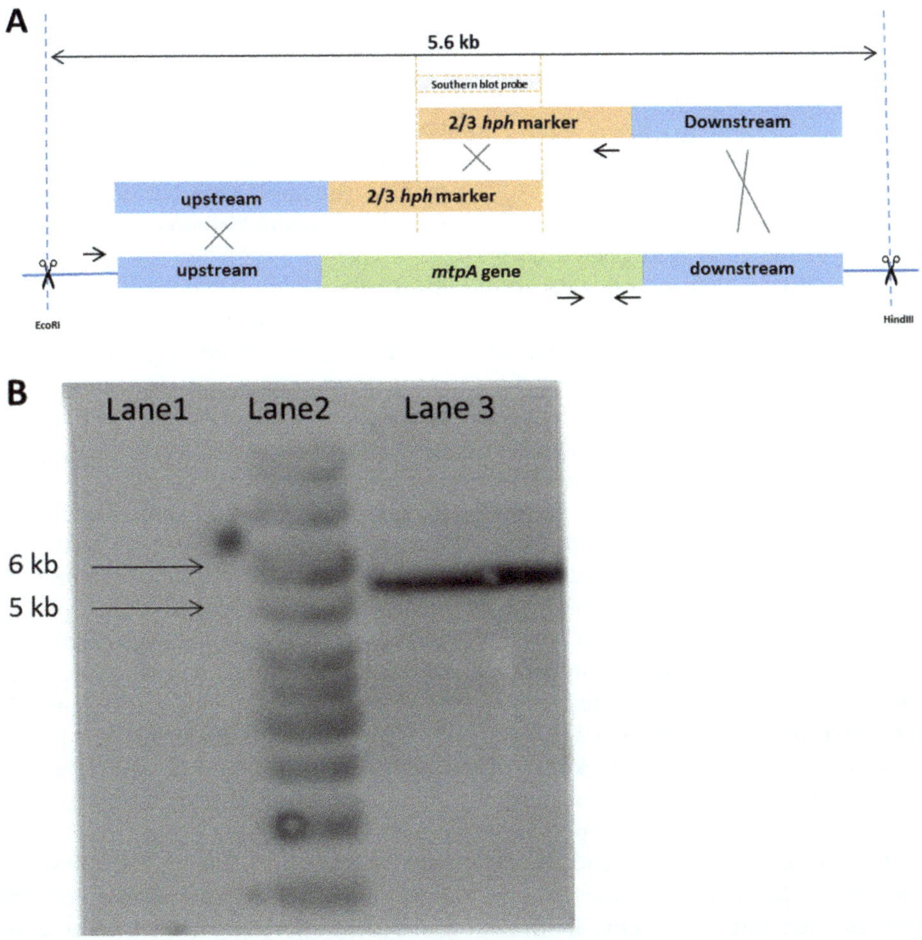


**Fig. 2** Verification of *mtpA* gene disruption in the transformant **a**) Disruption of *mtpA* gene with bipartite substrates in *A. carbonarius*. **b** Southern blotting analysis. Lane 1 hydridization of the probe to the genomic DNA fragment (from the wild-type) digested by *EcoRI* and *HindIII*, Lane 2, DNA ladder, Lane 3, hydridization of the probe to the genomic DNA fragment (from *mtpA-1* transformant) digested by *EcoRI* and *HindIII*

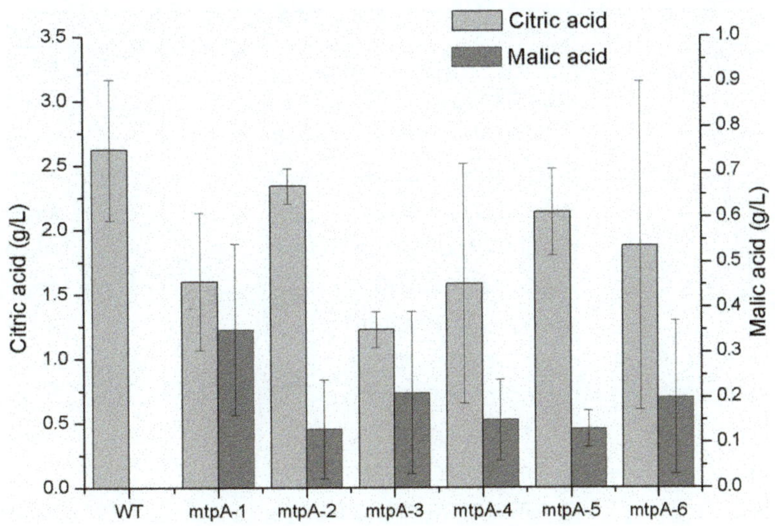


**Fig. 3** Organic acid production by *mtpA* transformants (Error bars shows standard error of the mean)

**Table 2** Effects of disrupting *mtpA* gene on organic acid production by *A. carbonarius* in shake flask fermentation

| Strains/Titer (g/L) | Glucose consumption | Citric acid | Malic acid |
|---|---|---|---|
| WT | 31.0 ± 1.27 | 2.6 ± 0.54 | n.d. |
| *mtpA-1* | 29.2 ± 7.83 | 1.6 ± 0.53 | 0.35 ± 0.19 |

(Note: Cultivations were performed in triplicates, showing standard error of the mean, n.d. - non-detectable level in HPLC)

the genes encoding each of the citric acid related mitochondrial transporters is similar in high and low citric acid production strains and conditions (Table 2).

In the study presented here, the effect of deletion of the putative oxaloacetate carrier MtpA may be somewhat moderate due to a rerouting of the organic acid fluxes by other mitochondrial dicarboxylate transporters that can transport malate from the cytosol to mitochondria, where malate is the precursor for oxaloacetate synthesis in the TCA cycle and thus for citric acid production. Since shaking cultures, as performed in this study, are prone to variation, we carried out replicate cultures to determine the statistical relevance of the results obtained. A further more definitive conclusion would require further investigations, including controlled fermentation experiments in different scales and under different optimized conditions or knockouts of *mtpA* in combination with other putative transporter genes. Also more detailed in vitro transporter specificity studies as performed for *A. luchuensis* CtpA and YhmA [19], could be performed for MtpA to determine kinetic transport parameters and substrate specificity. In these studies, the function of a transporter was defined in the presence or absence of substrates or even radioactive labeled substrates as was recently performed for the first time for a fungal mitochondrial transporter [30].

## Conclusions

A putative oxaloacetate transporter gene was identified in *A. carbonarius* and further investigated on its effects on citric and malic acid production. The *mtpA* knockout mutant obtained secreted malic acid at the expense of citric acid which was the only secreted organic acid in the wild type, in agreement with our original hypothesis for the effect of gene disruption on organic acid production.

## Methods

### Strains and growth conditions

*A. carbonarius* ITEM 5010 was used as the parental strain to construct transformants. All the fungal strains were cultivated as previously described in [31] . *Escherichia coli* strain DH5α was used as host for plasmid propagation.

### Genomic DNA extraction and plasmid construction

Genomic DNA extraction from freshly grown mycelia and measurement of DNA concentration were carried out as described in [31]. The knockout plasmid pSB4.1.1 containing *RP27-hph-βT* for hygromycin resistance, origin of replication and ampicillin resistance for growth in *E.coli* and a specific cassette facilitating simpleUSER cloning was constructed as described in [2]. All primers used in the study were designed with the primer software Primer3 [32] (Table 3). In the knock out plasmid pSB4.1.1-Antiporter upstream and downstream regions to the target gene were identified in the *A. carbonarius* genome provided by JGI. The upstream and downstream flanking regions with the size of 1 kb were selected for efficient homologous recombination [33] and amplified with primers containing uracil tails (primers 1–4, Table 3). Genomic DNA from *A. carbonarius* was used as template. The flanking regions of gene were amplified by PCR using

**Table 3** Primers used in the study

| Name | No. | Sequence (5′→3′) | Annotation |
|---|---|---|---|
| *mtpA* up-fw-U | 1 | GGGTTTAAUAGACATACCGTCGACCTTGG | Amplify upstream region of *mtpA* |
| *mtpA* up-rv-U | 2 | GGACTTAAUGAGGGTGAGTCTGGCAGAAG | Amplify upstream region of *mtpA* |
| *mtpA* do-fw-U | 3 | GGCATTAAUTCAGTTTTGCATGGTTGAGC | Amplify downstream region of *mtpA* |
| *mtpA* do-rv-U | 4 | GGTCTTAAUGCGGGTGGTATTCTCTGTGT | Amplify downstream region of *mtpA* |
| Bipart-1 rv | 5 | GATGTTGGCGACCTCGTATT | Amplify 1st bipartite fragment |
| Bipart-2 fw | 6 | GATGTAGGAGGGCGTGGATA | Amplify 2nd bipartite fragment |
| *mtpA* up-fw | 7 | AGACATACCGTCGACCTTGG | Amplify 1st bipartite fragment |
| *mtpA* do-rv | 8 | GCGGGTGGTATTCTCTGTGT | Amplify 2nd bipartite fragment |
| *mtpA* ko-ch-fw | 9 | GTCGCAAGCTTCAACTTTCC | Check for positive knockout of *mtpA* |
| *mtpA* ko-ch-rv | 10 | TATTGGAGAGCAAGGGATGG | Check for positive knockout of *mtpA* |
| Hph-Fw | 11 | GATGTAGGAGGGCGTGGATA | Amplify *hph* marker |
| Hph-Rv | 12 | GATGTTGGCGACCTCGTATT | Amplify *hph* marker |
| *mtpA*-fw | 13 | CAAGTTCTCCTTGAGTGAGTCG | Transcription analysis of *mtpA* |
| *mtpA*-rv | 14 | TCGACTGCCTTTACAAGACC | Transcription analysis of *mtpA* |

Pfu turbo cx polymerase (Agilent) and 65 °C as annealing temperature. The obtained PCR products were then ligated to the simpleUSER cassette in pSB4.1.1 that was already digested with the restriction enzyme *PacI* and the nicking enzyme *Nb.BbvCI* to create the complementary overhangs as previously described [2]. Through self-assembly the PCR fragments were cloned into the plasmid followed by transformation of *E. coli* with the plasmid for further propagation. Plasmid extraction with the GeneJET plasmid miniprep kit (ThermoFisher Scientific) was carried out according to the manufacturer's protocol. All plasmids were verified by Sanger sequencing using the sequencing service from StarSEQ (Mainz, Germany).

#### Protoplast transformation

Protoplasts were prepared from wild type *A. carbonarius* following the procedure described in [4]. Bipartite PCR fragments were made from the knockout plasmids pSB4.1.1-Antiporter and primer 5–8 (Table 3) and prepared for transformation as described in [31]. Transformation was done in 100 μl aliquots of protoplasts according to the method described by [4]. Transformants carrying hygromycin resistance were isolated and transferred to PDA plates containing hygromycin. All the transformants were preserved by mixing spores suspension with 20% glycerol (final concentration in cryostock) and stored at −80 °C. Validation of positive knockouts was carried out by PCR using primers 9–10 (Table 3).

#### Transcriptional analysis of *mtpA* gene

Transcription of the *mtpA* gene was analyzed by reverse-transcription PCR (RT-PCR). Total RNA purification and cDNA synthesis were prepared as previously described [5]. The cDNA was then used as template in PCR with *mtpA* gene specific primers no. 13–14 for transcription analysis.

#### Southern blotting of *mtpA* transformant

Southern blotting was carried out on a single *mtpA* transformant for confirmation of *mtpA* gene disruption. As shown in Fig. 2a, the hybridization probe was prepared by amplifying a part of the *hph* marker gene with primers 11–12 (Table 3) and labeled using a North2-South chemiluminescent detection kit (Thermo Fisher Scientific). Genomic DNA of the transformant was digested by *EcoRI* and *HindIII* and hybridized with the probe. Southern blotting was then performed using Whatman Turboblotter transfer system (GE healthcare life sciences) and detected by using a Pierce Chemiluminescent Nucleic acid detection module Kit (Thermo scientific). The blot was imaged in Thermo ECL imager.

#### Fermentation and shake flask cultivation condition

Fermentation was performed in 50 ml Erlenmeyer flasks in 10 ml media in triplicates. Fresh spore-suspension was added to 10 ml pre-culture medium (Yeast extract 3.6 g/l and peptone 10 g/l) to a final concentration of $2 \times 10^5$ spores/ml. Pre-culture was incubated at 25 °C, 200 rpm and 2.5 cm amplitude for 48 h (KS 4000 I control, IKA). For the pH controlled fermentation, the mycelia pellets from the pre-culture were transferred to Erlenmeyer flasks with cotton stoppers and 10 ml production media [34] containing 100 g/l glucose, 2 g/l $(NH_4)_2SO_4$, 0.15 g/l $KH_2PO_4$, 0.15 g/l $K_2HPO_4$, 0.1 g/l $MgSO_4{\cdot}7H_2O$, 0.1 g/l $CaCl_2{\cdot}2H_2O$, 0.005 g/l NaCl, 0.1 g/l $ZnSO_4$, 0.005 g/l $FeSO_4{\cdot}7H_2O$ and 30 g/l $CaCO_3$(for pH maintenance). Flasks were incubated at 25 °C, 200 rpm. The supernatant obtained from each fermentation culture was prepared as described by [35]. 1 ml sample from each flask was taken at day 5. 50 μl 50% $H_2SO_4$ was added and the mixture was heated to 80 °C and incubated for 15 min. The samples were then cooled down to the room temperature followed by centrifugation at 8000 rpm for 1 min. The supernatant was filtered through 0.45 μM HPLC-grade regenerated cellulose membrane filters for HPLC analysis. 250 μl of the filtrate were analyzed for the content of sugars and organic acids by HPLC (Dionex Ultimate 3000-LC system) with the conditions described in [31].

#### Statistical analysis

Comparison of results from triplicates were analyzed by *t*-test with a significance level of $p < 0.05$. Error bars on figures are standard error of the mean.

## Additional file

**Additional file 1: Figure S1.** Transcriptional analysis of the mtpA gene.

#### Abbreviations

CTP: Citrate transporter; HPLC: High-performance liquid chromatography; PCR: Polymerase chain reaction; rTCA branch: Reductive tricarboxylic acid branch; RT-PCR: Reverse transcription polymerase chain reaction; TCA cycle: Tricarboxylic acid cycle

#### Acknowledgements

Gitte Hinz-Berg and Sabine Malene Bjerregaard are thanked for excellent laboratory assistance.

#### Statement on methodology

In this study, description of methods on fungal cultivation, genomic DNA extraction, plasmid construction, protoplast transformation and HPLC analysis was simplified to avoid repetitive information that has been provided by our previous research paper. The detailed descriptions on these methods can be obtained from the reference [31].

#### Authors' contributions

LY, TL, ML, PP and PSL conceived the project and designed experiments. LY and TL performed the experiments and drafted the manuscript. AHH acquired and interpreted data for transcription analysis. AHH, ML, PP and PSL revised the manuscript. All authors read and approved the final manuscript.

#### Funding

The project was financed by Novozymes A/S and the Danish Strategic Research program MycoFuelChem (DSF grant no. 11-116803). The funders had no role in the design of this study, the collection, analysis, and interpretation of data or in preparation of the manuscript.

**Ethics approval and consent to participate**
The study does not contain any experiment with human participants or animals performed by any of the authors.

**Consent for publication**
Not applicable.

**Competing interests**
The co-author Peter J. Punt is a member of the editorial board (Section Editor) of this journal, and other authors declare that they have no competing interests.

**Author details**
[1]Department of Chemistry and Bioscience, Section for Sustainable Biotechnology, Aalborg University, A.C. Meyers Vaenge 15, DK-2450 Copenhagen, SV, Denmark. [2]Present address: AGC Biologics, Vandtaarnsvej 83B, DK-2860, Soeborg, Copenhagen, Denmark. [3]Dutch DNA Biotech BV, Padualaan 8, 3584CH, Utrecht, The Netherlands.

**References**

1. Gallezot P. Conversion of biomass to selected chemical products. Chem Soc Rev. 2012;41(4):1538–58.
2. Hansen NB, Lübeck M, Lübeck PS. Advancing USER cloning into simpleUSER and nicking cloning. J Microbiol Methods. 2014;96(1):42–9.
3. Yang L, Lübeck M, Lübeck PS. Effects of heterologous expression of phosphoenolpyruvate carboxykinase and phosphoenolpyruvate carboxylase on organic acid production in *Aspergillus carbonarius*. J Ind Microbiol Biotechnol. 2015;42(11):1533–45.
4. Yang L, Lübeck M, Lübeck PS. Deletion of glucose oxidase changes the pattern of organic acid production in *Aspergillus carbonarius*. AMB Express. 2014;4(1):54.
5. Yang L, Christakou E, Vang J, Lübeck M, Lübeck PS. Overexpression of a C4-dicarboxylate transporter is the key for rerouting citric acid to C4-dicarboxylic acid production in *Aspergillus carbonarius*. Microb Cell Factories. 2017;16(1):43.
6. Goldberg I, Rokem JS, Pines O. Organic acids: old metabolites, new themes. J Chem Technol Biotechnol. 2006;81(10):1601–11.
7. Papagianni M. Advances in citric acid fermentation by *Aspergillus niger*: biochemical aspects, membrane transport and modeling. Biotechnol Adv. 2007;25(3):244–63.
8. de Jongh WA, Nielsen J. Enhanced citrate production through gene insertion in *Aspergillus niger*. Metab Eng. 2008;10(2):87–96.
9. Blumhoff ML, Steiger MG, Mattanovich D, Sauer M. Targeting enzymes to the right compartment: metabolic engineering for itaconic acid production by *Aspergillus niger*. Metab Eng. 2013;19:26–32.
10. Li A, Pfelzer N, Zuijderwijk R, Punt P. Enhanced itaconic acid production in Aspergillus Niger using genetic modification and medium optimization. BMC Biotechnol. 2012;12.
11. Ruijter GJG, Panneman H, Xu D-, Visser J: Properties of *Aspergillus niger* citrate synthase and effects of citA overexpression on citric acid production. FEMS Microbiol Lett 2000, 184(1):35–40.
12. Rohr M, Kubicek CP. Regulatory aspects of citric acid fermentation by *Aspergillus niger*. Process Biochem. 1981;16:34.
13. Ruijter G, Kubicek C, Visser J: Production of organic acids by fungi. In *Industrial Applications*. Edited by Anonymous Springer; 2002:213–230.
14. Palmieri F, Stipani I, Quagliariello E, Klingenberg M. Kinetic study of the Tricarboxylate carrier in rat liver mitochondria. Eur J Biochem. 1972;26(4):587–94.
15. Conover TE. Does citrate transport supply both acetyl groups and NADPH for cytoplasmic fatty acid synthesis? Trends Biochem Sci. 1987;12(C):88–9.
16. Kaplan RS, Mayor JA, Wood DO. The mitochondrial tricarboxylate transport protein. cDNA cloning, primary structure, and comparison with other mitochondrial transport proteins. J Biol Chem. 1993;268(18):13682–90.
17. Bisaccia F, De Palma A, Dierks T, Krämer R, Palmieri F. Reaction mechanism of the reconstituted tricarboxylate carrier from rat liver mitochondria. Biochim Biophys Acta Bioenerg. 1993;1142(1–2):139–45.
18. Kirimura K, Kobayashi K, Ueda Y, Hattori T. Phenotypes of gene disruptants in relation to a putative mitochondrial malate–citrate shuttle protein in citric acid-producing *Aspergillus niger*. Biosci Biotechnol Biochem. 2016:1–10.
19. Kadooka C, Izumitsu K, Onoue M, Okutsu K, Yoshizaki Y, Takamine K, Goto M, Tamaki H, Futagami T. Mitochondrial citrate transporters CtpA and YhmA are involved in lysine biosynthesis in the white koji fungus, *Aspergillus luchuensis* mut. kawachii. bioRxiv. 2018:341370.
20. Hossain AH, Ter Beek A, Punt PJ. Itaconic acid degradation in *Aspergillus niger*: the role of unexpected bioconversion pathways. Fungal Biol Biotechnol. 2019;6(1):1.
21. Kirimura K, Kobayashi K, Yoshioka I. Decrease of citric acid produced by *Aspergihllus niger* through disruption of the gene encoding a putative mitochondrial citrate-oxoglutarate shuttle protein. Biosci Biotechnol Biochem. 2019:1–9.
22. Castegna A, Scarcia P, Agrimi G, Palmieri L, Rottensteiner H, Spera I, Germinario L, Palmieri F. Identification and functional characterization of a novel mitochondrial carrier for citrate and oxoglutarate in *Saccharomyces cerevisiae*. J Biol Chem. 2010;285(23):17359–70.
23. Werpy T, Petersen G: Top value added chemicals from biomass. Results of Screening for Potential Candidates from Sugars and Synthesis Gas 2004, 1.
24. Yang L, Lübeck M, Lübeck PS. *Aspergillus* as a versatile cell factory for organic acid production. Fungal Biol Rev. 2017;31(1):33–49.
25. El Moualij B, Duyckaerts C, Lamotte-Brasseur J, Sluse FE. Phylogenetic classification of the mitochondrial carrier family of *Saccharomyces cerevisiae*. Yeast. 1997;13(6):573–81.
26. Pel HJ, De Winde JH, Archer DB, Dyer PS, Hofmann G, Schaap PJ, Turner G, De Vries RP, Albang R, Albermann K, Andersen MR, Bendtsen JD, Benen JAE, Van Den Berg M, Breestraat S, Caddick MX, Contreras R, Cornell M, Coutinho PM, Danchin EGJ, Debets AJM, Dekker P, Van Dijck PWM, Van Dijk A, Dijkhuizen L, Driessen AJM, D'Enfert C, Geysens S, Goosen C, Groot GSP, De Groot PWJ, Guillemette T, Henrissat B, Herweijer M, Van Den Hombergh JPTW, Van Den Hondel CAMJJ, Van Der Heijden RTJM, Van Der Kaaij RM, Klis FM, Kools HJ, Kubicek CP, Van Kuyk PA, Lauber J, Lu X, Van Der Maarel MJEC, Meulenberg R, Menke H, Mortimer MA, Nielsen J, Oliver SG, Olsthoorn M, Pal K, Van Peij NNME, Ram AFJ, Rinas U, Roubos JA, Sagt CMJ, Schmoll M, Sun J, Ussery D, Varga J, Vervecken W, Van De Vondervoort PJJ, Wedler H, Wösten HAB, Zeng A-, Van Ooyen AJJ, Visser J, Stam H: Genome sequencing and analysis of the versatile cell factory *Aspergillus niger* CBS 513.88. Nat Biotechnol 2007, 25(2):221–231.
27. Aguilar-Pontes M, Brandl J, McDonnell E, Strasser K, Nguyen T, Riley R, Mondo S, Salamov A, Nybo JL, Vesth TC. The gold-standard genome of *Aspergillus niger* NRRL 3 enables a detailed view of the diversity of sugar catabolism in fungi. Stud Mycol. 2018;91:61–78.
28. Yin X, Shin H, Li J, Du G, Liu L, Chen J. Comparative genomics and transcriptome analysis of *Aspergillus niger* and metabolic engineering for citrate production. Sci Rep. 2017;7:41040.
29. Mortazavi A, Williams BA, McCue K, Schaeffer L, Wold B. Mapping and quantifying mammalian transcriptomes by RNA-Seq. Nat Methods. 2008;5(7):621.
30. Yuzbasheva EY, Agrimi G, Yuzbashev TV, Scarcia P, Vinogradova EB, Palmieri L, Shutov AV, Kosikhina IM, Palmieri F, Sineoky SP. The mitochondrial citrate carrier in *Yarrowia lipolytica*: its identification, characterization and functional significance for the production of citric acid. Metab Eng. 2019;54:264–74.
31. Linde T, Zoglowek M, Lübeck M, Frisvad JC, Lübeck PS. The global regulator LaeA controls production of citric acid and endoglucanases in *Aspergillus carbonarius*. J Ind Microbiol Biotechnol. 2016;43:1139–47.
32. Untergasser A, Cutcutache I, Koressaar T, Ye J, Faircloth BC, Remm M, Rozen SG. Primer3—new capabilities and interfaces. Nucleic Acids Res. 2012;40(15):e115.
33. Hynes MJ. Genetic transformation of filamentous fungi. J Genet. 1996;75(3):297–311.
34. Peleg Y, Stieglitz B, Goldberg I. Malic acid accumulation by *Aspergillus flavus* - I. biochemical aspects of acid biosynthesis. Appl Microbiol Biotechnol. 1988;28(1):69–75.
35. Battat E, Peleg Y, Bercovitz A, Rokem JS, Goldberg I. Optimization of L-malic acid production by *Aspergillus flavus* in a stirred fermentor. Biotechnol Bioeng. 1991;37(11):1108–16.

# Permissions

All chapters in this book were first published in BIOTECHNOLOGY, by BMC; hereby published with permission under the Creative Commons Attribution License or equivalent. Every chapter published in this book has been scrutinized by our experts. Their significance has been extensively debated. The topics covered herein carry significant findings which will fuel the growth of the discipline. They may even be implemented as practical applications or may be referred to as a beginning point for another development.

The contributors of this book come from diverse backgrounds, making this book a truly international effort. This book will bring forth new frontiers with its revolutionizing research information and detailed analysis of the nascent developments around the world.

We would like to thank all the contributing authors for lending their expertise to make the book truly unique. They have played a crucial role in the development of this book. Without their invaluable contributions this book wouldn't have been possible. They have made vital efforts to compile up to date information on the varied aspects of this subject to make this book a valuable addition to the collection of many professionals and students.

This book was conceptualized with the vision of imparting up-to-date information and advanced data in this field. To ensure the same, a matchless editorial board was set up. Every individual on the board went through rigorous rounds of assessment to prove their worth. After which they invested a large part of their time researching and compiling the most relevant data for our readers.

The editorial board has been involved in producing this book since its inception. They have spent rigorous hours researching and exploring the diverse topics which have resulted in the successful publishing of this book. They have passed on their knowledge of decades through this book. To expedite this challenging task, the publisher supported the team at every step. A small team of assistant editors was also appointed to further simplify the editing procedure and attain best results for the readers.

Apart from the editorial board, the designing team has also invested a significant amount of their time in understanding the subject and creating the most relevant covers. They scrutinized every image to scout for the most suitable representation of the subject and create an appropriate cover for the book.

The publishing team has been an ardent support to the editorial, designing and production team. Their endless efforts to recruit the best for this project, has resulted in the accomplishment of this book. They are a veteran in the field of academics and their pool of knowledge is as vast as their experience in printing. Their expertise and guidance has proved useful at every step. Their uncompromising quality standards have made this book an exceptional effort. Their encouragement from time to time has been an inspiration for everyone.

The publisher and the editorial board hope that this book will prove to be a valuable piece of knowledge for researchers, students, practitioners and scholars across the globe.

# List of Contributors

**Chu Chu, Zhaohui Yang, Li Yan, Chenyang Si, Yu Kang, Zhenzhen Chen, Yongchang Chen, Weizhi Ji and Yuyu Niu**
Yunnan Key Laboratory of Primate Biomedicine Research; Institute of Primate Translational Medicine, Kunming University of Science and Technology, Kunming 650500, China

**Jiayin Yang**
The Cardiology Division, Department of Medicine, Li Ka Shing Faculty of Medicine, University of Hong Kong, Hong Kong, SAR, China

**Xiang Tu, Meiying Xu, Jianjun Li, Enze Li, Rongfang Feng, Gang Zhao and Jun Guo**
Guangdong Institute of Microbiology, Guangdong Academy of Sciences, Guangzhou 510070, People's Republic of China
State Key Laboratory of Applied Microbiology Southern China, Guangzhou 510070, People's Republic of China
Guangdong Provincial Key Laboratory of Microbial Culture Collection and Application, Guangzhou 510070, People's Republic of China

**Shaobin Huang**
School of Environment and Energy, South China University of Technology, Guangzhou 510006, People's Republic of China

**Brent Fortune, Sizwe Mhlongo, Leonardo Joaquim van Zyl and Marla Trindade**
Institute for Microbial Biotechnology and Metagenomics, University of the Western Cape, Bellville, South Africa

**Robert Huddy and Mariette Smart**
Institute for Microbial Biotechnology and Metagenomics, University of the Western Cape, Bellville, South Africa
Centre for Bioprocess Engineering Research, University of Cape Town, Cape Town, Western Cape, South Africa

**Shugui Li**
Key Laboratory of Tobacco Biology and Processing, Tobacco Research Institute of Chinese Academy of Agricultural Sciences, Qingdao 266101, People's Republic of China

**Ge Zhang**
Key Laboratory of Tobacco Biology and Processing, Tobacco Research Institute of Chinese Academy of Agricultural Sciences, Qingdao 266101, People's Republic of China
CAS Key Laboratory of Biobased Materials, Qingdao Institute of Bioenergy and Bioprocess Technology, Chinese Academy of Sciences, Qingdao 266101, People's Republic of China
Haikou Cigar Research Institute, Hainan Provincial Branch of CNTC, Haikou 571100, People's Republic of China

**Haobao Liu**
Key Laboratory of Tobacco Biology and Processing, Tobacco Research Institute of Chinese Academy of Agricultural Sciences, Qingdao 266101, People's Republic of China
Haikou Cigar Research Institute, Hainan Provincial Branch of CNTC, Haikou 571100, People's Republic of China

**Fan Wang and Haibo Zhang**
CAS Key Laboratory of Biobased Materials, Qingdao Institute of Bioenergy and Bioprocess Technology, Chinese Academy of Sciences, Qingdao 266101, People's Republic of China

**Yuhua Xin and Zhong Shen**
Haikou Cigar Research Institute, Hainan Provincial Branch of CNTC, Haikou 571100, People's Republic of China

**Juan Wang and Ming Ma**
Haikou Cigar Research Institute, Hainan Provincial Branch of CNTC, Haikou 571100, People's Republic of China
China Tobacco Standardization Research Center, Zhengzhou Tobacco Research Institute of CNTC, Zhengzhou 450001, People's Republic of China

**Yingbo Xu**
Research and Development Center, China Tobacco Anhui Industrial Co., Ltd., Hefei 230088, People's Republic of China

**Francisco Cabrera**
Colegio de Ciencias de la Salud, Escuela de Medicina Veterinaria, Universidad San Francisco de Quito, 17-12-841 Quito, Ecuador
IRMB CHU Saint Eloi, University of Montpellier, France, 80 rue Augustin Fliche, 34295 Montpellier, cedex 5, France
Instituto de Investigaciones en Biomedicina, Universidad San Francisco de Quito, 17-12-841 Quito, Ecuador
Mito-Act Research Consortium, Quito, Ecuador

**Stephany Gallardo**
Colegio de Ciencias Biológicas y Ambientales, Escuela de Biotecnología, Universidad San Francisco de Quito, 17-12-841 Quito, Ecuador

**Mayra Ortega**
Colegio de Ciencias Biológicas y Ambientales, Escuela de Biotecnología, Universidad San Francisco de Quito, 17-12-841 Quito, Ecuador
Instituto de Investigaciones en Biomedicina, Universidad San Francisco de Quito, 17-12-841 Quito, Ecuador
Mito-Act Research Consortium, Quito, Ecuador

**Diego Barba, Lina Soto and Gabriela Peña**
Colegio de Ciencias de la Salud, Escuela de Medicina, Universidad San Francisco de Quito, 17-12-841 Quito, Ecuador

**Luis Alberto Pedroza**
Colegio de Ciencias de la Salud, Escuela de Medicina, Universidad San Francisco de Quito, 17-12-841 Quito, Ecuador
Baylor College of Medicine: Department of Pediatrics, Section of Immunology, Allergy, and Rheumatology, Baylor College of Medicine, Houston, TX, USA

**Christian Jorgensen**
IRMB CHU Saint Eloi, University of Montpellier, France, 80 rue Augustin Fliche, 34295 Montpellier, cedex 5, France
Mito-Act Research Consortium, Quito, Ecuador

**Eliseo Parra**
Laboratory of Nano-Regenerative Medicine, Faculty of Medicine, Universidad de Los Andes, Santiago, Chile
Consorcio Regenero, Chilean Consortium for Regenerative Medicine, Santiago, Chile
Cells for Cells, Santiago, Chile

**Maroun Khoury**
Laboratory of Nano-Regenerative Medicine, Faculty of Medicine, Universidad de Los Andes, Santiago, Chile
Consorcio Regenero, Chilean Consortium for Regenerative Medicine, Santiago, Chile
Cells for Cells, Santiago, Chile
Mito-Act Research Consortium, Quito, Ecuador

**Francesca Velarde**
Instituto de Investigaciones en Biomedicina, Universidad San Francisco de Quito, 17-12-841 Quito, Ecuador
Mito-Act Research Consortium, Quito, Ecuador

**Andrés Caicedo**
Colegio de Ciencias de la Salud, Escuela de Medicina, Universidad San Francisco de Quito, 17-12-841 Quito, Ecuador
Instituto de Investigaciones en Biomedicina, Universidad San Francisco de Quito, 17-12-841 Quito, Ecuador
Mito-Act Research Consortium, Quito, Ecuador
Sistemas Médicos, SIME, Universidad San Francisco de Quito, 17-12-841 Quito, Ecuador

**Wilasinee Puangtame, Thitikorn Chomtong, Suchada Srifa, Sukhita Sathitnaitham and Wasawat Dumrongthawatchai**
Special Research Unit in Microalgal Molecular Genetics and Functional Genomics (MMGFG), Department of Genetics, Faculty of Science, Kasetsart University, 50 Ngam Wong Wan road, Chatuchak, Bangkok 10900, Thailand

**Anongpat Suttangkakul, Anchalee Sirikhachornkit, Piyada Juntawong and Supachai Vuttipongchaikij**
Special Research Unit in Microalgal Molecular Genetics and Functional Genomics (MMGFG), Department of Genetics, Faculty of Science, Kasetsart University, 50 Ngam Wong Wan road, Chatuchak, Bangkok 10900, Thailand
Center of Advanced studies for Tropical Natural Resources, Kasetsart University, 50 Ngam Wong Wan road, Chatuchak, Bangkok 10900, Thailand

**Kanidtha Jariyachawalid**
PTT Research and Technology Institute, PTT Public Company Limited, Ayuthaya 13170, Thailand

**Casey M. Gries and David D. Lo**
Division of Biomedical Sciences, School of Medicine, University of California, Riverside, USA

**Rohith R. Mohan and Dimitrios Morikis**
Department of Bioengineering, University of California, Riverside, USA

**Kitson Liew, Mohd Firdaus Che Mat, Marini Marzuki, Norazlin Abdul Aziz, Tai-Lin Chu, Munirah Ahmad and Alan Soo-Beng Khoo**
Molecular Pathology Unit, Cancer Research Centre, Institute for Medical Research, National Institutes of Health (NIH Complex), Ministry of Health Malaysia, Level 4, Block C7, No: 1, Jalan Setia Murni U13/52, Section U13, Setia Alam, 40170 Shah Alam, 50588 Kuala Lumpur, Selangor, Malaysia

**Sin-Yeang Teow**
Molecular Pathology Unit, Cancer Research Centre, Institute for Medical Research, National Institutes of Health (NIH Complex), Ministry of Health Malaysia, Level 4, Block C7, No: 1, Jalan Setia Murni U13/52, Section U13, Setia Alam, 40170 Shah Alam, 50588 Kuala Lumpur, Selangor, Malaysia
Department of Medical Sciences, School of Healthcare and Medical Sciences, Sunway University, 47500 Bandar Sunway, Selangor Darul Ehsan, Malaysia

**Jong-Ju Park and Munkhzaya Byambaragchaa**
Animal Biotechnology, Graduate School of Future Convergence Technology,Institute of Genetic Engineering, Hankyong National University, Ansung 17579, Korea

**Kwan-Sik Min**
Animal Biotechnology, Graduate School of Future Convergence Technology, Institute of Genetic Engineering, Hankyong National University, Ansung 17579, Korea
Department of Animal Resource Science, Hankyong National University, Ansung 17579, Korea

**Myung-Hwa Kang**
Department of Food Science and Nutrition, Hoseo University, Asan 31499, Korea

**Régis P. Lemaitre, Aliona Bogdanova and Barbara Borgonovo**
Max Planck Institute of Molecular Cell Biology and Genetics, Pfotenhauerstrasse 108, 01307 Dresden, Germany

**David N. Drechsel**
Max Planck Institute of Molecular Cell Biology and Genetics, Pfotenhauerstrasse 108, 01307 Dresden, Germany
Research Institute of Molecular Pathology (IMP), Campus-Vienna-Biocenter 1, 1030 Vienna, Austria

**Jeffrey B. Woodruff**
Department of Cell Biology, Dept. of Biophysics, UT Southwestern Medical Center, Dallas, TX 75390, USA

**Song Chen Jiahua Shao, Peiliang Fu and Haishan Wu**
Department of Arthroplasty Surgery, Changzheng Hospital, Second Military Medical University, Shanghai 200003, China

**Zhenyu Xu**
State Key Laboratory for Diagnosis and Treatment of Infectious Diseases, and Collaborative Innovation Center for Diagnosis and Treatment of Infectious Diseases, School of Medicine, Zhejiang University, Hangzhou 310003, China

**Prassan Choudhary, Sudipta Das, Hillol Chakdar, Arjun Singh, Sanjay Kumar Goswami and Anil Kumar Saxena**
ICAR-National Bureau of Agriculturally Important Microorganisms (NBAIM), Mau, Uttar Pradesh 275103, India

**Seyed Mostafa Hallaji and Mohammad Kuroshkarim**
School of Environment, College of Engineering, University of Tehran, Tehran, Iran

**Seyede Parvin Moussavi**
Environmental Health Research Center, International Branch of Shahid Sadoughi University of Medical Sciences and Health Services, Yazd, Iran

**Zhenzhen Hao and Xiaoyun Su**
Key Laboratory for Feed Biotechnology of the Ministry of Agriculture, Feed Research Institute, Chinese Academy of Agricultural Sciences, No. 12 South Zhongguancun Street, Beijing 100081, China

**Maroua Omrane Benmrad and Sondes Mechri**
Laboratory of Microbial Biotechnology and Engineering Enzymes (LMBEE), Centre of Biotechnology of Sfax (CBS), University of Sfax, Road of Sidi Mansour Km 6, 3018 Sfax, Tunisia

**Nadia Zaraî Jaouadi, Mouna Ben Elhoul, Hatem Rekik, Samir Bejar and Bassem Jaouad**
Laboratory of Microbial Biotechnology and Engineering Enzymes (LMBEE), Centre of Biotechnology of Sfax (CBS), University of Sfax, Road of SidiMansour Km 6, 3018 Sfax, Tunisia
Biotech ECOZYM Start-up, Business Incubator, Centre of Biotechnology of Sfax (CBS), University of Sfax,Road of Sidi Mansour Km 6, 3018 Sfax, Tunisia

**Sami Sayadi**
Laboratory of Environmental Bioprocesses (LEBP), LMI COSYS-Med, Centre of Biotechnology of Sfax (CBS), University of Sfax, Road of Sidi Mansour Km 6, 3018 Sfax, Tunisia

**Nabil Kechaou**
Research Group of Agro-Food Processing Engineering (GP2A), Laboratory of Applied Fluid Mechanics, Process Engineering and Environment, National School of Engineers of Sfax (ENIS), University of Sfax, Road of Soukra Km 4, 3038 Sfax, Tunisia

**Magdalena Rakicka, Jakub Wolniak, Zbigniew Lazar and Waldemar Rymowicz**
Department of Biotechnology and Food Microbiology, Wroclaw University of Environmental and Life Sciences, 37 Chełmońskiego St, 51-630 Wroclaw, Poland

**Isabela T. Lourenço-Tessutti and Regina M. Amorim**
Embrapa Genetic Resources and Biotechnology, Brasília, DF, Brazil

**Elinea O. Freitas and Ana G. B. Leite**
Embrapa Genetic Resources and Biotechnology, Brasília, DF, Brazil
Federal University of Brasília, Brasília, DF, Brazil

**Bruno P. Melo**
Embrapa Genetic Resources and Biotechnology, Brasília, DF, Brazil
Federal University of Viçosa, Viçosa, MG, Brazil

**Fabrício B. M. Arraes**
Embrapa Genetic Resources and Biotechnology, Brasília, DF, Brazil
Federal University of Rio Grande do Sul, Porto Alegre, RS, Brazil

**Maria E. Lisei-de-Sá**
Embrapa Genetic Resources and Biotechnology, Brasília, DF, Brazil
Agricultural Research Company of Minas Gerais State, Uberaba, MG, Brazil

**Julia A. Costa and Maria F. Grossi-de-Sa**
Embrapa Genetic Resources and Biotechnology, Brasília, DF, Brazil
Catholic University of Brasilia - Post-Graduation Program in Genomic Sciences and Biotechnology, Brasília, DF, Brazil

**Muhammad Faheem**
Embrapa Genetic Resources and Biotechnology, Brasília, DF, Brazil
National University of Medical Sciences, Rawalpindi, Punjab, Pakistan

**Carolina V. Morgante**
Embrapa Genetic Resources and Biotechnology, Brasília, DF, Brazil
Embrapa Semi-Arid, Petrolina, PE, Brazil

**Elizabeth P. B. Fontes**
Federal University of Viçosa, Viçosa, MG, Brazil

**Márcio A. Ferreira**
Federal University of Rio de Janeiro, Rio Janeiro, RJ, Brazil

**Khadega A. Almaqdi, Rana Morsi, Bahia Alhayuti and Farah Alharthi**
Department of Chemistry, College of Science, UAE University, Al Ain, UAE

**S. Salman Ashraf**
Department of Chemistry, College of Arts and Sciences, Khalifa University, Abu Dhabi, UAE

**Siyu Chen, Jianjun Wang, Sheng Wu and Yong Tao**
CAS Key Laboratory of Microbial Physiological and Metabolic Engineering, Institute of Microbiology, Chinese Academy of Sciences, Beijing 100101, People's Republic of China

**Yingshuo Hou**
CAS Key Laboratory of Microbial Physiological and Metabolic Engineering,Institute of Microbiology, Chinese Academy of Sciences, Beijing 100101,People's Republic of China
University of Chinese Academy of Sciences, Beijing 100101, People's Republic of China

**Guizhen Liu**
Kaiping Genuine Biochemical Pharmaceutical Co. Ltd, Kaiping, People's Republic of China

**Lei Yang, Mette Lübeck and Peter S. Lübeck**
Department of Chemistry and Bioscience, Section for Sustainable Biotechnology, Aalborg University, A.C. Meyers Vaenge 15, DK-2450 Copenhagen, SV, Denmark

**Tore Linde**
Department of Chemistry and Bioscience, Section for Sustainable Biotechnology, Aalborg University, A.C. Meyers Vaenge 15, DK-2450 Copenhagen, SV, Denmark
AGC Biologics, Vandtaarnsvej 83B, DK-2860, Soeborg, Copenhagen, Denmark

**Abeer H. Hossain and Peter J. Punt**
Dutch DNA Biotech BV, Padualaan 8, 3584CH, Utrecht, The Netherlands

# Index

CPSIA information can be obtained
at www.ICGtesting.com
Printed in the USA
LVHW060853030122
707669LV00002B/114

9 781639 890811